DESIGN and ANALYSIS of CONTROL SYSTEMS

DESIGN and ANALYSIS of CONTROL SYSTEMS

Arthur G.O. Mutambara

CRC Press

Boca Raton   London   New York   Washington, D.C.

Library of Congress Cataloging-in-Publication Data

Mutambara, Arthur G.O.
    Design and analysis of control systems / Arthur G.O. Mutambara.
        p. cm.
    Includes bibliographical references (p.
    ISBN 0-8493-1898-X
    1. Automatic control. 2. System design. I. Title.
TJ213.M85 1999
629.8—dc21

99-25429
CIP

This book contains information obtained from authentic and highly regarded sources. Reprinted material is quoted with permission, and sources are indicated. A wide variety of references are listed. Reasonable efforts have been made to publish reliable data and information, but the author and the publisher cannot assume responsibility for the validity of all materials or for the consequences of their use.

Neither this book nor any part may be reproduced or transmitted in any form or by any means, electronic or mechanical, including photocopying, microfilming, and recording, or by any information storage or retrieval system, without prior permission in writing from the publisher.

The consent of CRC Press LLC does not extend to copying for general distribution, for promotion, for creating new works, or for resale. Specific permission must be obtained in writing from CRC Press LLC for such copying.

Direct all inquiries to CRC Press LLC, 2000 N.W. Corporate Blvd., Boca Raton, Florida 33431.

Trademark Notice: Product or corporate names may be trademarks or registered trademarks, and are used only for identification and explanation, without intent to infringe.

© 1999 by CRC Press LLC

No claim to original U.S. Government works
International Standard Book Number 0-8493-1898-X
Library of Congress Card Number 99-25429
Printed in the United States of America   2 3 4 5 6 7 8 9 0
Printed on acid-free paper

Preface

This book covers the principles and challenges involved in the design and analysis of control systems. A control system is an interconnection of interacting components forming a configuration that will provide a desired system response. Feedback is a part of the control system and consists of the process of measuring the controlled variable and using that information to influence the controlled variable. The material presented in this book is of practical interest because it provides insight into the fundamental nature of control systems and enables the reader to engineer effective controllers. The motivation of the book is to imbue the reader with skills to design and analyze implementable control algorithms for a wide range of real engineering applications.

The book is divided into nine chapters that can be grouped into three parts. The first part of the book (Chapters 1, 2, and 3) introduces the principles involved in control systems, discusses various applications, develops mathematical models for dynamic systems, and shows how the models are used to obtain and study the system response (behavior). This material is useful in a *Dynamic Systems II* class, which serves as a basis for a first course in *Control Systems*, whose material is covered in the second part of the book (Chapters 4, 5, and 6). This part of the book covers the basic characteristics of control systems, root locus design methods, and frequency-response design methods. The last part of the book (Chapters 7, 8, and 9) deals with subjects suitable for an *Advanced Control Systems* class: state space design methods, digital control systems, state and information space estimation, and nonlinear control systems.

The three integrated and overlapping classes that can be taught and studied using this book are suitable for both undergraduate and graduate students with no background in control systems. Control system practice cuts across aerospace, chemical, electrical, industrial, and mechanical engineering, therefore this text is useful in all these fields. Although written ostensibly as a classroom textbook, the book can also serve as an indispensable tool for self-study and reference for engineering professionals.

Key Features of the Book

- Covers the theoretical and practical principles involved in the design and analysis of control systems applicable to a broad spectrum of engineering disciplines.

- Develops mathematical models for different types of dynamic systems such as mechanical, electrical, electromechanical, mechatronic, thermal-fluid, chemical, hydraulic, and pneumatic systems.

- Illustrates how system models are used to study the system response (behavior), and how the models are employed to achieve system control.

- Presents standard control systems material in a systematic and easy-to-follow fashion.

- Introduces and develops advanced control systems concepts on a well established foundation of standard control system principles.

- Employs *"just-in-time learning"* where concepts are introduced and developed as they become relevant. This fosters a better understanding of the concepts and their engineering applications.

- Uses many worked examples, design problems, and a wide range of illustrative applications from aerospace, robotics, mechatronics, and manufacturing.

- Makes extensive use of MATLAB to clearly demonstrate control system concepts and their implementation.

- Includes descriptions of a number of new demonstrative modern control systems laboratory experiments such as *Magnetic Levitation*, the *Inverted Pendulum*, the *3-DOF Helicopter*, the *2-DOF Robot,* and the *Pendubot*.

Key Benefits for the User

- One textbook usable for three different but overlapping courses: *Dynamic Systems II*, *A First Course in Control Systems*, and *Advanced Control Systems*. It offers a more economical option than using three different textbooks.

- A textbook that emphasizes and embraces the multidisciplinary nature of control system engineering and employs *"just-in-time learning and teaching."* The user follows and learns the material better than from a conventional control system text.

- All the relevant principles and concepts (standard and advanced) involved in the control of dynamic systems are covered in one textbook.

- Detailed descriptions of a range of new control systems laboratory experiments that can be incorporated into the controls curriculum.

- The extensive use of worked examples and illustrations, and the text structure (just-in-time philosophy) make the book easy to read and understand. The practical examples used to demonstrate the concepts and show their application are from different fields of engineering. This promotes holistic engineering education.

- The use of MATLAB throughout the textbook helps to strengthen the grasping of concepts and solidify the development of a thorough understanding of computer implementation of control systems.

- A comprehensive appendix containing the essential basics of MATLAB, including laboratory exercises.

- There is a move to a new engineering curriculum (emphasizing the multidisciplinary nature of engineering practice and the need for just-in-time learning) sweeping throughout engineering schools in the USA, Europe, and Japan. This book is written with the specific objective of satisfying this new curriculum. None of the current books (the competition) satisfy the requirements of this new curriculum so well.

The Author

Arthur G.O. Mutambara is an assistant professor of robotics and mechatronics in the Mechanical Engineering Department at the joint Engineering College of Florida Agricultural and Mechanical University and Florida State University in Tallahassee. He has been a Visiting Research Fellow at the National Aeronautics and Space Administration (NASA) Jet Propulsion Laboratory in California (1994), at the Massachusetts Institute of Technology (MIT) in the Astronautics and Aeronautics Department (1995), and at the California Institute of Technology (1996). In 1997, he was a Visiting Research Scientist at the NASA Lewis Research Center in Cleveland, and, in the summer of 1998, a Visiting Research Professor at the Carnegie Mellon University Robotics Institute.

Professor Mutambara is currently the Martin Luther King, Jr., Visiting Assistant Professor of Aeronautics and Astronautics at MIT, and has served on both the Robotics Review Panel and the Dynamic Systems and Controls Panel for the U.S.A. National Science Foundation. In 1998, he authored *Decentralized Estimation and Control for Multisensor Systems,* a research book published by CRC Press.

In 1991, Professor Mutambara received his B.Sc. degree with honors in electrical engineering from the University of Zimbabwe. He later entered Oxford University as a Rhodes Scholar, where, in 1992, he earned an M.Sc. in computation and worked with the Programming Research Group. In 1995, he gained a Ph.D. in robotics from Merton College, Oxford University, where he worked with the Robotics Research Group.

Professor Mutambara's main research interests include multisensor fusion, estimation, decentralized control, mechatronics, and modular robotics. He teaches graduate and undergraduate courses in dynamic systems, control systems, robotics, and mechatronics. He is a member of the Institute of Electrical and Electronic Engineering (IEEE), the Institute of Electrical Engineering (IEE) and the British Computer Society (BCS).

Acknowledgments

The information presented in this book is an outgrowth of lecture material that I have used to teach integrated classes in dynamic systems, control systems and mechatronics at the FAMU-FSU college of engineering in Tallahassee, Florida. The examples and illustrations were derived from collaborative research that I have been involved in at various institutions in the United States. I greatly appreciate the comradeship, support and suggestions provided by my colleagues in these institutions.

I would like to acknowledge all my colleagues at FAMU and FSU for reviewing the book, classroom testing the material, and providing insightful comments. In particular, I would like to thank Professors Fred Foreman, Dexter Johnson (NASA Lewis), and Rodney Roberts for their detailed remarks and suggestions. Professor Mark Spong of the University of Illinois, Dr. Simon Julier of the Naval Research Laboratory, and Professor Paul Mason of the University of Florida also offered useful comments.

I am deeply indebted to all my undergraduate and graduate students for acting as an effective testbed for the book and providing a robust critique of the material. Their questions, comments, and extensive feedback helped me make the material more readable. In particular, I would like to thank three students who have also been my teaching assistants over the past two years: Majura Selekwa for participating in the initial gathering of some of the material in Chapters 7, 8, and 9; Sivaprasad Akasam and Anton Thomas for continuously reviewing the material and implementing some of the examples and problems in MATLAB. An incomplete list of other students who also carefully read various drafts of the chapters and made useful comments, corrections, and suggestions include Tameika Franklin, Andy Richardson, Saul Mtakula, Tinglun Song, Angel Jackson, Jamila Mack, Marwan Haik, Natalie Barrett, Blessing Mudavanhu, Kirc Savage, Madelyne Solomon, Marcnell Pierre, Tramone Curry, Keith Larson, Mauricio Ballesteros, Jack Tsai, David Randall, and Paul Mavima.

This book is dedicated to oppressed people throughout the world and their struggle for social justice and egalitarianism. Defeat is not on the agenda.

Contents

Chapter 1

An Introduction to Control Systems

1.1 Introduction

This chapter provides the background and motivation for the material presented in this book. The general principles and issues involved in the design and analysis of control systems are outlined and illustrated by using examples. Basic control system terms are introduced and defined. Open- and closed-loop control system configurations are introduced and appraised. Classes of control systems are discussed and a broad spectrum of control system examples from various applications are presented. Advanced applications of control systems are outlined in addition to a discussion about the nature of the control design process. Brief descriptions of a variety of experiments that can be used to illustrate the material presented in the book are also included. The chapter provides a book outline as a road map through the contents of the book.

Control is the process of causing a system variable to conform to some desired value or reference value. A system is any collection of interacting components for which there are cause-and-effect relationships among the variables. The components are connected so as to form a whole entity that has properties that are not present in the separate entities. Within this context, a *control system* is then defined as an interconnection of interacting components forming a system configuration that will provide a desired system response. Feedback is the process of measuring the controlled variable and using that information to influence the controlled variable. In order to understand the purpose of a control system, it is useful to examine examples of simple control systems. These simple systems incorporate the same ideas of control and feedback that are used in complex and advanced applications. Modern control engineering practice includes the use of control design strategies for improving manufacturing processes, efficient energy use and advanced automobile technology. These control principles are also used in rapid transit systems, advanced robotic systems, and in the emerging fields of mechatronics and micro-electromechanical systems (MEMS). Mechatronics involves the synergistic integration of mechanics, electronics, and computer science to produce optimum products and systems.

1.1.1   Background

The field of engineering is concerned with the understanding, usage, and control of natural materials and forces for the benefit of humankind. Control system engineering is a sub-aspect of engineering that is concerned with understanding and controlling segments of the environment, often called systems, to provide useful economic products for society. The twin goals of understanding and control are complementary because effective systems control requires that the systems be understood and modeled. Furthermore, control engineering must often consider the control of poorly understood systems such as chemical process systems. The present challenge to control engineers is the modeling and control of modern, complex, multidisciplinary systems such as traffic control systems, chemical processes, robotic systems, mechatronic systems, and MEMS. At the same time, the engineer has the opportunity to control many useful and intellectually challenging industrial automation systems. Perhaps the most characteristic quality of control engineering is the opportunity to control machines, industrial and economic processes for the benefit of society. Control engineering is based on the foundations of feedback theory and linear system analysis, and it integrates the concepts of network theory and communication theory.

Control engineering is ostensibly multidisciplinary. It is equally applicable to aeronautical, chemical, mechanical, environmental, civil, and electrical engineering. Feedback controllers are used in many different systems, from airplanes and rockets to chemical processing plants and semiconductor manufacturing. Quite often a single control system includes electrical, mechanical, and chemical components. The theory and practice of control systems are applicable to many disciplines other than engineering. As the understanding of the dynamics of business, social, and political systems increases, the ability to model and control these systems will also improve.

1.2   A Recent History of Control Systems

During the 1980s, the utilization of digital computers for control components became routine. The technology that allows these new control elements to perform accurate and rapid calculations was previously unavailable to control engineers. These computers are employed especially for process control systems in which many variables are measured and controlled simultaneously by the computer. With the advent of Sputnik (the first space vehicle) and the space age, another new impetus was imparted to control engineering. It became necessary to design complex, highly accurate control systems for missiles and space probes. Furthermore, the necessity to minimize the weight of satellites and to control them very accurately has spawned the important field of optimal control. Due to these requirements, the time-domain methods developed by Lyapunov, Minorsky and others have been embraced with great interest in the last two decades. Recent theories of optimal control developed by L. S. Pontryagin in the former Soviet Union and R. Bellman in the United States, and recent studies of robust systems have also contributed to the interest in time-domain methods. Control engineering must consider both the time-domain and the frequency-domain approaches simultaneously in the analysis and design of control systems.

Classical control theory, which deals only with single-input single-output (SISO) systems, is ineffectual for multiple-input-multiple-output (MIMO) systems. The availability of digital computers, from around 1965, made time-domain analysis of complex systems possible, leading to the development of modern control theory based on time-domain analysis and synthesis. This theory uses state variables to cope with the increased complexity of modern plants. These new methods also meet the stringent requirements on accuracy, weight, and cost in civilian, defense, space, and industrial applications. From 1965 to 1985, optimal control of both deterministic and stochastic systems, as well as adaptive and learning control of complex systems, were fully investigated. From 1980 to the present, developments in modern control theory have centered around robust control, H_∞ control, multisensor-based control, robust estimation, and associated topics. Now that digital computers have become cheaper and more compact, they are used as integral parts of control systems. Modern control theory has also started to find its way into such nonengineering systems as biological, biomedical, social, economic, and political systems.

1.2.1 Automatic Control

The control of an industrial process (manufacturing, production, and processing) by automatic rather than manual means is often called automation. Automation is the automatic operation or control of a process, device, or system. It is prevalent in chemical, electric power, paper, automotive, and steel industries, among others. The concept of automation is central to an industrial society. Automatic machines can be used to increase the productivity of a plant and to obtain high-quality products. Automatic control of machines and processes is utilized to produce a product within specified tolerances and to achieve high precision. Automatic control has played a vital role in the advancement of engineering and science. In addition to its extreme importance in space-vehicle, missile-guidance, robotic, and mechatronic systems, automatic control has become an important and integral part of modern manufacturing and industrial processes. For example, automatic control is essential in the numerical control of machine tools in the manufacturing industries, in the design of autopilot systems in aerospace industries, and in the design of vehicles in automobile industries. It is also essential in industrial operations that require the control of pressure, temperature, humidity, viscosity, and fluid flow. Due to the importance of automatic control as a means of attaining optimal performance of dynamic systems, improving productivity and relieving the drudgery of many routine repetitive manual operations, most engineers and scientists must now have a good understanding of this field. The current revolution in computer and information technology is causing an equally momentous social change: the expansion of information gathering and information processing as computers extend the reach of the human brain. Control systems are used to achieve increased productivity and improved performance of a device or system.

1.2.2   Multivariable Control

Due to the increasing complexity of the systems that must be controlled and the interest in achieving optimum performance, the importance of control system engineering has grown in the past decade. As the systems become more complex, the interrelationship of many controlled variables must be considered in the control scheme. This leads to control systems that have more than one feedback loop, i.e., *multi-loop control systems* as opposed to single-loop control systems (one feedback loop). Such systems are nontrivial and are much more challenging than single-loop control systems. Most of the standard control system theory applies to single-loop control only. Multiple control loops are needed whenever a plant has multiple sensors or multiple actuators. In this case, the interaction of every feedback loop with every other feedback loop must be accounted for. While many single-loop concepts hold in principle in the multi-loop case, the technicalities are much more involved. The performance benefits of multi-loop control, however, are often far more than one would expect from a collection of single-loop controllers. Such multivariable control is essential in multi-input multi-output (MIMO) systems, whereas single-loop control is sufficient for single-input single-output (SISO) systems.

1.3   The Basic Components of a Control System

A control system is an interconnection of components forming a system configuration that will provide a desired response. The basis for analysis of a system is provided by linear system theory, which assumes a cause-effect relationship for the components of a system. The input-output relationship represents the cause-and-effect relationship of the process, which in turn represents the processing of the input signal to provide an output signal variable, often with power amplification.

All control systems have a similar basic structure and consist of the same basic components:

- *Process (or Plant):* This is the main physical component of a control system as it is the component (a dynamic system) whose output is to be controlled. Usually, a mathematical model of the process is required for its behavior to be understood and then controlled.

- *Actuator:* A device that is used to physically influence the process. It is the muscle that receives the control signal from the controller and forces the plant to produce the desired output.

- *Controller:* An algorithm or mechanism that takes the error signal and generates the control signal required to drive the actuator. The controller is the *"brain"* of the control system and its design and analysis are the central motivation for the material presented in this book.

- *Sensor:* A device that measures the actual system output and produces the *measured output*. Sensors are not perfect; they are often affected by sensor errors and associated uncertainties. Hence there is always at least a slight difference between the actual and measured outputs.

- *Desired Output:* The desired value for the output of the process being controlled, and achieving this desired output is the objective of a control system.

- *Actual Output:* The actual state of the process that is to be controlled or influenced. It must be measured by a sensor and then compared with the desired output.

- *Comparator:* This component takes the desired output and the measured output as inputs and generates an error signal that is the difference between the desired and measured outputs. This error signal is sent to the controller.

- *Disturbance Signals or Noise:* These are signals that are external to the control system but affect the process. Examples of disturbance signals include heat losses, electromagnetic radiation, and vibrations. A good controller eliminates or minimizes the effects of disturbances, i.e., it manifests effective disturbance rejection.

These are the components that are common to all control systems irrespective of system complexity, nature of application, and type of dynamic system. However, depending on the type of control system, some components are not relevant. For example, the open-loop control system does not require both the sensor and the comparator.

1.4 Open-Loop Control vs. Closed-Loop Control

There are two main configurations for control systems:

- Open-loop

- Closed-loop

All control systems can be placed into these two general categories. In this section the general structures of these two configurations are presented and their applications, advantages, and disadvantages are discussed.

1.4.1 Open-Loop Control

An open-loop control system utilizes a controller and actuator to obtain the desired response without using any measurement of the actual system response (controlled variable) to influence the system. Thus the objective of an open-loop control system is to achieve the desired output by utilizing an actuating device to control the process directly without the use of feedback. The elements of an open-loop control system are shown in Figure 1.1.

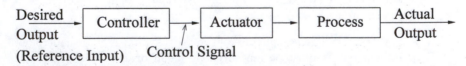

FIGURE 1.1
The Elements of an Open-Loop Control System

There are many applications where it is sufficient to use an open-loop control system. In particular, closed-loop control methods are unnecessary when system inputs and the plant model are known with total certainty, and there are no external disturbances. Also, open-loop systems are more convenient when obtaining the output measurement is either difficult or not economically feasible. For example, in a washer system, it would be quite expensive to provide a device that measures the quality of the washer output, i.e., cleanliness of the clothes.

However, there are several limitations of open-loop systems which include: slow system response, poor disturbance rejection, poor tracking under uncertainties, high sensitivity to system parameter errors (e.g., errors in plant or controller gains), and high sensitivity to changes in calibration errors (hence recalibration is usually necessary). A thorough comparison of the benefits and limitations of open- and closed-loop systems is presented later.

1.4.2 Closed-Loop Control

In contrast to an open-loop control system, a closed-loop control system utilizes a measure of the actual output to compare the actual output with the desired output response. The measure of the output is called the feedback signal. The elements of a general closed-loop feedback control system are shown in Figure 1.2. A closed-loop control system compares a measurement of the output with the desired output (reference or command input). The difference between the two quantities (the error signal) is then used to drive the output closer to the reference input through the controller and actuator.

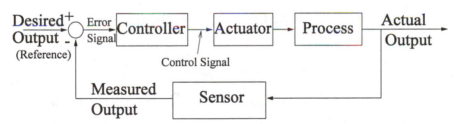

FIGURE 1.2
The Elements of a Closed-Loop Control System

Thus, a feedback control system is a control system that tends to maintain a prescribed relationship of one system variable to another by comparing functions of these variables and using the difference as a means of control. The system often uses a function of a prescribed relationship between the output and reference input to control the process. Often the difference between the output of the process under control and the reference input is amplified and used to control the process so that the difference is continually reduced. The feedback concept is the basis of control system analysis and design.

1.4.3   Advantages of Closed-Loop Systems

Closed-loop systems have the following advantages:

- Faster response to an input signal

- Effective disturbance rejection

- Better tracking of reference signals

- Low sensitivity to system parameter errors (e.g., errors in plant or controller gains)

- Low sensitivity to changes in calibration errors (recalibration is unnecessary)

- More accurate control of plant under disturbances and internal variations

- Effective and flexible control tuning by varying the control gain

- Used to stabilize systems that are inherently unstable in the open-loop form

1.4.4   Disadvantages of Closed-Loop Systems

The following are some of the disadvantages of closed-loop systems:

- Require the use of sensors which increase the system costs

- Involve more components which leads to more costs and complexity

- The power costs (due to high gains) are high

- More complex design, harder to build

- Sometimes obtaining the output measurement is either hard or not economically feasible

- Initial tuning is more difficult, especially if the bandwidth is narrow

- There is always a steady state error (with proportional controllers)

- The system tends to become unstable as the gain is increased beyond certain limits

- Closed-loop control methods are unnecessary when system inputs and the plant model are known with total certainty, and there are no external disturbances

- Closed-loop systems are not always controllable

Essentially, the advantages of closed-loop systems are the disadvantages of open-loop systems, and the disadvantages of closed-loop systems are the advantages of open-loop systems. The introduction of feedback enables the engineer to control a desired output and improve accuracy, but it requires attention to the issue of stability of response. Feedback is used for the purpose of reducing the error between the reference input and the system output. However, the significance of the effects of feedback in control systems is more complex. The reduction of system error is merely one of the many important effects that feedback may have upon a system. The effects of feedback on system performance characteristics such as stability, bandwidth, gain, disturbance, and sensitivity will be shown in Chapter 4.

In order to understand the effects of feedback on a control system, it is essential that the phenomenon be examined in a broad sense. When feedback is deliberately introduced for the purpose of control, its existence is easily identified. However, there are numerous situations where a physical system that is normally recognized as an inherently non-feedback system turns out to have feedback when it is observed from a certain perspective.

1.5 Examples of Control Systems

There are several examples of control systems (open and closed-loop) in biological systems, daily activities, and industrial operations. These systems could be manual, automatic, or semi-automatic and they are presented here to show the pervasiveness and usefulness of control system principles in general, and feedback control system principles in particular.

1.5.1 Manual Car Direction of Travel Control

Driving an automobile provides a good example for both manual and automatic control. The driving activity is a pleasant task when the automobile responds rapidly to the driver's command. Many cars have power steering and brakes, which utilize hydraulic amplifiers to amplify the forces used for braking and steering. An example of manual control occurs when the driver has a desired path (direction) for the car, observes the actual path of the car and then forces the car, using the steering wheel, to follow the desired path as closely as possible. Figure 1.3 shows a simple sketch of a car in a direction of travel under manual control.

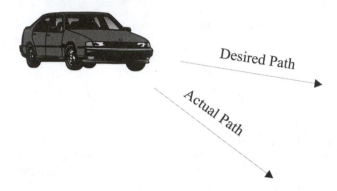

FIGURE 1.3
Control of a Car's Direction of Travel

An annotated block diagram of the car's manual steering control system is shown in Figure 1.4, illustrating the sensor, process, and actuator. The driver's eyes are the sensors that obtain a measurement of the car's actual direction. The driver then compares the desired course with the measured course and generates an error signal. This error signal is then used by the driver, who is the manual controller, to determine how to move the steering wheel (the actuator) so that the car (process) moves in the desired direction.

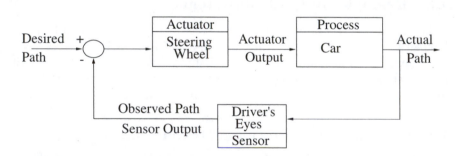

FIGURE 1.4
A Manual Closed-Loop Control System for a Car

A typical result of the direction of travel manual control is shown in Figure 1.5. The actual path tracks the desired path.

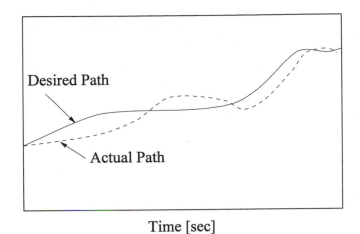

Time [sec]

FIGURE 1.5
Control of a Car's Path (Direction of Travel)

1.5.2   Cruise Control for a Car

Automobile cruise control provides a good example of automatic control. Most modern cars have a cruise control facility where a desired car speed is set, a speedometer measures the actual speed of the vehicle, and the difference between the two is used to drive an automatic controller (a programmed processor). This controller sends a signal to the engine throttle so that more fuel (or less) is burnt for the car to travel at the desired speed. Figure 1.6 shows a general car's cruise control system where the car is the process, the speedometer is the sensor and the engine is the actuator.

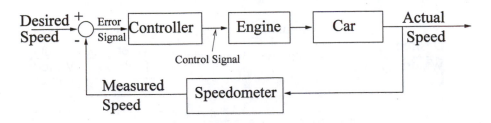

FIGURE 1.6
A Car Cruise Closed-Loop Control System

The car's cruise control system is very simple but very illustrative of control system phenomena. In Chapter 4, it is extensively modeled and quantitatively analyzed in both its open- and closed-loop forms.

1.5.3   Automatic Water Level Control

Water level control provides another simple platform to illustrate both manual and automatic control. Figure 1.7 shows a schematic diagram of a liquid level control system. The automatic controller maintains the water level by comparing the measured level with a desired level and correcting any error by adjusting the opening of the pneumatic valve.

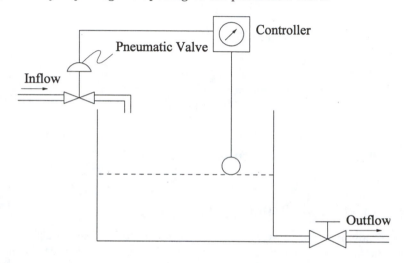

FIGURE 1.7
Automatic Water Level Control Mechanism

Figure 1.8 shows the corresponding block diagram of the water level control system where the water tank and its dynamics constitute the process, the float is the sensor, and the valve is the actuator. Thus, the mechanism automatically adjusts the water level until the desired level is achieved.

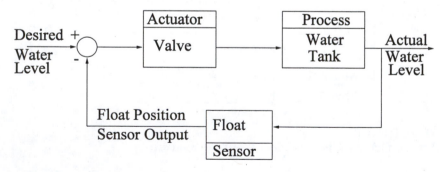

FIGURE 1.8
Block Diagram of an Automatic Water Level Control System

1.5.4   Manual Water Level Control

For the water level control system a manual control system can be used instead of an automatic control system. In this case, the human operator knows the desired water level, observes the water level, and uses the difference between the two to determine how to turn the pneumatic valve so that the desired water level is achieved. The control system block diagram for such a manual control system is shown in Figure 1.9.

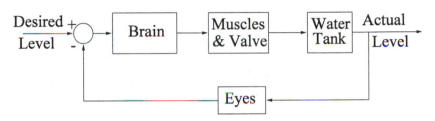

FIGURE 1.9
A Manual Water Level Closed-Loop Control System

The eyes of the human operator now constitute the sensor, the operator's brain is now the controller, and the actuator consists of the operator's muscles together with the pneumatic valve.

1.5.5   Turntable Speed Control

Another interesting example is the open-loop and closed-loop control of a turntable. Many modern devices use a turntable to rotate a disk at a constant speed. For example, a CD player, a computer disk drive, and a record player all require a constant speed of rotation in spite of motor wear and variation and other component changes. Figure 1.10 shows an open-loop (without feedback) turntable control system.

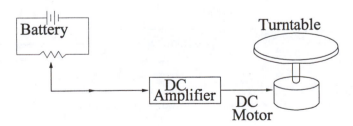

FIGURE 1.10
Turntable Speed Control: Open-Loop System

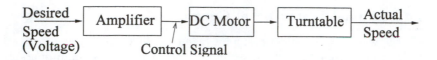

FIGURE 1.11
Turntable Speed Control System: Open-Loop System

The DC motor is the actuator because it provides a speed proportional to the applied motor voltage. For the input voltage to the motor, an amplifier is used because it can provide the required power. This system uses a battery source to provide a voltage that is proportional to the desired speed. This voltage is amplified and applied to the motor. The block diagram of the open-loop system identifying the controller (amplifier), actuator (DC motor), and the process (turntable) is shown in Figure 1.11.

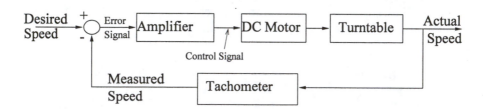

FIGURE 1.12
Turntable Speed Control: Closed-Loop System

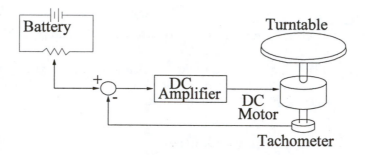

FIGURE 1.13
Turntable Speed Control: Closed-Loop System

In order to obtain a closed-loop turntable control system a sensor is required to measure the actual speed of the turntable. One such useful sensor is a tachometer that provides an output voltage proportional to the speed of its shaft. The measured speed is then compared with the desired speed and the difference between the two speeds is used to drive the controller (the amplifier). Thus the closed-loop feedback system takes the form shown in Figure 1.12. The block diagram model of the feedback system is shown in Figure 1.13. The error voltage is generated by the difference between the input voltage and the tachometer voltage. The feedback control system is better than the open-loop system in Figure 1.10 because the feedback system is faster, responds to errors, and works to reduce these errors. Detailed and quantitative discussions of the benefits of feedback control are presented in Chapter 4.

1.5.6 Blood Glucose Control

Another illustration of open- and closed-loop control systems can be obtained by considering blood glucose level monitoring, which is achieved by controlling insulin delivery. This is a common application of control engineering in the field of open-loop system drug delivery, in which mathematical models of the dose-effect relationship of the drugs are used. A drug-delivery system implanted in the body uses an open-loop system, since miniaturized glucose sensors are not yet available. The best solutions rely on individually programmable, pocket-sized insulin pumps that can deliver insulin according to a preset time history. More-complicated systems will use closed-loop control for the measured blood glucose levels.

The objective of a blood glucose control system is to design a system that regulates the blood glucose concentration in a diabetic person. The system must provide the insulin from a reservoir implanted within the diabetic person. Thus, the variable to be controlled is the blood glucose concentration. The specification for the control system is to provide a blood glucose level for the diabetic that closely approximates (tracks) the glucose level of a healthy person. An open-loop system would use a pre-programmed signal generator and miniature motor pump to regulate the insulin delivery rate as shown in Figure 1.14.

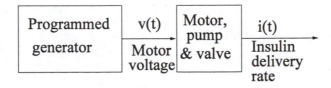

FIGURE 1.14
Blood Glucose Control: Open-Loop System

The feedback control system would use a sensor to measure the actual glucose level, compare this with the desired glucose level, and the error between the two levels to drive the amplifier. The amplifier then sends a control signal to the actuator (motor pump and valve) so that more (or less) insulin is delivered into the process (blood, body, and pancreas). Figure 1.15 depicts the block diagram of the blood glucose level closed-loop control system.

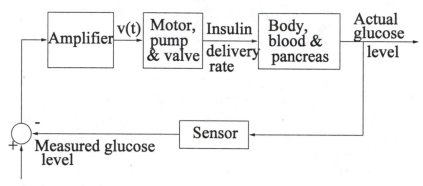

FIGURE 1.15
Blood Glucose Control: Closed-Loop System

1.5.7 Manual Control of a Sound Amplifier

Manual operation of a public address system presents another illustrative control system. The human operator is at the center of this system. The operator knows the desired sound level, his ears are the sensors that pick up a measure of the output sound, and then the operator compares the sound levels and appropriately uses the microphone, which in turn drives the amplifiers.

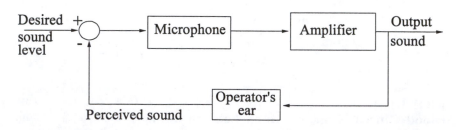

FIGURE 1.16
Manual Closed-Loop Control of a Sound Amplifier

1.5.8   Feedback in Social, Economic, and Political Systems

The principles of feedback and control are not restricted to engineering and biological systems. There are many applications and occurrences of feedback control concepts in the socioeconomic-political arena. Society is composed of many feedback systems and regulatory bodies that are controllers exerting the forces on society necessary to maintain a desired output. A simplified model of the national income feedback control system is shown in Figure 1.17 [7].

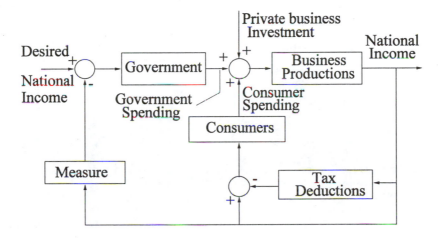

FIGURE 1.17
Application of the Principles of Feedback to Economic Systems

Such a control network fosters understanding of government control and expenditure. Other loops in the system are consumers and taxes. Although social, political, and economic feedback systems are not very rigorous, they do provide useful information and system understanding.

1.6   Classification of Control Systems

Feedback control systems can be classified in a number of ways, depending on the purpose of the classification. For instance, based on the method of analysis and design, control systems are classified as linear or nonlinear, and time-varying or time-invariant. When considering the type of signals used in the system, reference is often made to continuous-data and discrete-data systems, or modulated and unmodulated systems. Control systems are also often classified according to the purpose of the system.

For instance, a position-control system and a velocity-control system control the output variables according to ways implied by the names. The classification of a control system can also be defined according to the form of the open-loop transfer function. In general, there are many other ways of identifying control systems according to some special features of the system. It is important that some of these more common ways of classifying control systems are appreciated so that a proper perspective is gained before embarking on the analysis and design of these systems. The different classification strategies can be summarized as follows:

- Linear vs. nonlinear

- Time-variant vs. time-invariant

- Continuous-data vs. discrete-data

- Single-input single-output (SISO) vs. multiple-input multiple-output (MIMO)

- Regulator vs. tracking

- Purpose of control (e.g., position control or velocity control)

- Form of open-loop transfer function

Some of these classes are briefly discussed in the following sections.

1.6.1   Linear vs. Nonlinear Control Systems

This classification is made according to the methods of analysis and design. Strictly speaking, linear systems do not exist in practice, since all physical systems are nonlinear to some extent. Linear feedback control systems are idealized models fabricated by the analyst purely for the simplicity of analysis and design. When the magnitudes of signals in a control system are limited to ranges in which system components exhibit linear characteristics (i.e., the principle of superposition applies), the system is essentially linear. But when the magnitudes of signals are extended beyond the range of the linear operation, depending on the severity of the nonlinearity, the system should no longer be considered linear. For instance, amplifiers used in control systems often exhibit a saturation effect when their input signals become large; the magnetic field of a motor usually has saturation properties. Other common nonlinear effects found in control systems include the backlash or dead play between coupled gear members, nonlinear spring characteristics, and nonlinear friction or torque between moving members.

Quite often, nonlinear characteristics are intentionally introduced in a control system to improve its performance or provide more effective control. For instance, to achieve minimum-time control, an on-off (bang-bang or relay) type of controller is used in many missile or spacecraft control systems. Typically in these systems, jets are mounted on the sides of the vehicle to provide reaction torque for attitude control. These jets are often controlled in a full-on or full-off fashion, so a fixed amount of air is applied from a given jet for a certain time period to control the altitude of the space vehicle. For linear systems, there exists a wealth of analytical and graphical techniques for design and analysis purposes. A majority of the material in this book is devoted to the analysis and design of linear systems. Nonlinear systems, on the other hand, are usually difficult to treat mathematically, and there are no general methods available for solving a wide class of nonlinear systems. In the design of control systems, it is practical first to design the controller based on the linear-system model by neglecting the nonlinearities of the system. The designed controller is then applied to the nonlinear system model for evaluation or redesign by computer simulation.

1.6.2 Time-Invariant vs. Time-Variant Control Systems

When the parameters of a control system are stationary with respect to time during the operation of the system, the system is called a time-invariant system. In practice, most physical systems contain elements that drift or vary with time. For example, the winding resistance of an electric motor will vary when the motor is first being excited and its temperature is rising. Another example of a time-varying system is a guided-missile control system in which the mass of the missile decreases as the fuel on board is being consumed during flight. Although a time-varying system without nonlinearity is still a linear system, the analysis and design of this class of systems are usually much more complex than that of the linear time-invariant systems.

1.6.3 Continuous-Data vs. Discrete-Data Control Systems

A continuous-data system is one in which the signals are functions of the continuous time variable t. Among all continuous-data control systems, the signals may be further classified as *alternating current* (AC) or *direct current* (DC). Unlike the general definitions of these terms in electrical engineering, in an AC control system the signals are modulated by some form of modulation scheme, and in a DC control system not all the signals in the system are unidirectional, otherwise there would be no corrective control movement. A DC control system simply implies that the signals are unmodulated, but they are still AC signals according to the conventional definition.

Typical components of a DC control system include potentiometers, DC amplifiers, DC motors, and DC tachometers. AC control systems are used extensively in aircraft and missile control systems in which noise and disturbance often create problems. By using modulated AC control systems with carrier frequencies of 400 Hz or higher, the system will be less susceptible to low-frequency noise. Typical components of an AC control system include synchros, AC amplifiers, AC motors, gyroscopes, and accelerometers. In practice, not all control systems are strictly of the AC or DC type. A system may incorporate a mixture of AC and DC components, using modulators and demodulators to match the signals at various points in the system.

Discrete data control systems differ from continuous-data systems in that the signals at one or more points of the system are in the form of either a pulse train or a digital code. Usually, discrete-data control systems are subdivided into sampled-data and digital control systems. Sampled-data control systems refer to a more general class of discrete-data systems in which the signals are in the form of pulse data. A digital control system refers to the use of a digital computer or controller in the system, so that the signals are digitally coded, such as in binary code. In general, a sampled-data system receives data or information only intermittently at specific instants of time. A sampled-data system can also be classified as an AC system since the signal of the system is pulse modulated.

1.6.4   Regulator vs. Tracking Control Systems

A regulator control system is a control system whose desired output (the reference) is constant. Most control systems in industry, particularly the process industry, are of the regulator type. However, the desired output in many other systems is time-varying, i.e., the reference signal is time-variant and the controlled output is required to follow this signal as closely as possible. This situation characterizes the tracking problem.

1.7   Control System Design

Engineering design is both an art and a science that together constitute one of the main functions of an engineer. It is a complex process in which both creativity and analysis play major roles. Design is the process of conceiving or inventing the forms, parts, and details of a system to achieve a specified purpose. Design activity can be thought of as planning for the emergence of a particular product or system. Design is an innovative act whereby the engineer creatively uses knowledge and materials to specify the shape, function, and material content of a system. The steps can be summarized as follows:

- Determination of the need for a product or system

- Development of the specifications of the solution to the need

- Development and evaluation of various alternative solutions that meet these specifications

- Determination of the best solution to be designed in detail and fabricated

- Verification and validation of design

Any design process starts with the establishment of specifications. Specifications are statements that explicitly state the desired characteristic of device or product and clearly outline its functions. The design of technical systems aims to achieve appropriate design specifications and rests on four characteristics: complexity, trade-offs, design gaps, and risk. Complexity of design results from the wide range of tools, issues, and knowledge used in the process. The large number of factors to be considered illustrates the complexity of the design specification activity, not only in assigning these factors their relative importance in a particular design, but also in giving them substance either in numerical or written form, or both.

The design of control systems is a specific example of engineering design. Again, the goal of control engineering design is to obtain the configuration, specifications, and identification of the key parameters of a proposed system to meet an actual need. The first step in the design process is to establish the system goals. For example, the goal may be to control the velocity of a motor accurately. The second step is to identify the variables that are to be controlled, for example, the velocity of the motor. The third step is to write the specifications in terms of the accuracy that must be attained. This required accuracy of control then leads to the identification of a sensor to measure the controlled variable. The strategy is to proceed to the first attempt to configure a system that results in the desired control performance. This system configuration normally consists of a sensor, the process under control, an actuator, and a controller. The next step consists of identifying a candidate for the actuator. This depends on the process, but the actuation chosen must be capable of effectively adjusting the performance of the process. For example, if the objective is to control the speed of a rotating flywheel, a motor is selected as the actuator. The sensor, in this case, needs to be capable of accurately measuring the speed. The next step is then to obtain models for the actuator and sensor. This is followed by the design and selection of a controller, which often consists of a summing amplifier that compares the desired response and the actual response and then forwards this error-measurement signal to an amplifier. The final step in the design process is the adjustment of the parameters of the system in order to achieve the desired performance. The control system design process can be summarized as follows:

- Establishment of control goals

- Identification of the variables to be controlled

- Development of the specifications for the variables

- Establishment of the system configuration and identification of the actuator

- Design of the controller and selection of parameters to be adjusted

- Optimization parameter and analysis of performance

- Verification and validation of design

If the performance does not meet the specifications, then the configuration and the actuator are iterated. If the performance meets the specifications, the design is finalized and the parameters adjusted accordingly. The design is finalized and the results are then documented. If the performance still does not meet the specifications, there is a need to establish an improved system configuration and perhaps select an enhanced actuator and sensor. The design process is then repeated until the specifications are met, or until a decision is made that the specifications are too demanding and should be relaxed.

As an illustration, the performance specifications could describe how a closed-loop system should perform and will include:

- Effective disturbance rejection

- Fast and desirable system response

- Realistic actuator signals

- Low sensitivities to parameter variations

- Robustness

The control design process has been dramatically affected by the advent of powerful and inexpensive computers that characterize the information age, and the availability of effective control design and analysis software. For example, the Boeing 777, which incorporates the most advanced flight avionics of any U.S. commercial aircraft, was almost entirely computer-designed. Design verification of final designs is essential in safety-security critical systems (such as nuclear plants, ambulance systems, and surgery/emergency room systems) and high-fidelity computer simulations (advanced flight avionics). In many applications the certification of the control system in realistic simulations represents a significant cost in terms of money and time. The Boeing 777 test pilots flew about 2400 flights in high-fidelity simulations before the first aircraft was even built. Thus, system verification and validation constitute essential aspects of the design and analysis of control systems.

1.8   Advanced Applications of Control Systems

Control systems have advanced applications in the general areas of large scale systems, multisensor systems, space structures, manufacturing and flexible structures. Of particular interest are the specific fields of robotics, mechatronics, MEMS, and aerospace. In the following subsections a number of examples are discussed.

1.8.1   A Modular and Scalable Wheeled Mobile Robot

Robotics involves the study, design, and construction of multifunctional reprogrammable machines that perform tasks normally ascribed to human beings. A wheeled mobile robot (WMR) is an autonomous vehicle system whose mobility is provided by wheels. It is a multisensor and multiactuator robotic system, that has nonlinear kinematics and distributed means of acquiring information. A modular robotic vehicle has the same function as any conventional robot except that it is constructed from a small number of standard units. The benefits of modular technology include: *easier system design* and *maintenance*, *scalability* (the ease with which the number of modules is increased or decreased), *flexibility* of both design and application, *system survivability* (graceful degradation under system failure), *reduced system costs*, and *improved system reliability* (use of multiple sensors and redundancy). Each module has its own hardware and software, driven and steered units, sensors, communication links, power unit, kinematics, path planning, obstacle avoidance, sensor fusion, and control systems. There is no central processor on the vehicle. Vehicle kinematics and dynamics are invariably nonlinear and sensor observations are also not linearly dependent on sensed states. These kinematics, models and observation spaces must be distributed to the vehicle modules. A single WMR vehicle consists of three driven and steered units (DSUs), three battery units, power unit, communication units and sensors. The DSUs communicate by use of transputer architecture. There is no fixed axle; all three wheels are driven and steered. Thus the WMR is omni-directional.

Figure 1.18 shows a modular WMR system consisting of two coupled WMRs with mounted RCDD sonar trackers. Each differential sonar sensor rotates and tracks environment features. By tracking the range and bearing to these features over time, two or more sonars can provide continuous updates of vehicle position. This vehicle system effectively demonstrates the scalability, flexibility, and adaptability of a modular WMR system. It consists of 18 modular chassis frames, six wheels and four DSUs. Both WMRs use the same software (modular software design) on component units. The processors on the scaled vehicle include four transputers for the four DSUs, one transputer to run software and two more to run server processes.

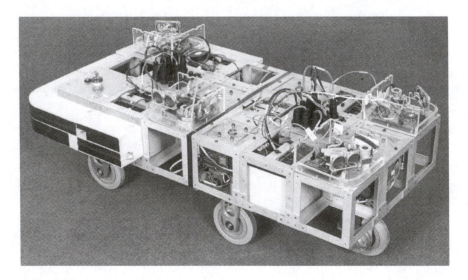

FIGURE 1.18
A Modular and Scalable WMR System

The vehicle system employs multiple sensors (sonar sensors, encoders, potentiometers) to measure its body position and orientation, wheel positions and velocities, obstacle locations and changes in the terrain. Sensor information from the modules is fused in a decentralized way and used to generate local control for each module. For the benefits of modularity to be functional and effective, fully decentralized and scalable multisensor fusion and control are mandatory.

A multisensor system such as the WMR system employs several sensors to obtain information in a real-world environment full of uncertainty and change. In order to provide autonomous navigation, path planning and obstacle avoidance for the WMR, it is essential that the information obtained from the sensors is interpreted and *combined* in such a way that a reliable, complete, and coherent description of the system is obtained. This is the *data fusion* problem. Multisensor fusion is the process by which information from many sensors is combined to yield an improved description of the observed system. The sensor fusion algorithms are then extended to optimal control algorithms for the WMR. The work presented in this book lays the foundation for the design and analysis of such multisensor fusion-based control systems.

1.8.2   The Mars Sojourner Rover

One robotic vehicle that fired many researchers' imagination in 1997 was the NASA Mars Pathfinder Mission's Sojourner Rover, which carried out exploration on Mars. The Prospector spacecraft containing the Rover landed on Mars on July 4th 1997. The Mars Pathfinder Rover team planned a vehicle traverse from the Rover Control Workstation at NASA (Jet Propulsion Laboratory) in Pasadena, California. Due to the speed-of-light time delay from Earth to Mars (11 minutes), and the constraint of a single uplink opportunity per day, the Rover was required to perform its daily operations autonomously. These activities included terrain navigation, rock inspection, terrain mapping, and response to contingencies.

During traverses the Rover used its look-ahead sensors (five laser stripe projectors and two CCD cameras) to detect and avoid rocks, dangerous slopes, and drop-off hazards, changing its path as needed before turning back toward its goal. Bumpers, articulation sensors, and accelerometers allowed the Rover to recognize other unsafe conditions. The hazard detection system could also be adopted to center the Rover on a target rock in preparation for deployment of its spectrometer. Other on-board experiments characterized soil mechanics, dust adherence, soil abrasiveness, and vehicle traverse performance. The Mars Rover is shown in Figure 1.19.

The capability of the Rover to operate in an unmodeled environment, choosing actions in response to sensor inputs to accomplish requested objectives, was unique among robotic space missions to date. As a complex and dynamic robotic vehicle characterized by a myriad of functions and different types of sensors while operating in an unmodeled and cluttered environment, the Sojourner Rover is an excellent example of a multisensor and multiactuator system.

Design modularity, decentralized estimation, and control provide certain advantages that would be relevant to the Rover. For example, if each wheel or unit is monitored and controlled by an independent mechanism, decentralized sensor processing and local control can permit the Rover to continue its mission even if one or more wheels/units are incapacitated.

FIGURE 1.19
The Mars Sojourner Rover (Photo Courtesy of NASA)

In addition, information from the various sensors will be efficiently utilized
to take optimal advantage of the redundancy inherent in the Rover's multiple sensors. The 23-pound Rover is solar-powered and can be controlled
from Earth by sending it path commands. The Rover control system can be
operated with or without feedback. The objective is to operate the Rover
with modest effects from disturbances (such as rocks) and with low sensitivity to changes in control gains and system parameter variations. Designing,
analyzing, and implementing robust adaptive and optimal control systems
for complex, compact, multisensor and multiactuator applications such as
the Rover provide the motivation for the material covered in this book.

1.8.3   Mars Surveyor 2001 Orbiter and Lander

The NASA Mars Surveyor 2001 is the next-generation spacecraft to be
sent to Mars and consists of an orbiter and lander launched separately on
Med-Lite launch vehicles. The Mars Surveyor 2001 Orbiter is scheduled
for launch on April 18, 2001. It will arrive at Mars on October 27, 2001,
if launched on schedule. The 2001 Orbiter will be the first to use the
atmosphere of Mars to slow down and directly capture a spacecraft into

orbit in one step, using a technique called aerocapture. It will then reach a circular mapping orbit within about one week after arrival. The Orbiter will carry two main scientific instruments, the Thermal Emission Imaging System (THEMIS) and the Gamma Ray Spectrometer (GRS). THEMIS will map the mineralogy and morphology of the Martian surface using a high-resolution camera and a thermal infrared imaging spectrometer. The GRS will achieve global mapping of the elemental composition of the surface and the abundance of hydrogen in the shallow subsurface. The gamma ray spectrometer was inherited from the lost Mars Observer mission. The 2001 Orbiter will also support communication with the Lander and Rover scheduled to arrive on Jan. 22, 2002. The orbiter is shown in Figure 1.20.

FIGURE 1.20
Mars Surveyor 2001 Orbiter (Photo Courtesy of NASA)

The Mars Surveyor 2001 Lander is scheduled for launch on April 10, 2001. It will land on Mars on Jan. 22, 2002, if launched on schedule. The 2001 Lander will carry an imager to take pictures of the surrounding terrain during its rocket-assisted descent to the surface. The descent imaging camera will provide images of the landing site for geological analyses, and will aid in planning for initial operations and traverses by the Rover. The 2001 Lander will also be a platform for instruments and technology experiments designed to provide key insights into decisions regarding successful and cost-effective human missions to Mars. Hardware on the Lander will be used for an *insitu* demonstration test of rocket propellant production using gases in the Martian atmosphere. Other equipment will characterize the Martian soil properties and surface radiation environment. The Lander is shown in Figure 1.21.

FIGURE 1.21
Mars Surveyor 2001 Lander (Photo Courtesy of NASA)

1.8.4   Deep Space 1

With the coming new millennium, NASA envisions an ambitious space exploration program through which the frontiers of the universe can be pushed back. Part of this vision is a challenge for engineers who must develop and design the control system for an extraordinary spacecraft of the future. The NASA *New Millennium Program*, with its advanced technology focus, is one of NASA's many efforts to develop and test an arsenal of cutting-edge technologies and concepts. Once proven to work, these technologies will be used by future missions to probe the universe. *Deep Space 1*, the spacecraft launched on October 24 1998 from Cape Canaveral, Florida, was the first in a series of deep space and Earth-orbiting missions that will be conducted to demonstrate new technologies in a space-borne testbed. Its mission to *"validate"* technologies pioneered the way for future spacecrafts to have an arsenal of technical capabilities for exploring the universe. As an experimental mission, Deep Space 1's primary goal was to test new technologies that had never been tested in space before: ion propulsion, autonomous systems, advanced microelectronics and telecommunications devices, and other exotic systems. The focus is on testing high-risk, advanced technologies in space with low-cost flights. Though testing technologies is the primary goal of the flights, they also allow an opportunity to collect scientifically valuable data. The idea was that most of the technologies on Deep Space 1 would have been completely validated during the first few months of flight, well before its encounter with an asteroid, where it would further test its new multisensor instrumentation, robust control system, and navigation technology. Figure 1.22 shows the Deep Space 1 spacecraft.

The spacecraft uses *ion propulsion* which is a technology which involves ionizing a gas to propel a craft. Instead of a spacecraft being propelled with standard chemicals, the gas xenon (which is like neon or helium, but heavier) is given an electrical charge, or ionized. It is then electrically accelerated to a speed of about 30 km/second. When xenon ions are emitted at such high speed as exhaust, they push the spacecraft in the opposite direction. The ultimate speed of a spacecraft using ion thrust depends upon how much propellant it carries. The same principle applies to chemical propulsion systems, although they are much less efficient. The ion propulsion system carries about 81.5 kilograms of xenon propellant, and it takes about 20 months of thrusting to use it all. It increases the speed of the spacecraft by about 4.5 kilometers per second, or about 10,000 miles per hour. Deep Space 4 is expected to use four ion engines to fly alongside a comet in 2004 so that it can land.

FIGURE 1.22
Deep Space 1 Spacecraft (Photo Courtesy of NASA)

1.8.5   The Experimental Unmanned Vehicle (XUV)

The motivation behind the USA Army Research Laboratory (ARL) DEMO III project is to provide 21st-century land forces with a family of highly mobile, multi-functional, intelligent, and unmanned ground vehicles. This work is part of what has been called *Force 21, or the Army after the Next Century.* The objective is to achieve a *"leap ahead"* capability across the spectrum of conflict in the digital battlefield of the future. The research work involves designing, verifying, validating, and building *experimental unmanned vehicles* (XUVs) such as the one shown in Figure 1.23. The mission is to use multiple, autonomous, and cooperating XUVs in battlefield environments that could be fraught with landmines, and chemical or biological weapons. In designing and evaluating the XUV, the performance goals can be grouped into five categories: mobility, planning, communications, control, and overall system performance. The mobility requirements include speed up to 40mph (daylight), 20mph (bad conditions), 10mph (night). A multi-thread, robust, adaptive, highly modular, and intelligent control system is required for the XUV. The competence of the vehicle behaviors is desired to be feasible at four interchangeable levels (which are dependent on the battle situation): *total autonomy, semi-autonomy, supervised autonomy,* and *tele-operation.* The work involves a myriad of technological issues such as intelligent subsystem architecture for robust and adaptive control, operator interface for nonautonomous behaviors, augmented machine vision, world model-based survivability behaviors, connection to the digital battlefield architecture, and extensive use of multiple sensors: FLIR, GHz radar, CCD camera, *mm* wave radar, laser, encoders.

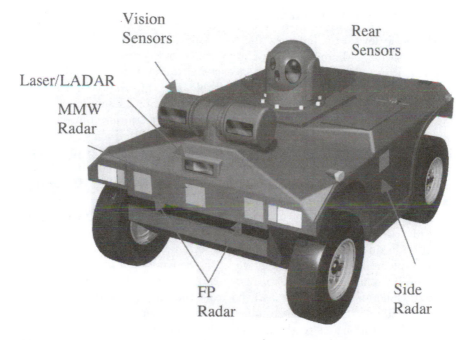

FIGURE 1.23
The Experimental Unmanned Vehicle (XUV) (Photo Courtesy of the USA ARL)

Some of the research problems in the XUV include the following: map sharing, information exchange, multisensor fusion, communication and coordination for multiple autonomous vehicles (XUVs) in battlefield environments. The map sharing capability is to use sensor information to augment the information available *a priori* from map data. In order to autonomously avoid these hazardous regions of terrain, multiple sensors and multisensor fusion algorithms will be integrated to detect and then to avoid these areas. Further issues addressed as part of the DEMO III project include requirements for different communication modes (e.g., vehicle-vehicle, operator-vehicle and center-vehicle), communication network topologies (and their reconfiguration), vehicle coordination outside communication range, and coordinated recovery from errors.

The XUV is essentially a wheeled mobile robot. Robotics can be understood as the *intelligent connection of perception to action.* The perception is achieved by the use of multiple sensors which send their information to an intelligent controller that generates intelligent instructions (control signals) and sends them to the actuator (motors and drives), which then executes the appropriate action.

1.8.6   Magnetic Levitation

Magnetic levitation provides the mechanism for several new modes of transportation. Magnetically levitated trains provide a high-speed, low-friction, low-noise alternative to conventional rail systems. The dynamics of magnetic levitation are inherently unstable, and thus require a controller to make the system behave in a useful manner. In fact, magnetic levitation systems are open-loop unstable, hence feedback control is required. Here the system is the mass of the train levitated by field-inducing coils. The input to the system is the current to the coils. The objective is to keep the train a safe distance above the coils. If the train is too high, it will leave the magnetic field and possibly veer to one side or the other. If the train is too low, it will touch the track with possibly disastrous results.

The response times in this system are fast, in the order of fractions of a second. Rather than using steel or rubber wheels, magnetically levitated trains ride on a nearly friction-free magnetic field. These trains provide a high-speed, low-friction alternative to the conventional metal-wheels-on-metal-rails configurations. The dynamics of the levitation system are inherently unstable, as can be deduced from the modeled thrust dynamics. Linearization leads to a transfer function between perturbations in the current and perturbations to the displacement from their nominal values and hence makes the use of feedback a very attractive option. Note that this system is inherently unstable; it has poles in the right-hand plane. A feedback control system can be designed by measuring the perturbations and passing the output through a controller and an actuator.

A magnetic levitation experiment, which dramatically demonstrates closed-loop levitation of permanent and ferromagnetic elements, is discussed later in the section that deals with control systems experiments.

1.8.7   Process Control: Metal Rolling Mill

Feedback control systems have been particularly successful in controlling manufacturing processes. Quantities such as temperature, flow, pressure, thickness, and chemical concentration are often precisely regulated in the presence of erratic behavior. In a steel-rolling mill, slabs of red-hot steel are rolled into thin sheets. Some new mills do a continuous pour of steel into sheets, but in either case, the goal is to keep the sheet thickness uniform and accurate. In addition, periodically the thickness must be adjusted in order to fulfill orders for different thicknesses. The response times in this system are relatively slow, from a few minutes to an hour.

1.9   Examples of Control System Experiments

This section outlines a number of new demonstrative, modern and practical control systems experiments that can be used to test and validate the material presented in this book. These experiments consist of real-time control design and implementation exercises that can be used in education and research. They are commercially available from such companies as *Educational Control Products* (ECP), *Quanser Consulting Incorporated*, and *Mechatronic Systems Incorporated*. The experiments are ideal for implementing and evaluating feedback strategies such as *proportional-integral-derivative (PID)*, *linear quadratic Gaussian (LQG)*, H_∞, *fuzzy logic*, *neural nets*, *adaptive* and *nonlinear* controllers. They are appropriate for all levels of university education and research, and include linear motion experiments that aptly utilize the concept of modularity. Modularity enables the reader to cost effectively employ the same power plant to perform experiments of varying complexity. By coupling the appropriate module to the plant, one achieves configurations ranging from simple position servo control to advanced MIMO systems such as the Seesaw/Pendulum. In addition, there are also available modular systems–specialty experiments including a 3-degrees of freedom (3-DOF) helicopter and earthquake simulating shaker table. Accessories include power modules and the MultiQ data acquisition. The experiments outlined are the *Inverted Pendulum, Magnetic Levitation, 2-DOF Helicopter, 3-DOF Helicopter, 2-DOF Robot* and *Pendubot*.

Most computer programs that are used to illustrate and demonstrate the design and analysis of control systems are implemented using MATLAB software. This is an interactive high-level programming language for numerical computation and data visualization. The basic principles and illustrative examples of this language are presented in Appendix B.

1.9.1   Inverted Pendulum

This unique ECP inverted pendulum design vividly demonstrates the need for and effectiveness of closed-loop control. It is not the conventional rod-on-cart inverted pendulum, but rather steers a horizontal rod in the presence of gravity to balance and control the vertical rod. As detailed analytically in the manual, the plant has both right-half plane poles and zeros as well as kinematic and gravitationally coupled nonlinearities. By adjusting mass properties, the characteristic roots may be varied to make the control problem range from being relatively simple to theoretically impossible. The inverted pendulum is shown in Figure 1.24.

FIGURE 1.24
The Inverted Pendulum Experiment (Photo Courtesy of ECP)

The mechanism includes removable and adjustable moment-arm counterweights on the vertical and horizontal rods for easy adjustment of plant dynamics. It features linear and rotary ball bearings at the joints for low friction and repeatable dynamic properties. This system is furnished with a set of experiments that provide for the identification of system dynamics and implement various control schemes for nonminimum phase and conditionally stable plants (right-half plane poles and zeros).

The Model 505 Inverted Pendulum apparatus is a unique mechanism that provides vivid demonstrations and challenging experiments for both undergraduate and graduate studies in controls. This novel design steers a horizontal sliding rod in the presence of gravity to balance and control the position of the vertical ("pendulum") rod. The mechanism is open-loop unstable (right-half plane pole) and non-minimum phase (right-half plane zero). As a result, feedback control is essential for stability and the structure of the controller must be selected carefully due the nonminimum phase characteristics.

1.9.2 2-DOF Helicopter Experiment

The Quanser 2-DOF (degrees of freedom) Helicopter experiment consists of a helicopter body with 2 motors equipped with propellers. The two propellers control the pitch and yaw of the body, which are measured using 4095 counts/rev encoders. The body is free to yaw an infinite number of times thanks to the slipring design. The slipring eliminates tangled wires and reduces the loading about the rotating axes to a minimum. The system is supplied with a joystick to compare human operator performance with computer control. The system is a highly coupled sixth order MIMO system. All mathematical models and controllers are supplied. The 2-DOF helicopter is shown in Figure 1.25.

The 2D flight simulator experiment consists of a helicopter model mounted on a fixed base. The helicopter model has two propellers driven by DC motors. The pitch propeller and the yaw propeller are used to control the pitch and yaw of the model. Motion about the two degrees of freedom is measured using two encoders. The purpose of the experiment is to design a controller that facilitates commanding a desired pitch and yaw angle. The performance of the system is also examined with an operator in the loop. A joystick is supplied that allows commanding the motors in an open-loop configuration in order to compare human operator performance with computer control. The joystick will also be used to provide the operator with a closed-loop controller that improves operator performance and makes the system easier to use. Electrical signals and power from the pitch encoder and the two motors are transmitted via a slipring. This allows for unlimited yaw and eliminates the possibility of wires tangling on the yaw axis.

A complete mathematical model including propeller dynamics and forces generated by the propellers is very difficult to obtain. The method used to design a controller is based on parameter estimation. Before parameter estimation can be performed however, a simple model that shows the relationships between the axes of motion and the inputs is developed. The pitch propeller is driven by a DC motor whose speed is controlled through the input voltage. The speed of rotation results in a force that acts normal

FIGURE 1.25
**2-DOF Helicopter Experiment (Photo Courtesy of Quanser Con-
sulting Inc.)**

to the body at a distance from the pitch axis. The rotation of the propeller
however, also causes a load torque on the motor shaft, which is in turn seen
at the yaw axis (parallel axis theorem).

- The dynamics of the rotating propellers will also affect the yaw and
  pitch axes but these effects are considered negligible.

- The response to a voltage input does not result in immediate response
  of the propeller speeds or output forces. The time constants are considered to be much faster than the body dynamics.

- Centrifugal forces during a yaw rotation will affect the pitch axis.

- Air turbulence around the body will result in unknown disturbances.

1.9.3   3-DOF Helicopter Experiment

The Quanser 3-DOF Helicopter experiment consists of a helicopter body with two motors equipped with propellers. The two propellers control the pitch and elevation of the body, which are measured using 4095 counts/rev encoders. The body is free to travel (circular motion) an infinite number of times thanks to the slipring design. The slipring eliminates tangled wires and reduces the loading about the rotating axes to a minimum. The system is supplied with a joystick to compare human operator performance with computer control. The system is a highly coupled eighth-order MIMO system. All mathematical models and controllers are supplied.

FIGURE 1.26
3-DOF Helicopter Experiment (Photo Courtesy of Quanser Consulting Inc.)

The 3-DOF Helicopter consists of a base upon which an arm is mounted as shown in Figure 1.26. The arm carries the helicopter body on one end and a counterweight on the other. The arm can pitch about an "elevation" axis as well as swivel about a vertical (travel) axis. Encoders mounted on these axes allow for measuring the elevation and travel of the arm. The helicopter body is free to swivel about a "pitch" axis. The pitch angle is measured via a third encoder. Two motors with propellers mounted on the helicopter body can generate a force proportional to the voltage applied to the motors. The force generated by the propellers can cause the helicopter body to lift off the ground. The purpose of the counterweight is to reduce the power requirements on the motors. The counterweight is adjusted such that applying about 1.5 volts to each motor results in hover. All electrical signals to and from the arm are transmitted via a slipring with eight contacts, thus eliminating the possibility of tangled wires and reducing the amount of friction and loading about the moving axes. The purpose of the experiment is to design a controller that facilitates commanding the helicopter body to a desired elevation and a desired travel rate or position.

The pitch axis is controlled by the difference of the forces generated by the propellers. If the force generated by the front motor is higher than the force generated by the back motor, the helicopter body will pitch up. The only way to apply a force in the travel direction is to pitch the body of the helicopter. This controller can also be designed as a second-order controller by selecting a desired peak time and damping ratio. The attractive feature in the structure of this controller is that we can limit the pitch command to not exceed a desired value. For example, we can limit pitch command to not exceed 20 degrees so that a "comfortable" ride is obtained. This limits the acceleration in the system.

1.9.4   2-DOF Robot Experiment

The Quanser 2-DOF Robot module converts two SRV-02 plants to an experiment in robotic control. Using the five bar mechanism, students can be taught the fundamentals of robotics such as *forward kinematics, inverse kinematics, singularities, force feedback control,* and *task teaching.* The 2-DOF robot is shown in Figure 1.27. This is the first of the three MIMO experiments that can be performed using the SRV-02. In order to perform this experiment, the following is required: 2x PA-0 1 03, 2x SRV-02 and one 2-DOF Robot module.

The 2-DOF Robot module consists of four links attached through joints and a support base to which two SRV-02s are mounted. The two SRV-02 plants are mounted to the support base using the eight supplied clamps. The purpose of the experiment is to design a controller that positions the end effector (joint 3) in a desired location and makes it track a previously learned trajectory.

1.9.5   Magnetic Levitation Experiment

ECP's unique MagLev plant dramatically demonstrates closed-loop levitation of permanent and ferromagnetic elements. The apparatus, with laser feedback and high flux magnetics, provides visually fascinating regulation and tracking demonstrations. The magnetic levitation experiment is shown in Figure 1.28. The system can be set up in both the open-loop stable and the unstable configurations. By using a repulsive field from the bottom

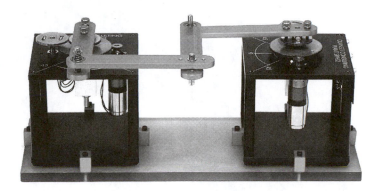

FIGURE 1.27
2-DOF Robot Experiment (Photo Courtesy of Quanser Consulting Inc.)

drive coil, a stable plant is effected, while the upper coil can be used in an attractive mode producing an open-loop unstable system. By adding a second magnet and driving both actuators, MIMO control is studied. The apparatus utilizes very high magnetic flux strengths to effect large displacements and thereby provide visually stimulating demonstrations.

The field interaction between the two magnets causes strong cross coupling and thus produces a true multivariable system. The inherent magnetic field nonlinearities can be inverted by using the provided real-time algorithms for linear control implementation or by considering the full system dynamics. An optional turntable accessory provides graphic demonstration of induced field levitation: the principal used in high-speed bullet trains. A set of experiments is included with this system that provides for the identification of system dynamics and implements various linear and nonlinear control schemes for the SISO and MIMO plant configurations.

1.9.6 The Pendubot

The Pendubot is a controls and robotics experimental platform produced by Mechatronic Systems, Inc. The name Pendubot is short for *Pendulum robot*. This name is derived from the fact that the Pendubot is like a two-link robot with the second motor removed, where the second link is unactuated and thus free to swing like a pendulum. In fact, the challenge in controlling the Pendubot is basically one of moving the first (or actuated) link to control the second (or passive) link. The Pendubot is, in some ways, similar to the classic inverted pendulum on a cart, where the linear motion

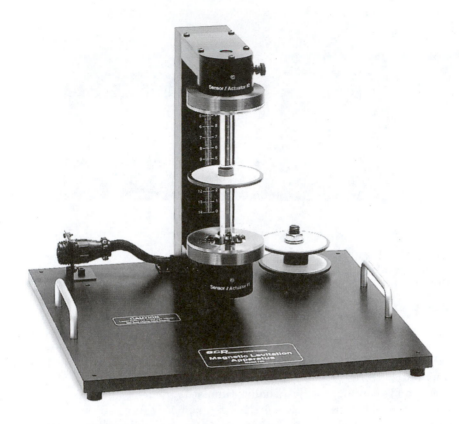

FIGURE 1.28
Magnetic Levitation Experiment (Photo Courtesy of ECP)

of the cart is used to balance the pendulum. The difference is that the Pendubot uses the rotational motion of the first link to balance the second (pendulum) link. Figure 1.29 shows a Pendubot.

This Pendubot is one of the newest and most advanced devices for controls education and research. It can be used for instruction at all levels; as simple demonstrations to motivate and instruct freshmen in the important concepts of dynamics and the systems approach to engineering design; in junior-level courses in linear control systems; in senior courses in mechatronics and real-time control; all the way to advanced graduate-level courses in nonlinear control theory. The Pendubot provides a platform where students can see first hand how linear models of nonlinear systems are, in general, operating point dependent, and what it means physically for a system to be uncontrollable.

FIGURE 1.29
The Pendubot (Photo Courtesy of Mechatronic Systems Inc.)

In addition, the Pendubot is a useful research platform in such areas as system identification, linear control, nonlinear control, optimal control, learning control, robust and adaptive control, fuzzy logic control, intelligent control, hybrid and switching control, gain scheduling, and other control paradigms. One can program the Pendubot for swing-up control, balancing, regulation and tracking, identification, gain scheduling, disturbance rejection, and friction compensation. The Pendubot can also be used in Robotics classes to illustrate and study kinematics (forward and inverse), robot dynamics, control of robots (position, velocity and force), and under-actuated robotic mechanisms.

The Pendubot consists of two rigid aluminum links, with the first link directly coupled to the shaft of a 90V permanent magnet DC motor mounted to a base. The motor mount and bearings are thus the support for the entire system. Link 1 also includes the bearing housing for joint two. Needle roller bearings riding on a ground shaft are used to construct the revolute joint for link 2. The shaft extends out both directions of the housing, allowing coupling to the second link and to an optical encoder mounted on link 1. The design gives both links full the 360 degrees of rotational motion. Two high-resolution optical encoders provide position information for feedback. One encoder is attached to the second joint and the other is attached to the motor. An advanced motion control 25A8 PWM servo amplifier is used to drive the motor.

1.10   Book Outline

This book is organized into nine chapters. **Chapter 1** introduces the concepts of control systems, discusses examples of control systems, and thus provides the motivation for the material covered in the book. In order to control a plant (or process), a mathematical model (description) of the plant is required. **Chapter 2** develops different techniques of obtaining and expressing mathematical descriptions for various physical systems, which include mechanical, electrical, electromechanical, thermal, chemical, and hydraulic systems. A system is any collection of interacting elements for which there are cause-and-effect relationships among the variables. The activity of mathematically capturing the behavior of physical systems is called system modeling. The chapter concentrates on dynamic systems, that is, systems whose variables are time-dependent. In most of the cases considered, not only will the excitations and responses vary with time but at any instant the derivatives of one or more variables will depend on the values of the system variables at that instant. Four forms of dynamic system models are presented, *the state-variable form, the input-output differential equation form, the transfer function form, and the block diagram form.* Methods for approximating a nonlinear system by a linear time-invariant model are developed. For time-varying or nonlinear systems that cannot be approximated by a linear and time-invariant model, computer solutions are employed. Derivation of models, conversion between models and analysis of models using MATLAB are presented.

Chapter 3 develops techniques for finding system responses for the dynamic systems modeled in Chapter 2. This activity is also called solving the model and involves using the mathematical model to determine certain features of the system cause-and-effect relationships. In order to design a control system, it is essential that the behavior of the plant (or process) is analyzed and understood. Three main mathematical approaches are used to obtain the system response: Direct solution of differential equations in the time domain, the use of the Laplace transform to solve differential equations in the frequency domain, and the deduction of system behavior from the system transfer function. Computer generation and analysis of system response using MATLAB are presented. The use of numerical methods, experimental time response data, and frequency-response data, instead of using analytical methods to determine the system response in circumstances where these methods are more feasible, are explored.

Chapter 4 introduces, develops, and analyzes the principles of feedback control systems and illustrates their characteristics and advantages. In several applications, there is a need to have automatic regulation and tracking. Quantities such as pressure, temperature, velocity, thickness, torque, and acceleration have to be maintained at desired levels. Feedback control is a convenient way in which these tasks can be accomplished. Two case studies, the cruise control system and the DC motor (both position and speed) control system are used as illustrative running examples throughout the chapter. The different types of controllers: Proportional (P), Proportional and Integral (PI), Proportional and Derivative (PD), Proportional and Integral and Derivative (PID) are discussed, together with their advantages and disadvantages. The concepts of system errors, tracking, disturbance rejection, and system type are covered. The notions of sensitivity, bounded input-bounded output (BIBO) stability, asymptotic internal stability, and Routh-Hurwitz stability are discussed and illustrated using examples.

Chapter 5 deals with the root locus design techniques, explains the procedure of creating root loci and outlines their uses. Definitions of the necessary terms are provided, including a step-by-step guide to constructing a root locus, and details of how to design and evaluate controllers using the root locus method. Given a feedback control system, the root locus illustrates how the poles of the closed-loop system vary with system parameters, in particular the closed-loop gain. Root locus is a powerful graphic method for analysis and design of control systems. Although this method is commonly used to study the effect of control gain variations, it can also be used to plot the roots of any polynomial expressed in the Evans root locus form. Most control systems work by regulating the system they are controlling around a desired operating point. The root locus method helps the designer of a control system to understand the stability and robustness properties of the controller at an operating point. Material presented in this chapter enables the reader to create a root locus and use the locus to understand the closed-loop system behavior given an open-loop system and a feedback controller. Case studies and examples that illustrate how to use the root locus for designing a control system are presented.

Chapter 6 seeks to investigate the steady state response of a dynamic system to sinusoidal inputs as the frequency varies. The design of feedback control systems in industry is accomplished by using frequency-response methods more often than any other method. Frequency-response design is popular primarily because it provides good designs in the face of uncertainty in the plant model. For example, for systems with poorly known or changing high-frequency resonances, the feedback compensation can be modified to alleviate the effects of those uncertainties. This modification is carried out more easily using frequency-response design than any other method. Another advantage of using frequency-response is the ease with which experimental information can be used for design purposes. The frequency-response design methods discussed in this chapter offer practical and important alternative approaches to the analysis and design of control systems. The main techniques covered include Bode plots, polar plots and Nyquist plots. The material presented empowers the reader with skills to hand draw these plots for a broad range of systems. Time-domain performance measures are developed in terms of the frequency response and then applied in system performance evaluation. These measures include gain margin, phase margin, and relative stability. The principles of compensation, lead and lag, are introduced and their applications discussed.

Chapter 7 discusses state space methods of analysis and design. The theoretical basics of the methods are discussed and followed by the actual methods. State space methods are analysis and design methods that use state variables, i.e., the analysis and the design are carried out in the state space. State space methods are somewhat simpler because they deal directly with the system states, which are first-order differential equations. Another advantage of these methods is their ability to handle multi-input multi-output (MIMO) systems. The chapter develops and demonstrates these advantages. The concepts of similarity transformations, observability and controllability, transfer function decomposition, and full state feedback control are discussed and illustrated using examples. Brief introductions to optimal control and estimator design are provided.

Chapter 8 addresses the issues involved in the design, analysis, and implementation of digital controllers. The rationale for digitization and using digital control is presented. The objective is to design and implement digital controllers such that the digitization and discretization effects of continuous time analog signals are either eliminated or minimized. First, the general characteristics of sampled data systems are introduced, and then an analysis of discrete time systems is presented including stability, root locus and frequency response analysis. The design and implementation of discrete time controllers is then discussed in detail. In particular, the discrete proportional and integral and derivative (PID) controller is developed and appraised. An overview of the hardware, software, and system integration issues is outlined.

In **Chapter 9** advanced topics and issues involved in the design and analysis of control systems are addressed. In particular, the subjects of discrete time estimation (both state space and information space), optimal stochastic control, and nonlinear control systems are presented. Adaptive control systems and robust control are briefly introduced. Multisensor fusion is the process by which information from a multitude of sensors is combined to yield a coherent description of the system under observation. General recursive estimation is presented and, in particular, the *Kalman filter* is discussed. A Bayesian approach to probabilistic information fusion is outlined and the notion and measures of information are defined. This leads to the derivation of the algebraic equivalent of the Kalman filter, the (linear) *Information filter*. State estimation for systems with nonlinearities is considered and the *extended* Kalman filter is treated. Linear information space is then extended to *nonlinear* information space by deriving the *extended* Information filter. The estimation techniques are then extended to LQG stochastic control problems, including systems involving nonlinearities, that is, the nonlinear stochastic control systems. This chapter also introduces the issues and concepts involved in the analysis and design of control systems for nonlinear dynamic systems.

Appendix A contains summaries of the properties of Laplace and $\mathcal{Z}$-transforms, including tables of key transforms. In **Appendix B** the basic principles and syntax of MATLAB are introduced and illustrated. Sample laboratory exercises are also provided.

1.11   Problems

Problem 1.1 *List the major advantages and disadvantages of closed-loop control systems with respect to open-loop systems.*

Problem 1.2 *The student-teacher learning process is inherently a feedback process intended to reduce the system error to a minimum. The desired output is knowledge being studied, and the student can be considered the process. Construct a feedback control system block diagram and identify the control blocks (sensor, actuator, process, controller, actual output, and desired output) of the system.*

Problem 1.3 *An engineering organizational system is composed of major groups such as management, research and development, preliminary design, experiments, product design and drafting, fabrication and assembling, and testing. These groups are interconnected to make up the whole operation.*

The system can be analyzed by reducing it to the most elementary set of components necessary that can provide the analytical detail required, and by representing the dynamic characteristics of each component by a set of simple equations. The dynamic performance of such a system can be determined from the relationship between progressive accomplishment and time.

Draw a functional block diagram showing an engineering organizational system.

Problem 1.4 *Many closed-loop and open-loop control systems can be found in homes.*

(a) List six such examples (three open-loop and three closed-loop).

(b) Construct feedback control system block diagrams for the six examples, and identify the control blocks (sensor, actuator, process, controller, actual output, and desired output).

Problem 1.5 *Give two examples of feedback control systems in which a human acts as a controller.*

Problem 1.6 *The following diagram depicts an automatic closed-loop system for paper moisture level control.*

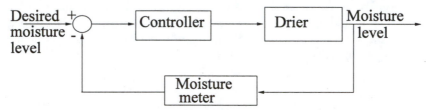

A Closed-Loop Control System for Paper Moisture

(a) Explain how this control system works.

(b) If the automatic controller is replaced by manual control, explain how the new system will function.

(c) Construct a feedback control system block diagram and identify the control blocks (sensor, actuator, process, controller, actual output, and desired output) of the new system.

Problem 1.7 *The following diagram depicts a closed-loop temperature control system.*

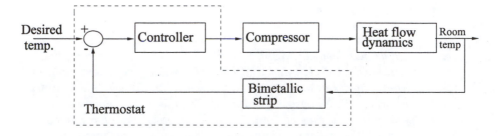

Room Temperature Control Using a Thermostat

(a) Explain how this control system works.

(b) If the automatic control is replaced by manual control, explain how the new system will function.

(c) Construct a feedback control system block diagram and identify the control blocks (sensor, actuator, process, controller, actual output, and de-·sired output) of the system.

Chapter 2

Modeling of Dynamic Systems

2.1 Introduction

In order to control a plant (or process), a mathematical model of the plant is required. This chapter develops different techniques of obtaining and expressing such mathematical descriptions for various physical systems, which include mechanical, electrical, electromechanical, thermal, chemical, and hydraulic systems. A system is any collection of interacting elements for which there are cause-and-effect relationships among the variables. This definition is necessarily general, in order to encompass a broad range of systems. The most important feature of the definition is that it indicates that the interactions among the variables must be taken into account in system modeling and analysis, rather than treating individual elements separately. A system can also be understood as a set of physical components connected so as to form a whole entity that has properties that are not present in the separate components. The activity of capturing in mathematical terms the behavior of physical systems is called system modeling.

This chapter concentrates on dynamic systems, that is, systems whose variables are time-dependent. In most of the cases considered, not only will the excitations and responses vary with time but at any instant the derivatives of one or more variables will depend on the values of the system variables at that instant. The basis for constructing a model of a system is rooted in physical laws such as conservation of energy, Newton's laws of motion and Kirchhoff's laws. These are the laws that the system elements and their interconnections are known to obey. The type of model sought will depend on both the objective of the engineer and the tools for analysis. If a pencil-and-paper analysis with parameters expressed in literal rather than numerical form is to be performed, a relatively simple model will be needed. To achieve this simplicity, the engineer should be prepared to neglect elements that do not play a dominant role in the system. On the other hand, if a computer is available for carrying out simulations of specific cases with parameters expressed in numerical form, a comprehensive mathematical model that includes descriptions of both primary and secondary effects might be appropriate. The modeling is restricted to lumped, continuous, non-quantized systems that can be described by sets of ordinary differential equations because well developed analytical techniques are available for solving linear ordinary differential equations with constant coefficients. Most of the examples will involve systems that are both linear and time-invariant. A method for approximating a nonlinear system by a linear time-invariant model will be developed. For time-varying or nonlinear systems that cannot be approximated by a linear and time-invariant model, one can resort to computer solutions. Four forms of dynamic system models are presented, the *state-variable form*, the *input-output differential equation form*, the *transfer function form* and the *block diagram form*.

2.1.1 Chapter Objectives

Upon completion of this chapter, the reader should be able to carry out the following tasks for a wide variety of dynamic systems such as mechanical (translational and rotational), electrical, thermal, chemical, and hydraulic systems:

- From the mathematical description of the system, construct a simplified version using idealized elements and define a suitable set of variables.

- Use the appropriate element and interconnection laws to obtain a mathematical model generally consisting of ordinary differential equations.

- If the model is nonlinear, determine the equilibrium conditions and, where appropriate, obtain a linearized model in terms of incremental variables.

- Arrange the equations that make up the model in readily usable forms such as the *state-variable form, input-output differential equation form, transfer function form* and *the block diagram.*

- Obtain system models from experimental data such as frequency response, transient response and stochastic steady state data.

- Compare and contrast the different dynamic system model forms and be able to convert from one form to the other.

2.2 Dynamic Systems

A dynamic system is a system whose variables are time-dependent, which means the inputs and outputs vary with time, and the derivatives of one or more variables will depend on the values of the system variables. There is a broad range of dynamic systems from different engineering disciplines. Such dynamic systems include the following:

- Mechanical (translational and rotational)

- Electrical

- Electromechanical

- Mechatronic

- Thermal and fluid

- Hydraulic and pneumatic

- Chemical and processing

- Civil and structural

- Manufacturing

Although these dynamic systems come from a wide spectrum of engineering fields, their modeling follows the same principles. What is specific to each type are the dynamic elements, the element laws and interconnection laws. Also, certain forms of the mathematical models are more useful in certain fields than others.

2.3 Dynamic System Models

The activity of capturing in mathematical terms the behavior of physical systems is called system mathematical modeling. The basis for constructing a system model consists of the physical laws, such as the conservation of energy, Newton's law and Kirchhoff's laws, which the system elements and their interconnections are known to obey. A dynamic system model is defined as the mathematical representation of the behavior of a dynamic system. The starting point in deriving such models is understanding the physics of the plant (the system being modeled), i.e., understanding what the dynamic elements are, the laws that govern them, how they affect each other and how the dynamic behavior is physically accomplished. Once this is mastered, a simplified diagrammatic representation of the system is constructed using idealized elements. In this way, a complex dynamic system can be represented by a simplified (albeit approximate) mathematical description. The representation is further broken into free-body diagrams or circuits to which element laws and interconnection laws are applied. The differential equations are then established using dynamic system laws such as Newton's laws (balance of forces or torques), Kirchhoff's current law (KCL) and Kirchhoff's voltage law (KVL) on the free-body diagrams and circuits.

2.3.1 Modeling Concepts

The issues involved in developing dynamic system models can be grouped as follows:

- Elements

- Element laws

- Interconnection laws

- Inputs and outputs

- State variables

- Free-body diagrams

- System model

2.4 Summary of the Model Derivation Procedure

The procedure of deriving models is similar for the different dynamic systems, and can be summarized as follows:

- Understand the physics and interaction of the elements of the dynamic system.

- Construct a simplified diagrammatic representation of the system using idealized elements.

- Apply element laws and interconnection laws.

- Draw the free-body or circuit diagrams.

- Identify or define the state variables, inputs, and outputs.

- Establish the system equations by using dynamic system laws such as D'Alembert's law, KCL and KVL.

- Obtain the desired form of the system model by manipulating the equations.

- If the model is nonlinear, determine the equilibrium conditions and obtain a linearized model.

2.4.1 Forms of the Dynamic System Model

There are several ways of expressing the differential equations that describe the dynamic behavior of a system. Put differently, the mathematical representation of a dynamic system can take different forms. The form of the dynamic system model employed depends on both the objective of the modeling and the available tools for analysis. Also, certain forms are more useful or practical for particular types of dynamic systems than others. There are four general forms of dynamic system models discussed in this book:

- State-variable matrix form

- Input-output differential equation form

- Transfer function form

- Block diagram form

2.5 Overview of Different Dynamic Systems

In this section an overview of different dynamic systems is carried out. The motivation is to identify the physical elements involved, the element laws, the interconnection laws, and the typical state variables in these systems. The following types of systems are outlined and reviewed: translation mechanical, rotational mechanical, electrical, electromechanical, hydraulic (fluid), and thermal systems.

2.5.1 Translational Mechanical Systems

In this section the variables, element laws, and interconnection laws for linear (translational mechanical) dynamic systems are presented. Translational systems are systems that can have only horizontal or/and vertical motion. For masses that can move vertically, the gravitational forces must be considered. Either force or displacement inputs can be applied to any part of the system. An applied force is a known function of time, but the motion of the body to which it is applied is not known at the beginning of a problem. Conversely, a displacement input moves some part of the system with a specified motion, but the force exerted by the external mechanism moving that part is normally not known. Displacements can be measured with respect to fixed reference positions or with respect to some other moving body. When relative displacements are used, it is important to keep in mind that the inertial force of a mass is always proportional to its absolute acceleration, not to its relative acceleration. To model a system, free-body diagrams are drawn, and the forces acting on every mass and junction point whose motion is unknown are summed up. The free-body diagram for a massless junction is drawn in the usual way, except that there is no inertial force [5].

The modeling issues specific to the modeling of translational systems are presented in this section. After introducing the variables to be used, the laws for the individual elements are presented in addition to the laws governing the interconnections of the elements. Next, the use of free-body diagrams as an aid in formulating the equations of the model is presented. Inputs consist of either the application of a known external force or the movement of a body with a known displacement. If the system contains an ideal pulley, then some parts can move horizontally and other parts vertically. Special situations, such as free-body diagrams for massless junctions and rules for the series or parallel combination of similar elements are also treated.

The modeling process can sometimes be simplified by replacing a series-parallel combination of stiffness or friction element by a single equivalent element. Special attention is given to linear systems that involve vertical motion. If displacements are measured from positions where the springs are neither stretched nor compressed, the gravitational forces must be included in the free-body diagrams for any masses that can move vertically.

2.5.1.1   State Variables

State variables of translational mechanical systems are the displacement $x(t)$, measured in meters (m), the velocity $v(t)$, which is measured in meters per second (m/s), the acceleration $a(t)$, measured in meters per second per second (m/s^2), and the force $f(t)$ which is measured in newtons (N).

$$v = \frac{dx}{dt}.$$

$$a = \frac{dv}{dt} = \frac{d^2x}{dt^2}.$$

In addition to the variables mentioned above, the energy w in joules (J) and power p in watts (w) are also defined variables.

$$p = fv.$$

$$p = \frac{dw}{dt}.$$

The energy supplied between time t_0 to t_1 is given by

$$\int_{t_0}^{t_1} p(\tau)d\tau,$$

and total energy supplied is obtained from

$$w(t) = w(t_0) + \int_{t_0}^{t} p(\tau)d\tau,$$

where $w(t_0)$ is the energy supplied up to time t_0.

2.5.1.2 Element and Interconnection Laws

The system model must incorporate both the element laws and the interconnection laws. The element laws involve displacements, velocities, and accelerations. Since the acceleration of a point is the derivative of the velocity, which in turn is the derivative of the displacement, all the element laws can be written in terms of x and its derivatives or in terms of $x(t)$, $v(t)$, and $a(t)$. It is important to indicate the assumed positive directions for displacements, velocities, and accelerations. The assumed positive directions for $a(t), v(t)$, and $x(t)$ will always be chosen to be the same, so it will not be necessary to indicate all three positive directions on the diagram. In this book, dots over the variables are used to denote derivatives with respect to time. For example,

$$\frac{dx}{dt} = \dot{x} \tag{2.1}$$

$$\frac{d}{dt}(mv) = f. \tag{2.2}$$

Equation 2.2 is Newton's second law and for a *constant* mass system it becomes,

$$m\frac{dv}{dt} = f. \tag{2.3}$$

The kinetic energy and the potential energy are given by

$$w_k = \frac{1}{2}mv^2$$

$$w_p = mgh.$$

D'Alembert's law, which will be formally developed later for both translational and rotational systems, is a special form of Newton's second law where an *inertial force* is defined as $-m\dfrac{dv}{dt}$ such that

$$\sum_{i=1}^{n}(f_{ext})_i - m\frac{dv}{dt} = 0, \tag{2.4}$$

where f_{ext} is an external force acting on the mass m. D'Alembert's law is applied to each mass or junction point whose velocity or acceleration is unknown beforehand. To do so, it is useful to draw a free-body diagram for each such mass or point, showing all external forces and the inertial force by arrows that define their positive senses. The element laws are used to express all forces except inputs in terms of displacements, velocities, and accelerations.

2.5.1.3 Friction

When two bodies slide over each other there is a frictional force $f_f(t)$ between them that is a function of the relative velocity Δv between the sliding surfaces.

$$
\begin{aligned}
f_f &= b\Delta v \\
&= b(v_2 - v_1), \tag{2.5}
\end{aligned}
$$

where b is the friction constant. A friction force that obeys such a linear relationship is modeled using viscous friction. The direction of the frictional force is such that it is in the opposite direction to the sliding motion. However, friction can be used to *transmit motion* from one moving mass to another mass with which it is in contact.

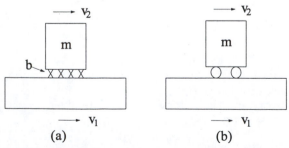

(a)   (b)

Sliding Bodies (a) with Friction (b) with Negligible Friction

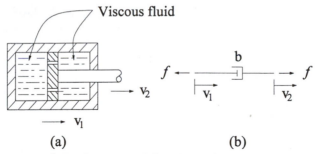

(a)   **(b)**

A Dashpot: Modeling Viscous Friction

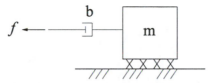

Force Transmitted in a Dashpot

The viscous friction described by Equation 2.5 is a linear element, for which the plot of f vs. Δv is a straight line passing through the origin, as shown in (a) in the following figure. In the same figure, (b) and (c) show examples of friction that obey nonlinear relationships: dry friction and drag friction, respectively.

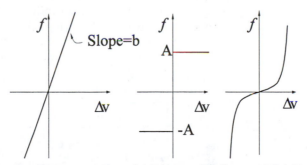

Friction Characteristics: (a) Linear, (b) Dry and (c) Drag

Dry friction is modeled by a force that is independent of the magnitude of the relative velocity, but dependent on the direction (sign) of the relative speed. Drag friction is caused by resistance to a body moving through a fluid such as wind resistance.

2.5.1.4   Spring Stiffness

When a mechanical element is subjected to a force f and goes through a change in length Δx, it can be characterized by a stiffness element. A common stiffness element is the spring and its characteristics shown in the following figure.

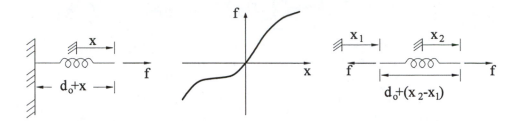

Characteristics of a Spring

For a linear spring, the force is given by

$$f = k\Delta x$$
$$= k(x_2 - x_1),$$

and the potential energy stored in the spring is

$$w_p = \frac{1}{2}k(\Delta x)^2.$$

A spring (stiffness) element is assumed to be massless. When a force f is applied to one side of a spring a force equal in magnitude but of opposite direction must be experienced on the other side. Thus, for the following figure the force f passes through the first spring and is exerted directly on the mass m.

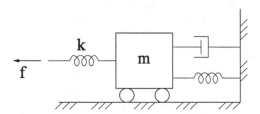

Force Transmitted Through a Spring

2.5.1.5   Interconnection Laws: D'Alembert's law

D'Alembert's law is developed from Newton's law for translational systems as follows:

$$\sum_{i=1}^{n}(f_{ext})_i = m\frac{dv}{dt}, \qquad (2.6)$$

where f_{ext} is an external force acting on a mass m moving with linear velocity $v(t)$. The term $-m\dfrac{dv}{dt}$ is called the *inertial force* and is always in the opposite direction to the direction of the mass's motion (the mass's displacement, velocity or acceleration). It follows that

$$\sum_{i=1}^{n}(f_{ext})_i - m\frac{dv}{dt} = 0$$

$$\sum_{i=1}^{n} f_i = 0. \tag{2.7}$$

This means that the sum of all the forces acting on a mass, including the inertial force, is zero. This is D'Alembert's law for translational mechanical systems.

2.5.1.6 The Law of Reaction Forces

In order to relate the forces exerted by the elements of friction and stiffness to the forces acting on a mass or junction point, Newton's third law of reaction forces is required. It states that for every force of an element on another, there is an equal and opposite *reaction force* on the first element. This is illustrated in the following diagram.

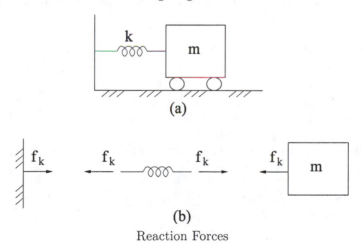

Reaction Forces

2.5.1.7 Series and Parallel Combination

Springs and dashpots can be placed in series or parallel. The following diagrams show two springs and two dashpots in series and parallel, respectively.

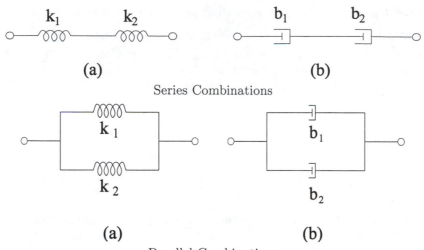

Series Combinations

Parallel Combinations

The equivalent stiffness for a parallel combination of springs is given by

$$k_{eq} = k_1 + k_2.$$

For a series combination it is given by

$$k_{eq} = \frac{k_1 k_2}{k_1 + k_2}.$$

For dashpot, the equivalent damping factor for a parallel combination is

$$b_{eq} = b_1 + b_2$$

and for a series combination it is

$$b_{eq} = \frac{b_1 b_2}{b_1 + b_2}.$$

Derivation of the these equations is left as an exercise for the reader. It is interesting to note that the equations are opposite to those of electrical resistors in the same configuration.

Series and parallel combinations can be mixed in the same configuration as shown in the following figure.

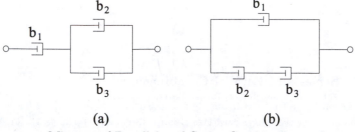

Mixture of Parallel and Series Combinations

The equivalent damping factor for configuration (a) is given by

$$b_{eq} = \frac{b_1(b_2 + b_3)}{b_1 + (b_2 + b_3)},$$

while that for configuration (b) is given by

$$b_{eq} = b_1 + \left(\frac{b_2 b_3}{b_2 + b_3} \right).$$

2.5.2 Rotational Mechanical Systems

Rotational mechanical systems are modeled using the same techniques as those for translational mechanical systems. The starting point is to introduce the three rotational elements that are analogs of mass, friction, and stiffness in translational systems. Two other elements, levers and gears, are characterized in a somewhat different way. The use of interconnection laws and free-body diagrams is very similar to their use for translational systems. All the four types of models: state-variable matrix form, output-input equation, transfer function, and block diagram are developed for rotational mechanical systems. Combined translational and rotational systems are also considered.

2.5.2.1 State Variables

For rotational mechanical systems, the symbols used for the variables are angular displacement $\theta(t)$ measured in radians (rad), angular velocity $\omega(t)$ measured in radians per second (rad/s), angular acceleration $\alpha(t)$ measured in radians per second squared (rad/s^2), and torque $T(t)$ measured in Newton-meters $(N.m)$.

$$\omega = \dot{\theta}$$
$$\alpha = \dot{\omega} = \ddot{\theta}.$$

The power supplied to the rotating body is

$$p = T\omega,$$

and the energy supplied up to time t is

$$w(t) = w(t_0) + \int_{t_0}^{t} p(\tau)d\tau.$$

2.5.2.2 Element and Interconnection Laws

The elements used to represent physical devices in rotational systems are moment of inertia, friction, stiffness, levers, and gears. Consideration is restricted to elements that rotate about fixed axes in an inertial reference frame.

The moment of inertia is given by

$$J = \int r^2 dm.$$

where r is the distance from the axis of reference and dm is the mass of the small element.

$$\frac{d}{dt}(J\omega) = T$$
$$J\dot{\omega} = T$$

Parallel axis theorem states that

$$J = J_0 + ma^2,$$

where a is the distance between the parallel axes and J_0 is the moment of inertia about the first (principal) axis.

2.5.2.3   Friction

A rotational friction element is one for which there is an algebraic relationship between the torque and the relative angular velocity between surfaces. Rotational viscous friction arises when two rotating bodies are separated by a film of oil. For a linear element, the curve must be a straight line passing through the origin. The power supplied to the friction element, is immediately lost to the mechanical system in the form of heat. The torque is proportional to the relative angular velocities as given by

$$T = b\Delta\omega$$
$$= b(\omega_2 - \omega_1).$$

2.5.2.4   Stiffness

Rotational stiffness is usually associated with a torsional spring, such as the main spring of a clock, or with a relatively thin shaft. It is an element for which there is an algebraic relationship between $T(t)$ and θ, and it is generally represented as shown in the following diagram. Since it is assumed that the moment of inertia of a stiffness element is either negligible or represented by a separate element, the torques exerted on the two ends of a stiffness element must be equal in magnitude and opposite in direction. For a linear torsional spring or flexible shaft,

$$T = k\Delta\theta.$$

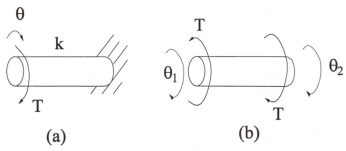

(a)   (b)

Rotational Stiffness

Potential energy is stored in a twisted stiffness element and is given by (for a linear spring or shaft),

$$W = \frac{1}{2}k\theta^2,$$

where $\theta(t)$ is the angular displacement.

2.5.2.5 The Lever

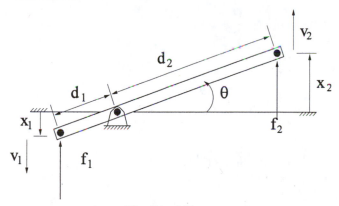

The Ideal Lever

An ideal lever is assumed to be a rigid bar that is pivoted at a point and has negligible mass, friction, momentum, and stored energy. In all the examples considered the pivot point is fixed. For small displacements

$$x_2 = \left(\frac{d_2}{d_1}\right)x_1.$$

By differentiating this equation the velocity is obtained

$$v_2 = \left(\frac{d_2}{d_1}\right)v_1.$$

By taking moments about the pivot it follows that

$$f_2 d_2 - f_1 d_1 = 0$$
$$f_2 = \left(\frac{d_1}{d_2}\right) f_1.$$

2.5.2.6   Gears

An ideal gear has no moment of inertia, no stored energy, no friction, and a perfect meshing of the teeth. Any inertia or bearing friction in an actual pair of gears can be represented by separate lumped elements in the free-body diagrams. The relative sizes of the two gears result in a proportionality constant for the angular displacements, angular velocities, and transmitted torques of the respective shafts. For purposes of analysis, it is convenient to visualize the pair of ideal gears as two circles that are tangent at the contact point and rotate without slipping. The spacing between teeth must be equal for each gear in a pair, so the radii of the gears are proportional to the number of teeth. Thus if r and n denote the radius and number of teeth, respectively, then a gear ratio N is defined as follows:

$$N = \frac{r_2}{r_1} = \frac{n_2}{n_1}$$
$$r_1 \theta_1 = r_2 \theta_2$$
$$\frac{\theta_1}{\theta_2} = \frac{r_2}{r_1} = N,$$

where θ_1 and θ_2 are the angular displacements for the gears. Differentiating the above equation leads to

$$\frac{\omega_1}{\omega_2} = \frac{r_2}{r_1} = N$$
$$\frac{T_2}{T_1} = -\frac{r_2}{r_1} = -N$$
$$T_1 \omega_1 + T_2 \omega_2 = 0.$$

2.5.2.7   Interconnection Laws

The interconnection laws for rotational systems involve laws for torques and angular displacements that are analogous to those for translational mechanical systems. The law governing reaction torques has an important modification when compared with the one governing reaction forces.

2.5.2.8 D'Alembert's Law

The D'Alembert's law is developed from Newton's law for rotational systems as follows:

$$\sum_{i=1}^{n} (T_{ext})_i = J \frac{d\omega}{dt} \tag{2.8}$$

where T_{ext} is an external torque acting on a body with constant moment of inertia J and rotating with angular velocity $\omega(t)$ about a fixed axis. The term $-J\frac{d\omega}{dt}$ is called the *inertial torque* and is always in the opposite direction to the direction of motion of the body. It follows that

$$\sum_{i=1}^{n} (T_{ext})_i - J \frac{d\omega}{dt} = 0$$

$$\sum_{i=1}^{n} T_i = 0. \tag{2.9}$$

This means that the sum of torques acting on a body, including the inertial torque, is zero. This is D'Alembert's law for rotational mechanical systems.

2.5.2.9 The Law of Reaction Torques

For bodies that are rotating about the same axis, any torque exerted by one element on another is accompanied by a reaction torque of equal magnitude and opposite direction on the first element.

2.5.2.10 Obtaining the System Model

The methods for using the element and interconnection laws to develop an appropriate mathematical model for a rotational system are the same as those discussed for translational mechanical systems. For each mass or junction point whose motion is unknown *a priori*, a free-body diagram is normally drawn showing all torques, including the inertial torque. All the torques except inputs are expressed in terms of angular displacements, velocities, or accelerations by means of the element laws. Then D'Alembert's law is applied. The state-variable matrix system is established, as will be discussed later.

2.5.3 Electrical Systems

Except at quite high frequencies, an interconnection of lumped elements and a very important portion of the applications of electrical phenomena can model a circuit by using ordinary differential equations and applying solution techniques discussed in this book. In this section, linear time-invariant circuits are considered using the same approach used for mechanical systems. The elements and interconnection laws are introduced and then combined to develop procedures for establishing the system model for a circuit.

2.5.3.1   State Variables

The variables most commonly used to describe the behavior of circuits are $v(t)$ voltage in volts (V) and current $i(t)$ in amperes (A). The related variable $q(t)$, charge in coulombs (C) is defined from

$$i = \frac{dq}{dt}$$

and

$$q(t) = q(t_0) + \int_{t_0}^{t} i(\tau)d\tau.$$

2.5.3.2   Element and Interconnection Laws

The elements in electrical circuits that will be considered include resistors, capacitors, inductors, and sources (both current and voltage). The first three of these are referred to as passive elements because, although they can store or dissipate energy that is present in the circuit, they cannot introduce additional energy. They are analogous to the dashpot, mass, and spring for mechanical systems. In contrast, sources are active elements that can introduce energy into the circuit and that serve as the inputs. They are analogous to the force or displacement inputs for mechanical systems.

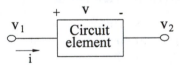

Denoting Circuit Voltages

A Resistor and its Variables

2.5.3.3   Resistor

A resistor is an element for which there is an algebraic relationship between the voltage across its terminals and the current through it, that is, an element that can be described by a curve of $v(t)$ vs. $i(t)$. A linear resistor is one for which the voltage and current are directly proportional to each other, that is, one described by Ohm's law:

$$v = Ri$$
$$i = \frac{1}{R}v.$$

2.5.3.4 Capacitor

A capacitor is an element that obeys an algebraic relationship between the voltage and the charge, where the charge is the integral of the current. The energy stored in the capacitor is a function of the voltage across its terminals and the characteristics of the capacitor are defined by the following equations:

$$q = Cv$$
$$i = C\frac{dv}{dt}$$
$$v(t) = v(t_0) + \frac{1}{C}\int_{t_o}^{t} i(\tau)d\tau.$$

The energy stored is given by

$$W = \frac{1}{2}Cv^2.$$

The following diagram shows the symbols of a capacitor and inductor.

(a)   (b)

Symbols of a Capacitor and an Inductor

2.5.3.5 Inductor

An inductor is an element for which there is an algebraic relationship between the voltage across its terminals and the derivative of the flux linkage.

The energy stored is a function of current.

$$v = \frac{d}{dt}(Li)$$

$$v = L\frac{di}{dt}$$

$$i(t) = i(t_0) + \frac{1}{L}\int_{t_o}^{t} v(\tau)d\tau$$

$$W = \frac{1}{2}Li^2.$$

To obtain the model of a circuit in state-variable form, an appropriate set of state variables is defined and then an equation for the derivative of each state variable in terms of only the state variables and inputs is derived. The choice of state variables is not unique, but they are normally related to the energy in each of the circuit energy-storing elements. Kirchhoff's voltage law (KVL) and Kirchhoff's current law (KCL) are used to obtain the circuit equations. All the techniques that have been discussed can still be used. The only basic difference is that the objective is to retain the variables $v_C(t)$ and $i_L(t)$ wherever they appear in the equations and to express other variables in terms of them.

2.5.3.6   Amplifiers and Controlled Sources

Some important types of electrical elements, unlike those in earlier sections, have more than two terminals to which external connections can be made. Controlled sources are considered in this section, and the frequently used operational amplifier receives special attention. Controlled sources arise in the models of transistors and other electronic devices. Rather than being independently specified, the values of such sources are proportional to the voltage or current somewhere else in the circuit. One purpose for which such devices are used is to amplify electrical signals, giving them sufficient power [5]. For example, to drive loud-speakers, instrumentation, or various electromechanical systems.

Ideal voltage and current amplifiers are shown in Figure 2.1 (a) and (b) [5]. The models for many common devices have two bottom terminals connected together. They may also include the added resistors shown in Figure 2.1 (c) and (d). For (c) to approach (a), R_n must be very large and R_o very small. In order for (d) to approach (b), R_n must be very small and R_o very large.

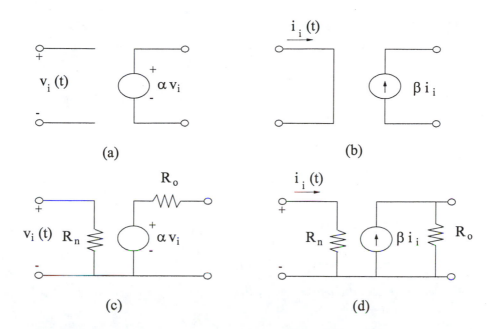

(a)   (b)   (c)   (d)

FIGURE 2.1
(a), (b) Ideal Voltage and Current Amplifiers. (c), (d) Amplifiers with Internal Resistance Causing Non-ideal Behavior.

Example 2.1 *Find the relationship between the input voltage $v_i(t)$ and the output voltage $v_o(t)$ in the following op-amp circuit.*

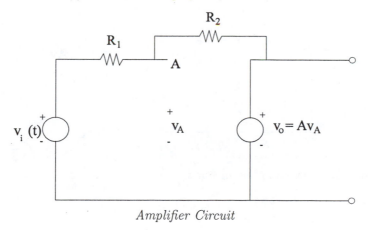

Amplifier Circuit

Solution 2.1 *Summing up the current at node A gives*

$$\frac{1}{R_1}[v_A - v_i(t)] + \frac{1}{R_2}(v_A - v_o) = 0$$

Since $v_o = Av_A$ it follows that

$$\left(\frac{1}{R_1} + \frac{1}{R_2} - \frac{A}{R_2}\right) v_A = \frac{1}{R_1} v_i(t).$$

Multiplying both sides of this equation by $R_1 R_2$, solving for v_A, and then setting $v_o = Av_A$ leads to

$$v_o = Av_A = \left[\frac{AR_2}{R_2 + (1-A)R_1}\right] v_i(t) = \left[\frac{R_2}{-R_1 + \frac{1}{A}(R_1 + R_2)}\right] v_i(t)$$

Note that for very large values of A,

$$v_o = -\frac{R_2}{R_1} v_i(t).$$

Under these conditions, the size of the voltage gain is determined solely by the ratio of the resistors.

Example 2.2 Find the relationship between the input voltage $v_i(t)$ and the output voltage $v_o(t)$ for the following op-amp circuit.

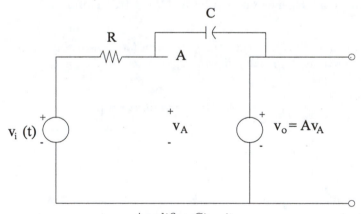

Amplifier Circuit

Solution 2.2 Summing the currents at node A gives

$$\frac{1}{R}[v_A - v_i(t)] + C(\dot{v}_A - \dot{v}_o) = 0.$$

Since $v_o = Av_A$, v_A can be replaced by v_o/A to give

$$\frac{1}{AR} v_o + \frac{C}{A}\dot{v}_o - C\dot{v}_o = \frac{1}{R} v_i(t).$$

Dividing both sides of the equation by C and rearranging terms, the input-output differential equation is obtained.

$$\left(1 - \frac{1}{A}\right)\dot{v}_o - \frac{1}{ARC}v_o = -\frac{1}{RC}v_i(t).$$

For very large values of A, this reduces to

$$\dot{v}_o = -\left(\frac{1}{RC}\right)v_i(t).$$

If there is no initial stored energy, then $v_o(0) = 0$, and

$$v_o = -\frac{1}{RC}\int_o^t v_i(\lambda)d\lambda.$$

The circuit is then called an integrator, because its output is proportional to the integral of the input.

2.5.3.7 The Operational Amplifier

The *operational amplifier* (often called an *op-amp*) is a particularly important building block in the electrical part of many modern systems. The device typically contains more than 20 transistors plus a number of resistors and capacitors, and it may have 10 or more external terminals [5]. However, its basic behavior is reasonably simple. There are two input terminals for time-varying signals and one output terminal. The symbol for the device and its equivalent circuits are shown in Figure 2.2. Complete physical descriptions can be found in electronics books and other textbooks [5], [7]. For an ideal op-amp, no current flows into the input terminals, but the output current is unknown. The op-amp gain is usually large enough so that the voltage between the two input terminals can be assumed to be zero.

The input terminals marked with the minus and plus signs are called the inverting and non-inverting terminals, respectively. The voltages are denoted with respect to the ground (which is at zero-volt reference) by v_A and v_B. One of the input terminals is often connected to the ground point, but this is not necessary. Typical values of r_n exceed $10^6\Omega$, and r_o is normally less than 100Ω. In most applications, the resistance r_n can be replaced by an open circuit, and r_o by a short circuit, leading to the simplified model as shown. Then no current can flow into the device from the left, and the output voltage is $v_o = A(v_A - v_B)$. The voltage amplification is extremely large, typically exceeding 10^5.

Note that the symbol in Figure 2.2 (a) does not show the ground point. In fact, the device itself does not have an external circuit that can be connected directly to ground. There are, however, terminals for the attachment

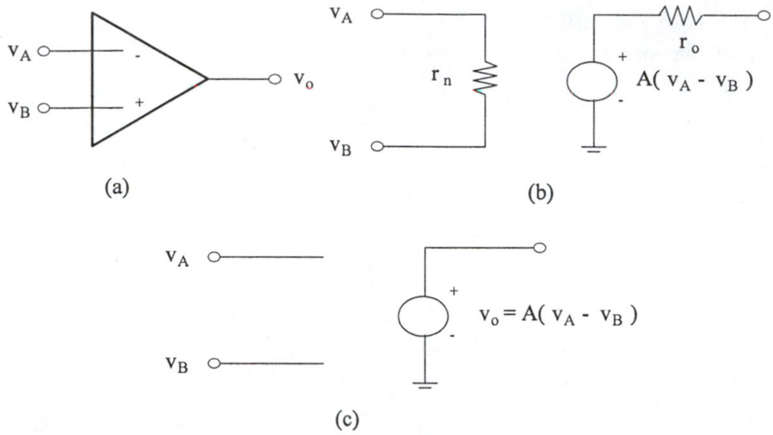

FIGURE 2.2
Operational Amplifier: (a) Schematic Diagram (b) Equivalent Circuit (c) Idealized Equivalent Circuit

of positive and negative bias voltages. The other ends of these constant voltages are connected to a common junction, which is the ground point that appears in (b) and (c) of the figure. Circuit diagrams involving op-amps must always show which of the other elements are connected to this external ground. Sometimes ground symbols appear in several different places on the diagram, in which case they all can be connected together. However, the diagrams for our examples will already have had this done. Because our interest is in the time-varying signals, the constant bias voltages are not normally shown on the circuit diagram.

Example 2.3 *Find the relationship between the input voltage $v_i(t)$ and the output voltage $v_o(t)$ for the circuit shown in the following diagram.*

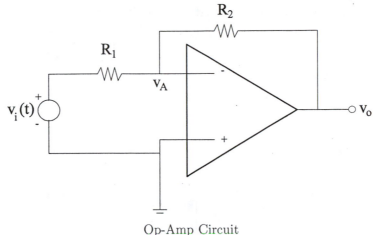

Op-Amp Circuit

Solution 2.3 *With the op-amp replaced by the ideal model, the circuits are equivalent to those done earlier except that* $v_o = -Av_A$ *rather than* $v_o = Av_A$. *The results are the same except that A is replaced by* $-A$.

$$v_o = \left(\frac{-R_2}{R_1 + \dfrac{1}{A}(R_1 + R_2)} \right) v_i(t),$$

which for very large values of A becomes

$$v_o = -\frac{R_2}{R_1} v_i(t).$$

Example 2.4 *Find the relationship between the input voltage* $v_i(t)$ *and the output voltage* $v_o(t)$ *for the circuit in the following diagram*

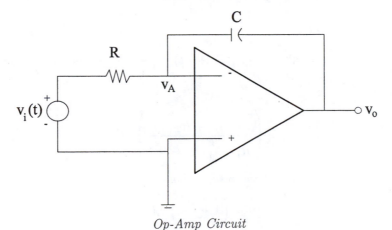

Op-Amp Circuit

Solution 2.4 *The problem is solved using similar approach.*

$$\left(1 + \frac{1}{A}\right)\dot{v}_o + \frac{1}{ARC}v_o = -\frac{1}{RC}v_i(t),$$

which for large values of A becomes

$$\dot{v}_o = -\frac{1}{RC}v_i(t).$$

Example 2.5 *Find an expression for the relationship between the input voltage $v_i(t)$ and the output voltage $v_o(t)$ for the circuit shown in the next diagram.*

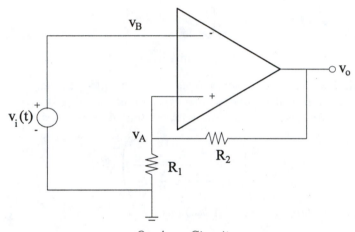

Op-Amp Circuit

Solution 2.5 *Since there is no current flowing into the terminals of the op-amp, the voltage divider rule can be used,*

$$v_A = \frac{R_1}{R_1 + R_2}v_o.$$

It then follows that

$$v_o = A[v_i(t) - v_A] = Av_i(t) - \left(\frac{AR_1}{R_1 + R_2}\right)v_o,$$

from which

$$v_o = \left(\frac{R_1 + R_2}{R_1 + \frac{1}{A}(R_1 + R_2)}\right)v_i(t).$$

For very large values of A the result becomes

$$v_o = \left(1 + \frac{R_2}{R_1}\right)v_i(t).$$

In the foregoing examples, note how the elements connected around the op-amp completely determine the behavior when the model with $A \to \infty$ is used to get a simpler expression. Since A is so large in practice, an easier method is often used to get the simpler expression directly. The output voltage of the device, given by $v_o = A(v_A - v_B)$, must be finite, so that the voltage $v_A - v_B$ between the input terminals must approach zero when A is very large. In practice, this voltage really is a tiny fraction of one volt, such as $0.1 mV$.

Assuming that the voltage difference $(v_A - v_B)$ is virtually zero, it is sometimes called the *virtual-short concept*, because the voltage across a short circuit is zero [5]. However, unlike a physical short circuit represented by an ideal wire (through which current could flow), it must be still assumed that no current flows into either of the input terminals. These two principles can be summarized as follows:

- Virtual short concept: $v_A - v_B = 0$, and if v_B is grounded then $v_A = v_B = 0$.

- No current flows into the op-amp.

Example 2.6 *Use the virtual-short concept to determine the relationship between the input voltage $v_i(t)$ and the output voltage $v_o(t)$ for the circuit shown below, when A is very large.*

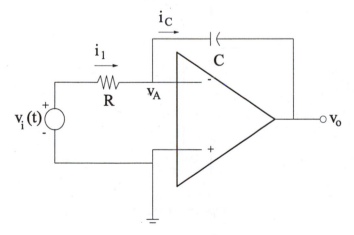

Op-Amp Circuit: Virtual-Short Principle

Solution 2.6 *The solution proceeds by applying KCL at A and using the two main principles associated with an op-amp: (1) no current flow into* the op-amp terminals, and (2) virtual-short concept, $v_A = v_B = 0$.

$$i_R = i_C \qquad \text{(no current flow into the op-amp terminals)}$$

$$\frac{v_i(t) - v_A}{R} = C\dot{v}_C$$

$$\frac{v_i(t) - v_A}{R} = C(\dot{v}_A - \dot{v}_o)$$

$$\frac{v_i(t)}{R} = -C\dot{v}_o \qquad \text{(virtual-short concept, } v_A = 0)$$

$$\dot{v}_o = -\frac{1}{RC}v_i(t).$$

This result is in agreement with the result from the earlier approach.

Example 2.7 *Use the virtual-short concept to determine the relationship between the input voltage $v_i(t)$ and the output voltage $v_o(t)$ for the circuit shown below, when A is very large.*

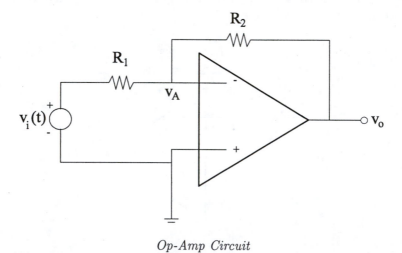

Op-Amp Circuit

Solution 2.7 *As in the previous example the solution proceeds by applying KCL at A while using two principles associated with an op-amp: (1) no current flow into the op-amp terminals, and (2) the virtual-short concept,*

$v_A = v_B = 0.$

$$i_{R_1} = i_{R_2} \qquad \textit{(no current flow into the op-amp terminals)}$$

$$\frac{v_i(t) - v_A}{R_1} = \frac{v_A - v_o}{R_2}$$

$$\frac{v_i(t)}{R_1} = \frac{-v_o}{R_2} \qquad \textit{(virtual-short concept, } v_A = 0 \textit{)}$$

$$v_o = -\frac{R_2}{R_1} v_i(t).$$

This result is in agreement with that from the earlier approach.

2.5.4 Electromechanical Systems

A wide variety of very useful devices is produced by combining electrical and mechanical elements. Among the electromechanical devices that will be considered are potentiometers, galvanometers, microphones, accelerometers, motors and generators. A detailed modeling process for the DC motor is presented in Chapter 4. A detailed example of an electromechanical system will be presented later in the chapter.

2.5.5 Pneumatic, Hydraulic and Fluid Systems

A hydraulic system is one in which liquids, generally considered incompressible, flow. Hydraulic systems commonly appear in chemical processes, automatic control systems, actuators and drive motors for manufacturing equipment. Such systems are usually interconnected to mechanical systems through pumps, valves, and movable pistons. A turbine driven by water and used for driving an electric generator is an example of a system with interacting hydraulic, mechanical, and electrical elements. The more general topic of fluid systems, which would include compressible fluids such as gases and air, will not be considered. An exact analysis of hydraulic systems is usually not feasible because of their distributed nature and the nonlinear character of the resistance to flow. For the dynamic analysis satisfactory results can be obtained by using lumped elements and linearizing the resulting nonlinear mathematical models. On the other hand, the design of chemical processes requires a more exact analysis wherein static, rather than dynamic, models are used. In most cases, hydraulic systems operate with the variables remaining close to a specific operating point. Thus models involving incremental variables are of importance. This fact is particularly helpful because such models are usually linear, although the model in terms of the total variables may be quite nonlinear. The principles involved in modeling fluid systems are presented below.

An orifice is a restriction in a fluid flow passage as illustrated below.

R

P_1 $\longrightarrow$ Q P_2

The Orifice: A Restriction in a Fluid Flow Passage

Such a restriction in the pipe presents hydraulic resistance to the flow of the fluid. The terms $p_1(t)$ and $p_2(t)$ represent the fluid pressure at the two ends (in N/m^2 or Pa), $Q(t)$ is the fluid volumetric flow rate (in m^3/s), and R is the hydraulic resistance (in Ns/m^5). From experimental research it has been established that the flow rate and the pressure drop across the orifice are related by

$$Q = k\sqrt{\Delta p} \quad \text{where } k \text{ is a constant}$$

and $\Delta p = p_1 - p_2$.

Graphically, this information can be represented as follows:

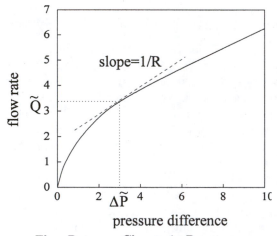

Flow Rate vs. Change in Pressure

Liquid storage exhibits hydraulic capacitance. Consider liquid stored in an open vessel of uniform cross-sectional area, A. The height of the liquid is h. Then the volume of liquid stored is given by

$$v = Ah.$$

The pressure at the base of the vessel is given by

$$p = \rho g h + p_a$$

$$p = \frac{\rho g}{A} v + p_a,$$

where p_a is the atmospheric pressure. Graphically, this information can be displayed as follows:

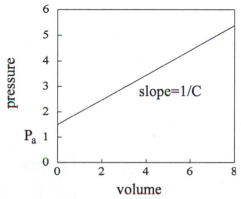

Pressure vs. Volume

C is the hydraulic capacitance of the vessel defined as the reciprocal of the pressure (p) vs. volume (v) curve, that is

$$C = \frac{A}{\rho g}$$

$$p = \frac{1}{C} v + p_a.$$

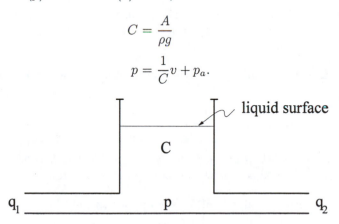

A Reservoir with an Inlet and an Outlet

The rate of liquid storage is given by

$$q_s = q_1 - q_2$$

$$\text{or} \quad \dot{v} = q_1 - q_2,$$

where $\dot{v}$ is the rate of change of volume. Hence, using the fact that

$$p = \frac{1}{C}v + p_a,$$

it follows that

$$\dot{p} = \frac{1}{C}\dot{v}$$
$$= \frac{1}{C}q_s.$$

Example 2.8 *Determine the equations that describe the height of the water in the tank shown below.*

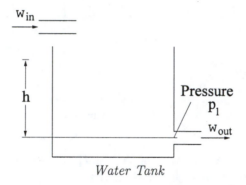

Water Tank

A is the area of the tank, ρ is the density of the water, $h = m/A\rho$ represents the height of the water and m is the mass of water in the tank.

Solution 2.8 *Using the equations derived above, the solution is obtained as*

$$\dot{h} = \frac{1}{A\rho}(w_{in} - w_{out}),$$

where A is area of the tank, ρ is the density of water, $h = m/A\rho$ is the height of the water, and m is the mass of water in the tank

Force equilibrium (balance of forces) applies to fluid flow systems as it does for mechanical systems. However, for fluid flow systems some forces may result from fluid pressure acting on a piston. The force f from the fluid is given by

$$f = pA,$$

where p is the fluid pressure and A is the area on which the fluid acts.

Example 2.9 *Determine the differential equation that describes the motion of the hydraulic piston shown below where there is a force F_D acting on the piston, and the pressure in the chamber is p.*

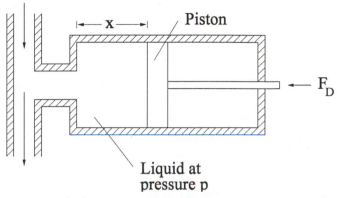

Hydraulic Piston

Solution 2.9 *The fluid flow laws apply directly, where the forces include the fluid pressure as well as the applied force. Hence the result is*

$$m\ddot{x} = Ap - F_D,$$

where A is the area of the piston, m is the mass of the piston, and $x(t)$ is the position of the piston.

For vessels with variable cross-sectional area, the pressure vs. volume curves are given below.

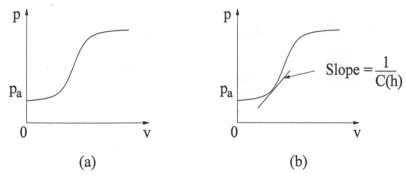

Pressure vs. Volume: Variable Cross-sectional Area $A(h)$

Hence the hydraulic capacitance is given by

$$C(h) = \frac{1}{\left(\dfrac{dp}{dv}\right)} = \frac{dv}{dp}$$

$$= \frac{dv}{dh}\frac{dh}{dp} \qquad \text{(chain rule of differentiation)}$$

$$= \frac{A(h)}{\rho g} \qquad \left(\text{from } \frac{dv}{dh} = A(h) \text{ and } \frac{dh}{dp} = \frac{1}{\rho g}\right).$$

From these equations, it can be deduced that for a vessel with constant area A the pressure equation reduces to

$$p = \frac{\rho g}{A}v + p_a,$$

which is as expected from previous discussions. The pressure vs. volume curve is also familiar from previous discussions and is shown below.

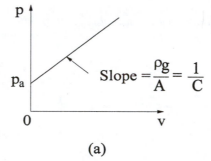

(a)

Pressure vs. Volume: Constant Area A

The hydraulic capacitance C is given by

$$C = \frac{A}{\rho g}.$$

The general rate of change of pressure at the base of the vessel is derived as follows

$$v(t) = v(0) + \int_0^t [w_{in}(\lambda) - w_{out}(\lambda)]d\lambda$$

$$\dot{v} = w_{in}(t) - w_{out}(t)$$

$$\dot{h} = \frac{1}{A(h)}[w_{in}(t) - w_{out}(t)]$$

$$\dot{p} = \frac{1}{C(h)}[w_{in}(t) - w_{out}(t)].$$

Fluid flows are common in many control systems components, one of the most common being the hydraulic actuators used to move the control surfaces on airplanes. The physical relations governing fluid flow are continuity, force equilibrium, and resistance.

In many cases of fluid-flow problems the flow is restricted either by a constriction in the path or by friction. The general form of the effect of resistance is given by the mass flow rate. Flow rates in between these extremes can yield intermediate values. The Reynolds number indicates the relative importance of inertial forces and viscous forces in the flow. It is proportional to a material's velocity and density and to the size of the restriction, and it is inversely proportional to the viscosity.

The basic variables in a hydraulic system are the flow rate and pressure. Other variables that are equivalent to the pressure at the bottom of a container are the volume of the liquid and the liquid's height. Since hydraulic systems are generally nonlinear, especially in the resistance to fluid flow, linearized models valid in the vicinity of an operating point are developed. Passive elements of hydraulic capacitance and hydraulic resistance are introduced in constructing such models. The former is associated with the potential energy of a fluid in a vessel, the latter with the energy dissipated when fluid flows through valves, orifices, and pipes.

2.5.6   Thermal Systems

Thermodynamics, heat transfer, and fluid dynamics are each subjects of complete textbooks. For purposes of generating dynamic models for use in control systems, the most important aspect of the physics is to represent the dynamic interaction between the variables. Experiments are usually required to determine the actual values of the parameters and thus to complete the dynamic model for purposes of control systems design. Thermal systems are systems in which the storage and flow of heat are involved. Their mathematical models are based on the fundamental laws of thermodynamics. Examples of thermal systems include a thermometer, an automobile engine's cooling system, an oven, and a refrigerator. Generally thermal systems are distributed, and thus they obey partial rather than ordinary differential equations. In this book, attention will be restricted to lumped mathematical models by making approximations where necessary. The purpose is to obtain linear ordinary differential equations that are capable of describing the dynamic response to a good approximation. The principles involved in modeling thermal systems are outlined below.

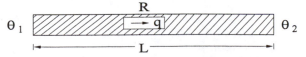

An Insulated Slab with Ends at Two Different Temperatures

The physical law for heat flow in a conductor is the Fourier law

$$\theta_1 - \theta_2 = Rq,$$

where θ is the temperature in *Kelvins* (K), q is the heat flow rate (J/s), and R is the thermal resistance (K/W). The temperatures and the heat flow may be time varying.

The figure below shows the cross section of a perfectly insulated conductor of mass m, which has a specific heat capacity c, and is initially at temperature θ_1.

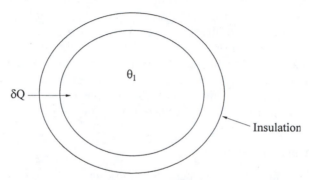

A Perfectly Insulated Conductor

The expression of the heat flow rate can be derived by considering a small quantity of heat δQ being injected into the mass in order to raise the temperature to θ_2 :

$$\delta Q = mc(\theta_1 - \theta_2)$$

$$\frac{dQ}{dt} = mc\frac{d\theta}{dt}$$

$$q = C\frac{d\theta}{dt},$$

where C is the thermal capacity of the mass $(C = mc)$.

Example 2.10 *An electric oven has total mass m and initially its ambient temperature is θ_a. The input heat flow rate is q_i and the oven temperature is θ. There is heat loss through the walls of the oven which have thermal resistance R. A diagram of the oven is shown below.*

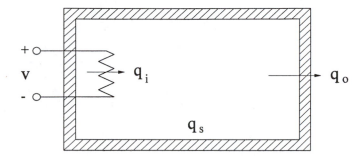

The terms $q_s(t)$ and $q_o(t)$ represent the rate of heat storage and the rate heat loss, respectively. Obtain the differential equation that describes the thermal behavior of the oven.

Solution 2.10 *The differential equation is obtained by first considering energy balance within the oven as follows:*

$$q_i - q_o = q_s$$

$$q_o = \frac{\theta - \theta_a}{R}$$

$$q_s = mc\frac{d\theta}{dt}$$

$$mc\frac{d\theta}{dt} + \frac{\theta - \theta_a}{R} = q_i$$

$$mc\frac{d\theta}{dt} + \frac{1}{R}\theta = q_i + \frac{1}{R}\theta_a.$$

This is the differential equation relating $\theta(t)$ to $q_i(t)$ and θ_a.

2.6 State-Variable Matrix Form

This form of the dynamic system model is achieved by first considering the system equations, choosing variables of interest and identifying the inputs and outputs. The system is then modeled by finding equations of the derivatives of the variables in terms of the variables and the inputs [5]. The outputs are also expressed in terms of variables and inputs. The state-variable matrix form is then developed by pulling out the coefficients of the state variables and inputs in order to obtain the matrices **A**, **B**, **C** and $\mathbf{D}$ such that

$$\dot{\mathbf{x}}(\mathbf{t}) = \mathbf{A}\mathbf{x}(\mathbf{t}) + \mathbf{B}\mathbf{u}(\mathbf{t}) \qquad (2.10)$$

$$\mathbf{y}(\mathbf{t}) = \mathbf{C}\mathbf{x}(\mathbf{t}) + \mathbf{D}\mathbf{u}(\mathbf{t}), \qquad (2.11)$$

where $\mathbf{x}(t)$ is a vector of state variables, $\dot{\mathbf{x}}(t)$ is a vector of the derivatives of the state variables, $\mathbf{u}(\mathbf{t})$ is a vector of inputs and $\mathbf{y}(\mathbf{t})$ is a vector of outputs. $\mathbf{A}$, $\mathbf{B}$, $\mathbf{C}$ and $\mathbf{D}$ are matrices of coefficients of the state variables and inputs and the general matrix elements are represented by a_{ij}, b_{ij}, c_{ij} and d_{ij}. For linear time-invariant systems the matrices $\mathbf{A}$, $\mathbf{B}$, $\mathbf{C}$ and $\mathbf{D}$ are fixed, which means that their elements are constants.

2.6.1   Choice of State Variables

The procedure of formulating the state variablesystem model begins with the selection of a set of state variables. This set of variables must completely describe the effect of the past history of the system and its response in the future. Although the choice of state variables is not unique, the state variables for dynamic systems are usually related to the energy stored in each of the system's energy-storing elements [5]. Since any energy that is initially stored in these elements can affect the response of the system at a later time, one state variable is normally associated with each of the independent energy storing elements. Hence, the number of independent state variables that can be chosen is equal to the number of independent energy storing elements in the system. The fact that the state variables are independent means it is impossible to express any state variable as an algebraic function of the remaining state variables and the inputs. A system with state variables chosen with this constraint is said to be minimal (or minimized) and the matrix $\mathbf{A}$ is full rank, and thus invertible. In some systems, the number of state variables is larger than the number of energy-storing elements because a particular interconnection of elements causes redundant variables or because there is need for a state variable that is not related to the storage of energy. The latter might occur when a particular variable or parameter in a system has to be monitored or controlled where this can only be achieved by using an extra state (redundant) variable. In such a state-variable matrix form the system is not minimal and matrix $\mathbf{A}$ is not full rank (i.e. not invertible). It is not possible to model a system using a fewer number of state variables than the number of independent energy storing elements in the system.

Example 2.11 *Two connected cars with an applied input force $u(t)$ and negligible rolling friction can be represented by a translational mechanical system as shown below.*

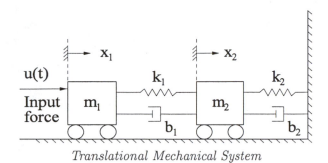

Translational Mechanical System

(a) Draw the free-body diagrams of this mechanical system.

(b) Write down the differential equations that represent the system.

(c) Explain why the vector of state variables should be chosen as

$$\mathbf{x}(t) = \begin{bmatrix} x_1 & v_1 & x_2 & v_2 \end{bmatrix}^T.$$

(d) Express these differential equations in the state-variable form, where the output is the frictional force in b_1.

Solution 2.11 *(a) The free-body diagrams for the translational mechanical system are drawn as shown below.*

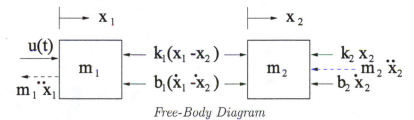

Free-Body Diagram

(b) Balancing the forces (D'Alembert's law) on the free-body diagrams of masses m_1 and m_2, respectively, gives

$$m_1 \ddot{x}_1 + b_1(\dot{x}_1 - \dot{x}_2) + k_1(x_1 - x_2) = u(t) \tag{2.12}$$

$$m_2 \ddot{x}_2 + k_2 x_2 + b_2 \dot{x}_2 = b_1(\dot{x}_1 - \dot{x}_2) + k_1(x_1 - x_2) \tag{2.13}$$

(c) There are four energy storing elements, two springs and two masses. The potential energy stored in a spring is a function of the displacement $x(t)$ and the kinetic energy stored in a moving mass is a function of the velocity $v(t)$ (which is equal to $\dot{x}(t)$). Hence, the state variables corresponding to the energy storing elements are given by

$$x(t) = \begin{bmatrix} x_1 & v_1 & x_2 & v_2 \end{bmatrix}^T.$$

(d) The state-variable form is achieved by obtaining expressions of the derivatives of these variables in terms of the variables and the input. Two of these expressions are simple:

$$\dot{x}_1 = v_1$$

$$\dot{x}_2 = v_2.$$

The other two expressions, for $\dot{v}_1$ and $\dot{v}_2$, are obtained by rearranging Equations 2.12 and 2.13.

$$m_1\ddot{x}_1 + b_1(\dot{x}_1 - \dot{x}_2) + k_1(x_1 - x_2) = u(t)$$

$$m_1\dot{v}_1 + b_1(v_1 - v_2) + k_1(x_1 - x_2) = u(t)$$

$$\dot{v}_1 = -\frac{b_1}{m_1}v_1 + \frac{b_1}{m_1}v_2 - \frac{k_1}{m_1}x_1 + \frac{k_1}{m_1}x_2$$

$$+\frac{1}{m_1}u(t)$$

$$m_2\ddot{x}_2 + k_2x_2 + b_2\dot{x}_2 = b_1(\dot{x}_1 - \dot{x}_2) + k_1(x_1 - x_2)$$

$$m_2\dot{v}_2 + k_2x_2 + b_2v_2 = b_1(v_1 - v_2) + k_1(x_1 - x_2)$$

$$\dot{v}_2 = \frac{b_1}{m_2}v_1 - \left(\frac{b_1}{m_2} + \frac{b_2}{m_2}\right)v_2 + \frac{k_1}{m_2}x_1 -$$

$$\left(\frac{k_1}{m_2} + \frac{k_2}{m_2}\right)x_2.$$

The results can be summarized as follows:

$$\dot{x}_1 = v_1$$

$$\dot{v}_1 = -\frac{b_1}{m_1}v_1 + \frac{b_1}{m_1}v_2 - \frac{k_1}{m_1}x_1 + \frac{k_1}{m_1}x_2 + \frac{1}{m_1}u(t)$$

$$\dot{x}_2 = v_2$$

$$\dot{v}_2 = \frac{b_1}{m_2}v_1 - \left(\frac{b_1}{m_2} + \frac{b_2}{m_2}\right)v_2 + \frac{k_1}{m_2}x_1 - \left(\frac{k_1}{m_2} + \frac{k_2}{m_2}\right)x_2.$$

The output, the frictional force in b_1, is given by

$$y(t) = b_1(v_1 - v_2)$$

$$= b_1v_1 - b_1v_2.$$

By extracting the coefficients of the variables and inputs, the state-variable matrix system is obtained.

$$
\begin{bmatrix} \dot{x}_1 \\ \dot{v}_1 \\ \dot{x}_2 \\ \dot{v}_2 \end{bmatrix} =
\begin{bmatrix}
0 & 1 & 0 & 0 \\
-\dfrac{k_1}{m_1} & -\dfrac{b_1}{m_1} & \dfrac{k_1}{m_1} & \dfrac{b_1}{m_1} \\
0 & 0 & 0 & 1 \\
\dfrac{k_1}{m_2} & \dfrac{b_1}{m_2} & -\dfrac{(k_1+k_2)}{m_2} & -\dfrac{(b_1+b_2)}{m_2}
\end{bmatrix}
\begin{bmatrix} x_1 \\ v_1 \\ x_2 \\ v_2 \end{bmatrix} +
\begin{bmatrix} 0 \\ \dfrac{1}{m_1} \\ 0 \\ 0 \end{bmatrix} u(t)
$$

$$
y(t) = \begin{bmatrix} 0 & b_1 & 0 & -b_1 \end{bmatrix}
\begin{bmatrix} x_1 \\ v_1 \\ x_2 \\ v_2 \end{bmatrix} + \begin{bmatrix} 0 \end{bmatrix} u(t).
$$

2.6.2   State-Variable Form for Nonlinear Systems

For linear time-invariant systems, all of the elements of $\mathbf{A}$, $\mathbf{B}$, $\mathbf{C}$ and $\mathbf{D}$ represented by a_{ij}, b_{ij}, c_{ij} and d_{ij} (coefficients of state variables and inputs) are constants. For linear systems whose parameters vary with time, some of these elements are functions of time. For nonlinear elements, some of the coefficients are functions of the state variables or inputs. Hence, in the general case of time-varying, nonlinear systems, the right-hand side of the state variable and output equations are nonlinear functions of the state variables, the inputs, and time such that,

$$
\dot{\mathbf{x}}(t) = \mathbf{f}(\mathbf{x}, \mathbf{u}, t) \tag{2.14}
$$
$$
\mathbf{y}(t) = \mathbf{g}(\mathbf{x}, \mathbf{u}, t). \tag{2.15}
$$

This means that for nonlinear systems the matrices $\mathbf{A}$, $\mathbf{B}$, $\mathbf{C}$ and $\mathbf{D}$ cannot be extracted, and state-variable form takes the general structure shown in Equations 2.14 and 2.15. Two other situations where the matrices cannot be extracted occur when derivatives of two independent state variables appear in the same state-variable equation, and when derivatives of the inputs appear in the state-variable equation. While these more general and more complicated systems are addressed in this chapter, emphasis is placed on linear time-invariant systems where the state-variable matrices can be easily extracted.

2.6.3   Characteristics of State-Variable Models

One of the characteristics of state-variable models is that they are easily captured by matrix notation and are thus amenable to the techniques of linear algebra. For example, any number of first-order state-variable equations can be represented by a single matrix differential equation merely by making the appropriate definitions. Furthermore, when one is dealing with complex multi-input, multi-output systems, using matrix concepts and properties leads to an understanding of system behavior that would be difficult to achieve otherwise. An additional advantage of using matrices is that one can apply many of the theoretical properties of matrices such as multiplication, evaluation of determinants, and inversion, to the study of dynamic systems once their models have been put into matrix form. The state-variable matrix model is obtained by pulling out the coefficients of the state variables and the inputs to obtain the coefficient matrices $\mathbf{A}, \mathbf{B}, \mathbf{C}$ and $\mathbf{D}$. These matrices can then be used in MATLAB (Appendix B) simulation to obtain the system response and design controllers to control the system response. State-variable equations are particularly convenient for complex multi-input, multi-output systems. They are often written in matrix form, and, in addition to their computational advantages, they can be used to obtain considerable insight into system behavior. The state variable concept has formed the basis for many of the theoretical developments in system analysis, in particular the state space methods discussed in Chapter 7.

2.6.4   Summary of the State-Variable Form Modeling

In the remainder of this section, a variety of examples illustrate the technique of deriving the mathematical model in state-variable form. The general approach employed in deriving the mathematical model in state-variable matrix form can be summarized as follows:

- Choose the state variables and identify the inputs and outputs.

- Draw free-body diagrams or circuit diagrams for the system elements.

- Obtain a set of differential equations by using dynamic system laws (mechanical, electrical, chemical, hydraulic, etc.)

- Manipulate the differential equations into state-variable form by expressing the derivative of each state variable as an algebraic function of the state variables and inputs. Take note of simple cases such as $\dot{x} = v$ where both x and v are state variables.

- Express the output variables as algebraic functions of the state variables and the inputs.

- Write the state variable and output equations in matrix form by pulling out the coefficients of the state variables and the inputs in order to obtain the matrices $\mathbf{A}, \mathbf{B}, \mathbf{C}$ and $\mathbf{D}$.

2.7   Input-Output Differential Equation Form

This form of the dynamic model is developed by expressing the system equations in terms of the inputs and the outputs, and their respective derivatives, while eliminating all other variables. This representation is thus in the form of input-output differential equations. For a system with one input $u(t)$ and one output $y(t)$, the output has the general form,

$$f(y, \dot{y}, \ddot{y}, ..., y^{(n)}) = f(u, \dot{u}, \ddot{u}, ..., u^{(m)})$$

$$a_0 y + a_1 \dot{y} + a_2 \ddot{y} + ... + a_n y^{(n)} = b_0 u + b_1 \dot{u} + b_2 \ddot{u} +$$
$$... + b_m u^{(m)}, \tag{2.16}$$

where

$$y^{(n)} = \frac{d^n y}{dt^n} \quad \text{and} \quad u^{(m)} = \frac{d^m u}{dt^m}.$$

For linear and time-invariant systems all the coefficients in Equation 2.16 are constants. Equation 2.17 shows a simple input-output differential equation with one input and one output

$$3y + \dot{y} + 2\ddot{y} = 4u + 7\dot{u}. \tag{2.17}$$

For systems with more than one output, the right side of the input-output differential equation will include additional input terms. If there are several outputs, there is need for a separate but similar equation for each output. For example, the following pair of equations represent a system with two outputs and three inputs,

$$3y_1 + \dot{y}_1 + 2\ddot{y}_1 = 4u_1 + u_2 + 7\dot{u}_3 \tag{2.18}$$

$$y_2 + \dot{y}_2 + 2\ddot{y}_2 = 4u_1 + 7u_2 + u_3. \tag{2.19}$$

Assuming the form of the input(s) is known, in the general case, each of the input-output differential equations involves only one unknown variable and its derivatives. Thus, unlike state-variable equations, each equation can be solved independently of the others. An input-output differential equation can be obtained by combining the equations from a state-variable model or more directly from a free-body diagram or circuit by labeling the variables in terms of the input and output variables.

2.7.1   Comparison with the State-Variable Form

For a first-order system, both forms of the system model involve a single first-order differential equation and are essentially identical. For higher-order systems, they are quite different. A set of n first-order differential equations in state-variable form must be solved for as a group, and the initial value of each state variable must be known in order to solve the set of n equations. An input-output differential equation of order n contains only one dependent variable, but there is need to know the initial values of that variable and its first $(n - 1)$ derivatives. In practice, finding the input-output differential equation and the associated initial conditions may require more effort than finding the information needed for a state variable solution. Using the state-variable equations has significant computational advantages when a computer solution is to be found. In fact, standard methods for solving a high-order, possibly nonlinear input-output differential equation numerically (using computers) usually require decomposition into a set of simultaneous first-order equations anyway. The analytical solution of input-output differential equations and of sets of state-variable equations is considered in Chapter 3. State-variable equations are particularly convenient for complex multi-input, multi-output systems, while the input-output differential equation form is very convenient when the transfer function is to be readily derived analytically.

2.8   Transfer Function Form

The models in the previous sections are either ordinary differential equations or algebraic equations of time-dependent quantities. They are called time-domain models. Another method of capturing the dynamic behavior of linear systems is through their frequency responses. Such models are called frequency-domain models. Laplace transformation is the key to frequency-domain modeling. The *Transfer Function Form* model of a dynamic system with an input $u(t)$ and output $y(t)$ is defined from the Laplace transforms of the input and output such that

$$H(s) = \frac{Y(s)}{U(s)}$$

$$= \frac{b_0 s^m + b_1 s^{m-1} + b_2 s^{m-2} + \ldots + b_m}{a_0 s^n + a_1 s^{n-1} + a_2 s^{n-2} + \ldots + a_n},$$

where $\{b_j\}$ and $\{a_i\}$ are constants, and $n \geq m$. The variable s is called the complex variable or the Laplace variable. In this book, both time-domain and frequency-domain models are employed, and it is important to be able to convert freely from one form to the other. In addition to modeling linear systems in the frequency domain, Laplace transformation is also used in engineering to solve the dynamic system ordinary differential equations, i.e., obtain the system response. This is the subject matter of Chapter 3.

2.8.1 Obtaining the Transfer Function

There are four ways of establishing the transfer function of a dynamic system

- Directly taking the Laplace transform of time-domain models (state variable or input-output differential equation)

- Using the s-operator

- Using transfer functions (obtained by employing Laplace transforms) of the components of the system to establish the overall system transfer function

- Using frequency response experimental data

2.8.2 Directly Taking Laplace Transforms

The transfer function model can be obtained directly from the time-domain models (when all the initial conditions are known) by taking the Laplace transforms throughout the differential equations. The time-domain models can either be in the state-variable form or in the input-output differential equation form. It is easier to start with the input-output differential equation form and hence it is advisable to convert the state-variable form to the input-output differential equation form, and then proceed to obtain the transfer function. The procedure can be summarized as follows:

- Consider the input-output differential equation model or convert the state-variable form of the model to an input-output differential equation form.

- Take the Laplace transforms of the input-output differential equation.

- Apply the initial conditions (usually set to zero).

- Obtain the Laplace transform of the output divided by the Laplace transform of the input.

Consider a system that has the input-output differential equation

$$a_1 y + a_2 \dot{y} + a_3 \ddot{y} = b_1 u + b_2 \dot{u},$$

where the input is $u(t)$ and output is $y(t)$. Taking Laplace transforms throughout the differential equation gives

$$a_1 Y(s) + a_2 \left[sY(s) - y(0) \right] + a_3 \left[s^2 Y(s) - sy(0) - \dot{y}(0) \right] = b_1 U(s)$$
$$+ b_2 \left[sU(s) - u(0) \right].$$

Setting all initial conditions to zero leads to

$$a_1 Y(s) + a_2 s Y(s) + a_3 s^2 Y(s) = b_1 U(s) + b_2 s U(s)$$

$$Y(s) \left[a_1 + a_2 s + a_3 s^2 \right] = U(s) \left[b_1 + b_2 s \right].$$

The system transfer function $H(s)$, defined as the Laplace transform of the output $Y(s)$ divided by the Laplace transform of the input $U(s)$ is then obtained as follows:

$$H(s) = \frac{Y(s)}{U(s)}$$

$$= \frac{b_1 + b_2 s}{a_1 + a_2 s + a_3 s^2}.$$

This is the transfer function of the dynamic system.

Consider a system with the following state-variable matrix model

$$\begin{bmatrix} \dot{x}_1 \\ \dot{x}_2 \end{bmatrix} = \begin{bmatrix} a_1 & a_2 \\ a_3 & 0 \end{bmatrix} \begin{bmatrix} x_1 \\ x_2 \end{bmatrix} + \begin{bmatrix} b_1 \\ b_2 \end{bmatrix} \mathbf{u} \qquad (2.20)$$

$$\mathbf{y} = \begin{bmatrix} c_1 & 0 \end{bmatrix} \begin{bmatrix} x_1 \\ x_2 \end{bmatrix} + [0]\, \mathbf{u}, \qquad (2.21)$$

where $x_1(t)$ and $x_2(t)$ are the state variables, the $u(t)$ is the input and $y(t)$ is the output. In order to obtain the transfer function form for this system, the input-output differential equation must be established first. This is done by eliminating all the state variables from the state-variable equations except for the output $y(t)$ (and its derivatives) and the input $u(t)$ (and its derivatives). From the matrix form, the state-variable equations are given by

$$\dot{x}_1 = a_1 x_1 + a_2 x_2 + b_1 u$$
$$\dot{x}_2 = a_3 x_1 + b_2 u \qquad (2.22)$$
$$y = c_1 x_1.$$

Elimination of the state variables proceeds as follows:

$$\dot{x}_1 = a_1 x_1 + a_2 x_2 + b_1 u$$
$$\implies \ddot{x}_1 = a_1 \dot{x}_1 + a_2 \dot{x}_2 + b_1 \dot{u} \qquad (2.23)$$
$$y = c_1 x_1$$
$$\implies x_1 = \frac{y}{c_1}.$$

Substituting for $\dot{x}_2$ (Equation 2.22) into Equation 2.23 gives

$$\ddot{x}_1 = a_1\dot{x}_1 + a_2\left(a_3x_1 + b_2u\right) + b_1\dot{u}. \tag{2.24}$$

Replacing x_1 by $\dfrac{y}{c_1}$ in Equation 2.24 leads to

$$\frac{\ddot{y}}{c_1} = a_1\frac{\dot{y}}{c_1} + a_2\left(a_3\frac{y}{c_1} + b_2u\right) + b_1\dot{u}$$

$$\ddot{y} = a_1\dot{y} + c_1a_2\left(a_3\frac{y}{c_1} + b_2u\right) + c_1b_1\dot{u}$$

$$\ddot{y} = a_1\dot{y} + a_2a_3y + c_1a_2b_2u + c_1b_1\dot{u}$$

$$\ddot{y} - a_1\dot{y} - a_2a_3y = c_1a_2b_2u + c_1b_1\dot{u}. \tag{2.25}$$

Equation 2.25 is the input-output differential equation for the system. Therefore, the transfer function of the system can be obtained by taking the Laplace transforms of this equation.

$$s^2Y(s) - sy(0) - \dot{y}(0) - a_1\left[sY(s) - y(0)\right] - a_2a_3Y(s) = c_1a_2b_2U(s) +$$
$$c_1b_1\left[sU(s) - u(0)\right].$$

Assuming zero initial conditions

$$s^2Y(s) - a_1sY(s) - a_2a_3Y(s) = c_1a_2b_2U(s) + c_1b_1sU(s)$$

$$Y(s)\left[s^2 - a_1s - a_2a_3\right] = U(s)\left[c_1a_2b_2 + c_1b_1s\right].$$

Hence, the system transfer function is given by

$$T(s) = \frac{Y(s)}{U(s)} = \frac{c_1b_1s + c_1a_2b_2}{s^2 - a_1s - a_2a_3}.$$

2.8.3   The s-Operator Method

Starting with the time-domain models (state-variable form or input-output differential equation form), use the differential operator identity

$$s \equiv \frac{d}{dt}.$$

This operator is used to convert the time domain models into the frequency domain, thus facilitating the derivation of the transfer function. The s-operator has also been called the p-operator. It is important to note that if the transfer function model of the system is given then the input-output differential equation is easily obtained using the same identity, however, the initial conditions will be lacking. The s-operator is also useful in the general reduction of simultaneous differential equations. It is often necessary to combine a set of differential equations involving more than one dependent variable into a single differential equation with a single dependent variable. An example of such an equation is the input-output differential equation model. This is very handy when it is not obvious how to eliminate the unwanted variables easily. This algebraic method provides a useful means of manipulating sets of differential equations with constant coefficients. It is important to note that s must operate on the variable or expression that follows it, and that it is not a variable or algebraic quantity itself.

Consider the state variable system in Equations 2.20 and 2.21, whose state-variable equations are given by

$$\dot{x}_1 = a_1 x_1 + a_2 x_2 + b_1 u$$
$$\dot{x}_2 = a_3 x_1 + b_2 u$$
$$y = c_1 x_1.$$

The three equations can be expressed in terms of the s-operator as follows:

$$sX_1 = a_1 X_1 + a_2 X_2 + b_1 U \qquad (2.26)$$

$$sX_2 = a_3 X_1 + b_2 U \qquad (2.27)$$

$$Y = c_1 X_1 \implies X_1 = \frac{Y}{c_1}. \qquad (2.28)$$

Pre-multiplying Equation 2.26 by s gives

$$s^2 X_1 = a_1 s X_1 + a_2 s X_2 + b_1 s U. \qquad (2.29)$$

Substituting for sX_2 (Equation 2.27) and X_1 (Equation 2.28) in Equation 2.29 leads to

$$s^2 X_1 = a_1 s X_1 + a_2 \left[a_3 X_1 + b_2 U \right] + b_1 s U$$

$$s^2 \frac{Y}{c_1} = a_1 s \frac{Y}{c_1} + a_2 \left[a_3 \frac{Y}{c_1} + b_2 U \right] + b_1 s U$$

$$Y \left[s^2 - a_1 s - a_2 a_3 \right] = U \left[c_1 b_1 s + c_1 a_2 b_2 \right]. \qquad (2.30)$$

Hence, the system transfer function is given by

$$T(s) = \frac{Y(s)}{U(s)} = \frac{c_1 b_1 s + c_1 a_2 b_2}{s^2 - a_1 s - a_2 a_3},$$

which is the same result as obtained by taking the direct Laplace transforms of the input-output Equation 2.25. Although the s-operator can be used to find the transfer function without first finding the input-output differential equation, it can also be used to get simplified time domain equations such as input-output differential equations. An expression of the input-output differential equation (Equation 2.25) can be obtained from Equation 2.30 by reversing the s-operator.

$$Y\left[s^2 - a_1 s - a_2 a_3\right] = U\left[c_1 b_1 s + c_1 a_2 b_2\right]$$

$$s^2 Y - a_1 s Y - a_2 a_3 Y = c_1 b_1 s U + c_1 a_2 b_2 U$$

$$\ddot{y} - a_1 \dot{y} - a_2 a_3 y = c_1 b_1 \dot{u} + c_1 a_2 b_2 u.$$

When the s-operator is used in this way, i.e., in the general reduction of simultaneous differential equations into a single differential equation with a single dependent variable, it is sometimes called the p-operator. Consider a system represented by the following pair of equations, where $y(t)$ is the output and $u(t)$ is the input.

$$\dot{x} + 2x + y = 3u \tag{2.31}$$

$$2\dot{x} + 5x - 2\dot{y} + 2y = 0. \tag{2.32}$$

The input-output differential equation and the system transfer function can be found by applying the s-operator. In terms of the s-operator, Equation 2.31 and 2.32 become

$$(s+2)X + Y = 3U$$

$$(2s+5)X + (-2s+2)Y = 0.$$

Pre-multiplying the first equation by $(2s+5)$ and pre-multiplying the second by $(s+2)$ gives

$$(2s+5)\left[(s+2)X + Y\right] = 3U(2s+5) \tag{2.33}$$

$$(s+2)\left[(2s+5)X + (-2s+2)Y\right] = 0. \tag{2.34}$$

Subtracting Equation 2.34 from Equation 2.33 gives

$$\left[(2s+5) - (s+2)(-2s+2)\right]Y = 3U(2s+5)$$

$$(2s^2 + 4s + 1)Y = U(6s + 15). \tag{2.35}$$

The input-output differential equation is obtained by reversing the s-operator in Equation 2.35. Hence,

$$2s^2 Y + 4sY + Y = 6sU + 15U$$

$$2\ddot{y} + 4y + y = 6\dot{u} + 15u.$$

The transfer function is obtained from Equation 2.35, thus

$$H(s) = \frac{Y(s)}{U(s)} = \frac{6s + 15}{2s^2 + 4s + 1}.$$

It is important to note that s-operator (or p-operator) has the same effect as taking the Laplace transform of a system.

2.8.4   The Component Transfer Function Method

In this approach, the overall transfer function is obtained by finding the transfer functions of each component of the system and then using the relevant physical interconnection relationships or laws to form a set of algebraic equations involving the component transfer functions. These frequency domain equations are then solved to obtain the overall transfer function relating the output in Laplace transforms to the input in Laplace transforms. The block diagram or the signal-flow diagram can also be used to establish the overall system transfer function from component transfer functions. As an example, consider a car's cruise control system, where a component transfer function $H_1(s)$ from the speed error $e(t)$ to the control signal $u(t)$ is given by

$$U(s) = KE(s)$$
$$\implies H_1(s) = \frac{U(s)}{E(s)} = K.$$

Another component transfer function $H_2(s)$ from the control signal $u(t)$ to the output speed $v(t)$ is given by

$$V(s) = \left[\frac{b}{ms + b} \right] U(s)$$
$$\implies H_2(s) = \frac{V(s)}{U(s)} = \frac{b}{ms + b}.$$

Therefore, the overall transfer function $H(s)$ from the speed error $e(t)$ to the output speed $v(t)$ can be obtained by

$$H(s) = \frac{V(s)}{E(s)}$$
$$= \frac{V(s)}{U(s)} \times \frac{U(s)}{E(s)}$$
$$= H_1(s)H_2(s)$$
$$= \frac{bK}{ms + b}.$$

Hence, in this case, the overall transfer function is obtained as a product of the component transfer functions. The following electrical circuit problem further illustrates the use of the component transfer function method.

Example 2.12 *In the RLC electrical circuit shown below, assuming zero initial conditions, find the following:*

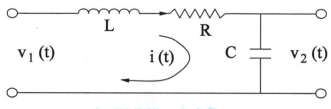

An RLC Electrical Circuit

(a) *The time domain equation relating $i(t)$ and $v_1(t)$.*

(b) *The time domain equation relating $i(t)$ and $v_2(t)$.*

(c) *The component transfer functions*

$$H_1(s) = \frac{V_1(s)}{I(s)} \quad and \quad H_2(s) = \frac{V_2(s)}{I(s)} \ .$$

(d) *The overall system transfer function*

$$H(s) = \frac{V_2(s)}{V_1(s)} \ .$$

Solution 2.12 *(a) The KVL is used to get the expressions for $v_1(t)$ and $v_2(t)$ as follows:*

$$v_1(t) = v_L + v_R + v_C$$

$$= L\frac{di}{dt} + Ri(t) + \frac{1}{C}\int i(t)dt \qquad (2.36)$$

$$v_2(t) = v_C$$

$$= \frac{1}{C}\int i(t)dt. \qquad (2.37)$$

(b) The transfer function $H_1(s)$ is obtained by taking the Laplace transforms

(assuming zero initial conditions) of Equation 2.36.

$$V_1(s) = sLI(s) + RI(s) + \frac{I(s)}{sC}$$

$$= I(s)\left[sL + R + \frac{1}{sC}\right]$$

$$H_1(s) = \frac{V_1(s)}{I(s)}$$

$$= sL + R + \frac{1}{sC} \; .$$

Similarly, the transfer function $H_2(s)$ is obtained by taking the Laplace transforms of Equation 2.37

$$V_2(s) = \frac{I(s)}{sC}$$

$$H_2(s) = \frac{V_2(s)}{I(s)}$$

$$= \frac{1}{sC} \; .$$

(c) The overall transfer function $H(s)$ is obtained from $H_1(s)$ and $H_2(s)$ as follows:

$$H(s) = \frac{V_2(s)}{V_1(s)}$$

$$= \frac{V_2(s)}{V_1(s)} \times \frac{I(s)}{I(s)} \qquad \text{multiplication by one}$$

$$= \frac{V_2(s)}{I(s)} \Big/ \frac{V_1(s)}{I(s)}$$

$$= \frac{H_2(s)}{H_1(s)}$$

$$= \frac{\dfrac{1}{sC}}{sL + R + \dfrac{1}{sC}}$$

$$= \frac{\dfrac{1}{LC}}{s^2 + \dfrac{R}{L}s + \dfrac{1}{LC}} \; .$$

In this example the overall transfer function is a ratio of the two component transfer functions.

2.8.5 The Transfer Function in Pole-Zero Factored Form

A special form of the transfer function of a dynamic system is the pole-zero factored form, where the numerator is expressed as the product of zero factors and the denominator as a product of pole factors, as opposed to being expressed as polynomials of s. The pole-zero form is derived from rewriting the transfer function as

$$H(s) = \frac{Y(s)}{U(s)}$$

$$= \frac{b_0 s^m + b_1 s^{m-1} + b_2 s^{m-2} + \dots + b_m}{a_0 s^n + a_1 s^{n-1} + a_2 s^{n-2} + \dots + a_n}$$

$$= \frac{b_0}{a_0} \left(\frac{s^m + \frac{b_1}{b_0} s^{m-1} + \frac{b_2}{b_0} b_2 s^{m-2} + \dots + \frac{b_m}{b_0}}{s^n + \frac{a_1}{a_0} s^{n-1} + \frac{a_2}{a_0} s^{n-2} + \dots + \frac{a_n}{a_0}} \right)$$

$$= \frac{b_0}{a_0} \left(\frac{(s - z_1)(s - z_2)\dots(s - z_m)}{(s - p_1)(s - p_2)\dots(s - p_n)} \right)$$

$$= K \frac{\Pi_{i=1}^{m}(s - z_i)}{\Pi_{j=1}^{n}(s - p_j)}, \tag{2.38}$$

where $K = \dfrac{b_0}{a_0}$. Equation 2.38 represents the pole-zero factored form of the transfer function and is useful because it clearly indicates the zeros (z_i) and poles (p_j) of the system. This form is very useful in determining system characteristics such as stability, and will be extensively used in later chapters.

2.9 Switching Between Different Model Forms

In this book emphasis is placed on both time-domain (state variable and input-output differential equation forms) and frequency-domain models (transfer function, pole-zero and block diagram forms). It is important to be able to convert from one model form to another as this enhances system understanding. Furthermore, depending on the system analysis, design of interest or specific application, one form may be more useful and informative than another. Consequently, the ability to analytically derive one form of the dynamic system model from another form is imperative. There are functions in MATLAB (Appendix B) that directly transform from one model form to another.

- ss2tf: This function converts from the state-variable form to the transfer function form

$$[\text{num, den}] = \text{ss2tf}\,(A, B, C, D),$$

  where *num* and *den* are the coefficients of the numerator polynomial and denominator polynomial of the system transfer function, i.e.,

$$H(s) = \frac{as^2 + bs + c}{ds^2 + es + f}$$

$$\text{num} = \begin{bmatrix} a & b & c \end{bmatrix} \quad \text{and} \quad \text{den} = \begin{bmatrix} d & e & f \end{bmatrix}.$$

  For example, the following state-variable matrix model can be converted to a transfer function model as follows:

$$\begin{bmatrix} \dot{x} \\ \dot{v} \end{bmatrix} = \begin{bmatrix} 0 & 1 \\ 0 & -\dfrac{1}{20} \end{bmatrix} \begin{bmatrix} x \\ v \end{bmatrix} + \begin{bmatrix} 0 \\ \dfrac{1}{20} \end{bmatrix} [v_r]$$

$$\begin{bmatrix} y \end{bmatrix} = \begin{bmatrix} 1 & 0 \end{bmatrix} \begin{bmatrix} x \\ v \end{bmatrix} + \begin{bmatrix} 0 \end{bmatrix} v_r.$$

$$[\text{num, den}] = \text{ss2tf}\,(A, B, C, D)$$

$$\text{num} = \begin{bmatrix} 0 & 0 & \dfrac{1}{20} \end{bmatrix} \quad \text{and} \quad \text{den} = \begin{bmatrix} 1 & \dfrac{1}{20} & 0 \end{bmatrix}$$

$$H(s) = \frac{\dfrac{1}{20}}{s^2 + \dfrac{s}{20}}$$

$$= \frac{\dfrac{1}{20}}{s(s + \dfrac{1}{20})}$$

$$= \frac{1}{20s^2 + s}.$$

- tf2ss: This function converts from the transfer function form to the state-variable form

$$[A, B, C, D] = \text{tf2ss}\,(\text{num, den})\,.$$

For example, MATLAB can be used to find the state-variable form of the following input-output differential equation where zero initial conditions are assumed.

$$\ddot{y} + 6\dot{y} + 25y = 9u + 3\dot{u}$$

First, the transfer function is obtained as follows:

$$\mathcal{L}\left[\ddot{y} + 6\dot{y} + 25y\right] = \mathcal{L}\left[9u + 3\dot{u}\right]$$

$$s^2 Y(s) + 6sY(s) + 25Y(s) = 9U(s) + 3sU(s)$$

$$\Longrightarrow H(s) = \frac{3s + 9}{s^2 + 6s + 25} \quad (2.39)$$

$$\text{num} = \begin{bmatrix} 0 & 3 & 9 \end{bmatrix} \quad \text{and} \quad \text{den} = \begin{bmatrix} 1 & 6 & 25 \end{bmatrix}$$

$$[A, B, C, D] = \text{tf2ss}\,(\text{num, den})$$

$$A = \begin{bmatrix} -6 & -25 \\ 1 & 0 \end{bmatrix}, \quad B = \begin{bmatrix} 1 \\ 0 \end{bmatrix} \quad (2.40)$$

$$C = \begin{bmatrix} 3 & 9 \end{bmatrix}, \quad D = \begin{bmatrix} 0 \end{bmatrix} \quad (2.41)$$

- ss2zp: This function converts from the state-variable form to the zero-pole form of the transfer function

$$[z, p, k] = \text{ss2zp}\,(A, B, C, D)\,.$$

- zp2ss: This function converts from the zero-pole form of the transfer function to the state-variable form

$$[A, B, C, D] = \text{zp2ss}\,(z, p, k)\,.$$

These last two functions can be illustrated by using the matrices in Equations 2.40 and 2.41.

$$[z, p, k] = \text{ss2zp}\,(A, B, C, D)$$

$$z = -3, \quad p = \begin{bmatrix} -3 + 4j \\ -3 - 4j \end{bmatrix}, \quad k = 3.$$

From these parameters, the transfer function can be deduced

$$H(s) = \frac{Y(s)}{U(s)}$$

$$= 3\frac{[s - (-3)]}{[s - (-3 + 4j)]\,[s - (-3 - 4j)]}$$

$$= \frac{3(s + 3)}{(s + 3 - 4j)(s + 3 + 4j)}$$

$$= \frac{3s + 9}{s^2 + 6s + 25}.$$

Note that this equation is the same as Equation 2.39, as expected.

2.10   Examples of Dynamic System Modeling

In this section, a number of detailed examples are presented to clearly elucidate the concepts of modeling.

Example 2.13 *The input to the translational mechanical system shown in the following diagram is the displacement $x_3(t)$ of the right end of the spring k_1. The displacement of m_2 relative to m_1 is x_2. The forces exerted by the springs are zero when $x_1 = x_2 = x_3 = 0$.*

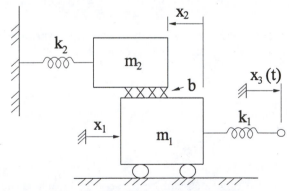

Translational Mechanical System

(a) Draw the free-body diagrams of the system
(b) Obtain the equations of motion.

Solution 2.13 *(a) The free-body diagrams are drawn as follows:*

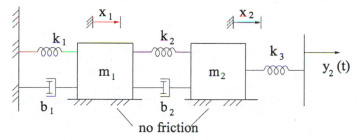

(b) The equations of motion are then obtain from the free-body diagrams.

$$m_1\ddot{x}_1 + b\dot{x}_2 - k_1[x_3(t) - x_1] = 0$$

$$m_1(\ddot{x}_1 - \ddot{x}_2) + k_2(x_1 - x_2) - b\dot{x}_2 = 0$$

Example 2.14 *Consider the mechanical system shown below, where the input is a displacement $y_2(t)$, and the output is the force in spring k_3.*

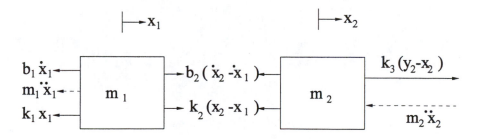

(a) Draw the free-body diagrams for the system.

(b) Obtain the equations of motion.

(d) Choose a minimum set of variables for the system and justify the choice.

(e) Express the equations of motion in state-variable matrix form (i.e. obtain A, B, C, D).

Solution 2.14 *(a) The forces acting on the masses are shown in the following free-body diagrams.*

Translational Mechanical System: Freebody Diagram

(b) *The equations of motion are obtained as:*

$$m_1\ddot{x}_1 + b_1\dot{x}_1 + k_1x_1 + b_2(\dot{x}_1 - \dot{x}_2) + k_2(x_1 - x_2) = 0 \qquad (2.42)$$

$$m_2\ddot{x}_2 + b_2(\dot{x}_2 - \dot{x}_1) + k_2(x_2 - x_1) + k_3(x_2 - y_2(t)) = 0 \qquad (2.43)$$

(c) *There are five energy storing elements (2 masses and 3 springs) but only four are independent, since m_2 and k_3 are dependent. Therefore, the state variables are chosen as*

$$\mathbf{x} = \begin{bmatrix} x_1 & \dot{x}_1 & x_2 & \dot{x}_2 \end{bmatrix}^T$$

where the input is $y_2(t)$.

(d) *Two of the equations of the derivatives of variables are simple*

$$\dot{x}_1 = \dot{x}_1$$
$$\dot{x}_2 = \dot{x}_2.$$

The other two equations are obtained by rearranging Equations 2.42 and 2.43,

$$\ddot{x}_1 = -\frac{k_1 + k_2}{m_1}x_1 - \frac{b_1 + b_2}{m_1}\dot{x}_1 + \frac{k_2}{m_1}x_2 + \frac{b_2}{m_1}\dot{x}_2$$

$$\ddot{x}_2 = \frac{k_2}{m_2}x_1 + \frac{b_2}{m_2}\dot{x}_1 - \frac{k_2 + k_3}{m_2}x_2 - \frac{b_2}{m_2}\dot{x}_2 + \frac{k_3}{m_2}y_2(t).$$

The output is the spring force in the spring k_3

$$y = k_3(x_2 - y_2(t))$$
$$= k_3x_2 - k_3y_2(t).$$

Therefore, the state variable system is given by

$$\begin{bmatrix} \dot{x}_1 \\ \ddot{x}_1 \\ \dot{x}_2 \\ \ddot{x}_2 \end{bmatrix} = \begin{bmatrix} 0 & 1 & 0 & 0 \\ -\left(\dfrac{k_1+k_2}{m_1}\right) & -\left(\dfrac{b_1+b_2}{m_1}\right) & \dfrac{k_2}{m_1} & \dfrac{b_2}{m_1} \\ 0 & 0 & 0 & 1 \\ \dfrac{k_2}{m_2} & \dfrac{b_2}{m_2} & -\left(\dfrac{k_2+k_3}{m_2}\right) & -\dfrac{b_2}{m_2} \end{bmatrix} \begin{bmatrix} x_1 \\ \dot{x}_1 \\ x_2 \\ \dot{x}_2 \end{bmatrix}$$

$$+ \begin{bmatrix} 0 \\ 0 \\ 0 \\ \dfrac{k_3}{m_2} \end{bmatrix} y_2(t),$$

where the output is given by

$$\mathbf{y} = \begin{bmatrix} 0 & 0 & k_3 & 0 \end{bmatrix} \begin{bmatrix} x_1 \\ \dot{x}_1 \\ x_2 \\ \dot{x}_2 \end{bmatrix} + \begin{bmatrix} -k_3 \end{bmatrix} \begin{bmatrix} y_2(t) \end{bmatrix}.$$

Example 2.15 *Consider the rotational mechanical system shown below.*

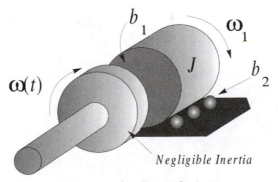

Diagram for Example 2.15

(a) *Draw the free-body diagrams of the system.*

(b) *Find the dynamic system model in the form of an input-output differential equation, where the input is angular speed $\omega(t)$ and the output is angular speed $\omega_1(t)$.*

Solution 2.15 *The free-body diagram for disk J is shown below.*

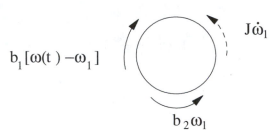

Free-Body Diagram for Example 2.15

From the free-body diagram of disk J the dynamic equation can be deduced

$$b_1 \left[\omega(t) - \omega_1 \right] = J\dot{\omega}_1 + b_2\omega_1$$

$$J\dot{\omega}_1 + (b_1 + b_2)\omega_1 = b_1\omega(t)$$

$$\dot{\omega}_1 + \frac{(b_1 + b_2)}{J}\omega_1 = \frac{b_1\omega(t)}{J}.$$

Example 2.16 *(a) Draw the free-body diagrams for the rotational mechanical system shown below.*

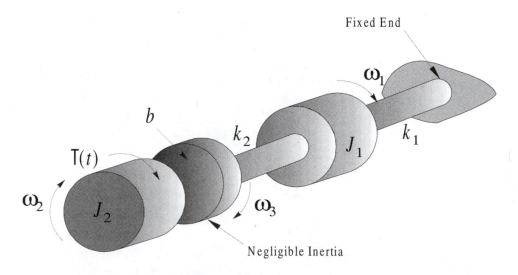

Rotational Mechanical System

(b) Explain why the state-variable vector should be chosen as

$$\mathbf{x}(t) = [\theta_1 \ \theta_3 \ \omega_1 \ \omega_2]^T.$$

(c) Express the dynamic equations in the state-variable matrix form (i.e. obtain A, B, C, D) where the input is the applied torque $T(t)$ and the output is the viscous frictional torque on disk J_2.

Solution 2.16 *(a) The free-body diagrams are given by*

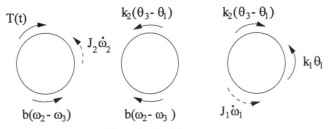

Free-Body Diagrams

(b) There are four independent energy storing elements; two shafts and two disks (with moment of inertia J_1 and J_2). The potential energy stored in a shaft depends on the angular displacement $\theta(t)$ and the kinetic energy stored in a rotating disk depends on the angular speed $\omega(t)$. Hence, the minimum set of state variables can be represented by

$$\mathbf{x}(t) = [\theta_1 \quad \theta_3 \quad \omega_1 \quad \omega_2]^T .$$

(c) From the free-body diagrams the equations of motion are given by

$$J_1\dot{\omega}_1 + (k_1 + k_2)\theta_1 - k_2\theta_3 = 0$$

$$J_2\dot{\omega}_2 + b\omega_2 - b\dot{\theta}_3 = T(t)$$

$$-k_2\theta_1 - b\omega_2 + b\dot{\theta}_3 + k_2\theta_3 = 0.$$

Adding the second and the third equation the following equation is obtained

$$-k_2\theta_1 + J_2\dot{\omega}_2 + k_2\theta_3 = T(t).$$

Rewriting these equations in terms of state variable gives

$$\dot{\theta}_1 = \omega_1$$

$$\dot{\omega}_1 = \frac{1}{J_1}\left[-(k_1 + k_2)\theta_1 + k_2\theta_3\right]$$

$$\dot{\omega}_2 = \frac{1}{J_2}[k_2\theta_1 - k_2\theta_3 + T(t)]$$

$$\dot{\theta}_3 = \frac{1}{b}[k_2\theta_1 + b\omega_2 - k_2\theta_3].$$

The output equation is given by

$$T_b = b(\omega_2 - \omega_3)$$
$$= b(\omega_2 - \dot{\theta}_3)$$
$$= b\omega_2 - [k_2\theta_1 + b\omega_2 - k_2\theta_3]$$
$$= -k_2\theta_1 + k_2\theta_3.$$

Putting the above equations in matrix form produces

$$\begin{bmatrix} \dot{\theta}_1 \\ \dot{\omega}_1 \\ \dot{\omega}_2 \\ \dot{\theta}_3 \end{bmatrix} = \begin{bmatrix} 0 & 1 & 0 & 0 \\ -\dfrac{(k_1 + k_2)}{J_1} & 0 & 0 & \dfrac{k_2}{J_1} \\ \dfrac{k_2}{J_2} & 0 & 0 & -\dfrac{k_2}{J_2} \\ \dfrac{k_2}{b} & 0 & 1 & -\dfrac{k_2}{b} \end{bmatrix} \begin{bmatrix} \theta_1 \\ \omega_1 \\ \omega_2 \\ \theta_3 \end{bmatrix} + \begin{bmatrix} 0 \\ 0 \\ \dfrac{1}{J_2} \\ 0 \end{bmatrix} [T]$$

$$T_b = \begin{bmatrix} -k_2 & 0 & 0 & k_2 \end{bmatrix} \begin{bmatrix} \theta_1 \\ \omega_1 \\ \omega_2 \\ \theta_3 \end{bmatrix} + [0][T].$$

Example 2.17 *(a) Draw the free-body diagram of the gear and shaft mechanism shown below.*

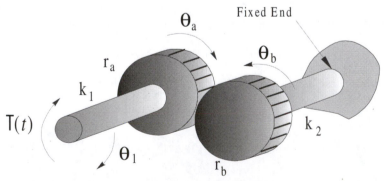

A Gear and Shaft Mechanism

(b) Find expressions for the gear ratio, N, and the angular displacements, $\theta_a(t)$ and $\theta_b(t)$.

(c) Find the equivalent stiffness constant k_{eq} such that the algebraic model of the system is given by

$$\theta_1 = \frac{T(t)}{k_{eq}},$$

where k_{eq} is a function of k_1, k_2, r_a, and r_b.

Solution 2.17 *(a) The free-body is shown here.*

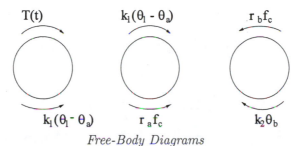

T(t) $k_1(\theta_1 - \theta_a)$ $r_b f_c$

$k_1(\theta_1 - \theta_a)$ $r_a f_c$ $k_2 \theta_b$

Free-Body Diagrams

(b) The gear ratio is given by

$$N = \frac{r_b}{r_a}$$

$$= \frac{\theta_a}{\theta_b}$$

$$\Rightarrow \theta_a = N\theta_b = \left(\frac{r_b}{r_a}\right)\theta_b. \tag{2.44}$$

From the free-body diagrams

$$T = k_1(\theta_1 - \theta_a) \qquad \text{(first free-body dgm)}$$

$$\Rightarrow \theta_1 = \frac{T}{k_1} + \theta_a \tag{2.45}$$

$$T = k_1(\theta_1 - \theta_a) = r_a f_c \qquad \text{(second free-body dgm)}$$

$$\Rightarrow f_c = \frac{T}{r_a} \tag{2.46}$$

$$r_b f_c = k_2 \theta_b \qquad \text{(third free-body dgm)}$$

$$\Longrightarrow \theta_b = \frac{r_b f_c}{k_2} = \frac{r_b}{k_2}\left(\frac{T}{r_a}\right) \qquad \text{(using Equation 2.46)}$$

$$\Longrightarrow \theta_b = \frac{r_b T}{r_a k_2}. \tag{2.47}$$

Substituting this expression of $\theta_b(t)$ (Equation 2.47) in Equation 2.44 leads to the expression for $\theta_a(t)$.

$$\theta_a = \left(\frac{r_b}{r_a}\right)\left(\frac{r_b T}{r_a k_2}\right)$$

$$= \left(\frac{r_b}{r_a}\right)^2 \frac{T}{k_2}. \qquad (2.48)$$

Substituting this expression for $\theta_a(t)$ in Equation 2.45 leads to

$$\theta_1 = \frac{T}{k_1} + \left(\frac{r_b}{r_a}\right)^2 \frac{T}{k_2}$$

$$= \left[\frac{1}{k_1} + \frac{\left(\frac{r_b}{r_a}\right)^2}{k_2}\right] T \equiv \frac{T}{k_{eq}}$$

$$\implies \frac{1}{k_{eq}} = \left[\frac{1}{k_1} + \frac{\left(\frac{r_b}{r_a}\right)^2}{k_2}\right]$$

$$\implies k_{eq} = \frac{k_1 k_2}{k_2 + k_1 \left(\frac{r_b}{r_a}\right)^2}.$$

This is the equivalent torsional stiffness k_{eq} for the gear and shaft system. It can also be expressed in terms of the gear ratio as

$$k_{eq} = \frac{k_1 k_2}{k_2 + k_1 N^2}.$$

Example 2.18 *In the following rotational mechanical system there is a driving torque $T(t)$ exerted on disk J_1 and a load torque $T_L(t)$ exerted on disk J_2.*

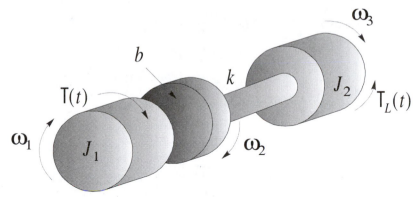

Rotational Mechanical System

(a) *Draw the free-body diagrams for the system.*

(b) *Choose the minimum number of state variables for this system and give a rationale for the choice.*

(c) *Express the dynamic equations in state-variable matrix form (i.e. obtain A, B, C, D) where both the applied torque $T(t)$ and the load torque $T_L(t)$ are considered as inputs. The output is the torque in the shaft k.*

Solution 2.18 (a) *The free-body diagrams are shown below.*

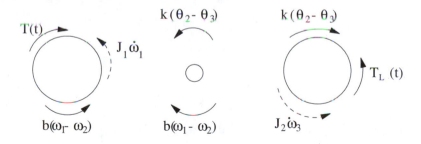

Free-Body Diagrams

(b) *There are three independent energy storing elements; one shaft and two disks (with moment of inertia J_1 and J_2). The potential energy stored in a shaft depends on the angular displacement $\theta(t)$ and the kinetic energy stored in a rotating disk depends on the angular speed $\omega(t)$. Hence, the minimum set of state variables can be represented by*

$$\mathbf{x}(t) = [\theta_r \quad \omega_1 \quad \omega_3]^T,$$

where $\theta_r(t)$ is the relative angular displacement between the two ends of the shaft,

$$\theta_r(t) = \theta_2 - \theta_3 \tag{2.49}$$

$$\implies \dot{\theta}_r(t) = \omega_2 - \omega_3. \tag{2.50}$$

(c) From the free-body diagrams the equations of motion are determined.

$$J_1\dot{\omega}_1 + b(\omega_1 - \omega_2) = T$$

$$b(\omega_1 - \omega_2) = k(\theta_2 - \theta_3)$$

$$k(\theta_2 - \theta_3) = T_L + J_2\dot{\omega}_3.$$

Replacing $(\theta_2 - \theta_3)$ by θ_r in these equations gives

$$J_1\dot{\omega}_1 + b(\omega_1 - \omega_2) = T \tag{2.51}$$

$$b(\omega_1 - \omega_2) = k\theta_r \tag{2.52}$$

$$k\theta_r = T_L + J_2\dot{\omega}_3. \tag{2.53}$$

Substituting Equation 2.52 into Equation 2.51 gives

$$J_1\dot{\omega}_1 + k\theta_r = T. \tag{2.54}$$

Also, from Equation 2.52, it follows that

$$\omega_2 = \omega_1 - \frac{k}{b}\theta_r. \tag{2.55}$$

Substituting Equation 2.55 into Equation 2.50 gives

$$\dot{\theta}_r = \omega_1 - \frac{k}{b}\theta_r - \omega_3. \tag{2.56}$$

Rearranging Equations 2.53, 2.54 and 2.56 produces the state variable equations.

$$\dot{\theta}_r = -\frac{k}{b}\theta_r + \omega_1 - \omega_3$$

$$\dot{\omega}_1 = -\frac{k}{J_1}\theta_r + \frac{T}{J_1}$$

$$\dot{\omega}_3 = \frac{k}{J_2}\theta_r - \frac{T_L}{J_2}.$$

The output equation is given by

$$T_k = k(\theta_2 - \theta_3)$$

$$= k\theta_r.$$

The state-variable matrix form is obtained by extracting the coefficients of the variables and inputs.

$$
\begin{bmatrix} \dot{\theta}_r \\ \dot{\omega}_1 \\ \dot{\omega}_3 \end{bmatrix} = \begin{bmatrix} -\dfrac{k}{b} & 1 & -1 \\[2mm] -\dfrac{k}{J_1} & 0 & 0 \\[2mm] \dfrac{k}{J_2} & 0 & 0 \end{bmatrix} \begin{bmatrix} \theta_r \\ \omega_1 \\ \omega_3 \end{bmatrix} + \begin{bmatrix} 0 & 0 \\[2mm] \dfrac{1}{J_1} & 0 \\[2mm] 0 & \dfrac{-1}{J_2} \end{bmatrix} \begin{bmatrix} T \\ T_L \end{bmatrix}
$$

$$
T_k = \begin{bmatrix} k & 0 & 0 \end{bmatrix} \begin{bmatrix} \theta_r \\ \omega_1 \\ \omega_3 \end{bmatrix} + \begin{bmatrix} 0 & 0 \end{bmatrix} \begin{bmatrix} T \\ T_L \end{bmatrix}.
$$

Example 2.19 *In the following electrical circuit the input is a current $i_i(t)$ and the output is a voltage $v_o(t)$.*

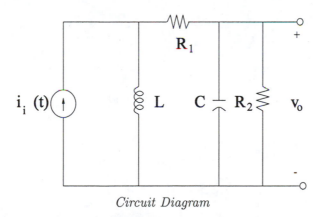

Circuit Diagram

(a) *Choose the state variables for the system.*
(b) *Derive the state-variable matrix model (i.e. obtain A, B, C, D).*
(c) *Find the input-output differential equation model for the circuit.*

Solution 2.19 *(a) There are two independent energy storing elements; a capacitor and an inductor. The energy stored in the capacitor is a function of the voltage $v_C(t)$ and that stored in the inductor depends on the current $i_L(t)$. Hence, the minimum set of variables is chosen as*

$$
\mathbf{x}(t) = \begin{bmatrix} i_L & v_C \end{bmatrix}^T,
$$

(b) The annotated circuit diagram with current directions is shown below.

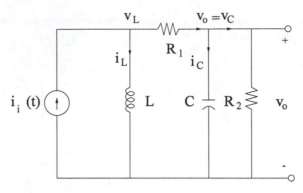

Annotated Circuit Diagram

From analyzing the inductor and capacitor it follows that

$$v_L = L\frac{di_L}{dt}$$

$$\Rightarrow \frac{di_L}{dt} = \frac{1}{L}v_L$$

$$i_C = C\frac{dv_C}{dt}$$

$$\Rightarrow \frac{dv_C}{dt} = \frac{1}{C}i_C.$$

Using KCL and KVL for the circuit

$$v_L = R_1(i_i - i_L) + v_C$$

$$\Rightarrow \frac{di_L}{dt} = \frac{1}{L}\left[R_1(i_i - i_L) + v_C\right] \tag{2.57}$$

$$i_C = i_{R_1} - i_o = (i_i - i_L) - \frac{v_C}{R_2}$$

$$\Rightarrow \frac{dv_C}{dt} = \frac{1}{C}\left[(i_i - i_L) - \frac{v_C}{R_2}\right]. \tag{2.58}$$

The output is given by

$$v_o = v_C.$$

Writing the above equations in state-variable matrix form produces

$$
\begin{bmatrix} \dot{i}_L \\ \dot{v}_C \end{bmatrix} = \begin{bmatrix} -\dfrac{R_1}{L} & \dfrac{1}{L} \\[2mm] -\dfrac{1}{C} & -\dfrac{1}{R_2 C} \end{bmatrix} \begin{bmatrix} i_L \\ v_C \end{bmatrix} + \begin{bmatrix} \dfrac{R_1}{L} \\[2mm] \dfrac{1}{C} \end{bmatrix} [i_i(t)]
$$

$$
[v_o] = \begin{bmatrix} 0 & 1 \end{bmatrix} \begin{bmatrix} i_L \\ v_C \end{bmatrix} + [0]\,[i_i(t)].
$$

(c) From Equations 2.57 and 2.58 the variables $i_L(t)$ and $\dfrac{di_L(t)}{dt}$ should be eliminated in order to obtain the input-output differential equation.

$$
\ddot{v}_o = \frac{1}{C}\left[\dot{i}_i - \dot{i}_L - \frac{\dot{v}_o}{R_2}\right]
$$

$$
= \frac{1}{C}\left[\dot{i}_i - \frac{1}{L}\{R_1(i_i - i_L) + v_o\} - \frac{\dot{v}_o}{R_2}\right] \qquad using \quad \dot{i}_L = \frac{1}{L}\{R_1(i_i - i_L) + v_o\}
$$

$$
= \frac{1}{C}\left[\dot{i}_i - \frac{1}{L}\{R_1(i_C + i_{R_2}) + v_o\} - \frac{\dot{v}_o}{R_2}\right] \qquad using \quad (i_i - i_L) = (i_C + i_{R_2})
$$

$$
= \frac{1}{C}\left[\dot{i}_i - \frac{1}{L}\{R_1\left(C\dot{v}_o + \frac{v_o}{R_2}\right) + v_o\} - \frac{\dot{v}_o}{R_2}\right]
$$

$$
= \frac{1}{C}\left[\dot{i}_i - \frac{R_1}{L}\left(C\dot{v}_o + \frac{v_o}{R_2} + \frac{v_o}{R_1}\right) - \frac{\dot{v}_o}{R_2}\right]
$$

$$
= \frac{1}{C}\left[\dot{i}_i - \frac{R_1}{L}\left(C\dot{v}_o + \left(\frac{R_1 + R_2}{R_1 R_2}\right)v_o\right) - \frac{\dot{v}_o}{R_2}\right]
$$

$$
== \frac{1}{C}\left[\dot{i}_i - \frac{R_1 C}{L}\dot{v}_o - \frac{R_1 + R_2}{LR_2}v_o - \frac{\dot{v}_o}{R_2}\right].
$$

Rearranging this equation gives the input-output differential equation for the electrical circuit as

$$
\ddot{v}_o + \left(\frac{R_1}{L} + \frac{1}{R_2 C}\right)\dot{v}_o + \left(\frac{R_1 + R_2}{R_2 CL}\right)v_o = \frac{\dot{i}_i}{C}.
$$

Example 2.20 *The differential equations for the inverted pendulum are defined as follows:*

$$
(I + M_p l^2)\ddot{\alpha} - M_p g l \alpha = M_p l \ddot{x}
$$

$$
(M_t + M_p)\ddot{x} + b\dot{x} - M_p l \ddot{\alpha} = u.
$$

(a) Why is it not possible to put these equations into state-variable form using state vector $\mathbf{x(t)} = [\alpha \ \ \dot{\alpha} \ \ x \ \ \dot{x}]^T$, where $u(t)$ is the input ?
(b) Write the equations in the form

$$\mathbf{E\dot{x}} = \mathbf{A'x} + \mathbf{B'u}$$

and identify the elements of $\mathbf{E}$, $\mathbf{A'}$ and $\mathbf{B'}$ (notice that $\mathbf{E}$ is a 4×4 matrix).
(c) Show how $\mathbf{A}$ and $\mathbf{B}$ for the standard state variable description of the inverted pendulum equations of motion can be computed.

Solution 2.20 *(a) It is not possible to put these equations into state-variable form because two derivatives of the variables, $\ddot{x}$ and $\ddot{\alpha}$, appear in the same equation.*
  (b) In order to write the equations in the form

$$\mathbf{E\dot{x}} = \mathbf{A'x} + \mathbf{B'u},$$

consider the set of state-variable equations (with terms involving derivatives of the variables on the left-hand side and those involving the variables on the right-hand side),

$$\dot{\alpha} = \dot{\alpha}$$

$$(I + M_p l^2)\ddot{\alpha} - M_p l\ddot{x} = M_p g l \alpha$$

$$\dot{x} = \dot{x}$$

$$-M_p l\ddot{\alpha} + (M_t + M_p)\ddot{x} = -b\dot{x} + u,$$

where the vector of state variables is given by

$$\mathbf{x(t)} = [\alpha \ \ \dot{\alpha} \ \ x \ \ \dot{x}]^T.$$

Pulling out the coefficients of the variable derivatives (on the left-hand side) and the coefficients of the variables and input (on the right-hand side) leads to

$$
\begin{bmatrix}
1 & 0 & 0 & 0 \\
0 & (I + M_p l^2) & 0 & -M_p l \\
0 & 0 & 1 & 0 \\
0 & -M_p l & 0 & (M_t + M_p)
\end{bmatrix}
\begin{bmatrix}
\dot{\alpha} \\ \ddot{\alpha} \\ \dot{x} \\ \ddot{x}
\end{bmatrix}
=
\begin{bmatrix}
0 & 1 & 0 & 0 \\
M_p g l & 0 & 0 & 0 \\
0 & 0 & 0 & 1 \\
0 & 0 & 0 & -b
\end{bmatrix}
\begin{bmatrix}
\alpha \\ \dot{\alpha} \\ x \\ \dot{x}
\end{bmatrix}
+
\begin{bmatrix}
0 \\ 0 \\ 0 \\ 1
\end{bmatrix}
u.
$$

(c)

$$\mathbf{E\dot{x}} = \mathbf{A'x} + \mathbf{B'u}$$

Pre-multiplying by E^{-1} the following equation is obtained

$$\mathbf{E^{-1}E\dot{x} = E^{-1}A'x + E^{-1}B'u}$$
$$\mathbf{\dot{x} = E^{-1}A'x + E^{-1}B'u} \tag{2.59}$$

The expressions for A and B are then obtained by comparing Equation 2.59 with the standard state-variable equation.

$$\mathbf{\dot{x} = Ax + Bu}$$
$$\mathbf{A = E^{-1}A'} \quad and \quad \mathbf{B = E^{-1}B'}.$$

Example 2.21 *(a) Write the dynamic equations for the circuit shown below.*

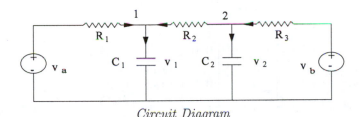

Circuit Diagram

(b) Put equations in the state-variable form where the voltages $v_a(t)$ and $v_b(t)$ are the inputs, and both $v_1(t)$ and $v_2(t)$ are the outputs of interest (i.e. obtain A, B, C, D). Choose the state variables as $v_1(t)$ and $v_2(t)$.

Solution 2.21 *(a) The dynamic equations are obtained by applying the KCL and manipulating the result. Employing the KCL at node 1 gives*

$$i_{C_1} = i_{R_1} + i_{R_2}$$
$$C_1\dot{v}_1 = \frac{v_a - v_1}{R_1} + \frac{v_2 - v_1}{R_2}$$

$$\Rightarrow \dot{v}_1 = -\left(\frac{1}{R_1 C_1} + \frac{1}{R_2 C_1}\right)v_1 + \frac{1}{R_2 C_1}v_2 + \frac{1}{R_1 C_1}v_a.$$

Applying the KCL at node 2 gives

$$i_{R_3} = i_{C_2} + i_{R_2}$$
$$i_{C_2} = i_{R_3} - i_{R_2}$$
$$C_2\dot{v}_2 = \frac{v_b - v_2}{R_3} - \frac{v_2 - v_1}{R_2}$$

$$\Rightarrow \dot{v}_2 = \frac{1}{R_2 C_2}v_1 - \left(\frac{1}{R_2 C_2} + \frac{1}{R_3 C_2}\right)v_2 + \frac{1}{R_3 C_2}v_b.$$

(b) Expressing these equations in state-variable matrix form is achieved by pulling out the coefficients of the variables and inputs.

$$
\begin{bmatrix} \dot{v}_1 \\ \dot{v}_2 \end{bmatrix} = \begin{bmatrix} -\left(\dfrac{1}{R_1 C_1} + \dfrac{1}{R_2 C_1}\right) & \dfrac{1}{R_2 C_1} \\[3mm] \dfrac{1}{R_2 C_2} & -\left(\dfrac{1}{R_2 C_2} + \dfrac{1}{R_3 C_2}\right) \end{bmatrix} \begin{bmatrix} v_1 \\ v_2 \end{bmatrix} +
$$

$$
\begin{bmatrix} \dfrac{1}{R_1 C_1} & 0 \\[3mm] 0 & \dfrac{1}{R_3 C_2} \end{bmatrix} \begin{bmatrix} v_a \\ v_b \end{bmatrix}
$$

$$
\begin{bmatrix} v_1 \\ v_2 \end{bmatrix} = \begin{bmatrix} 1 & 0 \\ 0 & 1 \end{bmatrix} \begin{bmatrix} v_1 \\ v_2 \end{bmatrix} + \begin{bmatrix} 0 & 0 \\ 0 & 0 \end{bmatrix} \begin{bmatrix} v_a \\ v_b \end{bmatrix}.
$$

Example 2.22 (a) Obtain the dynamic equations of the circuit shown below.

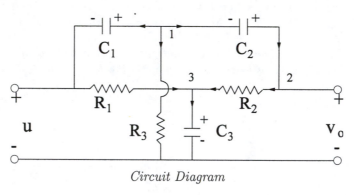

Circuit Diagram

(b) Explain why the state variables should be chosen as $v_{C_1}(t)$, $v_{C_2}(t)$, and $v_{C_3}(t)$.

(c) Express the equations in state-variable matrix form where the input is voltage $u(t)$ and the output is voltage $v_o(t)$.

Solution 2.22 (a) The output is given by using the KVL

$$
u + v_{C_1} + v_{C_2} - v_o = 0
$$

$$
v_o = v_{C_1} + v_{C_2} + u.
$$

Applying KCL at node 2,

$$-i_{C_2} = i_{R_2}$$

$$-C_2\ \dot{v}_{C_2} = \frac{v_0 - v_{C_3}}{R_2}$$

$$-C_2\ \dot{v}_{C_2} = \frac{u + v_{C_1} + v_{C_2} - v_{C_3}}{R_2}$$

$$\dot{v}_{C_2} = \frac{-u - v_{C_1} - v_{C_2} + v_{C_3}}{R_2 C_2}.$$

Applying KCL at node 1,

$$i_{C_1} + i_{R_3} + (-i_{C_2}) = 0 \qquad\qquad where\ -i_{C_2} = i_{R_2}$$

$$C_1\ \dot{v}_{C_1} + \frac{u + v_{C_1}}{R_3} + \frac{u + v_{C_1} + v_{C_2} - v_{c3}}{R_2} = 0$$

$$\dot{v}_{C_1} = \frac{-u - v_{C_1}}{R_3 C_1} + \frac{-u - v_{C_1} - v_{C_2} + v_{C_3}}{R_2 C_1}.$$

Applying KCL at node 3

$$i_{R_1} + i_{R_2} = i_{C_3}$$

$$\frac{u - v_{C_3}}{R_1} + \frac{u + v_{C_1} + v_{C_2} - v_{C_3}}{R_2} = C_3 \dot{v}_{C_3}$$

$$\dot{v}_{C_3} = \frac{u - v_{C_3}}{R_1 C_3} + \frac{u + v_{C_1} + v_{C_2} - v_{C_3}}{R_2 C_3}.$$

(b) There are three independent energy storing elements: three capacitors. The energy stored in the capacitor is a function of the voltage $v_C(t)$. Hence the minimum set of variables is chosen as

$$\mathbf{x(t)} = [v_{C_1}\quad v_{C_2}\quad v_{C_3}]^T.$$

(c) The state-variable matrix form is then obtained by extracting the co-efficients of the variables v_{C_1}, v_{C_2}, and v_{C_3}, and input u.

Example 2.23 *In the figure shown below, determine the input-output differential equation of the circuit.*

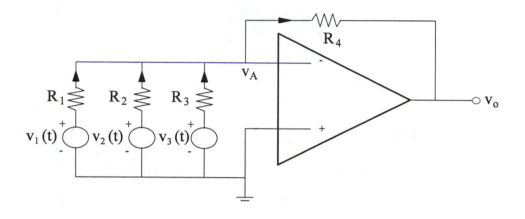

Solution 2.23 *Using the KCL and finding all the currents flowing from the sources (virtual ground) leads to*

$$i_{R_4} = i_{R_1} + i_{R_2} + i_{R_3}$$

$$\frac{v_A - v_o}{R_4} = \frac{v_1 - v_A}{R_1} + \frac{v_2 - v_A}{R_2} + \frac{v_3 - v_A}{R_3}$$

$$\frac{0 - v_0}{R_4} = \frac{v_1 - 0}{R_1} + \frac{v_2 - 0}{R_2} + \frac{v_3 - 0}{R_3}$$

$$v_o = -R_4 \left(\frac{v_1}{R_1} + \frac{v_2}{R_2} + \frac{v_3}{R_3} \right).$$

This means the circuit is a summer (it performs a summing function), since it adds up the input voltages.

Example 2.24 *Find the input-output differential equation describing the circuit shown in the following diagram*

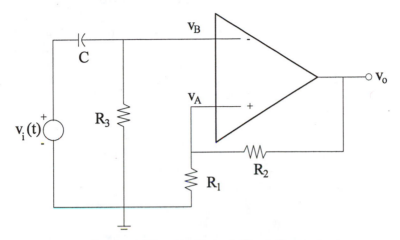

Op-Amp Circuit: Virtual-Short Concept

Solution 2.24 *Applying KCL at B gives*

$$i_C = i_{R_3}$$

$$C(\dot{v}_i - \dot{v}_B) = \frac{1}{R_3} v_B$$

$$C(\dot{v}_B - \dot{v}_i) + \frac{1}{R_3} v_B = 0.$$

By the virtual-short concept, $v_B = v_A$, which according to the voltage-divider rule can be written as $v_B = [R_1/(R_1 + R_2)]v_o$. Substituting this into the previous equation gives

$$\frac{R_1 C}{R_1 + R_2} \dot{v}_o - C\dot{v}_i + \frac{R_1}{(R_1 + R_2)R_3} v_o = 0$$

from which

$$R_1 R_3 C\dot{v}_o + R_1 v_o = (R_1 + R_2)R_3 C\dot{v}_i$$

or

$$\dot{v}_o + \frac{1}{R_3 C} v_o = \left(1 + \frac{R_2}{R_1}\right)\dot{v}_i.$$

This is the input-output equation for the op-amp circuit.

2.10.1   An Electromechanical System

Figure 2.3 shows a simplified model of a typical electromechanical system, that is, a system containing both moving mechanical parts and electrical components. The system represents a capacitor microphone, and consists of a parallel plate capacitor connected into an electric circuit. Capacitor plate 1 is rigidly fastened to the microphone frame. Sound waves pass through the mouthpiece and exert a force $f_s(t)$ on plate 2, which has mass m and is connected to the frame by a spring (k) and a damper (b). The

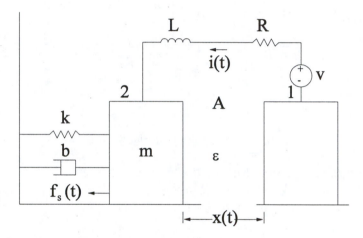

FIGURE 2.3
A Simple Electromechanical System

capacitance $C(x)$ is a function of the distance $x(t)$ between the plates

$$C(x) = \frac{\epsilon A}{x},$$

where ϵ is dielectric constant of the material between the plates, and A is the surface area of the plates. The charge q and the voltage e across the plates are related by

$$q = C(x)e.$$

The electric field in turn produces a force f_e, which opposes motion of the movable plate. This force is given by

$$f_e = \frac{q^2}{2\varepsilon A}.$$

Example 2.25 *Consider the electromechanical system in Figure 2.3.*

(a) Draw the electrical circuit and the free-body diagram that can be used to represent this system.

(b) Justify the choice of the vector of state variables as

$$x = [q \quad \dot{q} \quad x \quad \dot{x}]^T .$$

(c) Derive the state-variable model for this electromechanical system, where the output is the force f_e on the movable plate and the inputs are the voltage $v(t)$ and sound force $f_s(t)$

Solution 2.25 *(a) From Figure 2.3 the electrical circuit of the system can be deduced to be an RLC circuit with a variable capacitor $C(x)$ as shown in Figure 2.4.*

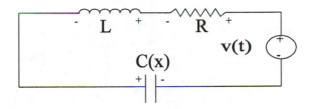

FIGURE 2.4
RLC Circuit for Electromechanical System

The free-body diagram is obtained by considering the forces acting on capacitor plate 2.

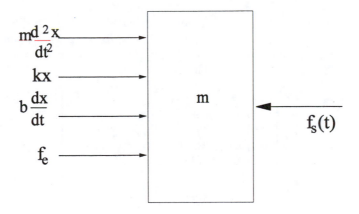

FIGURE 2.5
Free-Body Diagram: Electromechanical System

(b) There are four energy storing elements in the electromechanical system: a capacitor, an inductor, a spring, and a moving mass. The electrical energy stored in a capacitor is a function of the charge $q(t)$, the electrical energy stored in an inductor is a function of the current $\dot{q}(t)$, the potential energy stored in a spring is a function of the displacement $x(t)$, and the kinetic energy stored in a moving mass is a function of the velocity $\dot{x}(t)$. Hence, the minimum set of variables is given by

$$x = [q \quad \dot{q} \quad x \quad \dot{x}]^T.$$

(c) Employing the KVL in Figure 2.4 gives

$$v_C + v_L + v_R - v(t) = 0$$

$$v(t) = v_R + v_L + v_C$$

$$= Ri + L\frac{di}{dt} + \frac{1}{C(x)}\int i(t)dt$$

$$= R\frac{dq}{dt} + L\frac{d^2q}{dt^2} + \frac{1}{C(x)}\int \frac{dq}{dt}dt \quad \left(using \ i(t) = \frac{dq}{dt}\right)$$

$$= R\frac{dq}{dt} + L\frac{d^2q}{dt^2} + \frac{qx}{\epsilon A} \quad \left(using \ C(x) = \frac{\epsilon A}{x}\right).$$

Balancing the forces in the free-body diagram in Figure 2.5 gives

$$f_s(t) = m\ddot{x} + b\dot{x} + kx + f_e$$

$$= m\ddot{x} + b\dot{x} + kx + \frac{1}{2\epsilon A}q^2 \quad \left(using \ f_e = \frac{q^2}{2\epsilon A}\right).$$

This is a nonlinear system because of the terms qx and q^2, and hence, the matrices A and C cannot be extracted. The state variable system can be represented as follows:

$$\begin{bmatrix} \dot{q} \\ \ddot{q} \\ \dot{x} \\ \ddot{x} \end{bmatrix} = \begin{bmatrix} \dot{q} \\ -\frac{1}{L}\left(R\dot{q} + \frac{qx}{\epsilon A} - v\right) \\ \dot{x} \\ -\frac{1}{m}\left(b\dot{x} + kx + \frac{1}{2\epsilon A}q^2 - f_s\right) \end{bmatrix}$$

$$y = \frac{1}{2\epsilon A}q^2.$$

These equations are in the general state-variable form for nonlinear systems

$$\dot{x}(t) = f(x, u, t)$$
$$y(t) = g(x, u, t),$$

where the state variable and input vectors are given by

$$x(t) = [q \quad \dot{q} \quad x \quad \dot{x}]^T \quad and \quad u(t) = [v \quad f_s]^T.$$

Example 2.26 *The pulley system shown below is assumed to be ideal. Draw the free-body diagrams and obtain the modeling equations.*

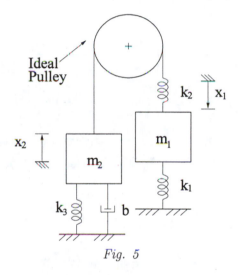

Fig. 5

Solution 2.26 *The free-body diagrams are shown in the following figures.*

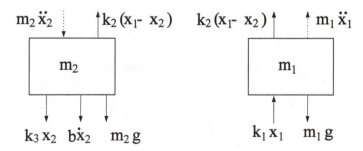

Note the inclusion of gravity (g) in the equations for the vertical system.

Example 2.27 *(a) Write the equations describing the series combination of the translational mechanical elements shown below.*

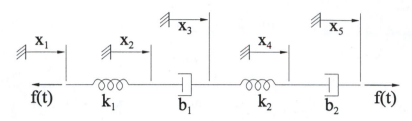

(b) Find the expressions for K_{eq} and B_{eq} (see the figure below) such that the motions of the ends of the combination are the same as those shown above.

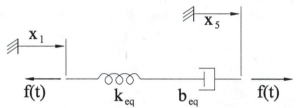

Solution 2.27 (a) The solution is obtained from analyzing the series as follows:

$$f(t) \quad 1 \quad k_1 \quad 2 \quad b_1 \quad 3 \quad k_2 \quad 4 \quad b_2 \quad 5 \quad f(t)$$

From (1-2)

$$f(t) = k_1(x_2 - x_1)$$

from (1-3)

$$f(t) = k_1(x_3 - x_1) + b_1(\dot{x}_3 - \dot{x}_1)$$

from (1-4)

$$f(t) = \frac{k_1 k_2}{k_1 + k_2}(x_4 - x_1) + b_1(\dot{x}_4 - \dot{x}_1)$$

from (1-5)

$$f(t) = \frac{k_1 k_2}{k_1 + k_2}(x_5 - x_1) + \frac{b_1 b_2}{b_1 + b_2}(\dot{x}_5 - \dot{x}_1)$$

(b)

$$f(t) \quad 1 \quad k_{eq} \quad b_{eq} \quad 5 \quad f(t)$$

from (1-5)

$$k_{eq} = \frac{k_1 k_2}{k_1 + k_2}; \qquad b_{eq} = \frac{b_1 b_2}{b_1 + b_2}$$

2.11   Linearization of Nonlinear Models

The differential equations describing the dynamic behavior of most practical systems are nonlinear. This is because in practice many elements of a dynamic system are inherently nonlinear and are only linear over a limited range of operating conditions. A nonlinear differential equation is one where the derivatives of the state variables have a nonlinear relationship to the states themselves and/or the inputs such that the differential equations cannot be written in the form

$$\dot{\mathbf{x}}(t) = \mathbf{Ax} + \mathbf{Bu}$$
$$\mathbf{y}(t) = \mathbf{Cx} + \mathbf{Du},$$

but rather in the general form

$$\dot{\mathbf{x}}(t) = \mathbf{f}(\mathbf{x}, \mathbf{u}, t)$$
$$\mathbf{y}(t) = \mathbf{g}(\mathbf{x}, \mathbf{u}, t).$$

The modeling, analysis, and control design are far easier for linear than for nonlinear systems. Real systems have all kinds of nonlinearities such as deadband, backlash, Coulomb friction, hysteresis, quantization, saturation, and kinematic nonlinearities. Thus, a controller designed for a linear system model to satisfy performance specifications may perform poorly when applied to the actual system. The trade-off here is between mathematical tractability of the linearized model and greater validity of a nonlinear model. When confronted with a mathematical model that contains nonlinearities, there are four approaches that can be used:

- Solve the nonlinear differential equations directly

- Small signal linearization (Linearization about an operating point)

- Linearization by feedback

- Obtain computer solutions of the response for specific cases of interest

The first alternative is possible only in specialized cases and will not be pursued. Linearization by feedback involves using part of the control effort to cancel the nonlinear terms and to design the remainder of the control based on linear theory. This technique is accomplished by subtracting the nonlinear terms out of the equations of motion and adding them to the control equation. The result is a linear system, provided the computer implementation of the control system has enough capability to compute the nonlinear terms fast enough. Linearization by feedback is popular in the field of robotics, where it is called the method of computed torque. It is also a research topic for control of aircraft and space shuttles.

2.11.1 Small Signal Linearization

The most popular and general approach employed in modeling nonlinear systems is small signal linearization. It is applicable for a broad range of systems and is extensively used in industry. Linearization is the process of finding a linear model that approximates a nonlinear one. If a small signal linear model is valid near an equilibrium and is stable, then there is a region (usually very small) containing the equilibrium within which the nonlinear system is stable. Thus one can make a linear model and design a linear control for it such that, in the neighborhood of the equilibrium, the design will be stable. Since a very important role of feedback control is to maintain the process variables near equilibrium, such small-signal linear models are a frequent starting point for control models.

The strategy is to determine the equilibrium points and then linearize the system. An equilibrium point is a state in which the system would remain if it were unperturbed by external disturbances. An equilibrium point can be unstable (an egg standing on its end), neutrally stable (a ball on a table), or stable (a book lying on a table). For a system under feedback control, an equilibrium point is called an operating point. A linearized model can be used to approximate a nonlinear system near an equilibrium point of the nonlinear system by a procedure called small signal linearization. The resulting linear system has an equilibrium point at zero that corresponds to the equilibrium point of the nonlinear system. While linearized models are only an approximation of the nonlinear system, they are convenient to analyze and they give considerable insight into the behavior of the nonlinear system near the equilibrium point. For example, if the zero equilibrium point of the linear system is stable, then the equilibrium of the nonlinear system is locally stable. The approximate linearized model of the system will be considered.

2.11.2 Linearization of Element Laws

A method is developed for linearizing an element law where the two variables, such as force and displacement, are not directly proportional. Next, the linearized element law is incorporated into the system model. Mechanical systems with nonlinear stiffness or friction elements are considered, and then nonlinear electrical systems are discussed. The object of linearization is to derive a linear model whose response will agree closely with that of the nonlinear model. Although the responses of the linear and nonlinear models will not match exactly and may differ significantly under some conditions, there will be a set of inputs and initial conditions for which the agreement is satisfactory. In this section the following are considered: the linearization of a single element law, a nonlinear operating point, a nominal value and incremental variable, graphical approach series-expansion approach.

2.11.3   Linearization of Models

The linearized element laws can now be incorporated into a system model, thus producing a linearized model from a nonlinear one. Starting with a nonlinear model, the procedure can be summarized as follows:

- Determine the operating point of the model by writing and solving the appropriate nonlinear algebraic equations. Select the proper operating point value if extraneous solutions also appear.

- Rewrite all linear terms in the mathematical model as the sum of their nominal and incremental variables, noting that the derivatives of constant terms are zero.

- Replace all nonlinear terms by the first two terms of their Taylor-series expansions, that is, the constant and linear terms.

- Using the algebraic equations defining the operating point, cancel the constant terms in the differential equations, leaving only linear terms involving incremental variables.

- Determine the initial conditions of all incremental variables in terms of the initial conditions of the variables in the nonlinear model.

The operating point of the system will be a condition of equilibrium in which each variable will be constant and equal to its nominal value and in which all derivatives will be zero. Inputs will take on their nominal values, which are typically selected to be their average values. For example, if a system input is

$$u(t) = A + B \sin \omega t,$$

the nominal value of the input would be used. Under these conditions, the differential equations reduce to algebraic equations that one can solve for the operating point, using a computer if necessary. In general, the coefficients involved in those terms that came from the expansion of nonlinear terms depend on the equilibrium conditions. Hence, a specific operating point must be found before the linearized model can be expressed in numerical form. The entire procedure will be illustrated by an example.

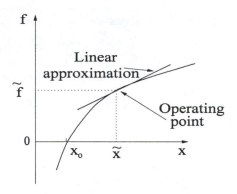

FIGURE 2.6
Nonlinear Spring Characteristic: Linear Approximation

2.11.4 Linearization Concepts

Linearization of the element law is carried with respect to an *operating point* as shown in the following diagram, which depicts the nonlinear spring characteristic (force $f(t)$ vs. displacement $x(t)$ curve). The operating point is a specific point on this curve and is denoted by the coordinates $(\tilde{x}, \tilde{f})$. Thus the displacement $x(t)$ can be expressed as

$$x(t) = \tilde{x} + \hat{x}(t),$$

where $\tilde{x}$ is a constant term called the *nominal value* of x, and $\hat{x}(t)$ is the *incremental variable* corresponding to x. Similarly, the force $f(t)$ is represented in the same way

$$f(t) = \tilde{f} + \hat{f}(t),$$

where

$$\tilde{f} = f(\tilde{x}).$$

With the necessary terms defined, element laws can be linearized graphically as shown in the previous figure. The tangent curve to the nonlinear law $f(x)$ is a good approximation to the nonlinear curve around the operating point. The slope of the tangent is given by

$$k = \frac{df}{dx}\Big|_{x=\tilde{x}}.$$

Hence the equation of the tangent is given by

$$f = \tilde{f} + k(x - \tilde{x})$$
$$\left(f - \tilde{f}\right) = k(x - \tilde{x})$$
$$\hat{f} = k\hat{x}.$$

Thus, the linearized element law can be expressed in terms of incremental variables, $\hat{f}(t)$ and $\hat{x}(t)$. The results are graphically displayed in the following diagram.

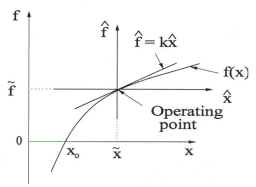

Nonlinear Spring Characteristic: Incremental Variable Coordinates

Example 2.28 *A nonlinear translational spring obeys the force-displacement law*

$$f(x) = |x|x.$$

Determine the linearized element law in numerical form for each of the following operating points

$$\tilde{x}_1 = -1, \quad \tilde{x}_2 = 0, \quad \tilde{x}_3 = 1, \quad \tilde{x}_4 = 2.$$

Solution 2.28 *The nonlinear law can be expressed as*

$$f(x) = \begin{cases} -x^2 & for \quad x < 0 \\ x^2 & for \quad x \geq 0. \end{cases}$$

The slope of the tangent at the operating point is given

$$k = \frac{df}{dx}\Big|_{x=\tilde{x}} = \begin{cases} -2\tilde{x} & for \ \tilde{x} < 0 \\ 2\tilde{x} & for \ \tilde{x} \geq 0 \end{cases}$$
$$= 2|\tilde{x}| \quad for \ all \ \tilde{x}.$$

Hence, the linearized element law in terms of incremental variables is given by

$$\hat{f} = 2|\tilde{x}|\hat{x},$$

where

$$k = 2|\tilde{x}|$$

describes the variable spring constant. For the four values of $\tilde{x}$ the results are summarized in the following figure and table:

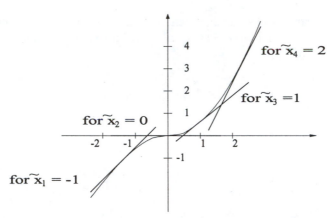

Nonlinear Spring Characteristic: Linear Approximations for the Four Values of $\tilde{x}$.

$$
\begin{bmatrix}
i & \tilde{x}_i & \tilde{f}_i & k_i \\
1 & -1 & -1 & 2 \\
2 & 0 & 0 & 0 \\
3 & 1 & 1 & 2 \\
4 & 2 & 4 & 4
\end{bmatrix}
$$

In terms of accuracy, the approximation is quite good for deviations up to 0.30 from the operating point.

2.12   Experimental Data Approach

Here, frequency response experimental data are used to construct the transfer function model. The techniques that are used form part of the important subject of system identification. There are several reasons for using experimental data to obtain a model of the dynamic system to be controlled. In the first place, the best theoretical model built from equations of motion is still only an approximation of the actual system. Sometimes, as in the case of a very rigid spacecraft, the theoretical model is extremely good. Other times, as with many chemical processes, such as paper making or metal working, the theoretical model is very approximate. In every case, before the final control design is done, it is important and prudent to verify the theoretical model with experimental data.

In situations where the theoretical model is especially complicated or the physics of the process is poorly understood, the only reliable information on which to base the control design is experimental data. Finally, the system is sometimes subject to on-line changes, which occur when the environment of the system changes. Examples include: an aircraft changing altitude or speed, a paper machine given a different composition of fiber, or a nonlinear system moving to a new operating point. On these occasions there is need to re-tune the controller by changing the control parameters. This requires a model for the new conditions, and experimental data are often the most effective, if not the only, information available for the new model.

There are four kinds of experimental data that can be used for generating a dynamic system model

- Transient response, such as that obtained from an impulse or a step input

- Frequency response data, which result from exciting the system with sinusoidal inputs at many frequencies

- Stochastic steady state information, as might come from flying an aircraft through turbulent weather or from some other natural source of randomness

- Pseudo-random-noise data, as may be generated in a digital computer

Each of these classes of experimental data has its own properties, advantages, and disadvantages.

2.12.1   Transient Response

Transient response data are quick and relatively easy to obtain. They are also often representative of the natural signals to which the system is subjected. Thus, a model derived from such data can be reliable for designing the control system. On the other hand, in order for the signal-to-noise ratio to be sufficiently high, the transient response must be highly noticeable. Hence, the method is rarely suitable for normal operations, so the data must be collected as part of special tests. A second disadvantage is that the data does not come in a form suitable for standard control systems designs, and some parts of the model, such as poles and zeros, must be computed from the data. This computation can be simple in special cases or complex in the general case.

2.12.2   Frequency Response

Frequency-response data (see Chapter 6) is simple to obtain but substantially more time-consuming than transient-response information. This is especially true if the time constants of the process are large, as often occurs in chemical processing industries. As with the transient-response data, it is important to have a good signal-to-noise ratio, hence obtaining frequency-response data can be very expensive. On the other hand, as will be seen in Chapter 6, frequency-response data are exactly in the right form for frequency-response design methods, so once the data have been obtained, the control design can proceed immediately.

2.12.3   Stochastic Steady State

Normal operating records from a natural stochastic environment at first appear to be an attractive basis for modeling systems since such records are by definition non-disruptive and inexpensive to obtain. Unfortunately, the quality of such data is inconsistent, tending to be worst when the control is best. This is because under these conditions the upsets are minimal and the signals are smooth. At such times, some or even most of the system dynamics are hardly excited. Since they contribute little to the system output, they will not be found in the model constructed to explain the signals. The result is a model that represents only part of the system and is sometimes unsuitable for control. In some instances, as is the case when trying to model the dynamics of the electroencephalogram (brain waves) of a sleeping or anesthetized person to locate the frequency and intensity of alpha waves, normal records are the only possibility. Usually they are the last choice for control purposes.

2.12.4   Pseudo-Random Noise (PRBS)

Finally, the pseudo-random signals that can be constructed using digital logic have much appeal. Especially interesting for model making is the pseudo-random binary signal (PRBS). The PRBS takes on the value $+A$ or $-A$ according to the output (1 or 0) of a feedback shift register. The feedback to the register is a binary sum of various states of the register that have been selected to make the output period (which must repeat itself in finite time) as long as possible. For example, with a register of 20 bits, 2^{11} (over a million) pulses are produced before the pattern repeats. Analysis beyond the scope of this text has revealed that the resulting signal is almost like a broad-band random signal. Yet this signal is entirely under the control of the engineer who can set the level (A) and the length (bits in the register) of the signal. The data obtained from tests with a PRBS must be analyzed by computer, and both special-purpose hardware and programs for general-purpose computers have been developed to perform this analysis. This approach has extensive applications.

2.12.5 Models from Transient-Response Data

In order to obtain a model from transient data it is assumed that a step response is available. If the transient is a simple combination of elementary transients, then a reasonable low-order model can be estimated using hand calculations. For example, the step response is monotonic and smooth.

2.12.6 Models from Other Data

A model can be generated using frequency-response data, which are obtained by exciting the system with a set of sinusoids and plotting. In Chapter 6 it will be shown how such plots can be used directly for design. Alternatively, one can use the frequency response to estimate the poles and zeros of a transfer function using straight-line asymptotes on a logarithmic plot. The construction of dynamic models from normal stochastic operating records or from the response to a PRBS can be based either on the concept of cross-correlation or on the least-squares fit of a discrete equivalent model, both topics in the field of system identification. They require substantial presentation and background that are beyond the scope of this text.

2.12.7 Pure Time Delay

When a tap is turned on at one end of a long hose-pipe it takes some time before the water appears at the far end. When the feed to a conveyor belt is changed it takes some time for the change to be observed at the other end. These are two examples of the phenomenon of pure-time delay, also called transportation lag or dead time. The waveforms of an arbitrary input to a pure-time delay of T seconds and the output are shown below. The output wave-form is exactly the input wave form shifted T seconds into the future. A pure-time delay can be modeled using Laplace transforms by

$$G(s) = e^{-sT}.$$

2.13 Problems

Problem 2.1 *For the following translational mechanical system, the springs are undeflected when $x_1 = x_2 = 0$.*

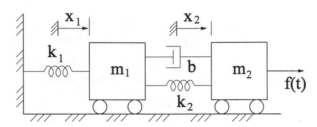

Translational Mechanical System

(a) Draw the free-body diagrams for the system.

(b) Write down the dynamic equations of motion.

(c) Choose the minimum set of state variables for the system and justify your choice.

(d) Develop the system state-variable matrix model (i.e. find A, B, C, D), where the input is force $f(t)$ and the output is the force in spring k_2.

Problem 2.2 *For the system shown below, the springs are undeflected when $x_1 = x_2 = 0$, and the input is force $f(t)$.*

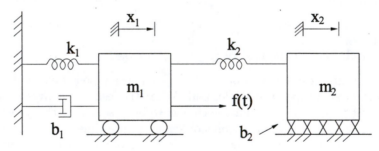

Translational Mechanical System

(a) Draw the free-body diagrams for the system.

(b) Write down the dynamic equations of motion.

(c) Choose the minimum set of state variables for the system and justify your choice.

(d) Develop the state-variable equations in matrix form (i.e. find A, B, C, D).

Problem 2.3 *For designing an automobile suspension, a two-mass system can be used for modeling as shown in the following diagram.*

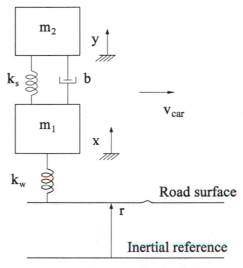

The Quarter-Car Model

This is called a quarter-car model because it comprises one of the four wheel suspensions. The car and wheel positions are denoted by $y(t)$ and $x(t)$ respectively. These displacements are from static equilibrium which corresponds to no inputs except gravity.

(a) Draw the free-body diagram of this system, assuming one-dimensional vertical motion of the mass above wheel.

(b) Write down the equations of motion for the automobile.

(c) Express these equations in a state-variable matrix form (A,B,C,D) using the following state-variable vector,

$$x(t) = [x \quad \dot{x} \quad y \quad \dot{y}]^T,$$

and justify this choice of state variables. Note that the car and wheel positions, $y(t)$ and $x(t)$, are the two outputs of the car system while the input is the unit step bump $r(t)$.

(d) Plot the position of the car and the wheel after the car hits a "unit bump" (i.e. $r(t)$ is a unit step) using MATLAB. Assume $m_1 = 10kg$, $m_2 = 250kg$, $k_w = 500,000N/m$, $k_s = 10,000N/m$. Find the value of b that you would prefer if you were a passenger in the car.

Problem 2.4 *The input to the translational mechanical system shown in the following diagram is the displacement $x_3(t)$ of the right end of the spring k_1. The absolute displacements of m_1 and m_2 are x_1 and x_2, respectively. The forces exerted by the springs are zero when $x_1 = x_2 = x_3 = 0$.*

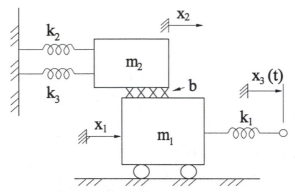

Translational Mechanical System

(a) Draw the free-body diagrams of the system.

(b) Obtain the system dynamic equations of motion.

Problem 2.5 *Consider the mechanical and the electrical systems shown in the following diagrams.*

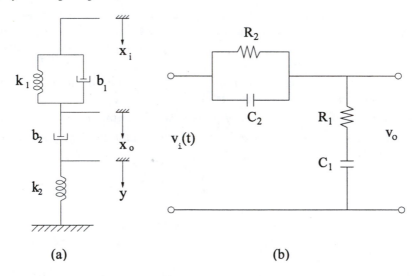

       (a)   **(b)**

(a) Obtain the transfer function of the mechanical system, where $x_i(t)$ is the input and $x_o(t)$ is the output.

(b) Obtain the transfer function of the electrical system, where $v_i(t)$ is the input and $v_o(t)$ is the output.

(c) Show that the transfer functions of the two systems have an identical form and are thus analogous.

Problem 2.6 *In the following mechanical system the input is a force $f(t)$ and the output is the spring force in k_2.*

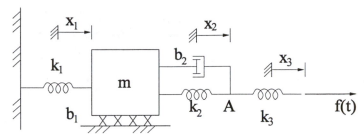

Translational Mechanical System

(a) Draw the free-body diagrams for the mass m and for the massless point A.

(b) Write down the equations of motion.

(c) Choose the minimum number of state variables for the system and develop the state-variable matrix model (i.e. obtain **A, B, C, D**).

(d) If the dashpot b_2 is removed show that

$$x_3 = \frac{1}{k_3}[(k_2 + k_3)x_2 - k_2 x_1].$$

*Choose a new set of independent state variables for this new system and develop the state-variable matrix model (i.e. obtain **A, B, C, D**).*

Problem 2.7 *In the following translational mechanical system the input is $f(t)$ and the output is the force in spring k_3.*

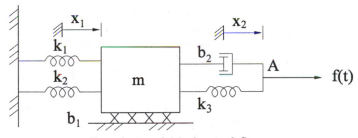

Translational Mechanical System

(a) Draw the free-body diagrams for the mass m and the massless point A.

(b) Write down the system equations of motion

(c) Choose the minimum set of state variables for the system and develop the state-variable matrix model (i.e. obtain A, B, C and D).

Problem 2.8 *For a certain dynamic system the state-variable equations for the position x_1 and velocity v_1 are given below:*

$$\dot{x}_1 = v_1$$

$$\dot{v}_1 = \frac{1}{m_1}[-(k_1 + k_2)x_1 - bv_1 + b\dot{x}_2 + k_2 x_2(t),$$

where $x_2(t)$ is the input and the output is x_1. Use the s-operator to obtain the input-output differential equation which relates the output x_1 and the input $x_2(t)$.

Problem 2.9 *Consider the following mechanical system.*

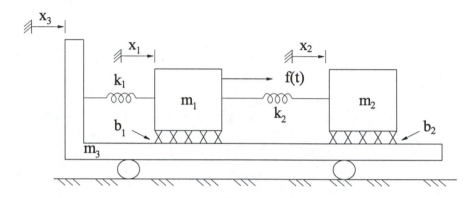

Translational Mechanical System

The forces exerted by the springs are zero when $x_1 = x_2 = x_3 = 0$. The input force is $f(t)$ and the absolute displacements of m_1, m_2 and m_3 are x_1, x_2 and x_3, respectively
   (a) Draw the free-body diagrams of the system.
   (b) Obtain the system dynamic equations of motion.
   (c) Choose a suitable vector of state variables and justify your choice.
   (d) Express the dynamic equations in the state-variable matrix form (A, B, C, D), where the output is the spring force in k_2.

Problem 2.10 *The following data is provided for the translational mechanical system considered in Problem 2.9.*

$$m_1 = 5kg$$
$$m_2 = 15kg$$
$$m_3 = 50kg$$
$$b_1 = 500N/m/s$$
$$b_2 = 600N/m/s$$
$$k_1 = 5,000N/m$$
$$k_2 = 10,000\dot{N}/m$$

Use MATLAB to find the following:
   (a) The output response (y) to an input force of $50N$
   (b) The response to an impulse force input.

(c) *If the values of b_1 and b_2 were variables choose values of b_1 and b_2 that will reduce oscillations, give practical settling time, and produce smooth responses.*

If the output was the platform displacement x_3, obtain the new matrices **C** *and* **D**.

(a) *Find the output response (y) to an input force of $50N$*

(b) *Find the response to an impulse force input.*

(c) *If the values of b_1 and b_2 were variables choose values of b_1 and b_2 that will reduce oscillations, and give practical settling time and smooth responses.*

Problem 2.11 *In some mechanical positioning systems the movement of a large object is controlled by manipulating a much smaller object that is mechanically coupled with it. The following diagram depicts such a system, where a force $u(t)$ is applied to a small mass m in order to position a larger mass M. The coupling between the objects is modeled by a spring constant with a damping coefficient b.*

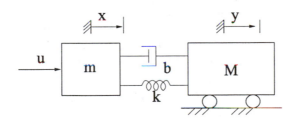

(a) *Draw the free-body diagrams of the system*

(b) *Write the equations of motion governing this system.*

(c) *Identify the appropriate state variables, and express the equations of motion in the state-variable matrix form (A, B, C, D).*

Problem 2.12 *In the following rotational mechanical system there is a driving torque $T(t)$ exerted on disk J_1 and a load torque $T_L(t)$ exerted on disk J_2. Both the applied torque $T(t)$ and the load torque $T_L(t)$ are considered as inputs and the output is the angular velocity $\omega_3(t)$.*

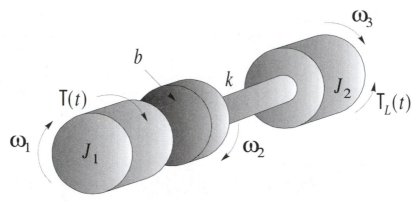

Rotational Mechanical System

(a) Draw the free-body diagrams for the system.

(b) Explain why the vector of state variables can be chosen as

$$\mathbf{x}(t) = [\,\theta_2 \quad \theta_3 \quad \omega_1 \quad \omega_3\,]^T .$$

(c) Express the dynamic equations in state-variable matrix form (i.e. obtain A, B, C, D).

(b) Use the s-operator method to express the dynamic equations in the form of an input-output differential equation.

Problem 2.13 *The diagram below shows a double pendulum system. Assume the displacement angles of the pendulums is small enough to ensure that the spring is always horizontal. The pendulum rods are taken to be massless, of length l, and the springs are attached 2/3 of the way down.*

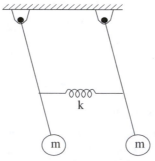

Double Pendulum

Derive two differential equations that model the motion of the double pendulum.

Problem 2.14 *Consider the RLC electrical circuit given below, where the input is current $i_1(t)$ and the output is voltage $v_o(t)$.*

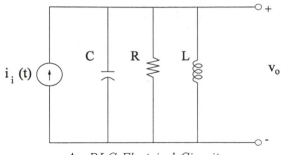

An RLC Electrical Circuit

(a) Find the state-variable model (A, B, C, D) for the circuit.

(b) Show that the input-output differential equation is given by

$$C\ddot{v}_o + \frac{1}{R}\dot{v}_o + \frac{1}{L}v_o = \dot{i}(t)$$

Problem 2.15 *In the figure below, determine the input-output differential equation of the circuit.*

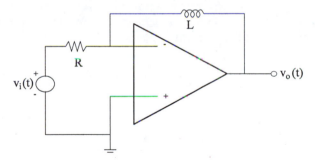

Problem 2.16 *Find the input-output differential equation relating v_o and $v_i(t)$ for the circuit shown below.*

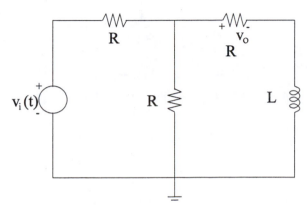

Problem 2.17 *(a) For the following circuit find the state-variable matrix model (A, B, C, D) where v_o is the output voltage and v_i is the input voltage.*

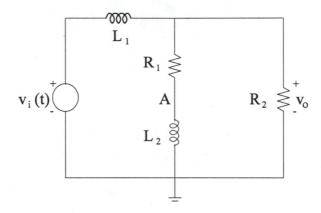

(b) Also find the input-output differential equation for the system.

Problem 2.18 *Consider the following circuit where $i_i(t)$ is the input current and $i_o(t)$ is the output current.*

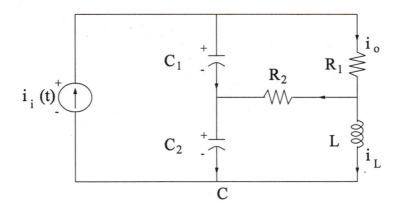

(a) Obtain the state-variable matrix model (A, B, C, D) for the circuit.
(b) Obtain the input-output differential equation for the circuit.

Problem 2.19 *For the following op-amp circuit derive the algebraic expression for the output voltage v_o in terms of the two input voltages $v_1(t)$ and $v_2(t)$. Indicate what mathematical operation the circuit performs.*

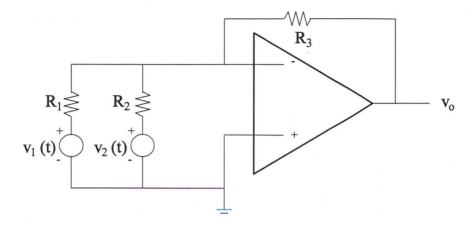

Problem 2.20 *(a) For the following op-amp circuit derive the input-output differential equation relating the output voltage v_o and the input voltage $v_i(t)$.*

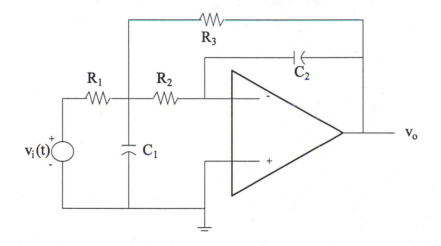

(b) Derive the circuit's state-variable matrix model (A, B, C, D), where the input is voltage $v_i(t)$ and there are two outputs: the voltage output v_o, and the current through R_3 (with positive sense to the right).

Problem 2.21 *Consider the following op-amp circuit where the output is voltage $v_o(t)$ and the input is voltage $v_i(t)$.*

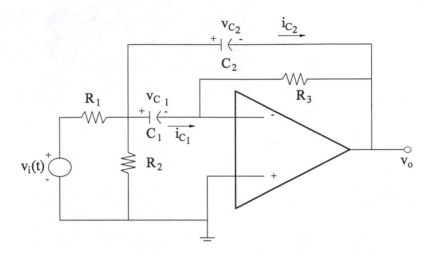

(a) Choose a set of state variables and justify your choice.

(b) Obtain the state-variable matrix model (A, B, C, D) for the system.

Chapter 3

Dynamic System Response

3.1 Introduction

In order to design a control system it is essential that the behavior of the plant (or process) be analyzed and understood. In the process of analyzing a system, two tasks must be performed: modeling the system and obtaining the dynamic system response. The first task was accomplished in Chapter 2 and the dynamic system response is obtained by solving the differential equations that constitute the system model. Once the system response is obtained, the function of a controller in a control system is then to influence the system response or behavior of the plant. The objective of this chapter is to develop techniques for finding the system responses for the dynamic systems modeled in Chapter 2. This activity is also called solving the model and involves using the mathematical model to determine certain features of the system cause-and-effect relationships.

Three main mathematical approaches are used to obtain the system response: direct solution of differential equations in the time domain, the use of the Laplace transform to solve differential equations in the frequency domain, and the deduction of system behavior from the system transfer function. The Laplace transform is a mathematical tool for transforming linear differential equations into an easier-to-manipulate algebraic form. In this domain, the differential equations are easily solved and the solutions are converted back into the time domain to give the system response. The transfer function was introduced in Chapter 2 as a modeling tool. The major advantage of this form of the dynamic system model is that the system response can be easily obtained from it. From the transfer function the system poles and zeros can be identified and these provide information about the characteristics of the system response. The location of the poles and zeros can then be manipulated to achieve certain desired characteristics or eliminate undesirable ones. In addition to the direct mathematical derivation of the transfer function, there are two visual tools that can be employed to derive it. The first is the block diagram, which was introduced in Chapter 2 and the other is the signal flow graph. The latter method consists of characterizing the system by a network of directed branches and associated gains (transfer functions) connected to nodes. Mason's rule is used to relate the graph to the algebra of the system simultaneous equations, thus determining the system transfer function.

Instead of using analytic methods to determine the system response in certain circumstances it is more feasible to use numerical methods. The type of equation involved in the model has a strong influence on the extent to which analytical methods can be used. For example, nonlinear differential equations not often solved in closed form, and the solution of partial differential equations is far more laborious than that of ordinary differential equations. Computers can be used to generate the responses to specific numerical cases for complex models. However, using a computer to solve a complex model has its limitations. Models used for computer studies should be chosen with the approximations encountered in numerical integration in mind and should be relatively insensitive to system parameters whose values are uncertain or subject to change. It must not be forgotten that the model being analyzed is only an approximate mathematical description of the system, not the physical system itself. When an analytical model is not available or is too complex to formulate the system behavior, it can be established from both experimental time-response data and frequency-response data. The development of models from experimental data is presented in Chapter 2.

3.1.1   Chapter Objectives

After finishing this chapter the reader should be able to accomplish a number of tasks. Given a mathematical model (or after deriving a model) for a dynamic system, the reader should be able to do the following:

- For the first or second-order system model, solve the differential equations directly to determine the system response

- For a dynamic model use the Laplace transform to
  a) find the complete time response

  b) determine the transfer function, its poles and zeros

  c) analyze stability, evaluate time constants, damping ratios, and undamped natural frequencies

- Use the transfer function to determine system response

- Use the Block diagram and Signal flow graph to determine the transfer function

- Find and analyze the impulse response, step response, and sinusoidal response

- In addition to using analytical methods, use numeric methods to obtain the system response of a linear or nonlinear model in numerical form

- Determine the system response from experimental data

3.2   Time Domain Solution of System Models

The dynamic response of a system can be determined directly in the time domain without any transformation. The starting point is expressing the system model in the input-output differential equation form, with all other variables eliminated. Consider the general n-th order model

$$a_0 y + a_1 \dot{y} + a_2 \ddot{y} + \dots + a_n y^{(n)} = b_0 u + b_1 \dot{u} + b_2 \ddot{u} +$$
$$\dots + b_m u^{(m)}, \tag{3.1}$$

where

$$y^{(n)} = \frac{d^n y}{dt^n} \quad \text{and} \quad u^{(m)} = \frac{d^m u}{dt^m}.$$

The terms on the right-hand side that involve the input and its derivatives constitute the forcing function

$$f(t) = b_0 u + b_1 \dot{u} + b_2 \ddot{u} + \dots + b_m u^{(m)}.$$

With this definition the input-output differential equation model can be expressed as

$$a_0 y + a_1 \dot{y} + a_2 \ddot{y} + \dots + a_n y^{(n)} = f(t). \tag{3.2}$$

The desired solution $y(t)$ for $t \geq 0$ must satisfy the input-output differential Equation 3.2.

3.2.1   Homogeneous Input-Output Equations

If $f(t) = 0$ in the input-output Equation 3.2 i.e., neither inputs nor their derivatives are present, then the equation is called a homogeneous differential equation,

$$a_0 y_h + a_1 \dot{y}_h + a_2 \ddot{y}_h + \dots + a_n y_h^{(n)} = 0. \tag{3.3}$$

The corresponding characteristic equation for this system is given by

$$a_0 + a_1 r + a_2 r^2 + \dots + a_n r^n = 0, \tag{3.4}$$

which has roots $r_1, r_2, r_3 \dots, r_n$. Hence, the solution of Equation 3.4 takes the form

$$y_h(t) = K_1 e^{r_1 t} + K_2 e^{r_2 t} + \dots + K_n e^{r_n t}$$
$$= \sum_{i=1}^{n} K_i e^{r_i t}, \tag{3.5}$$

where the terms K_i are real constants.

If two or more of the roots are identical (repeated roots) then Equation 3.5 is modified as follows:

$$y_h(t) = K_1 e^{r_1 t} + K_2 t e^{r_1 t} + K_3 e^{r_3 t} + \dots + K_n e^{r_n t}, \tag{3.6}$$

for $r_1 = r_2$. If any of the roots are complex conjugates ($r_1 = \alpha + j\beta$ and $r_2 = \alpha - j\beta$), the solution of Equation 3.4 can be expressed in three related forms

$$y_h(t) = K_1 e^{(\alpha + j\beta)t} + K_2 e^{(\alpha - j\beta)t}$$
$$= e^{\alpha t}(K_3 \cos \beta t + K_4 \sin \beta t)$$
$$= K e^{\beta t} \cos(\beta t + \phi),$$

where ϕ is a constant angle. Hence, in order to find the solution for any given homogeneous input-output differential equation with complex roots, the constants K_1, K_2, α and β must be determined (K_3, K_4, K and ϕ depend on the first four constants).

Example 3.1 *Consider a first-order system with no input such that the input-output differential equation is given by*

$$\dot{y} + 3y = 0.$$

Find its system response.

Solution 3.1 *The characteristic equation is given by*

$$r + 3 = 0 \Rightarrow r = -3$$
$$\implies y_h(t) = K e^{-3t}.$$

This is the system response, where K is a real constant, and thus the response represents a family of curves.

3.2.2 Nonhomogeneous Input-Output Equations

When the forcing function is not zero, i.e., $f(t) \neq 0$, the equation is called a nonhomogeneous input-output differential equation. Its solution $y(t)$ is called the **complete solution** (or general solution) and it consists of two parts; the **homogeneous solution** $y_h(t)$ and the **particular solution** $y_p(t)$ such that

$$y(t) = y_h(t) + y_p(t).$$

The homogeneous solution must satisfy the homogeneous equation,

$$a_0 y_h + a_1 \dot{y}_h + a_2 \ddot{y}_h + \dots + a_n y_h^{(n)} = 0,$$

while the particular solution must satisfy the entire differential equation,

$$a_0 y + a_1 \dot{y} + a_2 \ddot{y} + \dots + a_n y^{(n)} = f(t).$$

The procedure for obtaining the homogeneous solution $y_h(t)$ has already been explained. A general method of obtaining the $y_p(t)$ involves the variation of parameters procedure. When the forcing function $f(t)$ has only a finite number of different derivatives, the method of undetermined coefficients is sufficient. This method assumes that the form of $y_p(t)$ consists of terms similar to those in $f(t)$ and their derivatives. Each of the terms and derivatives is multiplied by coefficients that must be determined. Some common forms of particular solutions are listed in the table below.

$f(t)$	$y_p(t)$
β	b
$\alpha_1 t + \alpha_o$	$at + b$
$e^{\alpha t}$	$a e^{\alpha t}$
$\cos \omega t$	$a \cos \omega t + b \sin \omega t$
$\sin \omega t$	$a \cos \omega t + b \sin \omega t$

If $f(t)$ or one of its derivatives contains a term identical to a term in the homogeneous solution $y_h(t)$, the corresponding terms in the right-hand column of the table above should be multiplied by t. For example, if

$$y_h(t) = K_1 e^{-4t} + K_2 e^{-3t} \quad \text{and} \quad f(t) = 2e^{-3t},$$

then

$$y_p(t) = ate^{-3t},$$

should be used. If a term in $f(t)$ corresponds to a double root of the characteristic equation the normal form for $y_p(t)$ is multiplied by t^2. For example, if

$$y_h(t) = K_1 e^{-3t} + K_2 te^{-3t} \quad \text{and} \quad f(t) = 2e^{-3t},$$

then

$$y_p(t) = at^2 e^{-3t},$$

should be used. It is important to note that when solving for the complete or general solution

$$y(t) = y_h(t) + y_p(t),$$

the arbitrary constants K_1 through K_n cannot be evaluated until both $y_h(t)$ and $y_p(t)$ have been found.

The two parts of the complete solution have physical significance. The homogeneous solution $y_h(t)$ represents the natural behavior of the system where there are no external inputs i.e. $f(t) = 0$.

$$y_h(t) \equiv \text{Free response.}$$

The particular solution depends on the form of the forcing function.

$$y_p(t) \equiv \text{Forced response.}$$

Example 3.2 *Consider a first-order system with a forcing function $f(t)$ such that input-output differential equation is given by*

$$\dot{y} + 3y = f(t), \tag{3.7}$$

where $y(0) = 1$. Find the dynamic system response $y(t)$ for the following forcing functions:

(a) $f(t) = 5$
(b) $f(t) = 5\cos 2t$.

Solution 3.2 *The characteristic equation is obtained by inspecting the LHS of Equation 3.7,*

$$r + 3 = 0 \Rightarrow r = -3$$
$$\implies y_h(t) = Ke^{-3t}.$$

The value of K depends on $f(t)$ as well as $y(0)$ and hence it will be different for the two cases.

(a) For $f(t) = 5$, let the particular solution be given by

$$y_p(t) = b.$$

Substituting $f(t) = 5$ and $y_p(t) = b$ in Equation 3.7 leads to

$$\dot{y} + 3y = f(t)$$

$$\Longrightarrow \frac{db}{dt} + 3b = 5$$

$$\Longrightarrow 0 + 3b = 5$$

$$\Longrightarrow b = \frac{5}{3}.$$

The complete solution then becomes

$$y(t) = y_h(t) + y_p(t)$$

$$= Ke^{-3t} + \frac{5}{3}. \qquad (3.8)$$

The value of K is then obtained by using the initial condition, $y(0) = 1$, in Equation 3.8.

$$y(0) = Ke^{-3\times0} + \frac{5}{3} = 1$$

$$\Longrightarrow K + \frac{5}{3} = 1$$

$$\Longrightarrow K = -\frac{2}{3}.$$

Therefore, the system response is given by

$$y(t) = -\frac{2}{3}e^{-3t} + \frac{5}{3}.$$

(b) For the second case $f(t) = 5\cos 2t$, hence, the particular solution is of the form

$$y_P(t) = a\cos 2t + b\sin 2t.$$

Substituting $f(t) = 5\cos 2t$ and $y_p(t) = a\cos 2t + b\sin 2t$ in Equation 3.7 leads to

$$\dot{y} + 3y = f(t)$$

$$\frac{d}{dt}(a\cos 2t + b\sin 2t) + 3(a\cos 2t + b\sin 2t) = 5\cos 2t$$

$$-2a\sin 2t + 2b\cos 2t + 3a\cos 2t + 3b\sin 2t = 5\cos 2t$$

$$(3b - 2a)\sin 2t + (2b + 3a)\cos 2t = 5\cos 2t.$$

Collecting coefficients gives two simultaneous equations

$$(3b - 2a) = 0$$
$$(2b + 3a) = 5$$
$$\implies b = \frac{10}{13} \quad and \quad a = \frac{15}{13}.$$

The complete solution then becomes

$$y(t) = y_h(t) + y_p(t)$$
$$= Ke^{-3t} + \left(\frac{15}{13} \cos 2t + \frac{10}{13} \sin 2t \right).$$

The value of K is then obtained by using the initial condition, $y(0) = 1$. Hence,

$$y(0) = Ke^0 + \left(\frac{15}{13} \cos 0 + \frac{10}{13} \sin 0 \right) = 1$$
$$\implies K + \left(\frac{15}{13} \right) = 1$$
$$\implies K = -\frac{2}{13}.$$

Therefore,the system response is given by

$$y(t) = -\frac{2}{13}e^{-3t} + \left(\frac{15}{13} \cos 2t + \frac{10}{13} \sin 2t \right)$$
$$= -\frac{1}{13} \left(2e^{-3t} - 15 \cos 2t - 10 \sin 2t \right).$$

3.2.3 First-Order Systems

Consider a general first-order differential equation

$$\dot{y} + \frac{1}{\tau}y = f(t),$$

where τ is a real non-zero constant called the time constant. All first order differential equations can be expressed in this form. The characteristic equation is obtained by inspection from the homogeneous differential equation,

$$\dot{y} + \frac{1}{\tau}y = 0$$
$$\implies r + \frac{1}{\tau} = 0$$
$$\implies r = -\frac{1}{\tau}.$$

The homogeneous response is thus given by a family of curves of the form

$$y_h(t) = Ke^{-\frac{t}{\tau}},$$

where K is a real constant.

3.2.3.1 Stability Analysis

The stability of the family of curves can be established as follows:

$\tau > 0 \implies$ A stable system.

$\tau < 0 \implies$ An unstable system.

$\tau \longrightarrow \infty \implies$ A marginally stable system.

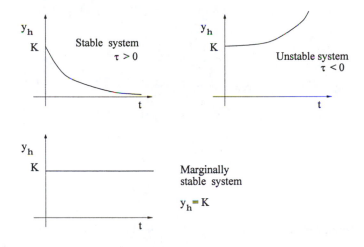

FIGURE 3.1
First-Order System Response: Stability Analysis

The complete dynamic system response can be interpreted as consisting of a transient response and a steady state response. For stable systems the transient part consists of terms that decay to zero as t approaches ∞. The steady state response is the part that remains after transient terms have decayed to zero. For the first-order system, as discussed before, the complete response is given by

$$y(t) = y_p(t) + y_h(t)$$

$$\text{where } y_h(t) = Ke^{-\frac{t}{\tau}}.$$

The homogeneous response $y_h(t)$ is also called the transient response, and if τ is positive, the system is stable. The particular solution $y_p(t)$ then represents the steady state response, where $y_p(t)$ can be a variety of functions (e.g., β, $a\cos\omega t + b\sin\omega t$ or $at + b$). However, if $y_p(t)$ is of the form $ate^{-t} + be^{-t}$, then it becomes part of the transient response. For stable systems whose $y_p(t)$ terms do not decay to zero the parts of the complete response are summarized as follows:

$$y_h(t) \equiv \text{ Transient response} \equiv \text{ Free response.}$$
$$y_p(t) \equiv \text{ Steady state response} \equiv \text{ Forced response.}$$

3.2.3.2   The Complete Response to a Constant Input

The system response for the special case where the input to a first-order system is a constant, can be simplified into a form that is easily obtained without explicitly finding $y_h(t)$ and $y_p(t)$ separately. Consider a first-order system with a constant input $f(t) = \beta$,

$$\dot{y} + \frac{1}{\tau}y = \beta. \tag{3.9}$$

The homogeneous and particular solutions are given by

$$y_h(t) = Ke^{-\frac{t}{\tau}} \text{ and } y_p(t) = \tau\beta,$$

such that

$$y = y_p(t) + y_h(t)$$
$$= \tau\beta + Ke^{-\frac{t}{\tau}}.$$

The particular solution $y_p(t)$ can be taken as the steady state response such that

$$y_p(t) = y_{ss}$$
$$\implies y(t) = y_{ss} + Ke^{-\frac{t}{\tau}}. \tag{3.10}$$

If the initial value of the response $y(0)$ is known, the constant K can be expressed in terms of it,

$$y(0) = y_{ss} + Ke^0$$
$$\Rightarrow K = y(0) - y_{ss}.$$

Substituting this expression of K in Equation 3.10 leads to

$$y(t) = y_{ss} + [y(0) - y_{ss}]e^{-\frac{t}{\tau}}. \tag{3.11}$$

Hence, for any first-order system with a constant input (or a constant forcing function) the system response can be obtained from Equation 3.11. The task is thus reduced to finding the steady state value y_{ss} and the time constant τ, where the initial value $y(0)$ is known. The term y_{ss} is obtained by setting the derivative $\dot{y}$ to zero, and τ is the inverse of the coefficient of $y(t)$ in the first-order Equation 3.9. Equation 3.11 is important because it allows the direct determination of the complete system response without establishing the homogeneous and particular solutions separately.

Example 3.3 *Find the dynamic system response for the spring mass damping system shown below, where $f(t) = \beta$, $x(0) = 0$ (i.e. there is no initial energy stored).*

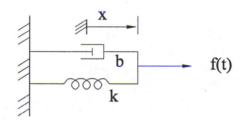

A First-Order Translational Mechanical System

Solution 3.3 *From the system free-body diagram*

$$b\dot{x} + kx = f(t)$$

$$\dot{x} + \frac{k}{b}x = \frac{1}{b}f(t)$$

$$\dot{x} + \frac{k}{b}x = \frac{\beta}{b}. \tag{3.12}$$

From Equation 3.12 the time constant τ and steady state value x_{ss} can be obtained.

$$\tau = \frac{b}{k}$$

$$0 + \frac{k}{b}x_{ss} = \frac{\beta}{b}$$

$$\Rightarrow x_{ss} = \frac{\beta}{k}.$$

Therefore, the system response can be obtained from Equation 3.11.

$$x(t) = x_{ss} + [x(0) - x_{ss}]e^{-\frac{t}{\tau}}$$

$$x(t) = \frac{\beta}{k} + \left[0 - \frac{\beta}{k}\right]e^{\frac{-kt}{b}}$$

$$= \frac{\beta}{k} - \frac{\beta}{k}e^{\frac{-kt}{b}} \qquad (\equiv y_h + y_p)$$

$$= \frac{\beta}{k}\left[1 - e^{\frac{-kt}{b}}\right]. \qquad (3.13)$$

Example 3.4 *Find the dynamic system response for the rotational mechanical system shown below. There is a constant input angular speed, $\omega(t) = \beta$, and zero initial angular speed, $\omega_1(0) = 0$ (i.e. there is no initial energy stored).*

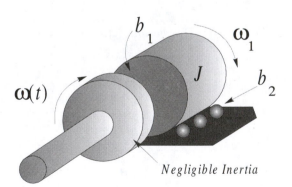

A First-Order Rotational Mechanical System

Solution 3.4 *From the system free-body diagrams*

$$J\dot{\omega}_1 + (b_1 + b_2)\omega_1 = b_1\omega(t)$$

$$\dot{\omega}_1 + \frac{(b_1 + b_2)}{J}\omega_1 = \frac{b_1\beta}{J} \qquad where \; \omega(t) = \beta.$$

From this equation the time constant τ and steady state value x_{ss} can be *obtained as follows:*

$$\tau = \frac{J}{b_1 + b_2}$$

$$0 + \frac{(b_1 + b_2)}{J}\omega_{1_{ss}} = \frac{b_1\beta}{J}$$

$$\implies \omega_{1_{ss}} = \frac{b_1\beta}{b_1 + b_2}.$$

Therefore, the system response can be obtained from Equation 3.11 as follows:

$$\omega_1(t) = \omega_{1ss} + [\omega_1(0) - \omega_{1ss}]e^{-\frac{t}{\tau}}$$

$$= \frac{b_1\beta}{b_1 + b_2} + \left(0 - \frac{b_1\beta}{b_1 + b_2}\right)e^{-\frac{t}{\tau}}$$

$$= \frac{b_1\beta}{b_1 + b_2} - \frac{b_1\beta}{b_1 + b_2}e^{-\frac{t}{\tau}} \qquad (\equiv y_h + y_p)$$

$$= \frac{b_1\beta}{b_1 + b_2}\left[1 - e^{-\frac{t}{\tau}}\right] \qquad where \;\; \tau = \frac{J}{b_1 + b_2}.$$

This is the system response of the rotational mechanical system.

Example 3.5 *Consider the electrical circuit shown below where $v_i(t) = \beta$, for $t \geq 0$. Find the dynamic system response $v_o(t)$.*

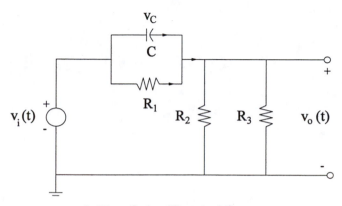

A First-Order Electrical System

Solution 3.5 *The starting point is determining the input-output differential equation model. This is done by using the KVL and KCL.*

$$i_c(t) + i_{R_1} = i_{R_T} \quad \text{where} \quad R_T = \frac{R_2 R_3}{R_2 + R_3} \quad \text{(parallel resistors)}$$

$$C(\dot{v}_i - \dot{v}_o) + \frac{1}{R_1}[v_i - v_o] = \frac{1}{R_T} v_o$$

$$C\dot{v}_o + \left(\frac{1}{R_1} + \frac{1}{R_T}\right) v_o = C\dot{v}_i + \frac{1}{R_1} v_i$$

$$\dot{v}_o + \left(\frac{1}{CR_1} + \frac{1}{CR_T}\right) v_o = \frac{\beta}{CR_1} \qquad (v_i = \beta \implies \dot{v}_i = 0) \qquad (3.14)$$

This is a first-order system with a constant input, and hence the system response is of the form

$$v_o(t) = v_{o_{ss}} + \left[v_o(0^+) - v_{o_{ss}}\right] e^{-\frac{t}{\tau}}, \qquad (3.15)$$

where the time constant is the inverse of the coefficient of $v_o(t)$ in Equation 3.14 and $v_{o_{ss}}$ is the steady state value obtained by setting the derivative $\dot{v}_o(t)$ to zero in the same equation. The term $v_o(0^+)$ represents the initial value of the output voltage approached from positive time.

$$\tau = \frac{R_1 R_T C}{R_1 + R_T}$$

$$v_{o_{ss}} = \frac{R_T \beta}{R_1 + R_T}.$$

The initial value $v_o(0^+)$ is obtained by using the KVL, the initial input voltage and the initial voltage across the capacitor.

$$v_i(t) - v_C(t) - v_o(t) = 0$$

$$\implies v_o(t) = v_i(t) - v_C(t)$$

$$\implies v_o(0^+) = v_i(0^+) - v_C(0^+)$$

$$\text{but} \quad v_C(0^+) = v_C(0^-) = 0 \quad \text{and} \quad v_i(0^+) = \beta$$

$$\implies v_o(0^+) = \beta - 0$$

$$\implies v_o(0^+) = \beta.$$

The dynamic system response is then obtained by substituting the determined expressions of τ, $v_{o_{ss}}$ and $v_o(0^+)$ in Equation 3.15.

3.2.4   Second-Order Systems

A general second-order system can be represented as follows:

$$\ddot{y} + a_1\dot{y} + a_o y = f(t)$$
$$\ddot{y} + 2\xi\omega_n\dot{y} + \omega_n^2 y = f(t),$$

where ω_n is the undamped natural frequency and ξ is the damping ratio. By inspection of these equations the characteristic equation can be determined.

$$r^2 + a_1 r + a_0 = 0$$
$$r^2 + 2\xi\omega_n r + \omega_n^2 = 0.$$

(a) If $\xi > 1 \implies$ roots are distinct and negative, y_h has two decaying exponential components.

(b) If $\xi = 1 \implies$ repeated root, $r = \omega_n$, $y_h(t) =$ terms of the form $e^{-\omega_n t}$ and $te^{-\omega_n t}$.

(c) If $0 < \xi < 1 \implies$ complex roots.

(d) The complex roots are given by

$$r_1, r_2 = -\xi\omega_n \pm j\omega_n\sqrt{1 - \xi^2}.$$

(e) With the complex roots the particular solution is given by

$$y_h = K_1 e^{r_1 t} + K_2 e^{r_2 t}$$
$$= e^{\alpha t}\left[K_1 \cos\beta t + K_2 \sin\beta t\right]$$
$$= K e^{\alpha t}\cos(\beta t + \phi).$$

(f) If $\xi < 0 \implies$ the system is unstable.

Example 3.6 *The system shown below obeys the differential equation*

$$(m_1 + m_2)\ddot{x} + (b_1 + b_2)\dot{x} + kx = f(t) - m_2 g,$$

where $x_1 = x_2 = x$.

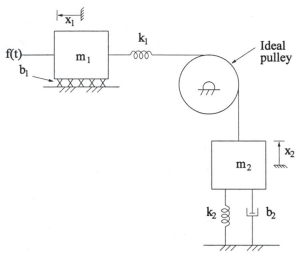

A Second-Order Translational Mechanical System

(a) *Verify that the differential equation is correct.*

(b) *Find the expressions for the damping coefficient ξ and the undamped natural frequency ω_n.*

(c) *Find the steady state response when the force is a step unit function*

Solution 3.6   (a) *Verification of the differential equation from the free-body diagrams*

$$m_1\ddot{x} + b_1\dot{x} + f_c = f(t)$$

$$m_2\ddot{x} + b_2\dot{x} + k_2x + m_2g - f_c = 0$$

$$(m_1 + m_2)\ddot{x} + (b_1 + b_2)\dot{x} + k_2x = f(t) - m_2g \qquad \text{(by adding the 2 equations).}$$

(b)

$$\ddot{x} + \frac{(b_1 + b_2)}{(m_1 + m_2)}\dot{x} + \frac{k_2}{(m_1 + m_2)}x = \frac{1}{(m_1 + m_2)}f(t) - \frac{m_2}{(m_1 + m_2)}g$$

$$\omega_n = \sqrt{\frac{k_2}{(m_1 + m_2)}}$$

$$\xi = \frac{1}{2\omega_n}\left(\frac{b_1 + b_2}{m_1 + m_2}\right)$$

$$= \frac{b_1 + b_2}{2\sqrt{m_1 + m_2}\sqrt{k_2}}.$$

(c)

$$x_{ss} = \left(\frac{m_1 + m_2}{k_2}\right)\left[\frac{1}{m_1 + m_2} - \frac{m_2 g}{m_1 + m_2}\right]$$

$$= \frac{1}{k_2}\left(1 - m_2 g\right).$$

3.2.5 Analogous Mechanical and Electrical Systems

Translational, rotational, and electrical systems can be shown to manifest the same dynamic behavior and hence their models can be used interchangeably. Consider the four second-order systems shown in Figures 3.2, 3.3, 3.4, and 3.5.

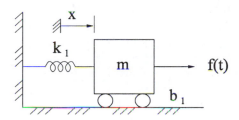

FIGURE 3.2
A Translational Mechanical System

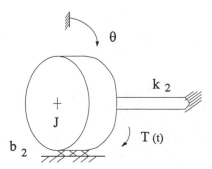

FIGURE 3.3
A Rotational Mechanical System

Using free-body diagrams and circuit laws (current and voltage) it can be shown that the input-output differential equations for the four systems

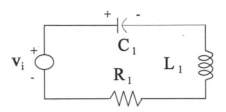

FIGURE 3.4
An RLC Series Circuit

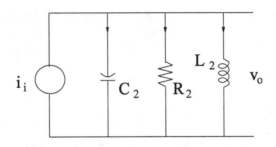

FIGURE 3.5
An RLC Parallel Circuit

are given by:

$$\ddot{x} + \frac{b_1}{m}\dot{x} + \frac{k_1}{m}x = \frac{1}{m}f(t)$$

$$\ddot{\theta} + \frac{b_2}{J}\dot{\theta} + \frac{k_2}{J}\theta = \frac{1}{J}T(t)$$

$$\ddot{i} + \frac{R_1}{L_1}\dot{i} + \frac{1}{L_1C_1}i = \frac{1}{L_1}\dot{v}_i(t)$$

$$\ddot{v}_o + \frac{1}{R_2C_2}\dot{v}_o + \frac{1}{L_2C_2}v_o = \frac{1}{C_2}\dot{i}(t).$$

The derivation of these models is left as an exercise for the reader. By comparing these models with the standard second-order model

$$\ddot{y} + 2\xi\omega_n\dot{y} + \omega_n^2 y = f(t),$$

the expressions for the respective natural frequency and damping ratio can be determined.

$$\omega_n = \sqrt{\frac{k_1}{m}} \quad \text{and} \quad \xi = \frac{b_1}{2\sqrt{mk_1}}$$

$$\omega_n = \sqrt{\frac{k_2}{J}} \quad \text{and} \quad \xi = \frac{b_2}{2\sqrt{Jk_2}}$$

$$\omega_n = \sqrt{\frac{1}{L_1 C_1}} \quad \text{and} \quad \xi = \frac{R_1}{2}\sqrt{\frac{C_1}{L_1}}$$

$$\omega_n = \sqrt{\frac{1}{L_2 C_2}} \quad \text{and} \quad \xi = \frac{1}{2R_2}\sqrt{\frac{L_2}{C_2}}.$$

The four mechanical and electrical systems are said to be analogous because they satisfy the same general second equation, which means that they manifest the similar dynamic properties.

3.2.6 Solution of State-Variable Matrix Equations

The matrix state-variable form of the dynamic system model can be solved in the time domain. Consider the general state-variable form

$$\dot{\mathbf{x}} = \mathbf{A}\mathbf{x} + \mathbf{B}\mathbf{u}$$
$$\mathbf{y} = \mathbf{C}\mathbf{x} + \mathbf{D}\mathbf{u}.$$

The homogeneous (zero input) response is obtained by setting the inputs to zero such that

$$\dot{\mathbf{x}} = \mathbf{A}\mathbf{x}$$
$$\mathbf{y} = \mathbf{C}\mathbf{x}.$$

The solutions or responses of the state variables are obtained from the equation

$$\mathbf{x(t)} = \phi(\mathbf{t})\mathbf{x(0)}$$

The function $\phi(\mathbf{t})$ is called the state-transition matrix defined by

$$\phi(\mathbf{t}) = \mathbf{e}^{At}$$
$$= \alpha_o \mathbf{I} + \alpha_1 \mathbf{A} + \alpha_2 \mathbf{A}^2 \ldots\ldots + \alpha_{n-1}\mathbf{A}^{n-1},$$

where α_o to α_{n-1} are scalar coefficients and $\mathbf{x(0)}$ is a vector of the initial values of the state variables. The characteristic values (or eigenvalues) of $\mathbf{A}$ are values of λ for which

$$|\lambda \mathbf{I} - \mathbf{A}| = 0.$$

This is called the system characteristic equation.

Example 3.7 *For a system that has two state variables such that* $\mathbf{x}(t) = [x_1 \ x_2]^T$, $\mathbf{x}(0) = [1 \ 1]^T$ *and*

$$A = \begin{bmatrix} 0 & 1 \\ -2 & -3 \end{bmatrix},$$

find the state variable responses $x_1(t)$ *and* $x_2(t)$.

Solution 3.7 *First, the eigenvalues are determined.*

$$|\lambda \mathbf{I} - \mathbf{A}| = 0$$

$$\Longrightarrow \left| \begin{bmatrix} \lambda & 0 \\ 0 & \lambda \end{bmatrix} - \begin{bmatrix} 0 & 1 \\ -2 & -3 \end{bmatrix} \right| = 0$$

$$\Longrightarrow \left| \begin{bmatrix} \lambda & -1 \\ 2 & (\lambda + 3) \end{bmatrix} \right| = 0$$

$$\Rightarrow \lambda(\lambda + 3) - (-1 \times 2) = 0$$

$$\Longrightarrow \lambda^2 + 3\lambda + 2 = 0$$

$$\Longrightarrow (\lambda + 1)(\lambda + 2) = 0$$

$$\Longrightarrow \lambda_1 = -1, \quad \lambda_2 = -2$$

The order of A is 2, which means the power series for e^{At} *contains only two terms, such that*

$$\mathbf{e}^{At} = \boldsymbol{\alpha}_o \mathbf{I} + \boldsymbol{\alpha}_1 \mathbf{A}. \qquad (3.16)$$

Replacing A in Equation 3.16 by the eigenvalues λ_1 *and* λ_2 *produces two scalar equations*

$$e^{-t} = \alpha_o - \alpha_1$$
$$e^{-2t} = \alpha_o - 2\alpha_1.$$

Solving for α_0 *and* α_1 *gives*

$$\alpha_o = 2e^{-t} - e^{-2t}$$
$$\alpha_1 = e^{-t} - e^{-2t}.$$

Substituting in the expression for $\mathbf{e}^{At}$ *leads to*

$$\mathbf{e}^{At} = \alpha_o \mathbf{I} + \alpha_1 \mathbf{A}$$

$$= \alpha_o \begin{bmatrix} 1 & 0 \\ 0 & 1 \end{bmatrix} + \alpha_1 \begin{bmatrix} 0 & 1 \\ -2 & -3 \end{bmatrix}$$

$$= \begin{bmatrix} \alpha_0 & 0 \\ 0 & \alpha_0 \end{bmatrix} + \begin{bmatrix} 0 & \alpha_1 \\ -2\alpha_1 & -3\alpha_1 \end{bmatrix}$$

$$= \begin{bmatrix} \alpha_0 & \alpha_1 \\ -2\alpha_1 & \alpha_o - 3\alpha_1 \end{bmatrix}$$

$$= \begin{bmatrix} 2e^{-t} - e^{-2t} & e^{-t} - e^{-2t} \\ -2\left(e^{-t} - e^{-2t}\right) & \left(2e^{-t} - e^{-2t}\right) - 3\left(e^{-t} - e^{-2t}\right) \end{bmatrix}$$

$$= \begin{bmatrix} 2e^{-t} - e^{-2t} & e^{-t} - e^{-2t} \\ -2\left(e^{-t} - e^{-2t}\right) & -e^{-t} + 2e^{-2t} \end{bmatrix}$$

Therefore, the system response (for the two states) is obtained by

$$\mathbf{x(t)} = \mathbf{e}^{At}\mathbf{x(0)}$$

$$= \begin{bmatrix} 2e^{-t} - e^{-2t} & e^{-t} - e^{-2t} \\ -2\left(e^{-t} - e^{-2t}\right) & -e^{-t} + 2e^{-2t} \end{bmatrix} \begin{bmatrix} 1 \\ 1 \end{bmatrix}$$

$$= \begin{bmatrix} 3e^{-t} - 2e^{-2t} \\ -3e^{-t} + 4e^{-2t} \end{bmatrix}.$$

Therefore, the responses of the two state variables are given by

$$x_1(t) = 3e^{-t} - 2e^{-2t}$$

$$x_2(t) = -3e^{-t} + 4e^{-2t}$$

Example 3.8 *A dynamic system is described by the state-variable equations* $\dot{\mathbf{x}} = \mathbf{A}\mathbf{x}$ *and* $\mathbf{y} = \mathbf{C}\mathbf{x}$, *where*

$$\mathbf{A} = \begin{bmatrix} 0 & 1 \\ 0 & -2 \end{bmatrix}, \quad \mathbf{C} = \begin{bmatrix} 3 & -1 \end{bmatrix}$$

and $\mathbf{x}(0) = \begin{bmatrix} 1 & 2 \end{bmatrix}^T$.

(a) Obtain the state-transition matrix $\phi(\mathbf{t})$.
(b) Find the state variable responses $x_1(t)$ *and* $x_2(t)$.
(c) Find the output response $y(t)$.
(d) For this system verify that

$$\phi(\mathbf{0}) = \mathbf{I}$$

and $\quad \phi^{-1}(\mathbf{t}) = \phi(-\mathbf{t}).$

Solution 3.8 *(a) The state-transition matrix is obtained as follows:*

$$\phi(\mathbf{t}) = e^{At} = \alpha_o \mathbf{I} + \alpha_1 \mathbf{A}.$$

where α_o *and* α_1 *are constants obtained from the eigenvalues as follows:*

$$|\lambda \mathbf{I} - \mathbf{A}| = 0$$

$$\lambda(\lambda + 2) = 0 \Rightarrow \lambda_1 = 0 \text{ and } \lambda_2 = -2$$

$$e^{0t} = \alpha_o I + \alpha_1 0$$

$$e^{-2t} = \alpha_o I - \alpha_1(-2)$$

$$\Longrightarrow \alpha_o = 1 \text{ and } \alpha_1 = \frac{1 - e^{-2t}}{2}.$$

Using these values of α_o *and* α_1 *the state-transition matrix* $\phi(t)$ *can now be determined.*

$$\phi(\mathbf{t}) = e^{At} = \alpha_o \mathbf{I} + \alpha_1 \mathbf{A}$$

$$\phi(t) = 1 \begin{bmatrix} 1 & 0 \\ 0 & 1 \end{bmatrix} + \frac{1 - e^{-2t}}{2} \begin{bmatrix} 0 & 1 \\ 0 & -2 \end{bmatrix}$$

$$= \begin{bmatrix} 1 & \dfrac{1 - e^{-2t}}{2} \\ 0 & e^{-2t} \end{bmatrix}$$

(b) The state variable responses are obtained as follows:

$$\mathbf{x(t)} = \boldsymbol{\phi}(\mathbf{t})\mathbf{x(0)}$$

$$= \begin{bmatrix} 1 & \dfrac{1-e^{-2t}}{2} \\ 0 & e^{-2t} \end{bmatrix} \begin{bmatrix} 1 \\ 2 \end{bmatrix}$$

$$= \begin{bmatrix} 1 + (1 - e^{-2t}) \\ 2e^{-2t} \end{bmatrix}$$

$$= \begin{bmatrix} 2 - e^{-2t} \\ 2e^{-2t} \end{bmatrix}.$$

(c) Now for the output response,

$$\mathbf{y(t)} = \mathbf{Cx(t)}$$

$$= \begin{bmatrix} 3 & -1 \end{bmatrix} \begin{bmatrix} 2 - e^{-2t} \\ 2e^{-2t} \end{bmatrix}$$

$$= 6 - 3e^{-2t} - 2e^{-2t}$$

$$= 6 - 5e^{-2t}$$

(d) The identities are verified by using the determined expression of $\phi(t)$.

(i)

$$\boldsymbol{\phi(0)} = \mathbf{I}$$

$$LHS = \phi(0)$$

$$= \begin{bmatrix} 1 & \dfrac{1-e^{-2t}}{2} \\ 0 & e^{-2t} \end{bmatrix}_{t=0}$$

$$= \begin{bmatrix} 1 & \dfrac{1-1}{2} \\ 0 & 1 \end{bmatrix}$$

$$= \begin{bmatrix} 1 & 0 \\ 0 & 1 \end{bmatrix}$$

$$= \mathbf{I}$$

$$= RHS.$$

(ii)

$$\phi^{-1}(\mathbf{t}) = \phi(-\mathbf{t})$$

$$LHS = \phi^{-1}(\mathbf{t})$$

$$= \frac{1}{(e^{-2t} - 0)} \begin{bmatrix} e^{-2t} & -\dfrac{\left(1 - e^{-2t}\right)}{2} \\ 0 & 1 \end{bmatrix}$$

$$= \begin{bmatrix} 1 & \dfrac{\left(1 - e^{2t}\right)}{2} \\ 0 & e^{2t} \end{bmatrix}$$

$$= \phi(-\mathbf{t})$$

$$= RHS.$$

Although these two properties have been verified for a specific system, they are valid for any state-transition matrix $\phi(\mathbf{t})$.

3.3   Frequency Domain Solution of Models

3.3.1   The Laplace Transform

Difficult mathematical problems can be transformed into equivalent problems that are easier to solve and the solution transformed back into the original framework. Such transformation techniques include Logarithms, Fourier series, the Fourier transform, the Laplace transform, and the $\mathcal{Z}$-transforms. The Laplace transform is a mathematical tool for transforming differential equations into an easier-to-manipulate algebraic form. In this domain the differential equations are easily solved and the solutions converted back into the time domain to give the system response. In this way, the Laplace transform is very useful in establishing and analyzing the system responses of linear dynamic systems by using relatively easily solved algebraic equations instead of the more difficult differential equations. The use of Laplace transforms to solve dynamic system models proceeds from two forms of the models; the input-output differential equation, or from the system transfer function.

3.3.1.1   Definitions

The Laplace transform converts a time function into a function of a complex variable that is denoted by s. The transform of a function of time $y(t)$ is represented by either $\mathcal{L}\left[y(t)\right]$ or $Y(s)$ such that

$$Y(s) = \mathcal{L}\left[y(t)\right],$$

where the symbol $\mathcal{L}$ stands for "*the Laplace transform of.*" One can think of the Laplace transform as providing a means of transforming a given problem from the time domain, where all variables are functions of t, to the complex-frequency domain, where all variables are functions of s. The Laplace transformation is defined as follows:

$$Y(s) = \mathcal{L}\left[y(t)\right]$$

$$= \int_{-\infty}^{\infty} y(t)e^{-st}dt.$$

In most applications only the one-sided Laplace transform is essential. This transform uses 0^- (a value just before $t = 0$) as the lower limit such that

$$Y(s) = \int_{0^-}^{\infty} y(t)e^{-st}dt.$$

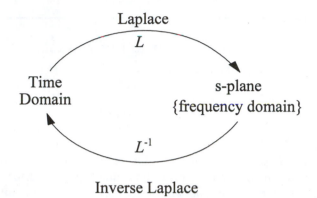

FIGURE 3.6
Laplace Transformation

Once the system model has been solved in the s-domain, it is converted back into the time domain to give the system response.

$$y(t) = \frac{1}{2\pi} \int_{\sigma-j\omega}^{\sigma+j\omega} Y(s)e^{st}ds.$$

In practice, this inverse Laplace transform equation is seldomly used because it is too complex. Instead, tables of simple inverse Laplace transforms are used. Such tables are in Appendix A. Figure 3.6 shows a summary of the Laplace transformation between the time and frequency domains.

3.3.2   Properties of Laplace Transforms

Two attributes of analytical techniques of linear time-invariant systems form the basis for their application to dynamic systems:

- A linear system response obeys the principle of superposition.

- The response of a linear constant system can be expressed as the convolution of the input with the unit impulse response of the system.

The principle of superposition states that if the system has an input that can be expressed as a sum of signals, then the response of the system can be expressed as the sum of the individual responses to the respective input signals. From the second property it follows immediately that the response of a linear time-invariant system to an exponential input is also exponential. This result is the principle reason for the usefulness of Fourier and

Laplace transforms in the study of linear constant systems. The properties of Laplace transforms can be summarized as follows:

- Superposition Principle

$$\mathcal{L}\{\alpha y_1(t) + \beta y_2(t)\} = \alpha Y_1(s) + \beta Y_2(s).$$

  Laplace transform are applied to linear systems and hence, they obey the principle of superposition.

- Convolution

$$y(t) = \int_{-\infty}^{\infty} u(\tau)h(t - \tau)d\tau$$

  where $u(t)$ is the input and $h(t)$ is the impulse response.

- Scaling Property

$$\mathcal{L}\{\alpha y(t)\} = \alpha Y(s).$$

  This is a special case of the superposition principle.

- Differentiation

$$\frac{d}{dt} = s \qquad \text{Differential operator}$$

$$\mathcal{L}\left[\frac{dy(t)}{dt}\right] = sY(s) - y(0)$$

$$\mathcal{L}\left[y^{(m)}(t)\right] = s^m Y(s) - s^{m-1}y(0) - s^{m-2}\dot{y}(0) - \dots - y^{(m-1)}(0).$$

- Integration

$$\int dt = \frac{1}{s} \qquad \text{Integral operator}$$

$$\mathcal{L}\left[\int y(t)dt\right] = \frac{Y(s)}{s}.$$

- Multiplication by Time

$$\mathcal{L}\left[ty(t)\right] = -\frac{dY(s)}{ds}.$$

  Multiplication of a function $y(t)$ by time corresponds to differentiating the negative of the Laplace transform, $-Y(s)$.

3.3.3   Laplace Transform of Some Key Functions

In order to build an understanding of Laplace transforms, the transforms of simple functions are derived in this section. Laplace transforms of some key functions are shown the following chart. Complete tables of the transforms and their properties are in Appendix A.

Number	$Y(s)$	$y(t),\ t \geqslant 0$
1	1	$\delta(t)$
2	$\dfrac{1}{s}$	$1(t)$
3	$\dfrac{1}{s^2}$	t
4	$\dfrac{m!}{s^{m+1}}$	t^m
5	$\dfrac{1}{s+a}$	e^{-at}
6	$\dfrac{1}{(s+a)^2}$	te^{-at}
7	$\dfrac{1}{(s+a)^m}$	$\dfrac{1}{(m-1)!}t^{m-1}e^{-at}$
8	$\dfrac{s+a}{(s+a)^2+b^2}$	$e^{-at}\cos bt$
9	$\dfrac{b}{(s+a)^2+b^2}$	$e^{-at}\sin bt$

(a) Step Function.

This is a function that is constant for $t \geq 0$ as shown in Figure 3.7 and Figure 3.8.

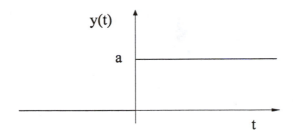

FIGURE 3.7
A Step Function

$$y(t) = a$$

$$Y(s) = \int_0^\infty (a)e^{-st}\,dt$$

$$= \frac{-ae^{-st}}{s}\Big|_0^\infty$$

$$= \frac{a}{s}.$$

(b) Ramp Function

This is represented as a line of constant gradient for $t \geq 0$ as shown in Figure 3.8.

$$y(t) = bt$$

$$Y(s) = \int_0^\infty (bt)e^{-st}\,dt$$

$$= \left[-\frac{bte^{-st}}{s} - \frac{be^{-st}}{s^2}\right]_0^\infty$$

$$= \frac{b}{s^2}.$$

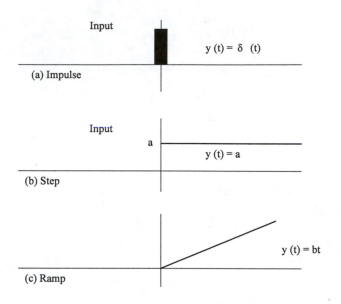

FIGURE 3.8
Impulse, Step, and Ramp Inputs

(c) Exponential Function

$$y(t) = e^{-at}$$

$$Y(s) = \int_0^\infty (e^{-at})e^{-st}dt$$

$$= \int_0^\infty e^{-(s+a)t}dt$$

$$= -\frac{e^{-(s+a)t}}{s+a}\Big|_0^\infty$$

$$= \frac{1}{s+a}.$$

(d) Impulse Function
This is a spike at the origin of $0\sec$ (0^- to 0^+) as shown in Figure 3.8.

$$y(t) = \delta(t)$$

$$Y(s) = \int_{0-}^{\infty} \delta(t)e^{-st}dt$$

$$= \int_{0-}^{0+} \delta(t)dt = 1.$$

(e) Sinusoid Function

$$y(t) = \sin \omega t$$

$$Y(s) = \int_{0}^{\infty} (\sin \omega t)e^{-st}dt$$

$$\text{using } \sin \omega t = \frac{e^{j\omega t} - e^{-j\omega t}}{2j}$$

$$Y(s) = \int_{0}^{\infty} \left(\frac{e^{j\omega t} - e^{-j\omega t}}{2j} \right) e^{-st}dt$$

$$= \frac{1}{2j} \int_{0}^{\infty} \left(e^{(j\omega - s)t} - e^{-(j\omega + s)t} \right) dt$$

$$= \frac{\omega}{s^2 + \omega^2}.$$

(f) Pure-Time Delay

Consider system $y(t)$ delayed by time T, such that the system is represented by $y_1 = y(t - T)$. Taking Laplace transforms leads to

$$y_1(t) = y(t - T)$$
$$\mathcal{L}\left[y_1(t)\right] = \mathcal{L}\left[y(t - T)\right]$$
$$Y_1(s) = e^{-sT}Y(s).$$

The function e^{-sT} represents the Laplace transform of a pure time delay function.

(g) Differentiation

$$\mathcal{L}\left[\frac{dy(t)}{dt}\right] = sY(s) - y(0)$$

For example, consider $y(t) = \sin \omega t$

$$\mathcal{L}\left[\frac{dy(t)}{dt}\right] = s\mathcal{L}(\sin \omega t) - \sin 0$$

$$= s\left[\frac{\omega}{s^2 + \omega^2}\right] - 0$$

$$= \frac{s\omega}{s^2 + \omega^2}$$

The result can be checked by taking the derivative first, and then taking the Laplace transforms.

$$\frac{dy(t)}{dt} = \frac{d(\sin \omega t)}{dt} = \omega \cos \omega t.$$

Therefore, $\mathcal{L}\left[\dfrac{dy(t)}{dt}\right] = \omega\mathcal{L}\left[\cos \omega t\right]$

$$= \frac{\omega s}{s^2 + \omega^2}. \quad \text{(same result as before)}$$

(h) Integration

$$\mathcal{L}\left[\int_0^t y(\tau)d\tau\right] = \frac{1}{s}F(s)$$

For example, consider $y(t) = \cos \omega t$

$$Y(s) = \frac{s}{s^2 + \omega^2}$$

$$\mathcal{L}\left[\int_0^t y(\tau)d\tau\right] = \frac{1}{s}\mathcal{L}(\cos \omega t)$$

$$= \frac{1}{s}\left[\frac{s}{s^2 + \omega^2}\right]$$

$$= \frac{1}{s^2 + \omega^2}.$$

The result can be checked by taking the integral first and then taking the

Laplace transforms as follows:

$$\int_0^t \cos \omega \tau d\tau = \frac{1}{\omega} \sin \omega t$$

Therefore, $\mathcal{L}\left[\frac{1}{\omega} \sin \omega t\right] = \frac{1}{\omega} \mathcal{L}(\sin \omega t)$

$$= \frac{1}{\omega}\left(\frac{\omega}{s^2 + \omega^2}\right)$$

$$= \frac{1}{s^2 + \omega^2} \qquad \text{(same result as before)}.$$

(i) The Differential Operator.

The s-operator can be considered as the differential operator.

$$s = \frac{d}{dt}$$

$$\Longrightarrow sY(s) = \frac{dy(t)}{dt}.$$

(j) The Integral Operator

The $\frac{1}{s}$ operator can be interpreted as the integral operator.

$$\frac{1}{s} = \int_0^t d\tau$$

$$\frac{Y(s)}{s} = \mathcal{L}\left[\int_0^t y(\tau)d\tau\right].$$

Example 3.9 *Find the Laplace transform of the function*

$$y(t) = \alpha \sin^2 t + \beta \cos^2 t.$$

Solution 3.9 *Using the half angle formula and rewriting the result leads to*

$$y(t) = \alpha \sin^2 t + \beta \cos^2 t$$

$$= \alpha \left(\frac{1}{2} - \frac{\cos 2t}{2}\right) + \beta \left(\frac{1}{2} + \frac{\cos 2t}{2}\right).$$

Therefore,

$$Y(s) = \alpha \mathcal{L}\left(\frac{1}{2} - \frac{\cos 2t}{2}\right) + \beta \mathcal{L}\left(\frac{1}{2} + \frac{\cos 2t}{2}\right).$$

From the Laplace tables it follows that

$$Y(s) = \alpha \left(\frac{1}{2s} - \frac{s}{2(s^2 + 4)} \right) + \beta \left(\frac{1}{2s} + \frac{s}{2(s^2 + 4)} \right).$$

Rearranging the terms

$$Y(s) = \frac{\alpha + \beta}{2s} + \frac{\beta - \alpha}{2} \left(\frac{s}{s^2 + 4} \right).$$

Use of Laplace transforms to solve dynamic system models proceeds from two forms of the models; the input-output differential equation, or from the system transfer function. The two approaches are related because the transfer function is defined as the output in Laplace divided by the input in Laplace. Both require taking inverse Laplace transforms of partial fractions. In the next subsection the techniques that are used to deal with partial fractions are discussed and then the determination of the system response by the two methods is developed.

3.3.4 Partial Fractions

Transfer functions can be expressed in terms of partial fractions, whose inverse Laplace transforms can be easily obtained from the Laplace transforms tables.

$$H(s) = \frac{Y(s)}{R(s)}$$

$$= \frac{b_0 s^m + b_1 s^{m-1} + b_2 s^{m-2} + \dots + b_m}{a_0 s^n + a_1 s^{n-1} + a_2 s^{n-2} + \dots + a_n}$$

$$= K \frac{\Pi_1^m (s - z_i)}{\Pi_1^n (s - p_j)}$$

$$= \frac{C_1}{s - p_1} + \frac{C_2}{s - p_2} + \dots + \frac{C_n}{s - p_n},$$

where $\{C_i\}$ are coefficients obtained by using the *cover up method*:

$$C_i = (s - p_i) H(s)|_{s=p_i}. \tag{3.17}$$

The nature of the partial fractions depend on the type of roots. There are three general categories of roots and they will be discussed through examples.

(a) Distinct Real Roots.

These roots consist of real numbers and they are all different, i.e., there are no repeated roots.

Example 3.10 *Express the following transfer function in terms of partial fractions*

$$H(s) = \frac{(s+7)(s+4)}{s(s+1)(s+3)}.$$

Solution 3.10 *The transfer function can be expressed in terms of three partial fractions*

$$H(s) = \frac{(s+7)(s+4)}{s(s+1)(s+3)}$$

$$= \frac{C_1}{s} + \frac{C_2}{s+1} + \frac{C_3}{s+3}.$$

The coefficients are obtained using the cover-up method.

$$C_1 = (s - p_1)H(s)|_{s=p_1}$$

$$= s \times \frac{(s+7)(s+4)}{s(s+1)(s+3)}\bigg|_{s=0}$$

$$= \frac{(s+7)(s+4)}{(s+1)(s+3)}\bigg|_{s=0}$$

$$= \frac{28}{3}.$$

$$C_2 = (s - p_2)H(s)|_{s=p_2}$$

$$= (s+1) \times \frac{(s+7)(s+4)}{s(s+1)(s+3)}\bigg|_{s=-1}$$

$$= \frac{(s+7)(s+4)}{s(s+3)}\bigg|_{s=-1}$$

$$= \frac{6 \times 3}{-1 \times 2}$$

$$= -9.$$

$$C_3 = (s - p_3)H(s)|_{s=p_3}$$

$$= (s + 3) \times \left. \frac{(s + 7)(s + 4)}{s(s + 1)(s + 3)} \right|_{s=-3}$$

$$= \left. \frac{(s + 7)(s + 4)}{s(s + 1)} \right|_{s=-3}$$

$$= \frac{4 \times 1}{-3 \times (-2)}$$

$$= \frac{2}{3}.$$

Therefore, the transfer function in partial fractions is given by

$$H(s) = \frac{\frac{28}{3}}{s} - \frac{9}{s + 1} + \frac{\frac{2}{3}}{s + 3}.$$

(b) Distinct Complex Roots

These occur when the roots consist of complex numbers and they are all different, i.e., there are no repeated roots.

Example 3.11 *Express the following transfer function in terms of partial fractions*

$$H(s) = \frac{2}{s(s^2 + s + 2)}.$$

Solution 3.11 *The transfer function can be expressed in terms of partial fractions as follows:*

$$H(s) = \frac{2}{s(s^2 + s + 2)}$$

$$= \frac{C_1}{s} + \frac{C_2 s + C_3}{s^2 + s + 2}$$

where C_1 is obtained by using the cover-up method

$$C_1 = (s - p_1)H(s)|_{s=p_1}$$

$$= s \times \left. \frac{2}{s(s^2 + s + 2)} \right|_{s=0}$$

$$= \left. \frac{2}{(s^2 + s + 2)} \right|_{s=0}$$

$$= 1.$$

The constants C_2 and C_3 are obtained by equating the original transfer function to its partial fraction form (with the computed value of C_1) and then comparing coefficients of the powers of s.

$$\frac{2}{s(s^2 + s + 2)} = \frac{1}{s} + \frac{C_2 s + C_3}{s^2 + s + 2}$$

$$\frac{2}{s(s^2 + s + 2)} = \frac{(s^2 + s + 2) + C_2 s^2 + s C_3}{s(s^2 + s + 2)}$$

$$2 = (1 + C_2) s^2 + (1 + C_3) s + 2$$

$$C_2 = -1 \quad and \quad C_3 = -1.$$

Therefore, the transfer function in partial fractions is given by

$$H(s) = \frac{1}{s} - \frac{s + 1}{s^2 + s + 2}.$$

Complex roots occur in conjugate pairs

$$p_1 = a + jb \quad and \quad p_1^* = a - jb.$$

Hence, the above transfer function (which has a conjugate pair of poles) can be expressed in terms of complex partial fractions as follows:

$$H(s) = \frac{2}{s(s^2 + s + 2)}$$

$$= \frac{2}{s(s - p_1)(s - p_1)}$$

$$= \frac{C_1}{s} + \frac{C_2}{s - p_1} + \frac{C_3}{s - p_1^*},$$

where

$$p_1 = -\frac{1}{2} + j\frac{\sqrt{7}}{2} \quad and \quad p_1^* = -\frac{1}{2} - j\frac{\sqrt{7}}{2}.$$

This is an alternative way of expressing a Laplace function that has complex roots in partial fractions. The constants C_1, C_2 and C_3 are then determined by the cover-up method. For any function with a complex conjugate pair of poles, it can be shown that the two constants C_3 and C_2^* are also complex conjugates, i.e.,

$$C_3 = C_2^*.$$

Hence, only one of them needs to be determined.

(c) Repeated Roots

A repeated root p_1, occurs when there are multiples of the factor $(s-p_1)$. The constant C_i for a factor $(s-p_1)$ with multiplicity k is given by

$$C_{k-i} = \frac{1}{i!}\left[\frac{d^i}{ds^i}\left[(s-p_1)^k\,Y(s)\right]\right]_{s=p_1},$$

where $i = 0, ..., k-1$.

Example 3.12 *Express the following transfer function in terms of partial fractions*

$$H(s) = \frac{s+4}{(s+1)(s+2)^2}$$

Solution 3.12 *The transfer function can be expressed in terms of three partial fractions*

$$H(s) = \frac{s+4}{(s+1)(s+2)^2}$$
$$= \frac{C_1}{s+1} + \frac{C_2}{s+2} + \frac{C_3}{(s+2)^2}$$

The coefficients are obtained as follows:

$$C_1 = (s-p_1)H(s)|_{s=p_1}$$
$$= (s+1) \times \frac{s+4}{(s+1)(s+2)^2}\bigg|_{s=-1}$$
$$= \frac{s+4}{(s+2)^2}\bigg|_{s=-1}$$
$$= 3.$$

$$C_2 = \frac{d}{ds}(s+2)^2 H(s)\big|_{s=-2}$$

$$= \frac{d}{ds}\left[(s+2)^2 \frac{s+4}{(s+1)(s+2)^2}\right]_{s=-2}$$

$$= \frac{d}{ds}\left[\frac{s+4}{s+1}\right]_{s=-2}$$

$$= \frac{(s+1)-(s+4)}{(s+1)^2}\bigg|_{s=-2}$$

$$= \frac{-3}{(s+1)^2}\bigg|_{s=-2}$$

$$= -3.$$

$$C_3 = (s-p_2)^2 H(s)\big|_{s=p2}$$

$$= (s+2)^2 \times \frac{s+4}{(s+1)(s+2)^2}\bigg|_{s=-2}$$

$$= \frac{s+4}{(s+1)}\bigg|_{s=-2}$$

$$= -2.$$

Therefore, the transfer function in partial fractions is given by

$$H(s) = \frac{3}{s+1} - \frac{3}{s+2} - \frac{2}{(s+2)^2}.$$

From these partial fraction forms of the Laplace transforms, the inverse Laplace transforms are obtained by comparing the partial fractions with those in the standard Laplace transform tables. Thus, for the partial fraction forms in Examples 3.10 and 3.12 the inverse Laplace transforms are obtained as follows:

$$H(s) = \frac{\frac{28}{3}}{s} - \frac{9}{s+1} + \frac{\frac{2}{3}}{s+3}$$

$$h(t) = \frac{28}{3} - 9e^{-t} + \frac{2}{3}e^{-3t}$$

$$H(s) = \frac{3}{s+1} - \frac{3}{s+2} - \frac{2}{(s+2)^2}$$

$$h(t) = 3e^{-t} - 3e^{-2t} - 2te^{-2t}.$$

For the partial fractions obtained in Example 3.11 (first approach), a special technique of *completing the square* is required to put the second partial fraction into a form where the inverse Laplace transform can be deduced from the Laplace transform tables.

3.3.4.1   Completing the Square

When the denominator of a Laplace transform function is expressed as a sum of two squares, the following equations (from Laplace Tables) can be used to get its inverse Laplace transform.

$$\mathcal{L}^{-1}\left[\frac{s+a}{(s+a)^2+b^2}\right] = e^{-at}\cos bt \tag{3.18}$$

$$\mathcal{L}^{-1}\left[\frac{b}{(s+a)^2+b^2}\right] = e^{-at}\sin bt \tag{3.19}$$

$$\mathcal{L}^{-1}\left[\frac{a^2+b^2}{s\left[(s+a)^2+b^2\right]}\right] = 1 - e^{-at}\left(\cos bt + \frac{a}{b}\sin bt\right). \tag{3.20}$$

The method of completing the square is a useful technique employed to express the denominator of a Laplace algebraic function as a sum of two squares. This method involves the two techniques of adding zero to an expression and multiplying an expression by one. It is very useful when dealing with second-order systems or higher-order systems that can be broken down into lower-order functions that include second-order ones. Consider the general second-order Laplace transform function given by

$$H(s) = \frac{ds+c}{s^2+as+b}. \tag{3.21}$$

where a, b, c and d are constants. The denominator can be expressed as a sum of two squares as follows:

$$s^2 + as + b = s^2 + as + \left(\frac{a^2}{4} - \frac{a^2}{4}\right) + b \qquad \text{Adding zero}$$

$$= \left(s^2 + as + \frac{a^2}{4}\right) - \frac{a^2}{4} + b$$

$$= \left(s + \frac{a}{2}\right)^2 + \left(b - \frac{a^2}{4}\right)$$

$$= \left(s + \frac{a}{2}\right)^2 + \left[\left(b - \frac{a^2}{4}\right)^{\frac{1}{2}}\right]^2. \tag{3.22}$$

This form of the denominator (Equation 3.22) is then used in Equation 3.21 and the numerator of the result is further manipulated to put the equation in the structure of the left-hand side of any of the Equations 3.18, 3.19 and 3.20.

$$H(s) = \frac{ds + c}{s^2 + as + b}$$

$$= \frac{ds + c}{\left(s + \frac{a}{2}\right)^2 + \left[\sqrt{\left(b - \frac{a^2}{4}\right)}\right]^2} \qquad \text{Completing the square}$$

$$= \frac{ds + c + \left(\frac{ad}{2} - \frac{ad}{2}\right)}{\left(s + \frac{a}{2}\right)^2 + \left[\sqrt{\left(b - \frac{a^2}{4}\right)}\right]^2} \qquad \text{Adding zero to the numerator}$$

$$= \frac{d\left(s + \frac{a}{2}\right) + \left(c - \frac{ad}{2}\right)}{\left(s + \frac{a}{2}\right)^2 + \left[\sqrt{\left(b - \frac{a^2}{4}\right)}\right]^2}$$

$$= \frac{d\left(s + \frac{a}{2}\right)}{\left(s + \frac{a}{2}\right)^2 + \left[\sqrt{\left(b - \frac{a^2}{4}\right)}\right]^2}$$

$$+ \frac{\left(c - \frac{ad}{2}\right) \frac{\sqrt{\left(b - \frac{a^2}{4}\right)}}{\sqrt{\left(b - \frac{a^2}{4}\right)}}}{\left(s + \frac{a}{2}\right)^2 + \left[\sqrt{\left(b - \frac{a^2}{4}\right)}\right]^2} \qquad \text{Multiplication by one}$$

$$= \frac{d\left(s + \frac{a}{2}\right)}{\left(s + \frac{a}{2}\right)^2 + \left[\sqrt{\left(b - \frac{a^2}{4}\right)}\right]^2} + \frac{\left[\frac{\left(c - \frac{ad}{2}\right)}{\sqrt{\left(b - \frac{a^2}{4}\right)}}\right]\sqrt{\left(b - \frac{a^2}{4}\right)}}{\left(s + \frac{a}{2}\right)^2 + \left[\sqrt{\left(b - \frac{a^2}{4}\right)}\right]^2}.$$

The function is now in the form of Equations 3.18 and 3.19 and the inverse Laplace transform can be found as follows:

$$h(t) = \mathcal{L}^{-1}\left[\frac{ds + c}{s^2 + as + b}\right] \tag{3.23}$$

$$= \mathcal{L}^{-1}\left[\frac{d\left(s + \frac{a}{2}\right)}{\left(s + \frac{a}{2}\right)^2 + \left[\sqrt{\left(b - \frac{a^2}{4}\right)}\right]^2} + \frac{\frac{\left(c - \frac{ad}{2}\right)}{\sqrt{\left(b - \frac{a^2}{4}\right)}}\sqrt{\left(b - \frac{a^2}{4}\right)}}{\left(s + \frac{a}{2}\right)^2 + \left[\sqrt{\left(b - \frac{a^2}{4}\right)}\right]^2}\right]$$

$$= de^{-\frac{a}{2}t}\cos\left[\sqrt{\left(b - \frac{a^2}{4}\right)}\right]t + \left[\frac{\left(c - \frac{ad}{2}\right)}{\sqrt{\left(b - \frac{a^2}{4}\right)}}\right]e^{-\frac{a}{2}t}\sin\left[\sqrt{\left(b - \frac{a^2}{4}\right)}\right]t.$$

This is the inverse Laplace transform of a general second-order function. Understanding the principles involved in this derivation is very important, as these principles are very useful in determining the inverse Laplace transforms of second-order systems in particular those that have complex roots.

Example 3.13 *Using partial fractions, find the time function $y(t)$ of the following Laplace functions.*
   (a)

$$Y(s) = \frac{2}{s(s + 2)}$$

   (b)

$$Y(s) = \frac{10}{s(s + 1)(s + 10)}$$

(c)

$$Y(s) = \frac{2s^2 + s + 1}{s^3 - 1}$$

(d)

$$Y(s) = \frac{3s + 2}{s^2 + 4s + 20}.$$

Solution 3.13 *In all the four cases, the given function is split into partial fractions and then inverse Laplace transformation is applied to get* $y(t)$.

(a)

$$Y(s) = \frac{2}{s(s + 2)}$$

$$= \frac{C_1}{s} - \frac{C_2}{s + 2}$$

$$= \frac{1}{s} - \frac{1}{s + 2}$$

$$y(t) = 1 - e^{-2t}.$$

(b)

$$Y(s) = \frac{10}{s(s + 1)(s + 10)}$$

$$= \frac{C_1}{s} - \frac{C_2}{s + 1} + \frac{C_3}{s + 10}$$

$$= \frac{1}{s} - \frac{\frac{10}{9}}{s + 1} + \frac{\frac{1}{9}}{s + 10}$$

$$y(t) = 1 - \frac{10}{9}e^{-t} + \frac{1}{9}e^{-10t}.$$

(c)

$$Y(s) = \frac{2s^2 + s + 1}{s^3 - 1}$$

$$= \frac{C_1}{s - 1} + \frac{C_2 s + C_3}{s^2 + s + 1}$$

$$= \frac{\frac{4}{3}}{s - 1} + \frac{\frac{2}{3}s + \frac{1}{3}}{s^2 + s + 1}.$$

The next step is completing the square of the denominator of the second term.

$$Y(s) = \frac{\frac{4}{3}}{s - 1} + \frac{\frac{2}{3}s + \frac{1}{3}}{\left(s + \frac{1}{2}\right)^2 - \frac{1}{4} + 1}$$

$$= \frac{\frac{4}{3}}{s - 1} + \frac{\frac{2}{3}s + \frac{1}{3}}{\left(s + \frac{1}{2}\right)^2 + \left(\sqrt{\frac{3}{4}}\right)^2}.$$

The last step is to express the numerator of the second term as a function of $\left(s + \frac{1}{2}\right)$ so that Equations 3.18 can be used to get the inverse Laplace transforms.

$$Y(s) = \frac{\frac{4}{3}}{s - 1} + \frac{\frac{2}{3}\left(s + \frac{1}{2}\right)}{\left(s + \frac{1}{2}\right)^2 + \left(\sqrt{\frac{3}{4}}\right)^2}$$

$$y(t) = \frac{4}{3}e^t + \frac{2}{3}e^{-\frac{t}{2}} \cos \frac{\sqrt{3}}{2}t.$$

(d) The first step is completing the square of the denominator.

$$Y(s) = \frac{3s + 2}{s^2 + 4s + 20}$$

$$= \frac{3s + 2}{(s + 2)^2 - 4 + 20}$$

$$= \frac{3s + 2}{(s + 2)^2 + 4^2}.$$

The next step is rearranging the numerator so that Equations 3.18 and 3.19 can be used to get the inverse Laplace transforms. This is done by expressing the numerator in terms of $(s + 2)$ and 2, by using the technique *of adding a zero.*

$$Y(s) = \frac{3s + 2 + (4 - 4)}{(s + 2)^2 + 4^2} \qquad adding\ a\ zero$$

$$= \frac{3(s + 2) - 4}{(s + 2)^2 + 4^2}$$

$$= \frac{3(s + 2)}{(s + 2)^2 + 4^2} - \frac{4}{(s + 2)^2 + 4^2}$$

$$y(t) = 3e^{-2t}\cos 4t - e^{-2t}\sin 4t$$

$$= e^{-2t}[3\cos 4t - \sin 4t].$$

3.4 Determination of the System Response

Now that the techniques of partial fractions have been discussed, Laplace transforms can be used to determine the dynamic response. This solution of dynamic system models proceeds from two forms of the models; the input-output differential equation, or from the system transfer function.

3.4.1 Using the Input-Output Equation

The procedure of obtaining the system response starting from the input-output differential equation can be summarized as follows:

- Obtain the input-output differential equation model.

- Apply Laplace transformation to this model.

- Solve the resulting algebraic equations for the Laplace transform of the output.

- Take the inverse Laplace transform of this output.

Example 3.14 *Consider the homogeneous input-output differential equation,*

$$\ddot{y} + y = 0, \tag{3.24}$$

where $y(0) = \alpha$ *and* $\dot{y}(0) = \beta$.

(a) Find the system response $y(t)$.
(b) How can this response be verified?

Solution 3.14 *(a) The system is determined by taking the Laplace transforms of the
   equation.*

$$\mathcal{L}\left[\ddot{y} + 2y\right] = \mathcal{L}\left[0\right]$$

$$\left[s^2 Y(s) - sy(0) - \dot{y}(0)\right] + Y(s) = 0$$

$$s^2 Y(s) - \alpha s - \beta + Y(s) = 0$$

$$Y(s)[s^2 + 1] = \alpha s + \beta$$

$$Y(s) = \frac{\alpha s + \beta}{s^2 + 1}$$

$$= \frac{\alpha s}{s^2 + 1} + \frac{\beta}{s^2 + 1}$$

$$y(t) = \mathcal{L}^{-1}[Y(s)]$$

$$= \alpha \cos t + \beta \sin t.$$

*(b) This result can be verified by showing that it satisfies Equation 3.24,
i.e.,*

$$LHS = \ddot{y} + y$$

$$= \frac{d^2\left[\alpha \cos t + \beta \sin t\right]}{dt^2} + \left[\alpha \cos t + \beta \sin t\right]$$

$$= -\alpha \cos t - \beta \sin t + \alpha \cos t + \beta \sin t$$

$$= 0$$

$$= RHS.$$

Example 3.15 *The car speed cruise control input-output differential equation is given by*

$$\dot{v} + \frac{b}{m}v = \frac{b}{m}v_r,$$

*for a step input $v_r(t)$. The car speed response was obtained by the direct
time domain determination as*

$$v(t) = v_r\left[1 - e^{-\frac{b}{m}t}\right].$$

Show that the Laplace transforms method produce the same result.

Solution 3.15 *The same response can be obtained using Laplace transforms as follows:*

$$\mathcal{L}\left[\dot{v} + \frac{b}{m}v\right] = \mathcal{L}\left[\frac{b}{m}v_r\right]$$

$$sV(s) - v(0) + \frac{b}{m}V(s) = \frac{\frac{b}{m}v_r}{s}$$

$$V(s)\left[s + \frac{b}{m}\right] = \frac{\frac{b}{m}v_r}{s} \qquad \textit{where } v(0) = 0$$

$$V(s) = \frac{bv_r}{m}\left[\frac{1}{s\left(s + \frac{b}{m}\right)}\right]$$

$$= \frac{bv_r}{m}\left[\frac{C_1}{s} + \frac{C_2}{s + \frac{b}{m}}\right] \qquad (3.25)$$

$$\textit{where} \qquad C_1 = \left.\frac{1}{s + \frac{b}{m}}\right|_{s=0} = \frac{m}{b}$$

$$\textit{and} \qquad C_2 = \left.\frac{1}{s}\right|_{s=-\frac{b}{m}} = -\frac{m}{b}.$$

Substituting these coefficients in Equation 3.25 leads to,

$$V(s) = \frac{bv_r}{m} \left[\frac{\frac{m}{b}}{s} - \frac{\frac{m}{b}}{s + \frac{b}{m}} \right]$$

$$= v_r \left[\frac{1}{s} - \frac{1}{s + \frac{b}{m}} \right]$$

$$v(t) = \mathcal{L}^{-1}\left[V(s) \right]$$

$$= v_r \left[1 - e^{-\frac{b}{m}t} \right]$$

Thus, the same result is obtained.

Example 3.16 *Consider a spring mass damper system with the following model*

$$m\ddot{y} + f\dot{y} + ky = u(t) \quad \text{where } u(t) \text{ is a constant, and } \dot{y}(0) = 0$$

(a) Find the Laplace transforms of the output $Y(s)$.
(b) For the specific case with the following data

$$m = 1, \ k = 2, \ f = 3, \ y(0) = 1 \ \text{and } u(t) = 0,$$

find the system response.

Solution 3.16 *(a) The Laplace of the output is obtained by taking the Laplace transforms of both sides of the equation.*

$$\mathcal{L}\left(m\ddot{y} + f\dot{y} + ky \right) = \mathcal{L}(u(t))$$

$$m\left[s^2 Y(s) - sy(0) - \dot{y}(0) \right] + f\left[sY(s) - y(0) \right] + kY(s) = U(s)$$

$$ms^2 Y(s) - msy(0) + fsY(s) - fy(0) + kY(s) = \frac{u}{s}$$

$$Y(s)\left[ms^2 + fs + k \right] = (ms + f)y(0) + \frac{u}{s}.$$

Therefore,

$$Y(s) = \frac{(ms+f)y(0) + \dfrac{u}{s}}{ms^2 + fs + k}$$

$$= \frac{ms^2 y(0) + sfy(0) + u}{s(ms^2 + fs + k)}.$$

(b) With the given data, the Laplace of the output then reduces to

$$Y(s) = \frac{s^2 + 3s}{s(s^2 + 3s + 2)}$$

$$= \frac{s+3}{(s+1)(s+2)}$$

$$= \frac{C_1}{s+1} + \frac{C_2}{s+2} \qquad (3.26)$$

$$C_1 = \frac{s+3}{s+2}\Big|_{s=-1} = 2$$

$$C_2 = \frac{s+3}{s+1}\Big|_{s=-2} = -1$$

Substituting these coefficients in Equation 3.26 leads to

$$Y(s) = \frac{2}{s+1} - \frac{1}{s+2}$$

$$y(t) = \mathcal{L}^{-1}[Y(s)]$$

$$= \mathcal{L}^{-1}\left(\frac{2}{s+1}\right) - \mathcal{L}^{-1}\left(\frac{1}{s+2}\right)$$

$$= 2e^{-t} - e^{-2t}$$

Example 3.17 *For the RC electrical circuit shown below, where $R = 2\Omega$, $C = 2F$, and $u(t)$ is a unit step input, find the system output $v_C(t)$.*

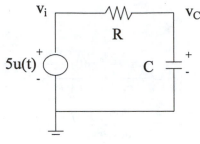

An RC Circuit

Solution 3.17 *The input-output differential equation model is obtained by using the KVL*

$$5u(t) - v_R - v_C(t) = 0$$
$$5u(t) - RC\dot{v}_C - v_C(t) = 0$$
$$4\dot{v}_C + v_C(t) = 5u(t).$$

The output is obtained by taking the Laplace transforms.

$$\mathcal{L}\left[4\dot{v}_C + v_C(t)\right] = \mathcal{L}\left[5u(t)\right]$$

$$4[sV_C(s) - v_C(0)] + V_C(s) = \frac{5}{s}$$

$$V_C(s)[1 + 4s] = \frac{5}{s} \qquad\qquad where\ v_C(0) = 0$$

$$V_C(s) = \frac{\dfrac{5}{4}}{s\left(s + \dfrac{1}{4}\right)}$$

$$= \frac{C_1}{s} + \frac{C_2}{s + \dfrac{1}{4}}$$

$$= \frac{5}{s} - \frac{5}{s + \dfrac{1}{4}}$$

$$v_C(t) = \mathcal{L}^{-1}\left[\frac{5}{s} - \frac{5}{s + \dfrac{1}{4}}\right]$$

$$= 5\left(1 - e^{-\frac{t}{4}}\right).$$

Example 3.18 *For the following nonhomogeneous input-output differential equations find their respective system response*

(a)

$$\ddot{y} + 4\dot{y} + 3y = u(t)$$
$$\text{where } y(0) = 1, \ \dot{y}(0) = 0 \ \text{and } u(t) = 2$$

(b)

$$\ddot{y} + 5\dot{y} + 4y = u(t)$$
$$\text{where } y(0) = \alpha, \ \dot{y}(0) = \beta \ \text{and } u(t) = 3$$

(c)

$$\ddot{y} + 5\dot{y} + 4y = u(t)$$
$$\text{where } y(0) = 0, \ \dot{y}(0) = 0 \ \text{and } u(t) = 2e^{-2t}$$

Solution 3.18 *The system responses are obtained by taking the Laplace transforms of the input-output differential equations.*
   (a)

$$\mathcal{L}\,[2] = \mathcal{L}\,[\ddot{y} + 4\dot{y} + 3y]$$

$$\frac{2}{s} = \left[s^2 Y(s) - sy(0) - \dot{y}(0)\right] + 4\left[sY(s) - y(0)\right] + 3Y(s)$$

$$\frac{2}{s} = \left[s^2 Y(s) - s\right] + 4\left[sY(s) - 1\right] + 3Y(s)$$

$$\frac{2}{s} = Y(s)[s^2 + 4s + 3] - (s+4)$$

$$Y(s) = \frac{s+4}{s^2 + 4s + 3} + \frac{2}{s(s^2 + 4s + 3)}$$

$$Y(s) = \left[\frac{C_1}{s+1} + \frac{C_2}{s+3}\right] + \left[\frac{C_3}{s+1} + \frac{C_4}{s+3} + \frac{C_5}{s}\right]$$

The coefficients are found by cover-up methods such that

$$Y(s) = \left[\frac{\frac{3}{2}}{s+1} + \frac{-\frac{1}{2}}{s+3}\right] + \left[\frac{-1}{s+1} + \frac{\frac{1}{3}}{s+3} + \frac{\frac{2}{3}}{s}\right]$$

$$y(t) = \mathcal{L}^{-1}[Y(s)]$$

$$= \left[\frac{3}{2}e^{-t} - \frac{1}{2}e^{-3t}\right] + \left[-e^{-t} + \frac{1}{3}e^{-3t}\right] + \frac{2}{3}$$

$$= \frac{1}{2}e^{-t} - \frac{1}{6}e^{-3t} + \frac{2}{3}.$$

(b)

$$\mathcal{L}[3] = \mathcal{L}[\ddot{y} + 5\dot{y} + 4y]$$

$$\frac{3}{s} = \left[s^2 Y(s) - s\alpha - \beta\right] + 5[sY(s) - \alpha] + 4Y(s)$$

$$\frac{3}{s} = Y(s)\left[s^2 + 5s + 4\right] - [s\alpha + \beta + 5\alpha]$$

$$Y(s) = \frac{s(s\alpha + \beta + 5\alpha) + 3}{s(s+1)(s+4)}$$

$$= \frac{C_1}{s} + \frac{C_2}{s+1} + \frac{C_3}{s+4}$$

The coefficients are then obtained and used to determine the system response.

$$Y(s) = \frac{\frac{3}{4}}{s} - \frac{\frac{3 - \beta - 4\alpha}{3}}{s+1} + \frac{\frac{3 - 4\alpha - 4\beta}{12}}{s+4}$$

$$y(t) = \frac{3}{4} - \frac{3 - \beta - 4\alpha}{3}e^{-t} + \frac{3 - 4\alpha - 4\beta}{12}e^{-4t}.$$

(c)

$$\mathcal{L}[\ddot{y} + 5\dot{y} + 4y] = \mathcal{L}\left[2e^{-2t}\right]$$

$$s^2 Y(s) + 5sY(s) + 4Y(s) = \frac{2}{s+2}$$

$$Y(s) = \frac{2}{(s+2)(s+1)(s+4)}$$

$$= \frac{-1}{s+2} + \frac{\frac{2}{3}}{s+1} + \frac{\frac{1}{3}}{s+4}$$

$$y(t) = \mathcal{L}^{-1}[Y(s)]$$

$$= -e^{-2t} + \frac{2}{3}e^{-t} + \frac{1}{3}e^{-4t}.$$

Example 3.19 *Consider a system that has the input-output differential equation*

$$a_1 y + a_2 \dot{y} + a_3 \ddot{y} = b_1 u + b_2 \dot{u},$$

where the input is $u(t)$ and output is $y(t)$ where all initial conditions are zero. Find the system response $y(t)$.

Solution 3.19 *Taking Laplace transforms throughout the differential equation gives*

$$a_1 Y(s) + a_2 [sY(s) - y(0)] + a_3 [s^2 Y(s) - sy(0) - \dot{y}(0)] = b_1 U(s)$$
$$+ b_2 [sU(s) - u(0)].$$

Setting all initial conditions to zero leads to

$$a_1 Y(s) + a_2 s Y(s) + a_3 s^2 Y(s) = b_1 U(s) + b_2 s U(s)$$

$$Y(s) \left[a_1 + a_2 s + a_3 s^2 \right] = U(s) \left[b_1 + b_2 s \right]$$

$$\implies Y(s) = \frac{b_1 + b_2 s}{a_1 + a_2 s + a_3 s^2} U(s)$$

$$\implies y(t) = \mathcal{L}^{-1} \left[\frac{b_1 + b_2 s}{a_1 + a_2 s + a_3 s^2} U(s) \right].$$

Given the specific form of the input $u(t)$, the actual expression of the response $y(t)$ can be determined.

3.5   Using the System Transfer Function

The procedure of obtaining the system response using the system transfer function is very similar to that used when the model is in the form of an input-output differential equation. The procedure can be summarized as follows:

- Obtain the system transfer function $H(s)$ (in Laplace transforms).

- Express the Laplace transform of the output $Y(s)$ in terms of the transfer function $H(s)$ and the Laplace transform of the input $U(s)$.

- Take the inverse Laplace transform of this output $Y(s)$.

These steps can be expressed in terms of equations as follows:

$$H(s) = \frac{Y(s)}{U(s)}$$
$$Y(s) = H(s)U(s)$$
$$y(t) = \mathcal{L}^{-1} \left[H(s)U(s) \right]. \tag{3.27}$$

This is the general system response of any system in terms of the transfer function and the input in Laplace transforms.

Example 3.20 *Consider the dynamic system whose input-output differential equation is given by*

$$\dot{y} + 2y = u(t),$$
$$\text{where } y(0) = 0,\ \dot{y}(0) = 0 \text{ and } u(t) = 2.$$

(a) Find the system transfer function $H(s)$.
(b) Find the system response $y(t)$.

Solution 3.20 *(a) The system transfer function is determined by taking the Laplace transforms of the equation.*

$$\mathcal{L}\left[\dot{y} + 2y\right] = \mathcal{L}\left[u(t)\right]$$

$$[sY(s) - y(0)] + 2Y(s) = U(s)$$

$$sY(s) + 2Y(s) = U(s)$$

$$Y(s)[s+2] = U(s)$$

$$\text{Therefore,} \qquad H(s) = \frac{Y(s)}{U(s)}$$

$$= \frac{1}{s+2}.$$

(b) The system response is obtained from the transfer function and the Laplace transform of the input.

$$y(t) = \mathcal{L}^{-1}\left[H(s)U(s)\right]$$

$$= \mathcal{L}^{-1}\left[\left(\frac{1}{s+2}\right)\left(\frac{2}{s}\right)\right]$$

$$= \mathcal{L}^{-1}\left[\frac{2}{s(s+2)}\right]$$

$$= \mathcal{L}^{-1}\left[\frac{1}{s} - \frac{1}{s+2}\right]$$

$$= 1 - e^{-2t}.$$

Example 3.21 *The car speed cruise control input-output differential equation is given by*

$$\dot{v} + \frac{b}{m}v = \frac{b}{m}v_r,$$

for a step input v_r.

(a) *Determine the system transfer function $H(s)$.*

(b) *Find the dynamic system response $v(t)$.*

Solution 3.21 (a) *The system transfer function is obtained by taking the Laplace transforms*

$$\mathcal{L}\left[\dot{v} + \frac{b}{m}v\right] = \mathcal{L}\left[\frac{b}{m}v_r\right]$$

$$[sV(s) - v(0)] + \frac{b}{m}V(s) = \frac{b}{m}V_r(s)$$

$$V(s)\left[s + \frac{b}{m}\right] = \frac{b}{m}V_r(s) \qquad where\ v(0) = 0$$

Therefore, $\qquad H(s) = \dfrac{V(s)}{V_r(s)}$

$$= \frac{\dfrac{b}{m}}{s + \dfrac{b}{m}}.$$

(b) *The system response is obtained from the transfer function and the*

Laplace transform of the input.

$$v(t) = \mathcal{L}^{-1}\left[H(s)V_r(s)\right]$$

$$= \mathcal{L}^{-1}\left[\left(\frac{\dfrac{b}{m}}{s+\dfrac{b}{m}}\right)\left(\frac{v_r}{s}\right)\right]$$

$$= \left(\frac{bv_r}{m}\right)\mathcal{L}^{-1}\left[\frac{1}{s\left(s+\dfrac{b}{m}\right)}\right]$$

$$= \left(\frac{bv_r}{m}\right)\mathcal{L}^{-1}\left[\frac{C_1}{s} + \frac{C_2}{s+\dfrac{b}{m}}\right].$$

The coefficients are obtained by the cover up method as

$$C_1 = \frac{m}{b} \quad and \quad C_2 = -\frac{m}{b}.$$

Hence,

$$v(t) = \left(\frac{bv_r}{m}\right)\mathcal{L}^{-1}\left[\frac{\dfrac{m}{b}}{s} - \frac{\dfrac{m}{b}}{s+\dfrac{b}{m}}\right]$$

$$= v_r\mathcal{L}^{-1}\left[\frac{1}{s} - \frac{1}{s+\dfrac{b}{m}}\right]$$

$$= v_r\left(1 - e^{-\frac{b}{m}t}\right).$$

These last two examples illustrate how the system response can be determined from the system transfer function. Two simple but common inputs that are useful in the study of the response of any dynamic system are the impulse and step inputs.

3.5.1   The Impulse Response (Natural Response)

The impulse response $h(t)$ is the output when the input is an impulse $\delta(t)$. Such a response for any system with a general transfer function $H(s)$ is derived from the general system response in Equation 3.27 by substituting for $U(s)$ with the Laplace transform of an impulse input given by $U(s) = 1$, such that

$$
\begin{aligned}
y(t) &= \mathcal{L}^{-1} Y(s) \\
&= \mathcal{L}^{-1} [H(s)U(s)] \\
y(t) &= \mathcal{L}^{-1} H(s) = h(t) \\
&\Longrightarrow h(t) = \mathcal{L}^{-1} H(s).
\end{aligned}
$$

This means that the impulse response of any system is the inverse Laplace transform of the system transfer function. This is a very important response and is also called the natural response. It is very significant because it is the simplest type of response that can be analyzed for any system in order to deduce system characteristics.

3.5.2   The Unit Step Response

The unit step response $y_u(t)$ of a system is the output when the input is a unit step function. Such a response for any system with a general transfer function $H(s)$ is derived from the general system response in Equation 3.27 by substituting for $U(s)$ with the Laplace transform of a unit step input given by $U(s) = \dfrac{1}{s}$, such that

$$
\begin{aligned}
y(t) &= \mathcal{L}^{-1} [H(s)U(s)] \\[2mm]
y(t) &= \mathcal{L}^{-1} \left[\frac{H(s)}{s} \right]
\end{aligned}
$$

$$
\Longrightarrow y_u(t) = \mathcal{L}^{-1} \left[\frac{H(s)}{s} \right].
$$

This is the general expression for the step response of any dynamic system.

3.5.3   The Impulse and Unit Step Responses: The Relationship

A relationship can be established between the impulse and step responses by comparing their expressions

$$
h(t) = \mathcal{L}^{-1} H(s) \quad \text{and} \quad y_u(t) = \mathcal{L}^{-1} \left[\frac{H(s)}{s} \right],
$$

and using the properties that s and $\frac{1}{s}$ represent the differential and integrator operators, respectively. Integrating the natural response leads to

$$\int h(t)dt = \mathcal{L}^{-1}\left[\frac{1}{s} \times H(s)\right]$$
$$= \mathcal{L}^{-1}\left[\frac{H(s)}{s}\right]$$
$$= y_u(t).$$

Alternatively, taking the derivative of the step response leads to

$$\frac{dy_u(t)}{dt} = \left[s \times \frac{H(s)}{s}\right]$$
$$= \mathcal{L}^{-1}H(s)$$
$$= h(t).$$

Thus, the relationship between responses can be summarized as follows:

$$y_u(t) = \int h(t)dt$$
$$h(t) = \frac{dy_u(t)}{dt}.$$

3.5.4   Final Value Theorem (FVT)

This theorem is used to compute the steady state value of a time function $y(t)$.

$$\lim_{t \to \infty} y(t) = \lim_{s \to 0} sY(s)$$
$$= \lim_{s \to 0} s\left[H(s)U(s)\right],$$

where $H(s)$ is the system transfer function and $U(s)$ is the Laplace transform of the input. This theorem only applies to stable systems since the final value of an unstable system is not defined. Consider two systems whose outputs in Laplace transforms are given by

$$Y_1(s) = \frac{4(s+2)}{s(s^2 + 2s + 20)}$$
$$Y_2(s) = \frac{7}{s(s-4)}.$$

The first system's steady state value is obtained by using FVT as follows:

$$\lim_{t \to \infty} y_1(t) = \lim_{s \to 0} sY_1(s)$$

$$= \lim_{s \to 0} s \left[\frac{4(s+2)}{s(s^2 + 2s + 20)} \right].$$

$$= \lim_{s \to 0} \frac{4(s+2)}{(s^2 + 2s + 20)}$$

$$= \frac{4 \times 2}{20}$$

$$= \frac{8}{20}.$$

For the second system, from the denominator it is clear that the system has a root in the right-hand plane (RHP) which means that it is an unstable system, hence the FVT is not defined. If the FVT is naively applied a steady state value of $-\frac{7}{4}$ is obtained, which is incorrect as the response grows without bound. In fact

$$y_2(t) = 7e^{4t} + 7,$$

which is clearly an unstable system growing without bound. System stability is discussed in detail in Chapters 4 and 5.

3.5.5 Initial Value Theorem (IVT)

This theorem is used to compute the initial value of a time function $y(t)$.

$$y(0) = \lim_{s \to \infty} sY(s)$$
$$= \lim_{s \to \infty} s\left[H(s)U(s)\right],$$

where $H(s)$ is the system transfer function and $U(s)$ is the Laplace of the input. The theorem is valid for all functions $Y(s)$ where the degree of the numerator polynomial m is less than that of the denominator n. Consider the Laplace function,

$$Y(s) = \frac{s^3 + 2s^2 + 4s + 7}{s^3 + 2s}$$

Attempting to apply the IVT leads to

$$y(0) = \lim_{s \to \infty} sY(s)$$

$$= \lim_{s \to \infty} s \left(\frac{s^3 + 2s^2 + 4s + 7}{s^3 + 2s} \right)$$

$$= \infty \quad \text{(undefined)}$$

The reason why the IVT is not defined for this system is because $m = n = 3$ in the Laplace function.

3.5.6   DC Gain

The DC gain is defined as the final value of the system response to a unit step input. It is obtained by employing the FVT when the input is a unit step function, i.e., $U(s) = \dfrac{1}{s}$.

$$\text{DC Gain} = \lim_{s \to 0} sY(s)$$

$$= \lim_{s \to 0} s\left[H(s)U(s)\right]$$

$$= \lim_{s \to 0} s\left[H(s)\frac{1}{s}\right]$$

Therefore, the DC Gain $= \lim_{s \to 0} H(s)$,

where $H(s)$ is the system transfer function. It is important to note that since the DC gain is defined in terms of the FVT when this theorem is not valid (e.g., an unstable system) then the DC gain is not defined. Consider two systems whose transfer functions are given by

$$H_1(s) = \frac{s+5}{(s+5)(s+7)(s+1)}$$

$$H_2(s) = \frac{3s+5}{(s+5)(s-7)}.$$

Find the DC gain for each of these systems. For the first system

$$\text{DC Gain} = \lim_{s \to 0} H_1(s)$$

$$= \lim_{s \to 0} \left[\frac{s+5}{(s+5)(s+7)(s+1)}\right]$$

$$= \frac{5}{5 \times 7 \times 1}$$

$$= \frac{1}{7}.$$

The second system is unstable because the characteristic equation has a root in the RHP. Hence the FVT is not valid which in turn makes the DC gain undefined.

3.6   First-Order Systems

A general first-order system is of the form

$$H(s) = \frac{1}{s + \sigma}$$
$$h(t) = e^{-\sigma t}$$
$$\sigma > 0 \equiv \text{ pole is located at } s < 0$$

$e^{-\sigma t}$ decays $\equiv$ impulse response is stable

$$\text{if } \sigma < 0 \Rightarrow s > 0 \equiv \text{ exponential growth } \rightarrow \text{ unstable}$$

Time constant $\tau = \dfrac{1}{\sigma} \equiv$ when response is $\dfrac{1}{e}$ times the initial value.

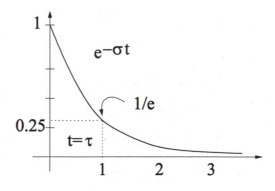

FIGURE 3.9
A First-Order System

3.7   Second-Order Systems

The transfer function of a second-order system takes the general form

$$H(s) = \frac{\omega_n^2}{s^2 + (2\xi\omega_n)\,s + \omega_n^2} \equiv \frac{b(s)}{a(s)}, \tag{3.28}$$

where the complex poles are given by

$$s = -\sigma \pm j\omega_d.$$

The relationship between the pole variables (σ and ω_d) and the system parameters (ξ and ω_n) can be established by expressing the characteristic equation in terms of the poles.

$$
\begin{aligned}
a(s) &= (s - s_1)(s - s_2) \\
&= [s - (-\sigma + j\omega_d)]\,[s - (-\sigma - j\omega_d)] \\
&= (s + \sigma - j\omega_d)(s + \sigma + j\omega_d) \\
&= (s + \sigma)^2 + \omega_d^2 \\
&= s^2 + 2\sigma s + \sigma^2 + \omega_d^2 \\
&= s^2 + (2\sigma)s + (\sigma^2 + \omega_d^2).
\end{aligned}
$$

Comparing the coefficients of this characteristic equation with those of the standard one in Equation 3.28 leads to

$$
\begin{aligned}
2\sigma &= 2\xi\omega_n \\
&\Longrightarrow \sigma = \xi\omega_n \tag{3.29} \\
\sigma^2 + \omega_d^2 &= \omega_n^2 \\
&\Longrightarrow (\xi\omega_n)^2 + \omega_d^2 = \omega_n^2 \\
&\Longrightarrow \omega_d^2 = \omega_n^2 - (\xi\omega_n)^2 \\
&\Longrightarrow \omega_d = \omega_n\sqrt{1 - \xi^2} \tag{3.30}
\end{aligned}
$$

$$\xi \equiv \text{ damping ratio}$$
$$\omega_n \equiv \text{ undamped natural frequency.}$$

The relationship can be summarized as

$$\sigma = \xi\omega_n \quad \text{and} \quad \omega_d = \omega_n\sqrt{1 - \xi^2}.$$

Therefore, the poles are located at a radius ω_n in the s-plane and at an angle θ where

$$\tan\theta = \frac{\sigma}{\omega_d}$$

$$= \frac{\xi\omega_n}{\omega_n\sqrt{1-\xi^2}}$$

$$\tan\theta = \frac{\xi}{\sqrt{1-\xi^2}} = \frac{\sin\theta}{\cos\theta}$$

$$\Rightarrow \sin\theta = \xi \quad \text{and} \quad \cos\theta = \sqrt{1-\xi^2}$$

$$\Rightarrow \theta = \sin^{-1}\xi.$$

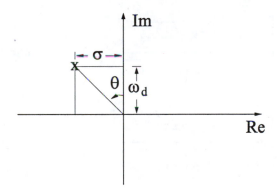

FIGURE 3.10
S-plane

3.7.1 Impulse and Step Responses

In this section, the general impulse and step responses for second-order systems are derived, plotted and compared. Consider the general second-order transfer function

$$H(s) = \frac{\omega_n^2}{s^2 + (2\xi\omega_n)\,s + \omega_n^2}.$$

The general second-order system impulse response $h(t)$ is determined as follows:

$$h(t) = \mathcal{L}^{-1} Y(s)$$

$$= \mathcal{L}^{-1} \left[H(s) U(s) \right]$$

$$= \mathcal{L}^{-1} H(s) \qquad\qquad (\text{where } U(s) = 1)$$

$$= \mathcal{L}^{-1} \left[\frac{\omega_n^2}{s^2 + (2\xi\omega_n)\, s + \omega_n^2} \right]$$

$$= \mathcal{L}^{-1} \left[\frac{\omega_n^2}{(s + \xi\omega_n)^2 + \left(\omega_n \sqrt{(1-\xi^2)} \right)^2} \right] \qquad \text{Completing the square}$$

$$= \mathcal{L}^{-1} \left[\frac{\omega_n^2 \times \dfrac{\sqrt{(1-\xi^2)}}{\sqrt{(1-\xi^2)}}}{(s + \xi\omega_n)^2 + \left(\omega_n \sqrt{(1-\xi^2)} \right)^2} \right] \qquad \text{Multiplication by one}$$

$$= \mathcal{L}^{-1} \left[\frac{\left(\dfrac{\omega_n}{\sqrt{1-\xi^2}} \right) \omega_n \sqrt{(1-\xi^2)}}{(s + \xi\omega_n)^2 + \left(\omega_n \sqrt{(1-\xi^2)} \right)^2} \right] \qquad \text{Rearranging}$$

$$= \frac{\omega_n}{\sqrt{1-\xi^2}} e^{-\xi\omega_n t} \sin\left(\omega_n \sqrt{(1-\xi^2)} \right) t \qquad \text{Using Equation 3.19}$$

$$= \frac{\omega_n}{\sqrt{1-\xi^2}} e^{-\sigma t} \sin \omega_d t.$$

It is important to note that the natural response of a second-order system is a special case of the more general response given in Equation 3.23, where $d = 0$, $c = b = w_n^2$ and $a = 2\xi\omega_n$.

The step response $y_u(t)$ of a general second-order system is obtained in a similar fashion.

$$y_u(t) = \mathcal{L}^{-1} Y(s)$$

$$= \mathcal{L}^{-1} \left[H(s)U(s) \right]$$

$$= \mathcal{L}^{-1} \left[\frac{H(s)}{s} \right]$$

$$= \mathcal{L}^{-1} \left[\frac{\omega_n^2}{s \left(s^2 + (2\xi\omega_n) s + \omega_n^2 \right)} \right]$$

$$= \mathcal{L}^{-1} \left[\frac{\omega_n^2}{s \left[(s + \xi\omega_n)^2 + \left(\omega_n \sqrt{(1 - \xi^2)} \right)^2 \right]} \right] \quad \text{Completing the square}$$

$$= \mathcal{L}^{-1} \left[\frac{\omega_n^2 + \left[(\xi\omega_n)^2 - (\xi\omega_n)^2 \right]}{s \left[(s + \xi\omega_n)^2 + \left(\omega_n \sqrt{(1 - \xi^2)} \right)^2 \right]} \right] \quad \text{Adding zero}$$

$$= \mathcal{L}^{-1} \left[\frac{(\xi\omega_n)^2 + \left(\omega_n \sqrt{(1 - \xi^2)} \right)^2}{s \left[(s + \xi\omega_n)^2 + \left(\omega_n \sqrt{(1 - \xi^2)} \right)^2 \right]} \right] \quad \text{Rearranging}$$

$$= 1 - e^{-\xi\omega_n t} \left[\cos \left(\omega_n \sqrt{(1 - \xi^2)} \right) t + \frac{\xi\omega_n \sin \left(\omega_n \sqrt{(1 - \xi^2)} \right) t}{\omega_n \sqrt{(1 - \xi^2)}} \right]$$

(by using Equation 3.20)

$$= 1 - e^{-\sigma t} \left[\cos \omega_d t + \frac{\sigma}{\omega_d} \sin \omega_d t \right].$$

The impulse and step responses are compared by analyzing their two responses,

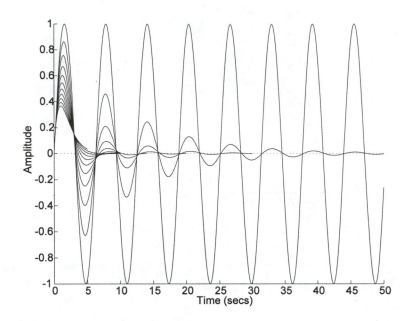

FIGURE 3.11
The Impulse Response for Different Values of ξ

$$h(t) = \frac{\omega_n}{\sqrt{1-\xi^2}} e^{-\sigma t} sin\omega_d t$$

$$y_u(t) = 1 - e^{-\sigma t} \left[\cos\omega_d t + \frac{\sigma}{\omega_d} \sin\omega_d t \right].$$

Plots of the impulse response and the step response for different values of ξ are shown in Figure 3.11 and Figure 3.12.

The transient characteristics of the two responses are similar and depend on the value of ξ.

(a) $\xi = 0 \implies$ Oscillations with constant amplitude of 1.

(b) $\xi = 1 \implies$ Least overshoot and fastest settling time.

(c) The steady state value or final value is 1 for the step response and 0 for the impulse response.

3.7.2   Stability for Second-Order System

Whether a system is stable or not can be established by studying its natural response. For a general second-order system this response is of the

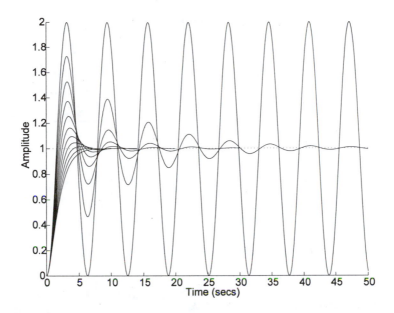

FIGURE 3.12
Step Response for Different Values of ξ

form

$$h(t) = \frac{\omega_n}{\sqrt{1-\xi^2}} e^{-\sigma t} \sin\omega_d t.$$

- The response envelope is defined by $e^{-\sigma t}$ and hence the real part of the pole σ determines the decay rate of the natural responses.

- If $\sigma < 0$, then the pole is in the RHP and the natural response grows without bound, which means the system is unstable.

- If $\sigma > 0$, then the pole is in the LHP and the natural response decays, which means the system is stable (Figure 3.13).

- If $\sigma = 0$, then the pole is on the imaginary axis and the natural response oscillates with a fixed amplitude, which means the system is marginally stable or has indeterminate stability.

Example 3.22 *(a) Discuss the relationship between the poles of the following system transfer function*

$$H(s) = \frac{2s+1}{s^2 + 2s + 5},$$

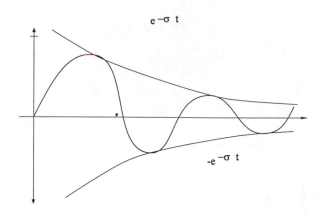

FIGURE 3.13
A Stable Second-Order System

and the corresponding impulse response.

   (b) Find the exact response.

Solution 3.22 *(a) Comparing the characteristic equation of the system with the standard second-order equation gives*

$$s^2 + 2\xi\omega_n s + \omega_n^2 = s^2 + 2s + 5.$$

Equating the coefficients gives

$$\omega_n^2 = 5 \Rightarrow \omega_n = \sqrt{5}$$

$$2\xi\omega_n = 2 \Rightarrow \xi = \frac{1}{\sqrt{5}}.$$

The poles are obtained directly from the characteristic equation as $(-1 \pm 2j)$ *or as follows:*

$$s_1, s_2 = -\sigma \pm j\omega_d$$

$$where \quad \sigma = \xi\omega_n \quad and \quad \omega_d = \omega_n\sqrt{1 - \xi^2}$$

$$\Longrightarrow s_1, s_2 = -1 \pm 2j.$$

(b) The exact impulse response is obtained by breaking the transfer function

into partial fractions and using the method of completing the square.

$$H(s) = \frac{2s + 1}{(s + 1)^2 + 2^2}$$

$$= \frac{2(s + 1)}{(s + 1)^2 + 2^2} - \frac{\left(\dfrac{1}{2}\right) \times 2}{(s + 1)^2 + 2^2}$$

$$h(t) = 2e^{-t} \cos 2t - \frac{1}{2}e^{-t} \sin 2t.$$

The impulse response can be plotted using MATLAB as follows:

$$num = [2 \quad 1];$$
$$den = [1 \quad 2 \quad 5];$$
$$impulse(num, den).$$

3.7.3   Characteristics of a Step Response

The characteristics of a general step response are given in Figure 3.14.

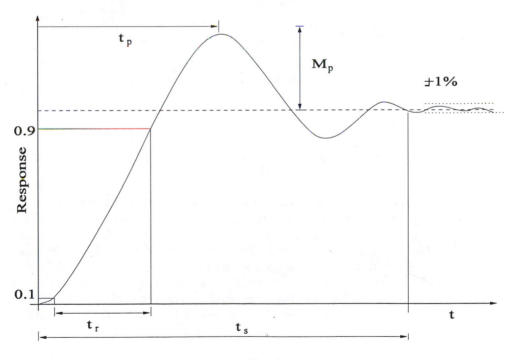

FIGURE 3.14
A General Response to a Step Input

- Rise time (t_r) : The time to reach vicinity of a new set point.

- Settling time (t_s) : Time for system transients to decay.

- Overshoot (M_p) : Maximum overshoot (as a % of final value).

- Peak time (t_p) : Time to reach the overshoot.

- The values of the parameters are summarized as follows:

$$t_r = \frac{1.8}{\omega_n}$$

$$t_p = \frac{\pi}{\omega_d}$$

$$M_p = e^{\frac{-\pi \xi}{\sqrt{1-\xi^2}}}$$

$$t_s = \frac{4.6}{\xi \omega_n} = \frac{4.6}{\sigma}.$$

- The effect of an extra pole is to increase the rise time.

3.7.4   Effects of Pole-Zero Location on the Response

The transfer function is used to analyze the response of the system it represents.

$$H(s) = \frac{b(s)}{a(s)} \equiv \frac{\text{a polynomial in } s}{\text{a polynomial in } s}.$$

Poles are the values of s such that $a(s) = 0$, zeros are values of s such that $b(s) = 0$. Assuming no common factors, when $a(s) = 0$, i.e., $H(s)$ is infinity, the roots of the equation are called poles of $H(s)$. The roots of the polynomial $b(s) = 0$ are called the zeros of the system. Systems with poles in the right-hand plane (RHP) are said to be *unstable*, while systems with zeros in the right-hand plane (RHP) are called *nonminimum phase* systems. The stability of dynamic systems and the significance of nonminimum phase systems will be discussed in detail in later chapters.

The effect of poles on the response of a system can be briefly established by considering the following transfer function,

$$H(s) = \frac{2s+1}{s^2 + 3s + 2} = \frac{b(s)}{a(s)},$$

where

$$b(s) = 2\left(s + \frac{1}{2}\right)$$

$$a(s) = s^2 + 3s + 2$$

$$= (s+1)(s+2).$$

This means that there are two poles at $s = -1$ and $s = -2$, and one zero at $s = -\frac{1}{2}$. The location of the poles ($\times$) and zero ($\circ$) are illustrated in the following diagram.

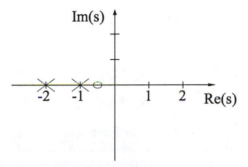

Location of Zero and Poles

Using partial fractions the natural response can be obtained as follows:

$$H(s) = \frac{2\left(s + \frac{1}{2}\right)}{(s+1)(s+2)}$$

$$H(s) = -\frac{1}{s+1} + \frac{3}{s+2}$$

$$h(t) = -e^{-t} + 3e^{-2t} \text{ for } t \geq 0$$

$$= 0 \text{ for } t < 0.$$

Thus, the natural response is determined by the location of poles. The fast pole is at $s = -2$ and the slow pole is at $s = -1$, where this is determined by the relative rate of signal decay:

$$3e^{-2t} \text{ dominates the early part.}$$

$$-e^{-t} \text{ dominates the later part.}$$

The sketches of the natural system response can be obtained by considering the numerator $(2s+1)$ and denominator (s^2+3s+2) in MATLAB as follows:

$num = [2\ 1];$

$den = [1\ 3\ 2];$
$impulse(num, den).$

Zeros exert their influence by modifying the coefficients of the exponential terms whose shape is decided by the poles. This is illustrated by considering a transfer function and its corresponding natural response.

$$H_1 = \frac{2}{(s+1)(s+2)} = \frac{2}{s+1} - \frac{2}{s+2}$$

$$\Longrightarrow h(t) = 2e^{-t} - 2e^{-2t}. \tag{3.31}$$

If a zero is introduce near the pole at $s = -1$, (at $s = -1.1$) the new transfer function and its corresponding natural response are given by

$$H_2 = \frac{2(s+1.1)}{1.1(s+1)(s+2)} = \frac{0.18}{s+1} + \frac{1.64}{s+2}$$

$$\Longrightarrow h(t) = 0.18e^{-t} + 1.64e^{-2t}. \tag{3.32}$$

Comparing the magnitudes of the coefficients of the exponential functions in Equations 3.31 and 3.32, it can be observed that the introduction of a zero at $s = -1.1$ dramatically reduces the effect of the pole at $s = -1$. In fact if a zero is placed exactly at $s = -1$, the term due to the pole -1 (i.e. $2e^{-t}$) completely vanishes. In general, a zero near a pole reduces the amount of that term in the total response.

Some of the effects of zero and pole patterns on the general step response presented in Figure 3.14 can be summarized as follows:

- In a second-order system with no finite zeros

$$t_r = \frac{1.8}{\omega_n}$$

$$M_p = 5\% \text{ for } \xi = 0.7$$

$$= 16\% \text{ for } \xi = 0.5$$

$$= 35\% \text{ for } \xi = 0.3$$

$$t_s = \frac{4.6}{\sigma}.$$

- An additional LHP zero will increase overshoot.

- An additional RHP zero will depress the overshoot (start in the opposite direction to that of the final steady state value).

- An additional pole in LHP will increase the rise time.

3.8 Examples: System Response (Laplace Transforms)

Example 3.23 *(a) Write down the equations that describe the following electrical circuit in state-variable form*

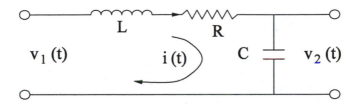

FIGURE 3.15
An RLC Electrical Circuit

(b) Express the model as a second-order differential equation in y(t).

Solution 3.23 *Choose the voltage across the capacitor y(t) and the current flowing through the inductor i(t), as the state variables.*

For the capacitor

$$i(t) = C\frac{dy}{dt} \implies \dot{y} = \frac{i(t)}{C}.$$

Applying the KVL for the circuit

$$u - L\frac{di}{dt} - Ri - y = 0$$

$$\frac{di}{dt} = \frac{u}{L} - \frac{R}{L}i - \frac{y}{L}.$$

The state-variable matrix system takes the form

$$\begin{bmatrix} \dot{i} \\ \dot{y} \end{bmatrix} = \begin{bmatrix} -\dfrac{R}{L} & -\dfrac{1}{L} \\ \dfrac{1}{C} & 0 \end{bmatrix} \begin{bmatrix} i \\ y \end{bmatrix} + \begin{bmatrix} \dfrac{1}{L} \\ 0 \end{bmatrix} u$$

$$y = \begin{bmatrix} 0 & 1 \end{bmatrix} \begin{bmatrix} i \\ y \end{bmatrix} + [0]\, u.$$

The input-output differential equation is given by

$$\frac{di}{dt} = \frac{u}{L} - \frac{R}{L}i - \frac{y}{L}$$

$$\implies \frac{d\,(C\,\dot{y})}{dt} = \frac{u}{L} - \frac{R}{L}(C\,\dot{y}) - \frac{y}{L} \quad where \quad i(t) = C\dot{y}$$

$$\implies \frac{Cd\,\dot{y}}{dt} + \left(\frac{CR}{L}\right)\dot{y} + \frac{y}{L} = \frac{u}{L}$$

$$\implies \frac{d\,\dot{y}}{dt} + \left(\frac{R}{L}\right)\dot{y} + \frac{y}{CL} = \frac{u}{CL}.$$

Given the values of $L = 1$, $R = 2$, $C = 1$ and using them leads to

$$\begin{bmatrix} \dot{i} \\ \dot{y} \end{bmatrix} = \begin{bmatrix} -2 & -1 \\ 1 & 0 \end{bmatrix} \begin{bmatrix} i \\ y \end{bmatrix} + \begin{bmatrix} 1 \\ 0 \end{bmatrix} u$$

$$y = \begin{bmatrix} 0 & 1 \end{bmatrix} \begin{bmatrix} i \\ y \end{bmatrix} + [0]\, u.$$

This is the state-variable form. The input-output differential equation reduces to

$$\frac{di}{dt} = -2i - y + u \quad where \quad i = \frac{dy}{dt}$$

$$\implies \ddot{y} = -2\dot{y} - y + u$$

$$\implies \ddot{y} + 2\dot{y} + y - u = 0.$$

Applying Laplace transform, assuming the following conditions: $u = 0$, $y(0) = 1$ and $\dot{y}(0) = 0$, leads to

$$s^2 Y(s) - sy(0) - \dot{y}(0) + 2\left[sY(s) - y(0)\right] + Y(s) = 0$$

$$s^2 Y(s) + 2sY(s) + Y(s) = sy(0) + \dot{y}(0) + 2y(0)$$

$$(s^2 + 2s + 1)Y(s) = s + 2$$

Therefore,

$$Y(s) = \frac{s+2}{s^2 + 2s + 1}$$

$$= \frac{s+2}{(s+1)^2}$$

$$= \frac{(s+1)+1}{(s+1)^2}$$

$$= \frac{1}{s+1} + \frac{1}{(s+1)^2}. \tag{3.33}$$

The time domain system response is obtained by taking the inverse Laplace transform of the function in Equation 3.33, and then plotting the result.

$$y(t) = \mathcal{L}^{-1}\left[\frac{1}{s+1} + \frac{1}{(s+1)^2}\right]$$

$$= e^{-t} + te^{-t} = e^{-t}(1+t).$$

Using MATLAB's initial command the same result can be obtained.

$$\begin{bmatrix} i_o \\ y_o \end{bmatrix} = \begin{bmatrix} C\dot{y}_o \\ y_o \end{bmatrix} = \begin{bmatrix} 0 \\ 1 \end{bmatrix}$$

The syntax for the MATLAB function is given by

initial(A, B, C, D, X_o),

where X_0 is the vector of initial values $[0 \ \ 1]^T$. The results are shown in the following diagram.

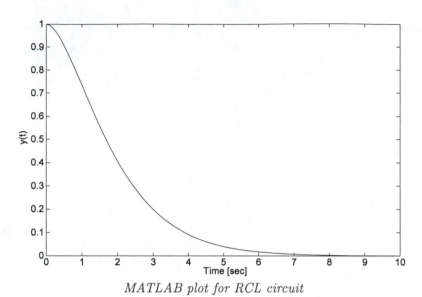

MATLAB plot for RCL circuit

Example 3.24 *Find the transfer function if the input and output are respectively given by the following equations*

$$u(t) = e^{-t}$$
$$y(t) = 2 - 3e^{-t} + e^{-2t}\cos 2t$$

Solution 3.24

$$U(s) = \frac{1}{s+1}$$

$$Y(s) = \frac{2}{s} - \frac{3}{s+1} + \frac{s+2}{(s+2)^2 + 4}$$

$$G(s) = \frac{Y(s)}{U(s)}$$

$$= \frac{\dfrac{2}{s} - \dfrac{3}{s+1} + \dfrac{s+2}{(s+2)^2 + 4}}{\dfrac{1}{s+1}}$$

$$= \frac{s^2 + 2s + 16}{s(s^2 + 4s + 8)}.$$

Example 3.25 *Consider the following RLC circuit and find:*

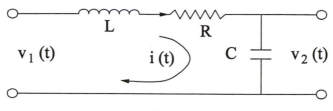

An RLC Electrical Circuit

(a) *The time domain equation relating $i(t)$ and $v_1(t)$.*

(b) *The time domain equation relating $i(t)$ and $v_2(t)$.*

(c) *The transfer function $V_2(s)/V_1(s)$, assuming zero initial conditions.*

(d) *The damping ratio ξ and the natural frequency ω_n.*

(e) *The value of R such that the overshoot, $M_p = 25\%$ where $v_1(t)$ is a unit step input, $L = 10mH$ and $C = 4\mu F$.*

Solution 3.25 *The KVL and KCL are used to set up the equations.*

(a) *From applying the KVL in the input loop leads to*

$$v_1(t) = L\frac{di}{dt} + Ri(t) + \frac{1}{C}\int i(t)dt.$$

(b)*From applying the KVL in the output loop leads to*

$$v_2(t) = \frac{1}{C}\int i(t)dt.$$

(c)*The transfer function $\dfrac{V_2(s)}{V_1(s)}$ is obtained by first taking the Laplace transforms of the expressions of $v_2(t)$ and $v_2(t)$.*

$$V_1(s) = \left(sL + R + \frac{1}{sC}\right)I(s)$$
$$V_2(s) = \frac{1}{sC}I(s).$$

The transfer function is the ratio of the transforms

$$H(s) = \frac{V_2(s)}{V_1(s)}$$

$$= \frac{\dfrac{1}{sC}I(s)}{\left(sL + R + \dfrac{1}{sC}\right)I(s)}$$

$$= \frac{\dfrac{1}{LC}}{s^2 + \dfrac{R}{L}s + \dfrac{1}{LC}} \tag{3.34}$$

$$\equiv \frac{\omega_n^2}{s^2 + 2\xi\omega_n s + \omega_n^2}. \tag{3.35}$$

(d) The damping ratio ξ and the natural frequency ω_n are obtained by comparing the coefficients of Equation 3.34 with those of the standard form of a general second-order system shown in Equation 3.35,

$$\omega_n^2 = \frac{1}{LC}$$

$$\Rightarrow \omega_n = \sqrt{\frac{1}{LC}}$$

$$2\xi\omega_n = \frac{R}{L}$$

$$\Rightarrow \xi = \frac{1}{2\omega_n} \times \frac{R}{L} = \frac{1}{2}\sqrt{LC}\frac{R}{L}$$

$$\Rightarrow \xi = \frac{R}{2}\sqrt{\frac{C}{L}}. \tag{3.36}$$

(iv) Given that the overshoot $M_p = 25\%$, a value of the damping ratio ξ can be obtained from the following expression for the overshoot

$$M_p = e^{-\frac{\pi\xi}{\sqrt{1-\xi^2}}}$$
$$\Rightarrow \xi \dot{=} 0.4.$$

The value of R is then determined by substituting this value of the damping *ratio in Equation 3.36. Hence, it follows that*

$$\xi = \frac{R}{2}\sqrt{\frac{C}{L}}$$

$$\Leftrightarrow R = 2\xi\sqrt{\frac{L}{C}} = (2)(0.4)\sqrt{\frac{10 \times 10^{-3}}{4 \times 10^{-6}}}$$

$$\Rightarrow R = 40\Omega.$$

Example 3.26 *Consider the following transfer functions for certain dynamic systems.*

(a)

$$H_1(s) = \frac{5s}{s^2 + 2s + 5}$$

(b)

$$H_2(s) = \frac{5}{s^2 + 2s + 5}$$

(c)

$$H_3(s) = \frac{\omega_n^2}{s^2 + 2\xi\omega_n + \omega_n^2}.$$

Find the output for each system if the input is as shown below.

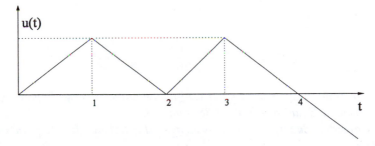

Solution 3.26 *(a) The input can be written in the form of a ramp function and a delayed ramp function using superposition principle*

$$u(t) = r(t) - 2r(t-1) + 2r(t-2) - 2r(t-3),$$

where $r(t) = t$ is the ramp function. Thus, only the response to a ramp function $y_r(t)$ has to be determined and the total response $y(t)$ is then expressed in terms of the ramp response using the superposition principle.

$$y(t) = y_r(t) - 2y_r(t-1) + 2y_r(t-2) - 2y_r(t-3).$$

In order to determine the ramp response $y_r(t)$ consider the general transfer function equation,

$$H(s) = \frac{Y(s)}{U(s)}$$

$$\Leftrightarrow Y(s) = H(s)U(s)$$

$$\Rightarrow Y_r(s) = H(s)R(s)$$

$$\Rightarrow Y_r(s) = H(s)\frac{1}{s^2}$$

$$\Leftrightarrow y_r(t) = \mathcal{L}^{-1}\left[H(s)\frac{1}{s^2}\right].$$

This is true for all the three transfer functions (a), (b), and (c), hence the corresponding ramp responses are given by the following equations,
  respectively.

$$y_r(t) = \mathcal{L}^{-1}\left[\frac{5s}{s^2 + 2s + 5} \times \frac{1}{s^2}\right] \tag{3.37}$$

$$y_r(t) = \mathcal{L}^{-1}\left[\frac{5}{s^2 + 2s + 5} \times \frac{1}{s^2}\right] \tag{3.38}$$

$$y_r(t) = \mathcal{L}^{-1}\left[\frac{\omega_n^2}{s^2 + 2\xi\omega_n + \omega_n^2} \times \frac{1}{s^2}\right]. \tag{3.39}$$

Thus, the remaining task is determining the three inverse Laplace transforms given in Equations 3.37, 3.38, and 3.39.
  (a) Consider Equation 3.37 and express the Laplace function in partial fractions

$$\frac{5s}{s^2 + 2s + 5} \times \frac{1}{s^2} = \frac{5}{s(s^2 + 2s + 5)}$$

$$= \frac{A}{s} + \frac{Bs + C}{s^2 + 2s + 5}. \tag{3.40}$$

The following coefficients are obtained,

$$A = 1, \quad B = -1, \quad C = -2.$$

Using these coefficients and then taking the inverse Laplace transform of Equation 3.40 leads to

$$y_r(t) = \mathcal{L}^{-1}\left[\frac{5s}{s^2 + 2s + 5} \times \frac{1}{s^2}\right]$$

$$= \mathcal{L}^{-1}\left[\frac{1}{s} - \frac{s+2}{s^2 + 2s + 5}\right]$$

$$= \mathcal{L}^{-1}\left[\frac{1}{s}\right] - \mathcal{L}^{-1}\left[\frac{s+2}{s^2 + 2s + 5}\right]$$

$$= \mathcal{L}^{-1}\left[\frac{1}{s}\right] - \mathcal{L}^{-1}\left[\frac{s+1}{s^2 + 2s + 5}\right] - \mathcal{L}^{-1}\left[\frac{1}{s^2 + 2s + 5}\right]$$

$$= \mathcal{L}^{-1}\left[\frac{1}{s}\right] - \mathcal{L}^{-1}\left[\frac{s+1}{(s+1)^2 + 4}\right] - \mathcal{L}^{-1}\left[\frac{1}{2} \times \frac{2}{(s+1)^2 + 4}\right]$$

$$= 1 - e^{-t}\cos 2t - \frac{1}{2}e^{-t}\sin 2t.$$

As explained above, the complete response $y(t)$ is then expressed in terms of this ramp response.

$$y(t) = y_r(t) - 2y_r(t-1) + 2y_r(t-2) - 2y_r(t-3)$$

(b) Consider Equation 3.38 and express the Laplace function in partial fractions

$$\frac{5}{s^2 + 2s + 5} \times \frac{1}{s^2} = \frac{5}{s^2(s^2 + 2s + 5)}$$

$$= \frac{C_1}{s} + \frac{C_2}{s^2} + \frac{C_3 s + C_4}{s^2 + 2s + 5}. \qquad (3.41)$$

The following coefficients are obtained,

$$C_1 = -\frac{2}{5}, \ C_2 = 1, C_3 = \frac{2}{5}, C_4 = -\frac{1}{5}.$$

These coefficients and the inverse Laplace transform of Equation 3.41 can be used to obtain the response $y_r(t)$. This is carried out as follows:

$$y_r(t) = \mathcal{L}^{-1} \left[\frac{-\frac{2}{5}}{s} + \frac{1}{s^2} + \frac{\frac{2}{5}s - \frac{1}{5}}{(s+1)^2 + 4} \right]$$

$$= \mathcal{L}^{-1} \left[\frac{-\frac{2}{5}}{s} + \frac{1}{s^2} + \frac{\frac{2}{5}(s+1) - \frac{2}{5} - \frac{1}{5}}{(s+1)^2 + 4} \right]$$

$$= \mathcal{L}^{-1} \left[\frac{-\frac{2}{5}}{s} + \frac{1}{s^2} + \frac{\frac{2}{5}(s+1)}{(s+1)^2 + 4} - \frac{-\frac{3}{5} \times \left(\frac{1}{2}\right) \times 2}{(s+1)^2 + 4} \right]$$

$$= -\frac{2}{5} + t + \frac{2}{5}e^{-t}\cos 2t - \frac{3}{10}e^t \sin 2t.$$

The complete response $y(t)$ is then expressed in terms of this ramp response.

$$y(t) = y_r(t) - 2y_r(t-1) + 2y_r(t-2) - 2y_r(t-3)$$

(c) Consider Equation 3.39 and express the Laplace function in partial fractions

$$\frac{\omega_n^2}{s^2 + 2\xi\omega_n + \omega_n^2} \times \frac{1}{s^2} = \frac{\omega_n^2}{s^2(s^2 + 2\xi\omega_n + \omega_n^2)}$$

$$= \frac{C_1}{s} + \frac{C_2}{s^2} + \frac{C_3s + C_4}{s^2 + 2\xi\omega_n + \omega_n^2}. \qquad (3.42)$$

The following coefficients are obtained

$$C_1 = -\frac{2\xi}{\omega_n}, \quad C_2 = 1, \quad C_3 = \frac{2\xi}{\omega_n}, \quad C_4 = 4\xi^2 - 1.$$

Using these coefficients and taking the inverse Laplace transform of Equa-

tion 3.42 gives the response $y_r(t)$. This is achieved as follows:

$$y_r(t) = \mathcal{L}^{-1}\left[-\frac{\dfrac{2\xi}{\omega_n}}{s} + \frac{1}{s^2} + \frac{\dfrac{2\xi}{\omega_n}s + (4\xi^2 - 1)}{s^2 + 2\xi\omega_n + \omega_n^2}\right]$$

$$= \mathcal{L}^{-1}\left[-\frac{\dfrac{2\xi}{\omega_n}}{s} + \frac{1}{s^2} + \frac{\dfrac{2\xi}{\omega_n}(s + \xi\omega_n) - \left(\dfrac{2\xi}{\omega_n} \times \xi\omega_n\right) + (4\xi^2 - 1)}{(s + \xi\omega_n)^2 + (\omega_n\sqrt{\xi^2 - 1})^2}\right]$$

$$= \mathcal{L}^{-1}\left[-\frac{\dfrac{2\xi}{\omega_n}}{s} + \frac{1}{s^2} + \frac{\dfrac{2\xi}{\omega_n}(s + \xi\omega_n) - 2\xi^2 + (4\xi^2 - 1)}{(s + \xi\omega_n)^2 + \left(\omega_n\sqrt{\xi^2 - 1}\right)^2}\right]$$

$$= \mathcal{L}^{-1}\left[-\frac{\dfrac{2\xi}{\omega_n}}{s} + \frac{1}{s^2} + \frac{\dfrac{2\xi}{\omega_n}(s + \xi\omega_n) + (2\xi^2 - 1)}{(s + \xi\omega_n)^2 + \left(\omega_n\sqrt{\xi^2 - 1}\right)^2}\right]$$

$$= \mathcal{L}^{-1}\left[-\frac{\dfrac{2\xi}{\omega_n}}{s} + \frac{1}{s^2} + \frac{\dfrac{2\xi}{\omega_n}(s + \xi\omega_n)}{(s + \xi\omega_n)^2 + \left(\omega_n\sqrt{\xi^2 - 1}\right)^2}\right]$$

$$+ \mathcal{L}^{-1}\left[\frac{(2\xi^2 - 1) \times \left(\dfrac{1}{(\omega_n\sqrt{\xi^2 - 1})}\right) \times \left(\omega_n\sqrt{\xi^2 - 1}\right)}{(s + \xi\omega_n)^2 + \left(\omega_n\sqrt{\xi^2 - 1}\right)^2}\right]$$

$$= -\frac{2\xi}{\omega_n} + t + \frac{2\xi}{\omega_n}e^{-\xi\omega_n t}\cos(\omega_n\sqrt{\xi^2 - 1})t$$

$$+ \frac{2\xi^2 - 1}{\omega_n\sqrt{\xi^2 - 1}}e^{-\xi\omega_n t}\sin(\omega_n\sqrt{\xi^2 - 1})t. \tag{3.43}$$

The complete response $y(t)$ is then expressed in terms of this ramp response,

$$y(t) = y_r(t) - 2y_r(t - 1) + 2y_r(t - 2) - 2y_r(t - 3),$$

where $y_r(t)$ is given in Equation 3.43.

3.9   Block Diagrams

A *block diagram* is an interconnection of symbols representing certain basic mathematical operations in such a way that the overall diagram obeys the system's mathematical model. In the diagram, the lines interconnecting the blocks represent the variables describing the system behavior, such as the input and state variables. Inspecting a block diagram of a system may provide new insight into the system's structure and behavior beyond that available from the differential equations themselves.[8]

In order to obtain the transfer function of a system, it is required to find the Laplace transform of the equations of motion and solve the resulting algebraic equations for the relationship between the input and the output. In many control systems the system equations can be written so that their components do not interact except by having the input of one part be the output of another part. In these cases, it is easy to draw a block diagram that represents the mathematical relationships. The transfer function of each component is placed in a box, and the input-output relationships between components are indicated by lines and arrows.

The general block diagram for a system with an input $u(t)$, an output $y(t)$ and transfer function $G(s)$ is shown in the following diagram.

U(s) ⟶ G(s) ⟶ Y(s)

Block Diagram Representation

The transfer function, the ratio (in Laplace transforms) of the output to the input, is given by

$$G(s) = \frac{Y(s)}{U(s)}.$$

Thus, the block can be represented by an electronic amplifier with the transfer function printed inside as illustrated in the preceding diagram.

3.9.1   Networks of Blocks

The blocks in a block diagram can be connected in different basic forms that include *series, parallel, negative feedback* and *positive feedback* networks. These basic forms constitute the building structures of complex block diagrams and it is essential to understand how their transfer functions are obtained. This knowledge can then be used to simplify complex block diagrams and determine their transfer functions.

Two blocks in series are represented as follows:

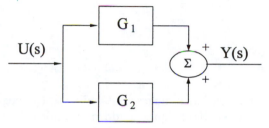

Blocks in Series

Hence, the overall transfer function is obtained from

$$G(s) = \frac{Y_2(s)}{U(s)}$$

$$= \frac{G_2(s)G_1(s)U(s)}{U(s)}$$

$$= G_2(s)G_1(s).$$

Two blocks in parallel are represented as follows:

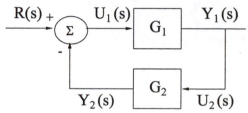

Blocks in Parallel

Hence, the overall transfer function is obtained from

$$G(s) = \frac{Y(s)}{U(s)}$$

$$= \frac{G_1(s)U(s) + G_2(s)U(s)}{U(s)}$$

$$= \frac{[G_1(s) + G_2(s)]U(s)}{U(s)}$$

$$= G_1(s) + G_2(s).$$

For negative feedback the block diagram takes the following form:

Negative Feedback Block Diagram

The closed-loop transfer function for the negative feedback system is then obtained as follows:

$$U_1(s) = R(s) - Y_2(s)$$

$$Y_2(s) = G_2(s)G_1(s)U_1(s)$$

$$Y_1(s) = G_1(s)U_1(s)$$

$$Y_1(s) = G_1(s)\left[R(s) - G_1(s)G_2(s)\frac{Y_1(s)}{G_1(s)}\right]$$

$$Y_1(s)\left[1 + G_1(s)G_2(s)\right] = G_1(s)R(s)$$

$$G(s) = \frac{Y_1(s)}{R(s)}$$

$$= \frac{G_1(s)}{1 + G_1(s)G_2(s)}.$$

The positive feedback block diagram is similar to the negative feedback one, except that the feedback signal is added and not subtracted.

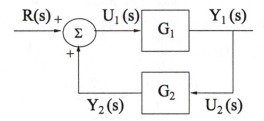

Positive Feedback Block Diagram

The closed-loop transfer function for the positive feedback block diagram is obtained in the same way as for the negative feedback one.

$$U_1(s) = R(s) + Y_2(s)$$

$$Y_2(s) = G_2(s)G_1(s)U_1(s)$$

$$Y_1(s) = G_1(s)U_1(s)$$

$$Y_1(s) = G_1(s)\left[R(s) + G_1(s)G_2(s)\frac{Y_1(s)}{G_{1(s)}}\right]$$

$$Y_1(s)\left[1 - G_1(s)G_2(s)\right] = G_1(s)R(s)$$

$$G(s) = \frac{Y_1(s)}{R(s)}$$

$$= \frac{G_1(s)}{1 - G_1(s)G_2(s)}.$$

It is instructive to note that for the positive and negative feedback block diagrams, the closed-loop transfer function has the same form

$$G(s) = \frac{G_1(s)}{1 \pm G_1(s)G_2(s)},$$

where the $+ve$ sign is for the negative feedback system and the $-ve$ sign is for the positive feedback system.

3.10 Simplifying Block Diagrams

When a system is modeled in a block diagram form the overall transfer function can be obtained by using block diagram simplification, which is often easier and more informative than algebraic manipulation, even though the methods are in every way equivalent. The central motivation behind simplifying a block diagram is the reduction of the complexity of the block diagram and obtaining the overall transfer function, while maintaining the same relationship among remaining variables. There are two main techniques that can be used to achieve this objective. The first method is *direct block diagram reduction* using block diagram algebra, and the other approach is based on *signal flow diagram analysis.*

Once the block diagrams have been simplified, the overall system transfer function can be obtained. If the block diagram is not available its structure can be established from component transfer functions. It is important to note that blocks can be connected in series only if the output of one block is not affected by the following block. If there are any loading effects between the components, it is necessary to combine these components into a single block. Any number of cascaded blocks representing non-loading components can be replaced by a single block, the transfer function of which is simply the product of the individual transfer functions.

3.10.1 Direct Block Diagram Reduction

A complicated block diagram involving many feedback loops can be simplified by a step-by-step rearrangement, using rules of block diagram algebra. This algebra depends on equivalence between simple block diagram configurations as shown in the following illustrations.

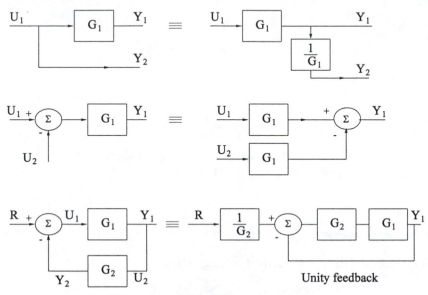

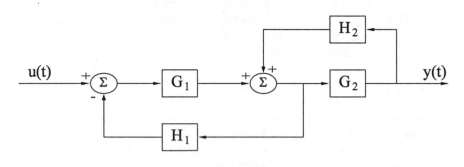

Examples of Block Diagram Algebra

The most important block diagram rules are summarized in Figure 3.16.

Example 3.27 *(a) Consider the following block diagram.*

Block Diagram

(a) Find the system transfer function.
(b) Choose system gains to make the transfer function unity.

Solution 3.27 *The block diagram is reduced as follows:*

	Original Block Diagrams	Equivalent Block Diagrams
1	$A \to \boxed{G} \xrightarrow{AG} \xrightarrow{+}(\Sigma)\xrightarrow{-B}$; output $AG - B$	$A \xrightarrow{+}(\Sigma) \to A - \dfrac{B}{G} \to \boxed{G} \to AG - B$; feedback $\dfrac{B}{G} \leftarrow \boxed{\dfrac{1}{G}} \leftarrow B$
2	$A \to \boxed{G} \to AG$; branch $\to AG$	$A \to \boxed{G} \to AG$; $A \to \boxed{G} \to AG$
3	$A \to \boxed{G} \to AG$; branch $\to A$	$A \to \boxed{G} \to AG$; $AG \to \boxed{\dfrac{1}{G}} \to A$
4	$A \xrightarrow{+}(\Sigma) \to \boxed{G_1} \to B$; feedback $\boxed{G_2}$	$A \to \boxed{\dfrac{1}{G_2}} \xrightarrow{+}(\Sigma) \to \boxed{G_2} \to \boxed{G_1} \to B$
5	$A \xrightarrow{+}(\Sigma) \to \boxed{G_1} \to B$; feedback $\boxed{G_2}$	$A \to \boxed{\dfrac{G_1}{1+G_1 G_2}} \to B$

FIGURE 3.16
Summary of Block Diagram Algebra

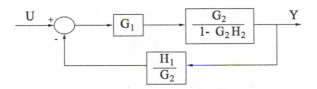

and finally to

$$U \to \boxed{\dfrac{\dfrac{G_1 G_2}{1-G_2 H_2}}{1+\dfrac{H_1}{G_2}\dfrac{G_1 G_2}{1-G_2 H_2}}} \to Y$$

Therefore, the transfer function is given by

$$\frac{Y(s)}{U(s)} = \frac{G_1 G_2}{1 - G_2 H_2 + H_1 G_1}.$$

(b) The design problem is to choose gains such that

$$\frac{Y(s)}{U(s)} = \frac{G_1 G_2}{1 - G_2 H_2 + H_1 G_1} = 1.$$

One set of gains that will satisfy this requirement is obtained by setting the gain products to 1, i.e.,

$$G_1 G_2 = G_2 H_2 = H_1 G_1 = 1.$$

For example, if G_1 is chosen to be 100, it then follows that

$$G_1 = 100 \Longrightarrow G_2 = \frac{1}{100} \text{ and } H_1 = \frac{1}{100}$$

$$G_2 = \frac{1}{100} \Longrightarrow H_2 = 100.$$

3.10.2   The Signal Flow Graph

The *signal flow graph* is another visual tool for representing causal relationships between components of the system. It is a simplified version of a block diagram introduced by S. J. Mason as a cause-and-effect representation of linear systems. In addition to the difference in physical appearances between the signal flow graph and the block diagram, the signal flow graph is constrained by more rigid mathematical rules. The signal flow graph is defined as a graphical means of portraying the input-output relationships between the variables of a set of linear algebraic equations. Consider a linear system described by a set of n algebraic equations such that,

$$y_j = \sum_{i=1}^{n} g_{ij} y_i \qquad \text{where } j = 1, 2, \ldots, n.$$

These n-equations are written in the form of cause-and effect relationships:

$$j\text{th effect} = \sum_{i=1}^{n} (\text{gain from } i \text{ to } j) \times (i\text{th cause})$$

or simply

$$\text{output} = \sum \text{gain} \times \text{input}.$$

This is the most important axiom in forming the set of algebraic equations for signal flow graphs. In the case where the system is represented by a set of differential equations, these equations must be first transformed into Laplace transform equations such that

$$Y_j(s) = \sum_{i=1}^{n} G_{ij}(s) Y_i(s) \quad \text{where } j = 1, 2, \ldots, n.$$

The basic elements of a signal flow graph can be summarized as follows:

- *Nodes*: Junction points that represent variables, internal signals, inputs or outputs

- *Paths (Branches)*: Segments connecting nodes, or sequences of connected blocks (route from one variable to another)

- *Forward path*: From input to output, such that no node is included more than once

- *Loop*: Closed path that returns to the starting node

- *Loop path*: The corresponding path

- *Path gain*: Product of component gains making up the path

- *Loop gain* : Gain associated with a loop path

Figures 3.17 and 3.18 illustrate block diagrams and their corresponding signal flow graphs.

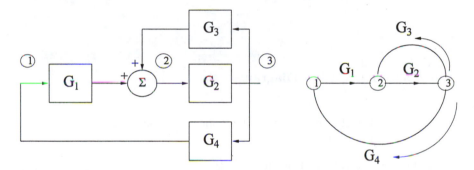

FIGURE 3.17
Block and Signal Flow Diagrams

3.10.3   Properties of Signal Flow Graphs

The properties of signal flow graphs can be summarized as follows:

- Signal flow graph analysis only applies to linear systems.

- The equations from which the graphs are drawn must be algebraic equations in the form of cause and effect.

- Nodes are used to represent variables and are normally arranged from left to right (from the input to the output) following a succession of cause-and-effect relationships.

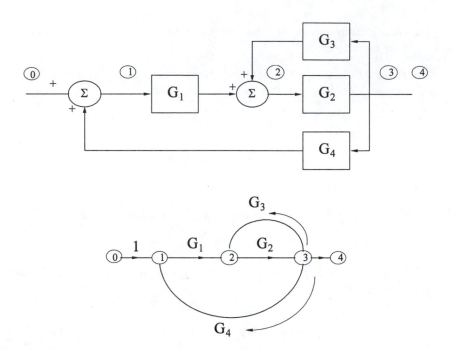

FIGURE 3.18
Block and Signal Flow Diagrams

- Signals travel along paths only in the direction described by the arrows of the paths.

- The path directing from node y_i to y_j represents the dependence of y_j upon y_i, but not the reverse.

- A signal y_i traveling along a path between y_i and y_j is multiplied by the gain of the path, g_{ij}, so that a signal $g_{ij}y_i$ is delivered at y_j.

3.10.4   Mason's Transfer Function Rule

Mason's transfer function rule states that, for a signal flow graph (or block diagram) with n forward paths and l loops, the transfer function between the input $r(t)$ and the output $y(t)$ is given by

$$T(s) = \frac{Y(s)}{R(s)}$$

$$= \frac{1}{\triangle} \sum_{i=1}^{n} G_i \triangle_i, \qquad (3.44)$$

where

$G_i \equiv$ *The path gain of the ith forward path.*

$\triangle \equiv$ *The system determinant*

$= 1 - \sum(all\ individual\ loop\ gains) +$

$\sum(gain\ products\ of\ all\ possible\ two\ non\text{-}touching\ loops) -$

$\sum(gain\ products\ of\ all\ possible\ three\ non\text{-}touching\ loops) + ...,$

$\triangle_i \equiv$ *The ith forward path determinant*

$=$ *The value of $\triangle$ for that part of the block diagram that does not touch the ith forward path.*

Due to the similarity between the block diagram and the signal flow graph, the transfer function formula given by Equation 3.44 can be applied to determine the input-output transfer function of either. In general, the transfer function formula can be applied directly to a block diagram. However, in complex systems, in order to identify all the loops and non-touching parts clearly, it is helpful to draw an equivalent signal flow graph to the block diagram before applying the gain formula.

In order to illustrate an equivalent signal flow graph of a block diagram and how the gain formula is applied to a block diagram, consider the block diagram shown in Figure 3.19 (a). The equivalent signal flow graph of the system is shown in Figure 3.19 (b). Notice that since a node on the signal flow graph is interpreted as a summing point of all incoming signals to the node, the negative feedbacks on the block diagram are represented by assigning negative gains to the feedback paths on the signal flow graph. The closed-loop transfer function of the system is obtained by applying the transfer function formula (Equation 3.44) to either the block diagram or the signal flow graph in Figure 3.19.

$$\frac{Y(s)}{R(s)} = \frac{G_1G_2G_3 + G_1G_4}{\Delta}$$

where

$$\Delta = 1 + G_1G_2H_1 + G_2G_3H_3 + G_1G_2G_3 + G_4H_2 + G_1G_4.$$

Therefore,

$$\frac{Y(s)}{R(s)} = \frac{G_1G_2G_3 + G_1G_4}{1 + G_1G_2H_1 + G_2G_3H_3 + G_1G_2G_3 + G_4H_2 + G_1G_4}.$$

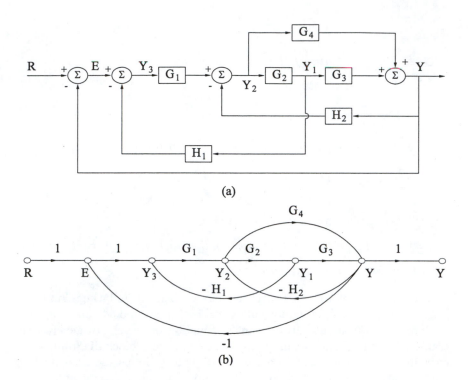

FIGURE 3.19
Block Diagram and Signal Flow Diagram

Similarly,

$$\frac{E(s)}{R(s)} = \frac{1 + G_1 G_2 H_1 + G_2 G_3 H_2 + G_4 H_2}{1 + G_1 G_2 H_1 + G_2 G_3 H_3 + G_1 G_2 G_3 + G_4 H_2 + G_1 G_4}$$

Therefore,

$$\frac{Y(s)}{E(s)} = \frac{Y(s)}{R(s)} \bigg/ \frac{E(s)}{R(s)}$$

$$= \frac{G_1 G_2 G_3 + G_1 G_4}{1 + G_1 G_2 H_1 + G_2 G_3 H_2 + G_4 H_2}.$$

3.11   Examples: Simplifying Block Diagrams

Example 3.28 *Determine the transfer function between $R(s)$ and $Y(s)$ in the following block diagram.*

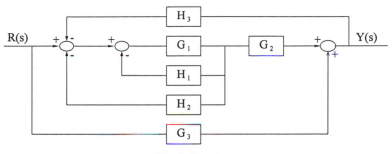

Block Diagram

Solution 3.28 *Simplify the internal closed-loop*

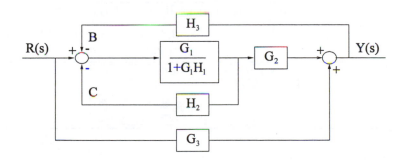

Add signal C, closed-loop and multiply before signal B

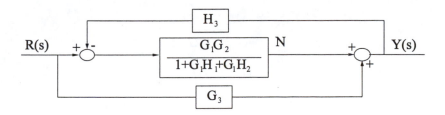

Move middle block N to the left side of the left summer.

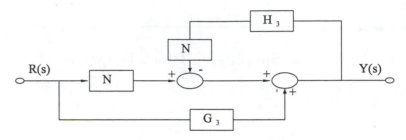

Now reverse the order of summers and close each block separately

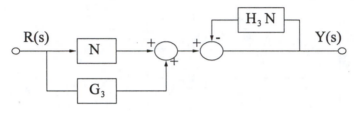

$$\frac{Y}{R} = (N + G_3)\left(\frac{1}{1 + NH_3}\right)$$

$$\frac{Y(s)}{R(s)} = \frac{G_1G_2 + G_3 + G_3G_1H_1 + G_3G_1H_2}{1 + G_1H_1 + G_1H_2 + G_1G_2H_3}.$$

The same result can be obtained by using Mason's rule. The starting point is establishing the forward path gains and loop gains.

Forward Path	Loops
$g_1 = G_1G_2$	$l_1 = -G_1H_1$
$g_2 = G_3$	$l_2 = -G_1H_2$
	$l_3 = -G_1G_2H_3$

$$\triangle = 1 - (-G_1H_1 - G_1H_2 - G_1G_2H_3)$$

$$= 1 + G_1H_1 + G_1H_2 + G_1G_2H_3$$

$$\triangle_1 = 1 - 0$$

$$\triangle_2 = 1 - (-G_1H_1 - G_1H_2)$$

$$= 1 + G_1H_1 + G_1H_2$$

$$\sum_{i=1}^{n} g_i\triangle_i = G_1G_2(1) + G_3(1 + G_1H_1 + G_1H_2)$$

$$= G_1G_2 + G_3 + G_3G_1H_1 + G_3G_1H_2$$

The overall transfer function is then obtained as follows:

$$\frac{Y(s)}{R(s)} = \frac{1}{\triangle} \sum_{i=1}^{n} g_i \triangle_i$$

$$= \frac{G_1 G_2 + G_3 + G_3 G_1 H_1 + G_3 G_1 H_2}{1 + G_1 H_1 + G_1 H_2 + G_1 G_2 H_3},$$

which is the same as the result obtained using block diagram algebra.

Example 3.29 *Find the transfer function of the system in the following block diagram.*

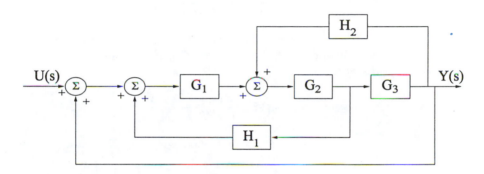

Block Diagram

Solution 3.29 *The block diagram is simplified as follows:*

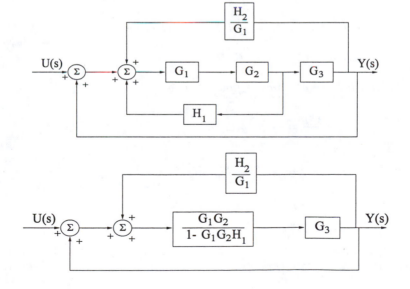

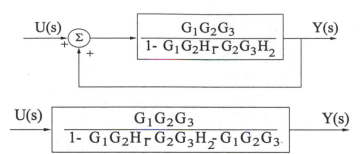

Therefore, the overall transfer function is given by

$$\frac{Y(s)}{U(s)} = \frac{G_1G_2G_3}{1 - G_1G_2H_1 - G_2G_3H_2 - G_1G_2G_3}.$$

The same result can be obtained using Mason's signal flow approach. The numerator of the overall transfer is the product of the transfer functions of the feed-forward path, i.e.,

$$num(s) = \sum_{i=1}^{n} g_i \triangle_i$$
$$= G_1G_2G_3.$$

The denominator $den(s)$ is given by

$$den(s) = \triangle$$

$$= 1 - \sum (\text{product of the transfer functions around each loop})$$

$$= 1 - (G_1G_2H_1 + G_2G_3H_2 + G_1G_2G_3)$$

$$= 1 - G_1G_2H_1 - G_2G_3H_2 - G_1G_2G_3.$$

Therefore, the overall transfer function is given by

$$\frac{Y(s)}{U(s)} = \frac{1}{\triangle} \sum_{i=1}^{n} g_i \triangle_i$$

$$= \frac{G_1G_2G_3}{1 - G_1G_2H_1 - G_2G_3H_2 - G_1G_2G_3}.$$

Notice that the positive feedback loop yields a negative term in the denominator.

Example 3.30 *Derive the transfer function, the input-output differential equation, and the state-variable matrix form for the following block diagram.*

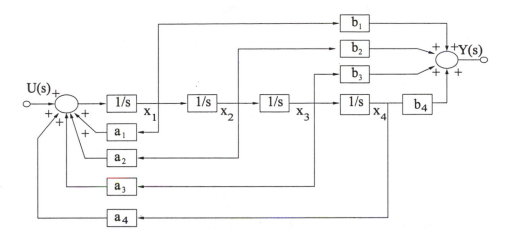

Block Diagram

Solution 3.30 *Forward path*

$$\frac{b_1}{s} + \frac{b_2}{s^2} + \frac{b_3}{s^3} + \frac{b_4}{s^4}$$

Feedback path

$$\frac{a_1}{s} + \frac{a_2}{s^2} + \frac{a_3}{s^3} + \frac{a_4}{s^4}$$

The general transfer function in terms of forward and feedback paths is given by

$$\frac{Y(s)}{U(s)} = \frac{Forward\ Path}{1 \pm\ Feedback\ Path},$$

where the +ve sign is for negative feedback and the −ve sign is for positive feedback. Since the system under consideration is a positive feedback one, the transfer function is obtained as follows:

$$\frac{Y(s)}{U(s)} = \frac{Forward\ Path}{1 -\ Feedback\ Path}$$

$$= \frac{\dfrac{b_1}{s} + \dfrac{b_2}{s^2} + \dfrac{b_3}{s^3} + \dfrac{b_4}{s^4}}{1 - (\dfrac{a_1}{s} + \dfrac{a_2}{s^2} + \dfrac{a_3}{s^3} + \dfrac{a_4}{s^4})}$$

$$= \frac{b_1 s^3 + b_2 s^2 + b_3 s + b_4}{s^4 - (a_1 s^3 + a_2 s^2 + a_3 s + a_4)}. \tag{3.45}$$

The differential equation is obtained from Equation 3.45.

$$Y(s)(s^4 - (a_1 s^3 + a_2 s^2 + a_3 s + a_4)) = U(s)(b_1 s^3 + b_2 s^2 + b_3 s + b_4)$$
$$\dddot{y} - a_1 \dddot{y} - a_2 \ddot{y} - a_3 \dot{y} - a_4 y = b_1 \dddot{u} + b_2 \ddot{u} + b_3 \dot{u} + b_4 u.$$

From the block diagram

$$\dot{x}_4 = x_3$$
$$\dot{x}_3 = x_2$$
$$\dot{x}_2 = x_1$$
$$\dot{x}_1 = u + a_1 x_1 + a_2 x_2 + a_3 x_3 + a_4 x_4$$
$$y = b_4 x_4 + b_3 x_3 + b_2 x_2 + b_1 x_1.$$

These equations can be expressed in matrix form as follows:

$$\dot{\mathbf{x}} = \begin{bmatrix} a_1 & a_2 & a_3 & a_4 \\ 1 & 0 & 0 & 0 \\ 0 & 1 & 0 & 0 \\ 0 & 0 & 1 & 0 \end{bmatrix} \begin{bmatrix} x_1 \\ x_2 \\ x_3 \\ x_4 \end{bmatrix} + \begin{bmatrix} 1 \\ 0 \\ 0 \\ 0 \end{bmatrix} u$$

$$y = \begin{bmatrix} b_1 & b_2 & b_3 & b_4 \end{bmatrix} \begin{bmatrix} x_1 \\ x_2 \\ x_3 \\ x_4 \end{bmatrix} + [0] u$$

Example 3.31 *Obtain a state space model for the following chemical processing plant.*

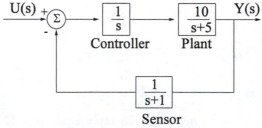

A Chemical Processing Plant

Solution 3.31 *The system involves one integrator and two delayed integrators. The output of each integrator or delayed integrator can be a state* *variable. Let the output of the plant be defined as x_1, the output of the controller as x_2, and the output of the sensor as x_3. It then follows that*

$$\frac{X_1(s)}{X_2(s)} = \frac{10}{s+5}$$

$$\frac{X_2(s)}{U(s) - X_3(s)} = \frac{1}{s}$$

$$\frac{X_3(s)}{X_1(s)} = \frac{1}{s+1}$$

$$Y(s) = X_1(s),$$

which can be rewritten as

$$sX_1(s) = -5X_1(s) + 10X_2(s)$$

$$sX_2(s) = -X_3(s) + U(s)$$

$$sX_3(s) = X_1(s) - X_3(s)$$

$$Y(s) = X_1(s).$$

By taking the inverse Laplace transforms of the preceding four equations, the state-variable equations are obtained as,

$$\dot{x}_1 = -5x_1 + 10x_2$$

$$\dot{x}_2 = -x_3 + u$$

$$\dot{x}_3 = x_1 - x_3$$

$$y = x_1.$$

Thus, a state space model of the system in the standard form is given by

$$\begin{bmatrix} \dot{x}_1 \\ \dot{x}_2 \\ \dot{x}_3 \end{bmatrix} = \begin{bmatrix} -5 & 10 & 0 \\ 0 & 0 & -1 \\ 1 & 0 & -1 \end{bmatrix} \begin{bmatrix} x_1 \\ x_2 \\ x_3 \end{bmatrix} + \begin{bmatrix} 0 \\ 1 \\ 0 \end{bmatrix} u$$

$$y = \begin{bmatrix} 1 & 0 & 0 \end{bmatrix} \begin{bmatrix} x_1 \\ x_2 \\ x_3 \end{bmatrix}.$$

It is important to know that this is not the only state space representation of the system, i.e., it is not a unique representation. Many other state space representations are possible. However, the minimum number of state variables is the same in any state space representation of the same system. In the present system, the number of state variables is three, regardless of what variables are chosen as state variables.

3.12   Problems

Problem 3.1 *Find and sketch the response of a dynamic system described by the equation*

$$6\dot{y} + y = f(t),$$

with the initial condition $y(0) = 4$ *when the input is:*
  (a) $f(t) = 10$
  (b) $f(t) = 5e^{-\frac{t}{2}}$
  (c) $f(t) = 10 + 5e^{-\frac{t}{2}}$
  (d) $f(t) = \sin t + \cos t$

Problem 3.2 *Use the definition and properties of Laplace transforms to find the Laplace transform of each of the following functions:*
  (a)

$$y_1(t) = te^{-at} + e^{-at}\cos \omega t$$

  (b)

$$y_2(t) = te^{-2t}\cos 3t + t^2 \sin 2t$$

  (c)

$$y_3(t) = \frac{d}{dt}(t^2 e^{-t}) + \int_0^t \lambda^2 e^{-\lambda}d\lambda$$

Problem 3.3 *Use partial fraction expansions to find the inverse Laplace transforms (the time system responses) of the following functions*
  (a)

$$Y_1(s) = \frac{5}{s(s+1)(s+10)}$$

  (b)

$$Y_2(s) = \frac{2s}{s^2 + 8s + 16}$$

  (c)

$$Y_3(s) = \frac{s^3 + 2s + 4}{s^4 - 16}$$

Problem 3.4 *(a) Find the inverse Laplace transform of the function*

$$Y(s) = \frac{3s^2 + 2s + 2}{(s+2)(s^2 + 2s + 5)},$$

by using a complete partial fraction expansion.

(b) Find the inverse Laplace transform of the same function by using the method of completing the square.

Problem 3.5 *(a) Show that the following rotational mechanical system is represented by the differential equation*

$$\dot{\omega}_1 + \left[\frac{b_1 + b_2}{J} \right] \omega_1 = \frac{b_1}{J} \omega(t).$$

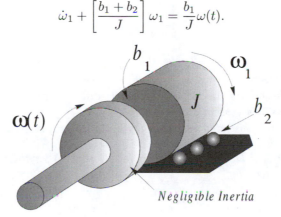

Rotational Mechanical System

(b) Given the following information: $\omega(t) = 10$, $b_1 = b_2 = 1$, $\omega_1(0) = 0$ and $J = 2$, find τ, the time constant and $\omega_{1_{ss}}$, the steady state speed.

(c) Deduce the system response (solution to the differential equation).

(d) Solve the same differential equation in (a) with the data in (b) by using Laplace transforms.

(e) Plot the system response and confirm your plot using the IVT and FVT.

Problem 3.6 *In the following mechanical system a velocity input, $v(t) = 6$ for $t \geq 0$, is applied as illustrated.*

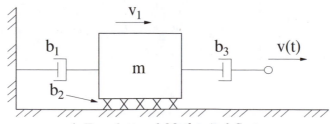

A Translational Mechanical System

(a) *Write the system's differential equation in terms of the velocity $v_1(t)$.*

(b) *What is the time constant τ and the steady state velocity $v_{1_{ss}}$ for the system?*

(c) *Given the information: $v_1(0) = 10$, $b_1 = b_2 = b_3 = 1$ and $m = 1$, obtain the system response.*

(d) *Sketch the system response.*

(e) *Solve the same differential equation in (a) with the data in (c), by using Laplace transforms.*

Problem 3.7 *(a) Derive the transfer function $H(s)$ for the general first-order system given by*

$$\dot{y} + \sigma y = ku,$$

where $y(t)$ is the output, $u(t)$ is the input, and all initial conditions are assumed to be zero.

(b) *Find the natural response of the first-order system.*

(c) *When is this system stable?*

(d) *Obtain the unit step response from the natural response in (b).*

Problem 3.8 *(a) Derive the transfer function $H(s)$ for the general second-order system given by*

$$\ddot{y} + 2\xi\omega_n\dot{y} + \omega_n^2 y = \omega_n^2 u,$$

where $y(t)$ is the output, $u(t)$ is the input, and all initial conditions are assumed to be zero.

(b) *Find the natural response of the second-order system.*

(c) *When is this system stable?*

(d) *Explain how the unit step response can be obtained from the natural response in (b).*

Problem 3.9 *Consider the RLC electrical circuit given below, where the input is current $i_i(t)$ and the output is voltage $v_o(t)$.*

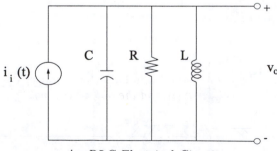

An RLC Electrical Circuit

(a) *The input-output differential equation for the circuit is given by*

$$C\ddot{v}_o + \frac{1}{R}\dot{v}_o + \frac{1}{L}v_o = \dot{i}(t).$$

Given that $R = \frac{1}{4}\Omega$, $L = \frac{1}{3}H$, $C = 1F$ *and assuming zero initial conditions, find the system transfer function* $H(s)$.

(b) *Find the impulse (natural) response of the system.*

(c) *Find the unit step response of the system.*

(d) *What is the relationship between the natural and unit step responses?*

Problem 3.10 *Consider the following diagram where the input is force* $f(t)$.

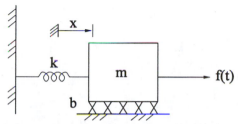

Translational Mechanical System

(a) *Obtain a differential equation in* x *that describes the behavior of the system.*

(b) *Use the following information:* $x(0) = 1$, $\dot{x}(0) = 0$, $m = 1$, $b = 4$, $k = 3$ *and* $f(t) = 9$, *use Laplace transforms to show that*

$$X(s) = \frac{s^2 + 4s + 9}{s(s^2 + 4s + 3)}.$$

(c) *Find the system response* $x(t)$ *and plot it.*

(d) *Use the IVT and FVT to check your plot.*

(e) *Given the following new information:* $x(0) = 0$, $\dot{x}(0) = 0$, $m = 1$, $b = 2$, $k = 5$ *and* $f(t) \equiv$ *unit impulse, show that*

$$X(s) = \frac{1}{s^2 + 2s + 5}.$$

Use the method of completing the square to find the response, $x(t)$ *and plot the system response.*

Problem 3.11 *The input-output differential equation of an electrical circuit whose input is a voltage* $v_i(t)$ *and the output is a voltage* $v_o(t)$, *is given by*

$$C\ddot{v}_o + \left(\frac{1}{R_1} + \frac{1}{R_2}\right)\dot{v}_o + \frac{1}{L}v_o = \frac{1}{R_1}\dot{v}_i.$$

(a) Find the expressions for the damping ratio ξ, and the undamped natural frequency ω_n.

(b) Given the information: $R_1 = 1\Omega$, $R_2 = \frac{1}{2}\Omega$, $C = 1F$, $L = \frac{1}{2}H$, $v_i(t) = 2t$ and assuming zero initial conditions, use Laplace transforms to solve the differential equation, i.e., obtain the system response.

(c) Plot the system response.

(d) Illustrate that the initial and final values of the response are the same as those obtained by using the IVT and FVT.

Problem 3.12 *Verify that the transfer function for the circuit shown below is given by*

$$H(s) = \frac{V_o(s)}{V_i(s)} = \frac{s^2 + 2s + 1}{s^2 + 4s + 4}.$$

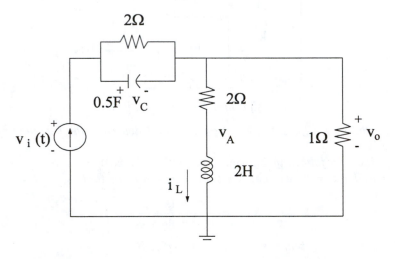

(b) Find the unit impulse response for the circuit.

(c) Find the unit step response for the circuit.

Problem 3.13 *Apply the initial and final value theorems to find $y(0^+)$ and $y(\infty)$ for each of the following transforms. If a theorem is not applicable to a particular transform, explain why this is so.*

(a)

$$Y(s) = \frac{s^3 + 2s + 4}{s(s+1)^2(s+2)}$$

(b)

$$Y(s) = \frac{4s^2 + 10s + 10}{s^3 + 2s^2 + 5s}$$

(c)

$$Y(s) = \frac{3(s^3 + 2s^2 + 4s + 1)}{s(s+3)^2}$$

(d)

$$Y(s) = \frac{s^3 - 4s}{(s+1)(s^2 + 4s + 4)}$$

Problem 3.14 *(a) Assuming zero initial conditions, find the transfer function for a system that obeys the equation*

$$\ddot{y} + 4\dot{y} + 4y = u(t).$$

(b) From the transfer function obtain the unit step response of the system.

(c) From the transfer function obtain the impulse response of the system.

(d) Differentiate the answer to part (a) and compare the result with the impulse response obtained in (c).

Problem 3.15 *A dynamic system is described by the state-variable equations* $\dot{\mathbf{x}} = \mathbf{A}\mathbf{x}$ *and* $\mathbf{y} = \mathbf{C}\mathbf{x}$, *where*

$$\mathbf{A} = \begin{bmatrix} 1 & 1 \\ 0 & -2 \end{bmatrix}, \quad \mathbf{C} = \begin{bmatrix} 2 & -1 \end{bmatrix}$$

and $\mathbf{x}(0) = \begin{bmatrix} 1 & 2 \end{bmatrix}^T$.

(a) Obtain the state-transition matrix $\phi(\mathbf{t})$.

(b) Find the state variable responses $x_1(t)$ *and* $x_2(t)$.

(c) Find the output response $y(t)$.

(d) For this system verify that:

$$\phi(\mathbf{0}) = \mathbf{I}$$
$$\phi^{-1}(\mathbf{t}) = \phi(-\mathbf{t})$$
$$\phi(\mathbf{1})\phi(\mathbf{2}) = \phi(\mathbf{3}).$$

Problem 3.16 *Simplify the following block diagram and obtain its overall transfer function,* $Y(s)/R(s)$.

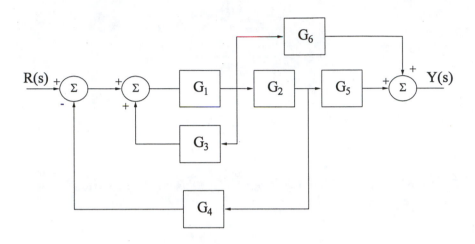

Block Diagram

Problem 3.17 *Show that the transfer functions obtained by using block diagram algebra and Mason's rule for the following block diagram are the same.*

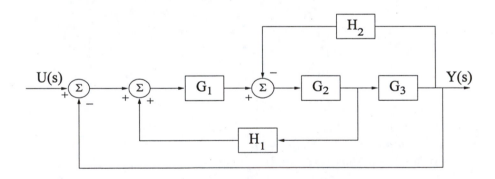

Block Diagram

Problem 3.18 *Simplify the following block diagram and obtain the closed-loop transfer function $Y(s)/R(s)$.*

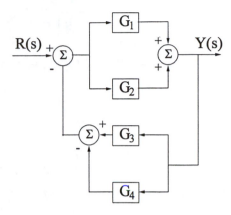

Block Diagram

Problem 3.19 *Simplify the following block diagram and obtain the closed-loop transfer function $Y(s)/R(s)$.*

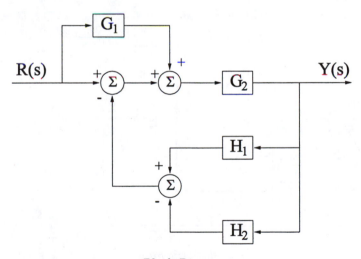

Block Diagram

Problem 3.20 *Simplify the following block diagram and obtain the closed-loop transfer function $Y(s)/R(s)$.*

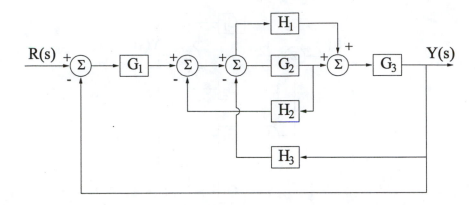

Block Diagram

Problem 3.21 *Derive the transfer function, the input-output differential equation and the state-variable matrix form for the following block diagram.*

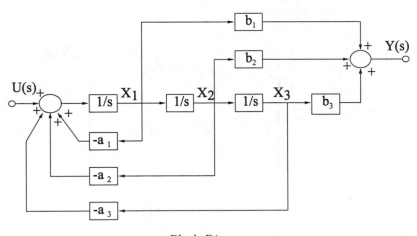

Block Diagram

Problem 3.22 *Use block diagram algebra or Mason's rule to determine the transfer function $Y(s)/R(s)$ for the following block diagram.*

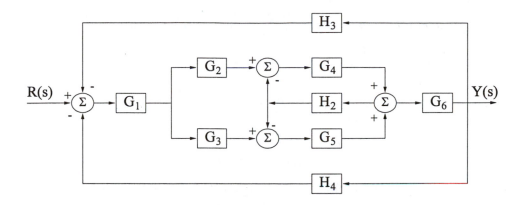

Block Diagram

Problem 3.23 *The following diagram shows a control system with conditional feedback. The transfer function $G(s)$ denotes the controlled process, and $D(s)$ and $H(s)$ are the controller transfer functions.*

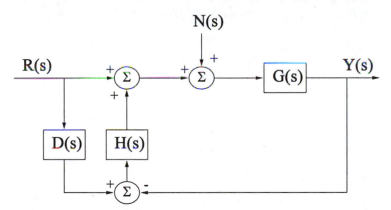

Conditional Feedback Control System

(a) Derive the transfer functions $Y(s)/R(s)|_{N=0}$ and $Y(s)/N(s)|_{R=0}$.

(b) Find $Y(s)/R(s)|_{N=0}$ when $D(s) = G(s)$.

(c) Let

$$G(s) = D(s) = \frac{100}{(s+1)(s+5)},$$

and find the output response $y(t)$ when $N(s) = 0$ and $r(t) = 1(t)$ (a unit step function).

(d) With $G(s)$ and $D(s)$ as given in part (c), select among the following choices of $H(s)$ such that when $n(t) = 1(t)$ and $r(t) = 0$, the steady state

value of $y(t)$ is equal to zero. (There may be multiple answers.)

$$H(s) = \frac{10}{s(s+1)} \qquad\qquad H(s) = \frac{10}{(s+1)(s+2)}$$

$$H(s) = \frac{10(s+1)}{(s+2)} \quad H(s) = \frac{K}{s^n} \ (n = positive\ integer,\ select\ n)$$

It is important to note that the poles of the closed-loop system must all be in the left half s-plane for the final value theorem to be valid.

Problem 3.24 (a) Draw a state diagram for the following state equations:

$$\frac{dx_1(t)}{dt} = -2x_1(t) + 3x_2(t)$$

$$\frac{dx_2(t)}{dt} = -5x_1(t) - 5x_2(t) + 2r(t)$$

(b) Find the characteristic equation of the system.
(c) Find the transfer functions $X_1(s)/R(s)$ and $X_2(s)/R(s)$.

Chapter 4

Characteristics of Feedback Control Systems

4.1 Introduction

The purpose of this chapter is to introduce the principles of feedback control systems and illustrate their characteristics and advantages. In several applications, there is a need to have automatic regulation or tracking. Quantities such as pressure, temperature, velocity, thickness, torque, and acceleration have to be maintained at desired levels. Feedback control is a convenient way in which these tasks can be accomplished. Control is the process of causing a system variable to conform to some desired value or reference value. A system is any collection of interacting components for which there are cause-and-effect relationships among the variables. The components are connected so as to form a whole entity that has properties that are not present in the separate entities. Within this context, a *control system* is then defined as an interconnection of interacting components forming a system configuration that will provide a desired system response.

Chapter 2 discussed the modeling of dynamic systems, while Chapter 3 dealt with obtaining the system response from the models. In this chapter the objective is to influence the dynamic system response by using feedback control. Feedback is the process of measuring the controlled variable and using that information to influence the controlled variable. Modern control engineering practice includes the use of control design strategies to improve manufacturing processes, the efficiency of energy use, advanced automobile control, and rapid transit systems. Feedback controllers are used in many different systems, from airplanes and rockets to chemical processing plants and semiconductor manufacturing. A feedback controller can be used to stabilize a system that is unstable in an open-loop configuration.

In this chapter, two case studies, the car cruise control system and the DC motor (both position and speed) control system are used to study, compare, and contrast the chief characteristics of open- and closed-loop control systems. The different types of controllers: Proportional (P), Proportional and Integral (PI), Proportional and Derivative (PD), Proportional and Integral and Derivative (PID) are discussed, together with their advantages and limitations. The concepts of system error, tracking, disturbance rejection, and system type are covered. The notions of sensitivity, bounded input-bounded output stability, asymptotic internal stability, and Routh-Hurwitz stability are discussed and illustrated using practical examples.

4.2   Open-Loop Control vs. Closed-Loop Control

4.2.1   Open-Loop Control

An open-loop control system utilizes a controller and actuator to obtain the desired response without monitoring the actual system response (controlled variable). This means that the objective of an open-loop control system is to achieve the desired output by utilizing an actuating device to control the process directly without the use of feedback. The elements of an open-loop control system are shown in Figure 4.1 and the block diagram representation is in Figure 4.2.

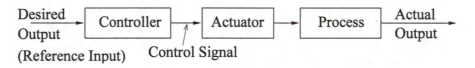

FIGURE 4.1
The Elements of an Open Loop Control System

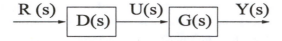

FIGURE 4.2
General Open Loop Control Block Diagram

$R(s)$ is the Laplace transform of the input $r(t)$, $Y(s)$ is the Laplace transform of the output $y(t)$, $D(s)$ is the controller transfer function, $G(s)$ is the plant transfer function, and $U(s)$ is the Laplace transform of the control signal $u(t)$.

4.2.2 Closed-Loop Control

In contrast to an open-loop control system, a closed-loop control system utilizes an additional measure of the actual output to compare the actual output with the desired output response. The measure of the output is called the feedback signal. The elements of a general closed-loop feedback control system are shown in Figure 4.3. A closed-loop control system compares a measurement of the output with the desired input (reference or command input). The difference between the two quantities is then used to drive the output closer to the reference input through the controller and actuator. The general block diagram model of a closed-loop control system is shown in Figure 4.4.

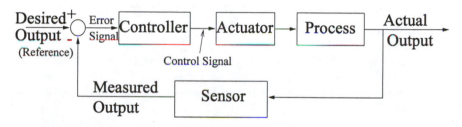

FIGURE 4.3
The Elements of a Closed Loop Control System

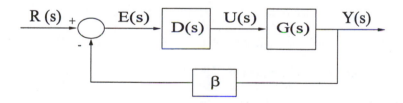

FIGURE 4.4
General Closed Loop Control Block Diagram

$R(s)$ is the Laplace transform of the input $r(t)$, $Y(s)$ is the Laplace transform of the output $y(t)$, $D(s)$ is the controller transfer function, $G(s)$ is the plant transfer function, $U(s)$ is the Laplace transform of the control signal $u(t)$, $E(s)$ is the Laplace transform of the error signal $e(t)$, and β is the sensor gain (or transfer function). For unit feedback control systems, $\beta = 1$.

4.2.3 Advantages of Closed-Loop Systems

Closed-loop systems have the following advantages:

- Faster response to an input signal

- Effective disturbance rejection

- Better tracking of reference signals

- Low sensitivity to system parameter errors (e.g., errors in plant or controller gains)

- Low sensitivity to changes in calibration errors (recalibration is unnecessary)

- More accurate control of plant under disturbances and internal variations

- Effective and flexible control tuning by varying the control gain

- Used to stabilize systems that are inherently unstable in the open-loop form

4.2.4 Disadvantages of Closed-Loop Systems

The following are some of the disadvantages of closed-loop systems:

- Require the use of sensors which increase the system costs

- Involve more components which leads to more costs and complexity

- The power costs (gain) are high

- More complex design, harder to build

- Less convenient when obtaining the output measurement is either hard or not economically feasible

- Initial tuning is more difficult, in particular if the bandwidth is narrow

- There is always a steady state error (with proportional controllers)

- Tend to become unstable as the gain is increased beyond certain limits

- Unnecessary when system inputs and the plant model are known with total certainty, and there are no external disturbances

- Not always controllable

4.2.5 Examples of Open- and Closed-Loop Systems

A variety of control systems (open and closed-loop) were outlined in Chapter 1. These include blood glucose control, manual car direction control, turntable speed control, automatic water level control and room temperature control. Two running examples are used in the next sections to illustrate and quantitatively compare open and closed-loop systems: Speed cruise control design for a car and control of a DC motor.

4.3 Car Cruise Control System (Open-Loop)

The model of the cruise control system for a car can be established in a relatively simple fashion. The car is modeled as a single translational mass where the rotational inertia of the wheels is neglected. The input to the car system is the applied engine force $f(t)$, which produces horizontal motion $x(t)$ at speed $v(t)$ and acceleration $\dot{v}(t)$. The frictional force $bv(t)$, which is proportional to the car's speed, opposes the car's motion. The model of the car's cruise control system is thus reduced to a simple mass and damper system as shown in Figure 4.5.

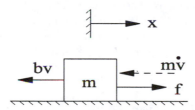

FIGURE 4.5
Car Cruise Control: A Single Mass Model

Using Newton's second law, the car's equation of motion is given by

$$m\dot{v} + bv = f$$
$$\dot{v} + \frac{b}{m}v = \frac{f}{m}. \tag{4.1}$$

The applied force $f(t)$ can be related to the desired speed or (reference speed) $v_r(t)$ by considering the steady state conditions, i.e., when all derivatives are equal to zero. The reference speed is equal to the steady state speed $v_{ss}(t)$, which is also the maximum attainable speed when a force $f(t)$ is applied to the car.

$$\frac{b}{m}v_{ss} = \frac{f}{m}$$

$$\Leftrightarrow v_{ss} = \frac{f}{b} = v_r.$$

For example, an applied force of $500N$, where $b = 50N \sec /m$, would mean that the reference speed v_r is $10m/\sec$.

4.3.1   Input-Output Form

The input-output Equation 4.1 can be rewritten with the reference speed $v_r(t)$ as the input as follows:

$$\dot{v} + \frac{b}{m}v = \frac{b}{m}v_r$$

$$\Leftrightarrow m\dot{v} + bv = bv_r. \tag{4.2}$$

This is a much more intuitive representation of the input-output differential equation as it clearly relates the desired output $v_r(t)$ and the actual output $v(t)$, where the two quantities are of the same type and dimension. Most textbooks consider Equation 4.1 as the input-output differential equation, but this does not make much sense when one has to design a car's cruise control system (open-loop or closed-loop). This is because under such a representation the system will have to be designed in terms of a reference force input $f(t)$ (as the desired output) while the variable being controlled (the actual output) is the car speed $v(t)$. Although this is technically correct, it is very unintuitive.

4.3.2   Transfer Function Form

The transfer function is obtained by taking the Laplace transform of the input-output Equation 4.2. When finding the transfer function, zero initial conditions are assumed.

$$\dot{v} + \frac{b}{m}v = \frac{b}{m}v_r$$

$$\implies sV(s) + \frac{b}{m}V(s) = \frac{b}{m}V_r(s)$$

$$\implies V(s)\left(s + \frac{b}{m}\right) = \frac{b}{m}V_r(s)$$

$$\implies \frac{V(s)}{V_r(s)} = \frac{\dfrac{b}{m}}{s + \dfrac{b}{m}}. \tag{4.3}$$

This is the open-loop transfer function of the system with respect to the reference or desired car speed.

4.3.3 Block Diagram Form

The block diagram model of a system (open-loop or closed-loop) can be derived from analyzing segments of the transfer function. Any general open-loop system can be represented as shown in Figure 4.6.

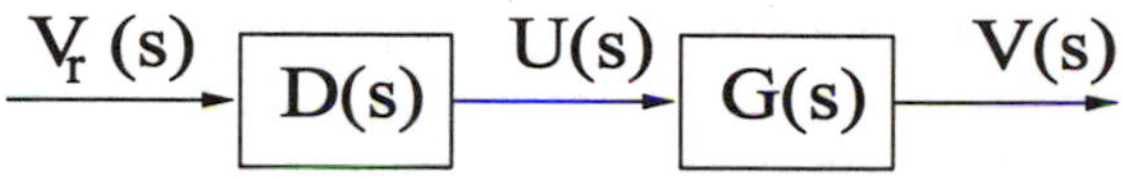

FIGURE 4.6
General Open-Loop Control System

The specific block diagram form is determined by pairing up elements of this generic diagram with the corresponding segments of the specific transfer function being considered. In the car's cruise control problem, the input is the reference car speed $v_r(t)$ and the actual car speed $v(t)$ is the output. The plant gain $G(s)$ is obtained from the transfer function in Equation 4.3 as

$$G(s) = \frac{\dfrac{b}{m}}{s + \dfrac{b}{m}}.$$

The controller is a unit proportional controller i.e. $D(s) = K = 1$, where

$$U(s) = D(s)V_r(s).$$

Hence, the block model for the open-loop car's cruise control system takes the form shown in Figure 4.7.

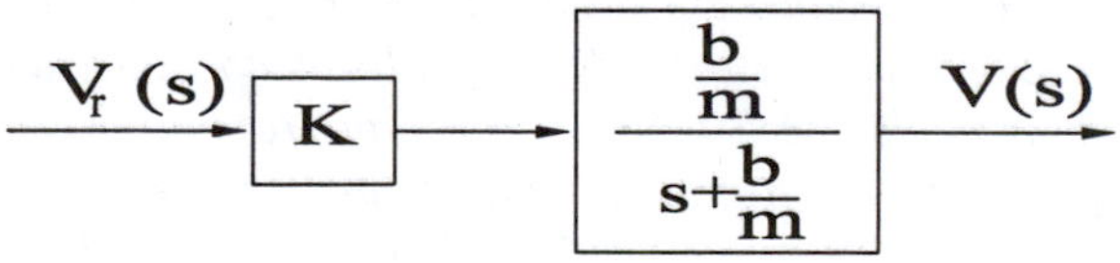

FIGURE 4.7
Open-Loop Cruise Control

4.3.4 State-Variable Form

The system has one independent energy storing element (the mass), therefore the minimum number of variables is one, and this variable is chosen as $v(t)$, the car speed.

$$\dot{x} = v$$
$$\dot{v} = -\frac{b}{m}v + \frac{b}{m}v_r.$$

The state-variable matrix form (with the output as the car velocity) is then given by

$$\dot{v} = \left[-\frac{b}{m}\right]v + \left[\frac{b}{m}\right]v_r$$

$$y = [1]\,v + [0]\,v_r.$$

If the car position $x(t)$ is also of interest, for example in a situation where car position control is an objective, then an extra redundant variable $x(t)$ is required to model the system. The state-variable matrix form (with both position and velocity as outputs) becomes

$$\begin{bmatrix} \dot{x} \\ \dot{v} \end{bmatrix} = \begin{bmatrix} 0 & 1 \\ 0 & -\frac{b}{m} \end{bmatrix} \begin{bmatrix} x \\ v \end{bmatrix} + \begin{bmatrix} 0 \\ \frac{b}{m} \end{bmatrix} [v_r] \tag{4.4}$$

$$\begin{bmatrix} y_1 \\ y_2 \end{bmatrix} = \begin{bmatrix} 1 & 0 \\ 0 & 1 \end{bmatrix} \begin{bmatrix} x \\ v \end{bmatrix} + \begin{bmatrix} 0 \\ 0 \end{bmatrix} v_r. \tag{4.5}$$

Example 4.1 *Open-Loop Cruise Control System Design: Consider a car whose mass, $m = 1000kg$ and $b = 50N\sec/m$, where the reference (input) speed of interest is $10m/\sec$. Design requirements: The desired speed of the car is of 10 m/s (22 m.p.h.). An automobile should be able to accelerate up to that speed in less than 5 seconds. Since this is only a cruise control system, a 10% overshoot on the velocity will not do much damage. A 2% steady state error is also acceptable for the same reason. Keeping the above in mind, the following design criteria for this problem is proposed: rise time < 5 sec, overshoot $< 10\%$, and steady state error $< 2\%$. The step function in MATLAB calculates the response to a step input (1m/\sec). Hence, the input matrix $\boldsymbol{B}$ will have to be multiplied by the magnitude of the reference speed, $10m/\sec$. As a result the MATLAB command is given by*
   step$(A,v_r*B,C,D,1,t)$.

Solution 4.1 *The state-variable matrices are obtain from Equations 4.4 and 4.5 and t is the time of observation.*

$$
A = \begin{bmatrix} 0 & 1 \\ 0 & -\dfrac{b}{m} \end{bmatrix} , \ B = \begin{bmatrix} 0 \\ \dfrac{b}{m} \end{bmatrix} , \ C = \begin{bmatrix} 1 & 0 \\ 0 & 1 \end{bmatrix} , \ D = \begin{bmatrix} 0 \\ 0 \end{bmatrix} . \quad (4.6)
$$

Figures 4.8 and 4.9 show the MATLAB plots of the position and speed responses of the car with respect to time.

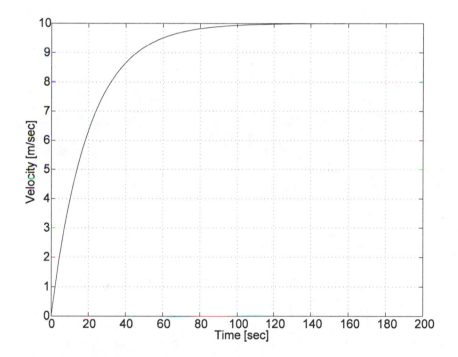

FIGURE 4.8
Car Cruise Control (Open-Loop System)

Figure 4.8 shows the open-loop car speed response. The car achieves the desired speed of 10m/ sec in about 110 sec (the settling time) without any steady state error nor overshoot. The settling time is too large and does not satisfy the rise time criterion of less than 5 seconds. However, the overshoot requirement (less than 10%) and the steady state error criterion (less than 2%) are satisfied. Figure 4.9 shows the car position response, which, after the settling time of 110 sec is a linear curve of gradient 10m/sec.

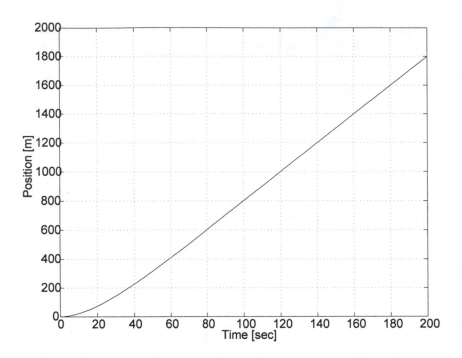

FIGURE 4.9
Car Cruise Control (Open-Loop System)

4.4   Car Cruise Control System (Closed-Loop)

The car's cruise control system considered so far has been open-loop.
There is no use of a sensor and hence there is no feedback of sensor information to correct errors in the car speed. A closed-loop car's cruise control
system engages a speed measuring sensor, the speedometer, to measure the
actual car speed. This measured speed is then compared with the desired
car speed and the difference between them is then sent to a controller that
seeks to reduce this system error. Consider the open-loop cruise control
model

$$\dot{v} + \frac{b}{m}v = \frac{b}{m}v_r.$$

In the closed-loop system the speed error, $[v_r(t)\text{-}v(t)]$, is used as the input
into the proportional controller K, instead of the reference speed, v_r. Thus,

the closed-loop model takes the form

$$\dot{v} + \frac{b}{m}v = \frac{b}{m}K(v_r - v). \tag{4.7}$$

4.4.1 Input-Output Form

The input to the whole system is still $v_r(t)$ while the output is still $v(t)$ and hence the input-output differential equation form is obtained by rearranging Equation 4.7.

$$\dot{v} + \frac{b}{m}v + \frac{bK}{m}v = \frac{b}{m}Kv_r$$

$$\dot{v} + \left(\frac{b}{m} + \frac{b}{m}K \right)v = \frac{b}{m}Kv_r. \tag{4.8}$$

4.4.2 Transfer Function Form

The transfer function form is obtained by taking the Laplace transform of the input-output Equation 4.8, while assuming zero initial conditions.

$$m\dot{v} + (b + bK)v = bKv_r.$$

Taking Laplace transforms it follows that

$$smV(s) + (b + bK)V(s) = bKV_r(s)$$

$$\implies \frac{V(s)}{V_r(s)} = \frac{bK}{ms + (b + bK)} \; . \tag{4.9}$$

This is the transfer function form of the closed-loop car's cruise control system with respect to the desired car speed.

4.4.3 Block Diagram Form

The block diagram form of the closed-loop system can be obtained in the same way as done in the open-loop system, that is, by comparing segments of the transfer function with a generic closed-loop block diagram. Any *closed-loop (negative feedback)* system can be represented by Figure 4.10, where β represents a sensor that measures the actual output.

The measured output is compared with the desired output, and the difference between the two is used to drive the controller $D(s)$. From the closed-loop equation

$$\dot{v} + \frac{b}{m}v = \frac{b}{m}K(v_r - v)$$

$$\Rightarrow \frac{m}{b}\dot{v} + v = K(v_r - v).$$

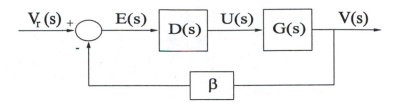

FIGURE 4.10
General Closed Loop Block Diagram

The block diagram is established by taking the Laplace transforms of this equation and then comparing the elements of the resulting transfer function with the generic block diagram in Figure 4.10.

$$\frac{m}{b}sV(s) + V(s) = K\left[V_r(s) - V(s)\right]$$

$$\Rightarrow V(s)\left[\frac{ms + b}{b}\right] = K\left[V_r(s) - V(s)\right],$$

where the speed error and the control signals in Laplace transforms are, respectively, given by

$$E(s) = V_r(s) - V(s) \tag{4.10}$$

$$U(s) = KE(s). \tag{4.11}$$

The output speed is thus obtained from the control signal by the equation

$$V(s) = \left[\frac{b}{ms + b}\right]U(s)$$

$$\Rightarrow V(s) = \left[\frac{\frac{b}{m}}{s + \frac{b}{m}}\right]U(s). \tag{4.12}$$

Putting the Equations 4.10, 4.11 and 4.12 in block diagram form produces Figure 4.11.

Comparing this closed-loop block diagram for the cruise control system with that of the corresponding open-loop system in Figure 4.7, it can be seen that the only difference between them is the *unit negative* feedback loop. In such a control system the exact actual output is compared with the desired output, and the difference between the two is used to drive the controller. The assumption is that a perfect sensor ($\beta = 1$) is used to measure the output. From the closed-loop block diagram the closed-loop transfer function can be deduced directly.

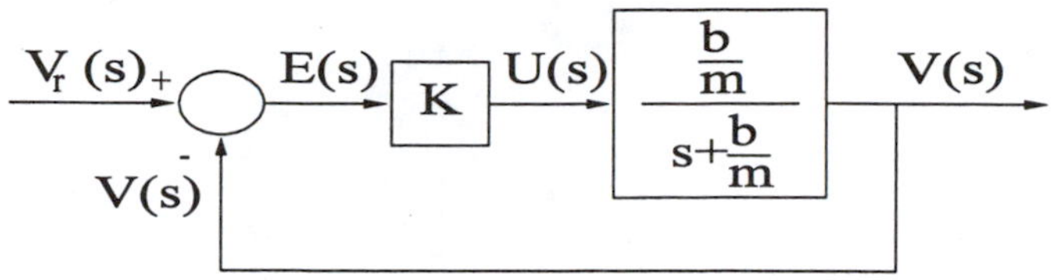

FIGURE 4.11
Closed Loop Cruise Control System

$$\frac{V(s)}{V_r(s)} = \frac{K\dfrac{\dfrac{b}{m}}{s + \dfrac{b}{m}}}{1 + K\dfrac{\dfrac{b}{m}}{s + \dfrac{b}{m}}}$$

$$\Rightarrow \frac{V(s)}{V_r(s)} = \frac{bK}{ms + (b + bK)}.$$

This equation is the same as Equation 4.9, thus effectively validating the block diagram shown in Figure 4.11.

4.4.4 State-Variable Form

Considering a single state variable, the car speed $v(t)$, the state-variable form is obtained from the closed-loop input-output differential equation,

$$\dot{v} + \left(\frac{b}{m} + \frac{b}{m}K\right)v = \frac{b}{m}Kv_r$$

$$[\dot{v}] = \left[-\frac{b(1 + K)}{m}\right][v] + \left[\frac{bK}{m}\right]v_r$$

$$y = [1]\,v + [0]\,v_r.$$

If motor position monitoring is also required then both $x(t)$ and $v(t)$ are chosen as state variables. In this case, there is a redundant state variable $x(t)$. Thus the state-variable matrix form of the position control system model (where both position and velocity are outputs) is represented as follows:

$$\begin{bmatrix} \dot{x} \\ \dot{v} \end{bmatrix} = \begin{bmatrix} 0 & 1 \\ 0 & -\dfrac{b(1+K)}{m} \end{bmatrix} \begin{bmatrix} x \\ v \end{bmatrix} + \begin{bmatrix} 0 \\ \dfrac{bK}{m} \end{bmatrix} [v_r]$$

$$\begin{bmatrix} y_1 \\ y_2 \end{bmatrix} = \begin{bmatrix} 1 & 0 \\ 0 & 1 \end{bmatrix} \begin{bmatrix} x \\ v \end{bmatrix} + \begin{bmatrix} 0 \\ 0 \end{bmatrix} v_r$$

Example 4.2 *Consider a car of similar characteristics and the same design requirements as those used for the open-loop system, i.e., $m = 1000kg$ and $b = 50N \sec /m$, where the reference (input) speed of interest is $10m/\sec$. The step function in MATLAB calculates the response to a step input $(1m/\sec)$. Hence, matrix **B** will correspondingly have to be multiplied by the magnitude reference speed $10m/\sec$. The MATLAB command is the same as that in the open-loop system, but the system matrices are different, as shown below*

Solution 4.2 $step(A, v_r {}^*B, C, D, 1, t)$

$$A = \begin{bmatrix} 0 & 1 \\ 0 & -\dfrac{b(1+K)}{m} \end{bmatrix}, \; B = \begin{bmatrix} 0 \\ \dfrac{bK}{m} \end{bmatrix}, \; C = \begin{bmatrix} 1 & 0 \\ 0 & 1 \end{bmatrix}, \; D = \begin{bmatrix} 0 \\ 0 \end{bmatrix} \qquad (4.13)$$

   Figures 4.12 and 4.13and show the MATLAB plots of the position and speed responses of the car with respect to time.

   In Figure 4.12 shows the car speed response for proportional control gain of $K = 20$. There is a steady state error of 5% from the desired speed of $10m/sec$, a settling time of 5sec and no overshoot. There is a dramatic improvement from the open-loop system on settling time (110sec to 5sec). The disadvantage of the proportional closed-loop control system with respect to the open-loop one is that there is always a steady state error, whereas there is no steady state error in the open-loop system. Although large values of K give low steady state errors, they involve high energy consumption, large (impractical) car accelerations and might lead to instabilities. Although it is desirable to reduce the steady state error, there has to be a trade-off between this objective and these negative tendencies. From the plot, it is clear that the closed-loop system satisfies all the three design requirements (rise-time, overshoot and steady state). Figure 4.13 shows the corresponding car position response, which, after the settling time (5sec), is a linear curve of gradient $9.5m/sec$.

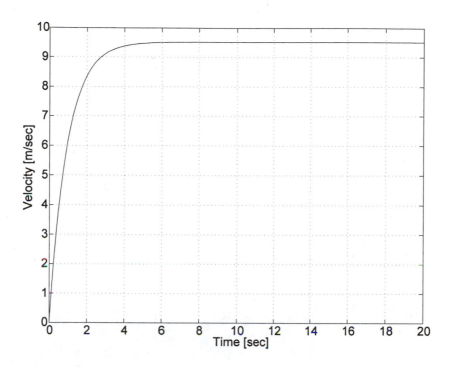

FIGURE 4.12
Car Cruise Control (Closed Loop System)

4.5 DC Motor Speed Control (Open-Loop)

A DC motor is a good example of a an electromechanical system, that is, a system that contains mechanical and electrical components interconnected to provide a composite function. It is a common actuator in a variety of engineering systems, including mechatronics and robotics. The DC motor directly provides rotary motion and, when coupled with cylinders and cables, it can provide translational motion. The electric circuit of the armature and the free-body diagram of the rotor are shown in Figure 4.14.

The modeling equations are obtained by using electrical circuit laws (KVL, KCL *etc*) and basic mechanics (Newton's laws). The armature is driven by the circuit shown in Figure 4.14 and the motor torque T is related to the armature current $i(t)$ by a constant factor K, while the rotor and shaft are assumed to be rigid. Hence, by summing up the torques on the rotor's free-body diagram in Figure 4.14, an expression for the current $i(t)$ in terms of the motor angular speed and acceleration is obtained as

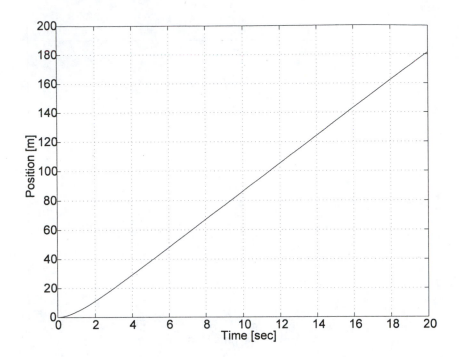

FIGURE 4.13
Car Cruise Control (Closed Loop Control)

follows:

$$J\ddot{\theta} + b\dot{\theta} \;\; = \;\; T = Ki$$

$$\Longleftrightarrow i = \frac{J\ddot{\theta} + b\dot{\theta}}{K}. \tag{4.14}$$

The back emf is related to the angular speed $\dot{\theta}(t)$ such that

$$L\frac{di}{dt} + Ri = v - K\dot{\theta}, \tag{4.15}$$

where $v(t)$ is the supply voltage. Equations 4.14 and 4.15 are the core
equations that can be used to produce different forms of the mathematical
model for a DC motor.

4.5.1   Input-Output Form

In this form of the model the system is expressed in terms of the input
voltage $v(t)$ and its derivatives, and the output angular speed $\omega(t)$ and its

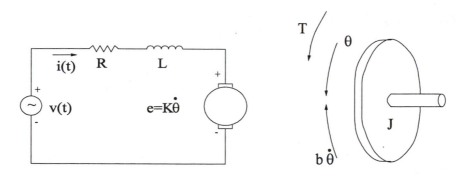

FIGURE 4.14
Circuit of the Armature and the Rotor Free-Body Diagram

derivatives, while eliminating all other variables.

$$L\frac{d}{dt}\left(\frac{J\ddot{\theta}+b\dot{\theta}}{K}\right) + R\left(\frac{J\ddot{\theta}+b\dot{\theta}}{K}\right) = v - K\dot{\theta}$$

$$\frac{JL}{K}\dddot{\theta} + \frac{Lb}{K}\ddot{\theta} + \frac{RJ}{K}\ddot{\theta} + \frac{Rb}{K}\dot{\theta} = v - K\dot{\theta}$$

$$\frac{JL}{K}\ddot{\omega} + \frac{Lb}{K}\dot{\omega} + \frac{RJ}{K}\dot{\omega} + \frac{Rb}{K}\omega = v - K\omega$$

$$\frac{JL}{K}\ddot{\omega} + \frac{Lb}{K}\dot{\omega} + \frac{RJ}{K}\dot{\omega} + \frac{Rb}{K}\omega + K\omega = v$$

$$JL\ddot{\omega} + Lb\dot{\omega} + RJ\dot{\omega} + Rb\omega + K^2\omega = Kv$$

$$(JL)\ddot{\omega} + (Lb + RJ)\dot{\omega} + (Rb + K^2)\omega = Kv. \qquad (4.16)$$

Equation 4.16 is the input-output differential equation model where the input is the voltage $v(t)$ and the output is the angular speed $\omega(t)$. Now, since the objective is to design a motor speed control system, it will be more useful to express the input-output differential equation with the desired (reference) motor speed $\omega_r(t)$ as the input. This reference speed (maximum attainable speed) can be deduced from Equation 4.16 as the steady state speed, which is a function of the input voltage, where all the derivatives are set to zero, i.e., $\ddot{\omega} = \dot{\omega} = 0$.

$$(Rb + K^2)\omega_{ss} = Kv$$

$$\Leftrightarrow w_r = w_{ss} = \frac{Kv}{(Rb + K^2)}$$

$$\Leftrightarrow v = \frac{(Rb + K^2)w_r}{K} . \tag{4.17}$$

Substituting this expression of $v(t)$ in Equation 4.16 gives the input-output differential equation model with respect to the desired DC motor speed.

$$(JL)\ddot{\omega} + (Lb + RJ)\dot{\omega} + (Rb + K^2)\omega = (Rb + K^2)w_r. \tag{4.18}$$

This is a much more useful form, as it depicts as the input the desired motor speed $\omega(t)$, where the motor speed $\omega(t)$ is the variable being controlled or monitored. Most textbooks consider Equation 4.16 as the input-output differential equation, which does not make much sense when one has to design a motor speed control system (open-loop or closed-loop). This is because under such a representation (Equation 4.16) the system will have to be designed in terms of a reference voltage input $v(t)$ (as the desired output) while the variable being controlled (the actual output) is the motor angular speed $\omega(t)$. While this is not incorrect and can be done, it is counter-intuitive and unnecessary.

4.5.2   Transfer Function Form

Two transfer functions can be obtained, one with respect to the input voltage and the other with respect to the desired motor speed. The transfer functions are obtained by taking the Laplace transform of the input-output Equations 4.16 and 4.18. First, the transfer function with respect to the input voltage is considered.

$$KV(s) = \left[s^2(JL) + s(Lb + RJ) + (Rb + K^2)\right]\Omega(s)$$

$$\implies \frac{\Omega(s)}{V(s)} = \frac{K}{(JL)s^2 + (Lb + RJ)s + (Rb + K^2)} .$$

Alternatively, this transfer function can be obtained by first finding the Laplace transforms of the angular speed $\Omega(s)$ and the voltage $V(s)$, and then finding the ratio,

$$\frac{\Omega(s)}{V(s)} = \frac{s\Theta(s)}{V(s)} \quad \text{(from } \omega = \dot{\theta}\text{).} \tag{4.19}$$

The expression for $V(s)$ in terms of $\Theta(s)$ is obtained by using the Laplace transform of the motor circuit Equation 4.15 and then substituting for the

Laplace transform of the motor current $I(s)$. The Laplace transform of Equation 4.15 gives

$$(Ls + R)I(s) = V(s) - Ks\Theta(s)$$
$$\Rightarrow V(s) = (sL + R)I(s) + Ks\Theta(s). \qquad (4.20)$$

The Laplace transform of the motor current $I(s)$ is obtained from Equation 4.14 as follows:

$$s(Js + b)\Theta(s) = KI(s)$$
$$\Longrightarrow I(s) = \frac{s(Js + b)\Theta(s)}{K}.$$

Substituting this expression of $I(s)$ in Equation 4.20 leads to

$$V(s) = (sL + R)\left(\frac{s(Js + b)\Theta(s)}{K}\right) + Ks\Theta(s)$$

$$= s\Theta(s)\left[(sL + R)\left(\frac{Js + b}{K}\right) + K\right]$$

$$= s\Theta(s)\left[\frac{(sL + R)(Js + b) + K^2}{K}\right]$$

$$\Longrightarrow \frac{\Omega(s)}{V(s)} = \frac{s\Theta(s)}{V(s)} = \frac{K}{(sL + R)(Js + b) + K^2}$$

$$\Leftrightarrow \frac{\Omega(s)}{V(s)} = \frac{K}{(JL)s^2 + (Lb + RJ)s + (Rb + K^2)}. \qquad (4.21)$$

Similarly the transfer function with respect to the desired motor speed $w_r(t)$ can be obtained by the two methods employed above. Taking Laplace transforms of the input-output Equation 4.18 gives

$$\frac{\Omega(s)}{\Omega_r(s)} = \frac{Rb + K^2}{(JL)s^2 + (Lb + RJ)s + (Rb + K^2)}. \qquad (4.22)$$

This transfer function is more intuitive and relevant to the DC motor speed control than that given in Equation 4.21. However, it is important to note that Equation 4.21 and Equation 4.22 have the same characteristic equation, which means that they are essentially describing the same dynamic system. From Equation 4.22, the block diagram forms of both open-loop and closed-loop speed control systems with the desired motor speed as reference can be easily derived.

4.5.3   Block Diagram Form (Open-Loop)

The block diagram model of a system can be derived by analyzing segments of the transfer function in Equation 4.22. Any general open-loop
system can be represented as shown in Figure 4.6. The specific block diagram form is determined by identifying what the elements of this generic
diagram correspond to in the particular system under consideration. In
the motor speed control problem, the input is the reference car speed $\omega_r(t)$
and the actual car speed $v(t)$ is the output. The plant function $G(s)$ is
obtained from the transfer function in Equation 4.22. The controller is a
unit proportional controller i.e. $D(s) = K_p = 1$. Thus, the block diagram
form for the motor speed control system is obtained as illustrated below in
Figure 4.15.

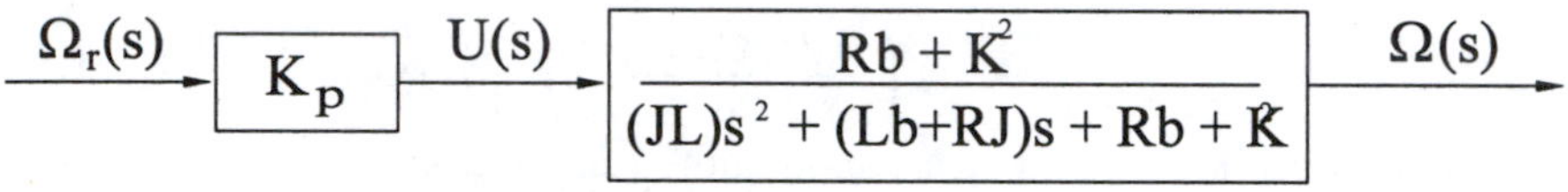

$$\Omega_r(s) \longrightarrow \boxed{K_p} \xrightarrow{U(s)} \boxed{\dfrac{Rb + K^2}{(JL)s^2 + (Lb+RJ)s + Rb + \check{K}}} \longrightarrow \Omega(s)$$

FIGURE 4.15
Open-Loop Motor Speed Control

4.5.4   State-Variable Form

There are two independent energy storing elements and hence two variables, the armature current $i(t)$ and the motor angular speed $\omega(t)$ are chosen. From the motor Equations 4.14 and 4.15 it follows that

$$\ddot{\theta} = -\frac{b}{J}\dot{\theta} + \frac{K}{J}i$$

$$\Leftrightarrow \dot{w} = -\frac{b}{J}w + \frac{K}{J}i$$

$$\dot{i} = -\frac{K}{L}\dot{\theta} - \frac{R}{L}i + \frac{1}{L}v$$

$$\Leftrightarrow \dot{i} = -\frac{K}{L}w - \frac{R}{L}i + \frac{1}{L}v.$$

Extracting the coefficients of the state variables and the input $v(t)$, produces the state-variable form

$$\begin{bmatrix} \dot{\omega} \\ \dot{i} \end{bmatrix} = \begin{bmatrix} -\dfrac{b}{J} & \dfrac{K}{J} \\[3mm] -\dfrac{K}{L} & -\dfrac{R}{L} \end{bmatrix} \begin{bmatrix} w \\ i \end{bmatrix} + \begin{bmatrix} 0 \\[2mm] \dfrac{1}{L} \end{bmatrix} [v]$$

$$y = \begin{bmatrix} 1 & 0 \end{bmatrix} \begin{bmatrix} \omega \\ i \end{bmatrix} + [0]\,[v].$$

This is the state-variable matrix system with respect to the input voltage $v(t)$. The state variable system using the desired motor speed as the input is obtained by substituting for $v(t)$ in the above matrix system, i.e., replacing it by the expression

$$v = \frac{(Rb + K^2)\omega_r}{K}.$$

The state-variable form, which is now more meaningful, becomes

$$\begin{bmatrix} \dot{\omega} \\ \dot{i} \end{bmatrix} = \begin{bmatrix} -\dfrac{b}{J} & \dfrac{K}{J} \\[3mm] -\dfrac{K}{L} & -\dfrac{R}{L} \end{bmatrix} \begin{bmatrix} w \\ i \end{bmatrix} + \begin{bmatrix} 0 \\[2mm] \dfrac{Rb + K^2}{LK} \end{bmatrix} [\omega_r] \qquad (4.23)$$

$$y = \begin{bmatrix} 1 & 0 \end{bmatrix} \begin{bmatrix} \omega \\ i \end{bmatrix} + [0]\,[\omega_r]. \qquad (4.24)$$

This is a much more intuitive and practical state-variable form of the motor speed control.

4.6 DC Motor Position Control (Open-Loop)

In order to achieve motor position control, an extra state variable θ is required in addition to the variables ω and i. From the motor Equations 4.14 and 4.15 it follows that

$$J\ddot{\theta} + b\dot{\theta} = Ki$$

$$L\frac{di}{dt} + Ri = v - K\dot{\theta}.$$

4.6.1   Input-Output Form

This form is obtained by proceeding just as for the speed control case but expressing the equations in terms of the angular displacement θ (and its derivatives) and not ω (and its derivatives). In particular from Equation 4.16 it follows that

$$(JL)\ddot{\omega} + (Lb + RJ)\dot{\omega} + (Rb + K^2)\omega = Kv$$

$$\Longleftrightarrow (JL)\dddot{\theta} + (Lb + RJ)\ddot{\theta} + (Rb + K^2)\dot{\theta} = Kv. \qquad (4.25)$$

Similarly, with respect to the desired speed the input-output differential equation is given by

$$(JL)\dddot{\theta} + (Lb + RJ)\ddot{\theta} + (Rb + K^2)\dot{\theta} = (Rb + K^2)\omega_r. \qquad (4.26)$$

4.6.2   Transfer Function

The transfer function can be obtained from that of the speed control system. From Equation 4.19 it follows that

$$\frac{\Omega(s)}{V(s)} = \frac{s\Theta(s)}{V(s)} \Leftrightarrow$$

$$\frac{\Theta(s)}{V(s)} = \frac{\Omega(s)}{sV(s)} = \frac{1}{s} \text{ (Speed control transfer function)}$$

$$= \frac{K}{s[(sL + R)(Js + b) + K^2]}$$

$$= \frac{K}{s\left[(JL)s^2 + (Lb + RJ)s + (Rb + K^2)\right]}.$$

The transfer equation with respect to the desired speed is obtained in a similar fashion.

$$\frac{\Omega(s)}{\Omega_r(s)} = \frac{s\Theta(s)}{\Omega_r(s)} \Leftrightarrow$$

$$\frac{\Theta(s)}{\Omega_r(s)} = \frac{\Omega(s)}{s\Omega_r(s)} = \frac{1}{s} \text{ (Speed control transfer function)}$$

$$= \frac{Rb + K^2}{s[(sL + R)(Js + b) + K^2]}$$

$$= \frac{Rb + K^2}{s\left[(JL)s^2 + (Lb + RJ)s + (Rb + K^2)\right]}.$$

4.6.3 State-Variable Form

In order to achieve motor position control an extra state variable $\theta(t)$ is required in addition to the variables $\omega(t)$ and $i(t)$.

$$\dot{\theta} = \omega$$

$$\ddot{\theta} = \dot{\omega} = -\frac{b}{J}\dot{\theta} + \frac{K}{J}i$$

$$\dot{i} = -\frac{K}{L}\omega - \frac{R}{L}i + \frac{Rb + K^2}{LK}\omega_r. \tag{4.27}$$

$$\begin{bmatrix} \dot{\theta} \\ \dot{\omega} \\ \dot{i} \end{bmatrix} = \begin{bmatrix} 0 & 1 & 0 \\ 0 & -\frac{b}{J} & \frac{K}{J} \\ 0 & -\frac{K}{L} & -\frac{R}{L} \end{bmatrix} \begin{bmatrix} \theta \\ \omega \\ i \end{bmatrix} + \begin{bmatrix} 0 \\ 0 \\ \frac{Rb + K^2}{LK} \end{bmatrix} [\omega_r] \tag{4.28}$$

$$y = \begin{bmatrix} 1 & 0 & 0 \end{bmatrix} \begin{bmatrix} \theta \\ \omega \\ i \end{bmatrix} + [0][\omega_r]. \tag{4.29}$$

If it is desired to control both the motor angular speed and its position, then the output Equation 4.29 is expanded to include the speed as follows:

$$\begin{bmatrix} y_1 \\ y_2 \end{bmatrix} = \begin{bmatrix} 1 & 0 & 0 \\ 0 & 1 & 0 \end{bmatrix} \begin{bmatrix} \theta \\ \omega \\ i \end{bmatrix} + \begin{bmatrix} 0 \\ 0 \end{bmatrix} [\omega_r]. \tag{4.30}$$

It is important to note that there are two independent energy storing elements in the DC motor system, the inductor (electrical) and the rotor (mechanical). Hence, when motor position control is required, the number of variables is greater than the number of independent energy storing elements. Thus, the third variable θ is a redundant state variable, which means the number of variables is not minimized. This fact is amply manifested by observing that the rows of matrix A in Equation 4.28 are not independent, whereas those of A in Equation 4.23 are independent. This means that for a system involving redundant states the rank of A is less than the number of variables, whereas for a system without redundant states the rank of A is the same as the number of variables. Put differently, A is not invertible for a system with redundant variables.

Example 4.3 *The system matrices are different as shown below. In this example, the following values for the physical parameters are assumed. These values were derived by experiment from an actual motor.*

moment of inertia of the rotor $J = 0.01 kg.m^2/s^2$

damping ratio of the mechanical system $b = 0.1 N ms$

electromotive force constant $K = 0.01 Nm/Amp$

electric resistance $R = 1\Omega$

electric inductance $L = 0.5H$

input (ω_r): desired motor angular speed

output (θ): position of shaft

Solution 4.3 *The system matrices for both speed and position control are given below*

$$
A = \begin{bmatrix} 0 & 1 & 0 \\ 0 & -\dfrac{b}{J} & \dfrac{K}{J} \\ 0 & -\dfrac{K}{L} & -\dfrac{R}{L} \end{bmatrix}, \ B = \begin{bmatrix} 0 \\ 0 \\ \dfrac{Rb + K^2}{LK} \end{bmatrix}, \ C = \begin{bmatrix} 1 & 0 & 0 \\ 0 & 1 & 0 \end{bmatrix}, \ D = \begin{bmatrix} 0 \\ 0 \end{bmatrix}.
$$

$$(4.31)$$

    Figure 4.16 and Figure 4.17 show the MATLAB plots of the speed and position responses of the motor with respect to time.

    Figure 4.16 shows the open-loop motor speed response. The motor achieves the desired speed of $1 rad/ \sec$ in about $5 \sec$ (the settling time) without any steady state error or overshoot. The settling time is too large and does not satisfy the rise time criterion of less than $1sec$. However, the overshoot requirement (less than 20%) and the steady state error criterion (less than 5%) are satisfied. Figure 4.17 shows the car position response which after the settling time of $5sec$ is a linear curve of gradient $1 rad/sec$.

4.7   DC Motor Speed Control (Closed-Loop)

4.7.1   Input-Output Form

    Consider the open-loop motor speed control input-output differential equation model,

$$(JL)\ddot{\omega} + (Lb + RJ)\dot{\omega} + (Rb + K^2)\omega = (Rb + K^2)w_r.$$

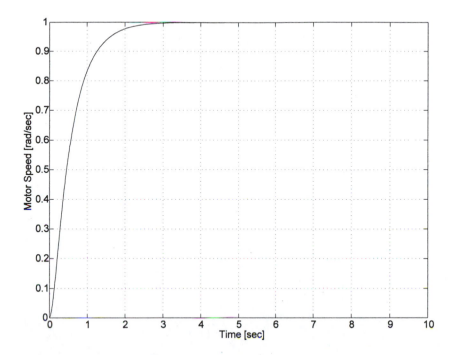

FIGURE 4.16
DC Motor Speed Control (Open-Loop)

In the closed-loop system, the motor speed error, $\omega_r(t) - \omega(t)$, is used as the input into the proportional controller K_p, instead of the reference speed, $\omega_r(t)$. The input to the system is still $\omega_r(t)$ while the output is still $v(t)$. Thus the closed-loop input-output differential equation model takes the form

$$(JL)\ddot{\omega} + (Lb + RJ)\dot{\omega} + (Rb + K^2)\omega = (Rb + K^2)K_P(\omega_r - \omega)$$

$$(JL)\ddot{\omega} + (Lb + RJ)\dot{\omega} + (Rb + K^2)(1 + K_P)\omega = (Rb + K^2)K_P\omega_r. \quad (4.32)$$

4.7.2 Transfer Function Form

The transfer function form is obtained by taking the Laplace transform of the input-output Equation 4.32 while assuming zero initial conditions.

$$\frac{\Omega(s)}{\Omega_r(s)} = \frac{K_P(Rb + K^2)}{(JL)s^2 + (Lb + RJ)s + (Rb + K^2)(1 + K_P)}. \quad (4.33)$$

From this transfer function equation the block diagram of the closed-loop speed control system with the desired motor speed as reference can be derived.

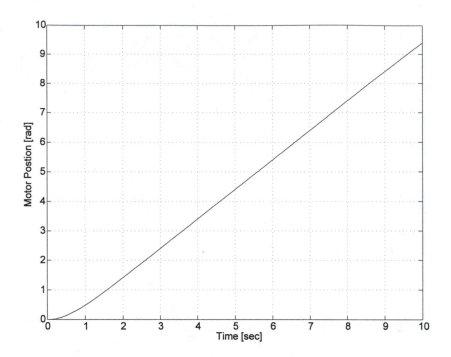

FIGURE 4.17
DC Motor Position Control (Open-Loop)

4.7.3 Block Diagram Form (Closed-Loop)

The block diagram form of the closed-loop system can be obtained in the same way as was done in the open-loop system by comparing segments of the transfer function with a generic closed-loop block diagram. The input is the reference motor speed $\omega_r(t)$ and the output is the actual motor speed $\omega(t)$. Any closed-loop motor speed control system can be represented by Figure 4.18. From the closed-loop input-output differential equation,

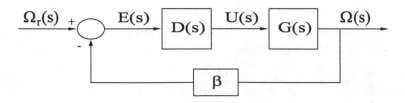

FIGURE 4.18
General Closed-Loop Control Block Diagram

$$(JL)\ddot{\omega} + (Lb + RJ)\dot{\omega} + (Rb + K^2)\omega = (Rb + K^2)K_P(w_r - w).$$

The block diagram is established by taking the Laplace transform of this equation and then comparing the elements of the resulting transfer function with the generic block diagram in Figure 4.18.

$$(JL)s^2\Omega(s) + (Lb + RJ)s\Omega(s) + (Rb + K^2)\Omega(s) = (Rb + K^2) \times$$
$$K_P[\Omega_r(s) - \Omega(s)]$$
$$\Omega(s)[(JL)s^2 + (Lb + RJ)s + (Rb + K^2)] = K_P(Rb + K^2)[\Omega_r(s) - \Omega(s)],$$

where the speed error and the control signals in Laplace transforms are given, respectively, as

$$E(s) = \Omega_r(s) - \Omega(s) \tag{4.34}$$

$$U(s) = K_P E(s). \tag{4.35}$$

The output speed is thus obtained from the control signal by the equation

$$\Omega(s) = \left[\frac{(Rb + K^2)}{(JL)s^2 + (Lb + RJ)s + (Rb + K^2)}\right] U(s). \tag{4.36}$$

Putting the Equations 4.34, 4.35, and 4.36 in block diagram form (with unit negative feedback, i.e., $\beta = 1$) produces Figure 4.19.

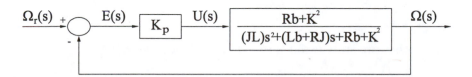

FIGURE 4.19
Closed-Loop Control Block Diagram (Proportional Control)

Compare this closed-loop block diagram for the motor speed control system with that of the corresponding open-loop system in Figure 4.15. It can be seen that the only difference between them is the unit negative feedback loop. From the closed-loop block diagram the closed-loop transfer function can be deduced directly.

$$\frac{\Omega(s)}{\Omega_r(s)} = \frac{K_P G(s)}{1 + K_P G(s)}$$

$$= \frac{\left[K_P \left[\dfrac{(Rb + K^2)}{(JL)s^2 + (Lb + RJ)s + (Rb + K^2)} \right] \right]}{\left[1 + K_P \left[\dfrac{(Rb + K^2)}{(JL)s^2 + (Lb + RJ)s + (Rb + K^2)} \right] \right]}$$

$$\Rightarrow \frac{\Omega(s)}{\Omega_r(s)} = \frac{K_P(Rb + K^2)}{(JL)s^2 + (Lb + RJ)s + (Rb + K^2)(1 + K_P)}.$$

Comparing this equation with Equation 4.33, it is clear that the two are the same, thus validating the block diagram.

4.7.4   State-Variable Form

The state-variable form for the closed-loop control is established in the same way as that for the open-loop system. The key difference is that the desired speed $w_r(t)$ in the state variable Equation 4.27 is replaced by $K_p[w_r(t) - w(t)]$ such that,

$$\dot{i} = -\frac{K}{L}w - \frac{R}{L}i + \frac{Rb + K^2}{LK}K_P[w_r - w]$$

$$\Leftrightarrow \dot{i} = -\left[\frac{K}{L} + \frac{K_P(Rb + K^2)}{LK} \right] w - \frac{R}{L}i + \frac{K_P(Rb + K^2)}{LK}w_r.$$

The other two state-variable equations are not affected,

$$\dot{\theta} = w$$

$$\dot{w} = -\frac{b}{J}w + \frac{K}{J}i.$$

Hence the closed-loop control system state-variable model, with both the motor angular speed $w_r(t)$ and angular position $\theta(t)$ as outputs, and the desired speed as the input, is given by the following two equations;

$$\begin{bmatrix} \dot{\theta} \\ \dot{\omega} \\ \dot{i} \end{bmatrix} = \begin{bmatrix} 0 & 1 & 0 \\ 0 & -\dfrac{b}{J} & \dfrac{K}{J} \\ 0 & -\left[\dfrac{K}{L} + \dfrac{K_P(Rb + K^2)}{LK}\right] & -\dfrac{R}{L} \end{bmatrix} \begin{bmatrix} \theta \\ \omega \\ i \end{bmatrix}$$

$$+ \begin{bmatrix} 0 \\ 0 \\ \dfrac{K_P(Rb + K^2)}{LK} \end{bmatrix} [\omega_r]$$

$$\begin{bmatrix} y_1 \\ y_2 \end{bmatrix} = \begin{bmatrix} 1 & 0 & 0 \\ 0 & 1 & 0 \end{bmatrix} \begin{bmatrix} \theta \\ \omega \\ i \end{bmatrix} + \begin{bmatrix} 0 \\ 0 \end{bmatrix} [\omega_r].$$

Example 4.4 *The system matrices are different as shown below. In this example, the following values for the physical parameters are assumed. These values were derived by experiment from an actual motor.*

moment of inertia of the rotor $J = 0.01 kgm^2/s^2$

damping ratio of the mechanical system $b = 0.1 Nms$

electromotive force constant $K = 0.01 Nm/Amp$

electric resistance $R = 1\Omega$

electric inductance $L = 0.5H$

input (ω_r): desired motor angular speed

output (θ): position of shaft

Solution 4.4 *The system matrices for both speed and position control can be obtained from the previous section. They are given by the following matrices:*

$$A = \begin{bmatrix} 0 & 1 & 0 \\ 0 & -\dfrac{b}{J} & \dfrac{K}{J} \\ 0 & -\left[\dfrac{K}{L} + \dfrac{K_P(Rb + K^2)}{LK}\right] & -\dfrac{R}{L} \end{bmatrix} , \ B = \begin{bmatrix} 0 \\ 0 \\ \dfrac{K_P(Rb + K^2)}{LK} \end{bmatrix}$$

$$C = \begin{bmatrix} 1 & 0 & 0 \\ 0 & 1 & 0 \end{bmatrix} , \ D = \begin{bmatrix} 0 \\ 0 \end{bmatrix} .$$

Figure 4.16 and Figure 4.17 show the MATLAB plots of the speed and position responses of the motor with respect to time.

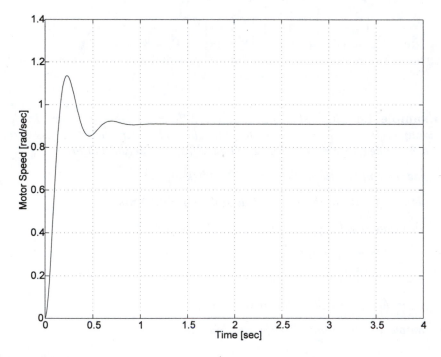

FIGURE 4.20
DC Motor Speed Closed-Loop Control (Proportional)

There is a steady state error of 10% from the desired motor speed of 1 rad/s, an overshoot of about 20% and a settling time of 1 second. A dramatic improvement is obtained over the open-loop system on settling time (3 sec to 1 sec). The disadvantage of the proportional closed-loop control system with respect to the open-loop one is that there is always a steady state error whereas there is no steady state error in the open-loop system. As the value of the proportional controller is increased, the steady state error is reduced, but there are practical limits to the increase of K_p. If this gain is too big this may lead to instability and impractical settling time (too short). Also, a large gain means excessive energy consumption. The motor position response is a curve, which, after the settling time of 1 sec is a linear curve of gradient 0.9 rad/sec.

4.8   Modeling of PID Controllers

Proportional feedback control can reduce error responses to disturbances, however, it still allows a non-zero steady state error. In addition, proportional feedback increases the speed of response but has a much larger transient overshoot. When the controller includes a term proportional to the integral of the error, then the steady state error can be eliminated, though this comes at the expense of the further deterioration in the dynamic response. Finally, addition of a term proportional to the derivative of the error can damp the dynamic response. Combined, these three kinds of control form the classical PID controller, which is widely used in the process industries and whose tuning rules have an interesting history. The establishment of the three-term controller is considered term by term.

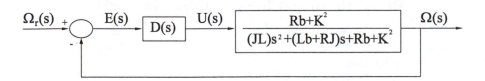

FIGURE 4.21
Closed-Loop Motor Speed Control (PID)

In the closed-loop system already discussed (Figure 4.21) the proportional controller K_P is replaced by a general controller $D(s)$ which can be a P, PI, PD or PID controller. The individual control components can be analyzed and their representation established.

4.8.1   Proportional Controller (P)

The proportional controller has already been developed in previous sections. The derivation of its model is presented here for completeness. When the feedback control signal is made to be linearly proportional to the error in the measured output, it is called proportional feedback. The general form of proportional control model is obtained as follows:

$$u(t) = K_P\left[r(t) - y(t)\right]$$
$$= K_P e(t)$$
$$U(s) = K_P E(s) \qquad \text{where} \quad E(s) = R(s) - Y(s)$$
$$D(s) = \frac{U(s)}{E(s)}$$
$$= K_P.$$

The proportional controller can be viewed as an amplifier with a "knob" to adjust the gain up or down. The system with proportional control may have a steady state offset (or drop) in response to a constant reference input and may not be entirely capable of rejecting a constant disturbance. For higher-order systems, large values of the proportional feedback gain will typically lead to instability. For most systems there is an upper limit on the proportional feedback gain in order to achieve a well-damped stable response, and this limit may still have an unacceptable steady state error. Therefore, there is a limit on how much the errors can be reduced by using proportional feedback only. One way to improve the steady state accuracy of control without adding extremely high proportional gains is to introduce integral control, which is discussed in the following section.

4.8.2   Proportional and Integral Controller (PI)

The primary reason for integral control is to reduce or eliminate constant steady state errors, but this benefit typically comes at the cost of worse transient response.

Integral feedback has the form

$$u(t) = K_P\left[r(t) - y(t)\right] + K_I \int_0^t \left[r(t) - y(t)\right] dt$$

$$= K_P e(t) + K_I \int_0^t e(t) dt$$

$$U(s) = K_P E(s) + \frac{K_I}{s} E(s)$$

$$= E(s)\left(K_P + \frac{K_I}{s}\right)$$

$$D(s) = \frac{U(s)}{E(s)}$$

$$= K_P + \frac{K_I}{s} \tag{4.37}$$

$$= \frac{K_P s + K_I}{s}. \tag{4.38}$$

This is how a generic PI controller is modeled, where the two parameters or gains K_P, and K_I are chosen to give the desired system dynamics. These gains are normally known as the proportional and integral gains, respectively. This feedback controller has the primary virtue that it can provide a finite value of control signal with no error signal input $e(t)$. This comes about because $u(t)$ is a function of all past values of $e(t)$ rather than just the current value, as in the proportional case. Therefore, past errors e "charge up" the integrator to some value that will remain, even if the error becomes zero and stays there.

Several limitations of proportional control are resolved by integral control. The steady state response to this class of load disturbance is completely eliminated. Thus, as long as the system remains stable, the system output equals the desired output regardless of the value of K_P. The final concern is with the dynamic response. If the designer wishes to increase the dynamic speed of response with large integral gain, then the response becomes very oscillatory. A way to avoid this behavior in some cases is to use both proportional and integral control at the same time.

In general, even though integral control improves the steady state tracking response, it has the effect of slowing down the response while the overshoot is kept unchanged. With both proportional and integral control, it can be seen that by choosing K_P and K_I the designer has independent control over two of the three terms in the characteristic terms and can provide better transient response than can be done with integral control alone.

4.8.3   Proportional and Derivative Controller (PD)

Derivative feedback (also called as rate feedback) has the form

$$K_D \frac{d}{dt} \left[r(t) - y(t) \right].$$

It is used in conjunction with proportional and/or integral feedback to increase the damping and generally improve the stability of the system. In practice, pure derivative feedback is not practical to implement, however, its approximations can be implemented. Another reason derivative feedback is not used by itself is that if the error signal $e(t)$ remains constant, then the output of the derivative controller would be zero and a proportional or integral term would be needed to provide a control signal at this time. In some cases, proportional and derivative control are combined to yield the Proportional and Derivative (PD) controller. The PD controller transfer function model is obtained as follows:

$$u(t) = K_P \left[r(t) - y(t) \right] + K_D \frac{d}{dt} \left[r(t) - y(t) \right]$$

$$= K_P e(t) + K_D \frac{de(t)}{dt}$$

$$U(s) = K_P E(s) + K_D s E(s)$$

$$= E(s) \left(K_P + K_D s \right)$$

$$D(s) = \frac{U(s)}{E(s)}$$

$$= K_P + K_D s.$$

In the derivative control the correction depends on the rate of change of error. As a result, a controller with derivative control exhibits an anticipatory response: Proportional-derivative behavior leads the proportion-only action by $\frac{1}{K_D}$ seconds. Derivative control may be introduced into the feedback loop in two ways, to the tachometer in a DC motor, or as a part of a dynamic compensator in the forward loop. In both cases the closed-loop characteristic equation is the same but the zeros from $r(t)$ to $y(t)$ are, of course, different; also, with the derivative in feedback, the reference is not differentiated, which may be a desirable result.

4.8.4   Proportional, Integral & Derivative Controller (PID)

For control over steady state and transient errors all the three control strategies discussed so far should be combined to get proportional-integral-derivative (PID) control. Here the control signal is a linear combination of

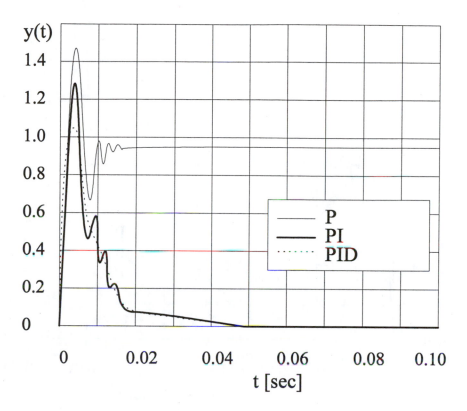

FIGURE 4.22
Transient Responses to a Step Disturbance

the error, the time integral of the error, and the time rate of change of the error. All three gain constants are adjustable. The PID controller contains all three control components (proportional, derivative, and integral). Its transfer function model can be derived as follows:

$$u(t) = K_P\left[r(t) - y(t)\right] + K_D \frac{d}{dt}\left[r(t) - y(t)\right] + K_I \int_0^t \left[r(t) - y(t)\right] dt$$

$$= K_P e(t) + K_D \frac{de(t)}{dt} + K_I \int_0^t e(t) dt$$

$$U(s) = K_P E(s) + K_D s E(s) + \frac{K_I}{s} E(s)$$

$$= E(s)\left(K_P + \frac{K_I}{s} + K_D s\right)$$

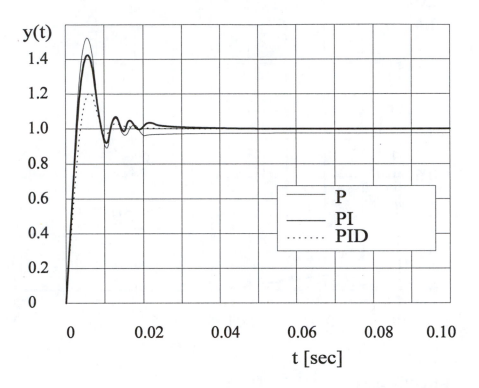

FIGURE 4.23
Transient Responses to a Step Reference Input

Therefore,

$$D(s) = \frac{U(s)}{E(s)}$$

$$= K_P + \frac{K_I}{s} + K_D s \tag{4.39}$$

$$= \frac{K_D s^2 + K_P s + K_I}{s}. \tag{4.40}$$

Thus, the corresponding block diagram representation takes the form shown in Figure 4.24. This is how a generic PID controller is modeled where the three parameters or gains K_P, K_I, and K_D are chosen to give the desired system dynamics. These gains are normally known respectively as the proportional, integral, and derivative gains.

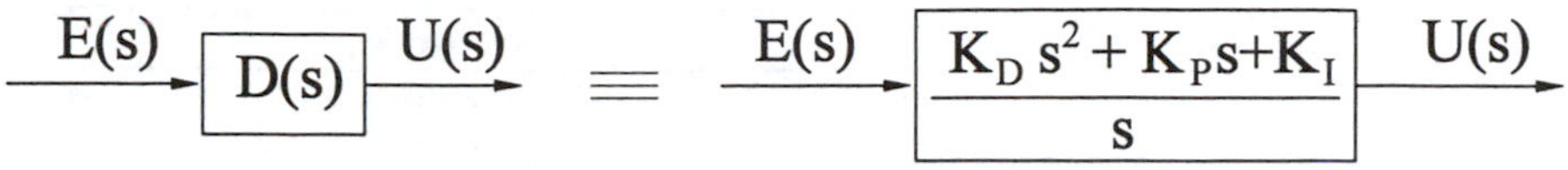

$$\xrightarrow{\;E(s)\;}\boxed{D(s)}\xrightarrow{\;U(s)\;}\;\;\equiv\;\;\xrightarrow{\;E(s)\;}\boxed{\dfrac{K_D s^2 + K_P s + K_I}{s}}\xrightarrow{\;U(s)\;}$$

FIGURE 4.24
PID Controller: Block Diagram Form

In order to design a particular control loop the engineer merely has to adjust the constants K_P, K_I and K_D in the above equation to arrive at acceptable performance. This adjustment process is called tuning the controller. Increasing K_P and K_I tends to reduce system errors but may not be capable of also producing adequate stability, while increasing K_D tends to improve stability. For example, the characteristic equation will have three coefficients and three parameters (K_P, K_I and K_D), and thus in theory the poles of such a system can be set wherever desired. The combination of the three control components in this system yields complete control over the system dynamics. The PID controller provides both an acceptable degree of error reduction and an acceptable stability and damping. PID controllers are so effective that PID control is standard in processing industries such as petroleum refining, papermaking, and metalworking.

An alternative and convenient form of the PID controller is obtained by expressing the three controller gains (K_P, K_I and K_D) in terms of one controller gain and two time constants (K_P, T_I and T_D) as follows:

$$D(s) = K_P + \frac{K_I}{s} + K_D s \tag{4.41}$$

$$= K_P\left(1 + \frac{1}{T_I s} + T_D s\right), \tag{4.42}$$

where the three parameters (K_P, T_I and T_D) now give a complete description of the PID controller. K_P is the *proportional gain (as before)*, T_I is the *integral (reset) time constant*, and T_D is the *derivative time constant*. The two PID controller models given in Equations 4.41 and 4.42 are used interchangeably where,

$$K_I = \frac{K_P}{T_I} \text{ and } K_D = K_P T_D.$$

The alternative models for the PI and PD controllers are easily deduced

Controller	Benefits	Drawbacks
P (K_P)	simple cheap	steady state error large overshoot & settling time poor transient response prone to instability (large gains)
I (K_I)	eliminates steady state error	poor damping large overshoot & settling time poor transient response reduces stability
D (K_D)	increases damping reduces overshoot reduces settling time improved transients improves stability	steady state error cannot be used alone

Table 4.1   Summary of PID Controller Characteristics

from Equation 4.42 as follows:

$$D(s) = K_P + \frac{K_I}{s} \qquad \text{(PI)}$$

$$= K_P \left(1 + \frac{1}{T_I s} \right) \qquad (4.43)$$

$$D(s) = K_P + K_D s \qquad \text{(PD)}$$

$$= K_P \left(1 + T_D s \right). \qquad (4.44)$$

Using the models of PID controllers that have been exhaustively developed in this section, the characteristics of the different types of these controllers (P, PI, PD and PID) can be studied, illustrated, and verified. This is achieved by implementing the controllers in MATLAB for systems with simple reference inputs and disturbances such impulse, step, and ramp functions

4.8.5   Summary of PID Controller Characteristics

The benefits and limitations of the three components of a PID controller (P, I and D) are summarized by studying the system responses of a system to a unit step disturbance (Figure 4.22) and unit step reference input (Figure 4.23), when three controllers; P, PI and PID, are used. The effects of the three components are deduced from the plots in Figure 4.22 and 4.23 and summarized in Table 4.1.

Example 4.5 *Consider the DC closed-loop motor control system considered in previous examples. Instead of the K_P use the $D(s) = PI$, PD or*

PID. The MATLAB code is shown in a later section.

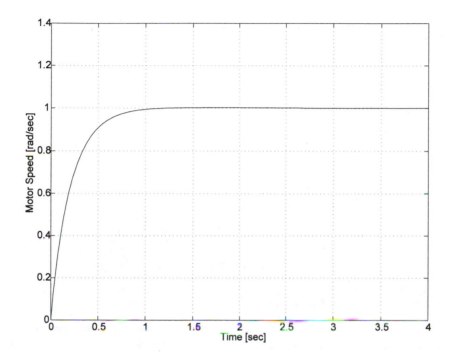

FIGURE 4.25
DC Motor Speed Closed-Loop Control (PID)

Solution 4.5 *The overshoot is reduced to 0%, the response time is reduced to 1sec, the steady state error is reduced to 0%. There is also improved stability, damping, and general system response. The motor position response is a curve, which, after the settling time of 1 sec, is a linear curve of gradient 1rad/sec (the desired motor speed). These results demonstrably illustrate the benefits of the three elements of the PID controller and how they compensate for each other's limitations. The state steady error introduced by the proportional controller (P), is eliminated by the integrator (I). The poor system transient response and poor stability introduced by the integrator (I) are resolved by the derivative controller (D), which improves both system damping and stability.*

4.9   MATLAB Implementation

4.9.1   State-Variable Form

If the state-variable matrices (A, B, C, D) are available for the open-loop and closed-loop systems, then the system response of the system to different input functions such as step, impulse, and ramp are easily obtained in MATLAB (Appendix B). For example, with a step reference input, $w_r(t) =$ constant, the system responses with respect to time t are obtained as follows:

$$y_{ol} = step(A_{ol}, w_r * B_{ol}, C_{ol}, D_{ol}, 1, t);$$
$$y_{cl} = step(A_{cl}, w_r * B_{cl}, C_{cl}, D_{cl}, 1, t);$$

where the number 1 indicates that there is one input. In most complicated closed-loop systems the state-variable matrices are not readily available. In these cases it is easier to use the transfer function model.

4.9.2   Transfer Function

From the open-loop or closed-loop transfer function, which is expressed as a ratio of two polynomials in s, the numerator and denominator consisting of coefficients of these polynomials are obtained: *numol* and *denol* for open-loop systems, *numcl* and *dencl* for closed-loop systems. These are then used to simulate the response of the system to input different functions such as step, impulse and ramp. For example, for a step input the open-loop system response is obtained as follows:

$$T_{ol} = D(s)G(s) = \frac{n(s)}{m(s)} \Rightarrow \frac{\text{numol}}{\text{denol}}$$

$$y_{ol} = step(numol, denol, t).$$

Using the closed-loop transfer function, the closed-loop system response for a step input is similarly computed.

$$T_{cl} = \frac{D(s)G(s)}{1 + \beta D(s)G(s)} = \frac{q(s)}{p(s)} \Rightarrow \frac{\text{numcl}}{\text{dencl}}$$

$$y_{cl} = step(numcl, dencl, t).$$

It is not necessary to completely establish the open-loop or closed-loop transfer function before MATLAB implementation. MATLAB can be used to compute these from the Controller $D(s)$, plant gain $G(s)$, by using the principle of convolution. Unity feedback is assumed ($\beta = 1$). The plant gain $G(s)$ is represented by

$$G(s) = \frac{b(s)}{a(s)} \Rightarrow \frac{\text{num}}{\text{den}}.$$

The controller $D(s)$ is represented by the block diagram in Figure 4.24 and hence its transfer function can be expressed as follows:

$$D(s) = \frac{K_D s^2 + K_P s + K_I}{s} = \frac{g(s)}{f(s)} \Rightarrow \frac{\text{numcon}}{\text{dencon}}.$$

Therefore, $[numol, denol]$ can be calculated from $[num, den]$ and $[numcon, dencon]$, and then $[numcl, dencl]$ from $[numol, denol]$. Hence the general sequence of MATLAB commands will be as follows:

num=[- - -];
den=[- - -];
numcon=[K_D, K_P, K_I];
dencon=[1 0];
numol=conv(num,numcon);
denol=conv(den,dencon);
[numcl,dencl]=cloop(numol,denol);
y_ol=step(numol,denol,t);
y_cl=step(numcl,dencl,t);.

4.9.3 Sample MATLAB Code: Motor Speed PID Control

The sample MATLAB code given below is that of the closed-loop control system for the DC motor discussed in section using a PID controller. The results of this implementation have already been presented. This code helps to illustrate how control systems can be implemented using MATLAB.

% DC Motor Speed Closed-Loop Control (PID)
t=input('Input the amount of time to look at in sec:');
J=0.01;   b=0.1;   K=0.01;   R=1;   L=2;
K_P=10; K_I=5; K_D=1;
num=[(R*b)+(K^2)];
den=[(J*L) ((J*R)+(L*b)) ((b*R)+K^2)];
numcon=[K_D, K_P, K_I];
dencon=[1 0];
numol=conv(num,numcon);
denol=conv(den,dencon);
[numcl,dencl]=cloop(numol,denol);
t = 0.:0.01:t;
y=step(numcl,dencl,t);

title('DC Motor Speed Closed-Loop Control (PID) (PID)')
plot(t,y),grid
xlabel('Time [sec]')
ylabel('Motor Speed [rad/sec]').

4.10   Tuning of PID Controllers

Methods are available to develop a controller that will meet steady state
and transient specifications for both tracking input references and rejecting disturbances. These methods require that control of the process use
complete dynamic models in the form of equations of motion or transfer
functions. Ziegler and Nichols gave two methods for tuning the controller
for such a model.

4.10.1   Quarter Decay Ratio Method

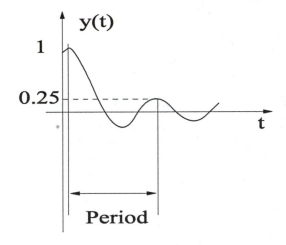

FIGURE 4.26
Quarter Decay Ratio

In the first method, the choice of controller parameters is based on a
decay ratio of approximately 0.25. This means that a dominant transient
decays to a quarter of its value after one period of oscillation as shown
in Figure 4.26. A quarter decay corresponds to $\xi = 0.21$ and is a good
compromise between quick response and adequate stability margins. A
large number of process control systems exhibit a process reaction curve.

Type of Controller	Optimum Gain
P	$K_P = 1/RL$
PI	$\begin{cases} K_P = 0.9/RL \\ K_I = 0.27/RL^2 \end{cases}$
PID	$\begin{cases} K_P = 1.2/RL \\ K_I = 0.6/RL^2 \\ K_D = 0.6/R \end{cases}$

Table 4.2 PID Controller Tuning: Quarter Decay Ratio

The slope of the curve is represented by R, and the intersection line with the time axis identifies the time delay $L = t_d$. The equations are simulated for the system on an analog computer and the control parameters adjusted until the transients showed a decay of 25% in one period. The regulator parameters suggested by Ziegler and Nichols for the common controller terms are shown in Table 4.2. It is important to note that the general expression for PID controller used is given by

$$D(s) = K_P + \frac{K_I}{s} + K_D s.$$

4.10.2 Stability Limit Method

In the second method the criteria for adjusting the parameters are based on evaluating the system at the limit of stability rather than on taking a step response. The proportional gain is increased until continuous oscillations are observed, that is, until the system becomes marginally stable. The

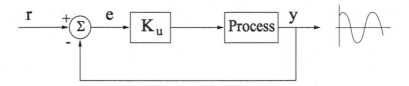

FIGURE 4.27
Determination of Ultimate Gain and Period

corresponding gain K_u (also called the ultimate gain) and the period of

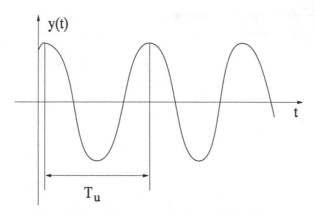

FIGURE 4.28
Marginally Stable System

oscillation T_u (also called the ultimate period) are determined as shown in Figure 4.27. Figure 4.28 shows a marginally stable system, and T_u should be measured when the amplitude of oscillation is quite small. Then the final parameters are chosen as shown in Table 4.3.

Experience has shown that the controller setting according to Ziegler-Nichols rules provide a good closed-loop response for many systems. The process operator can do the final tuning of the controller iteratively to yield satisfactory control.

4.11   Steady State Tracking and System Type

System types can be defined with respect to a reference $r(t)$ or a disturbance $w(t)$.

Consider the generic feedback system shown in Figure 4.29. Let the transfer function from the reference $r(t)$ to the output $y(t)$ be $T_r(s)$, and the transfer function from the disturbance $w(t)$ to the output $y(t)$ be $T_w(s)$. For a generic reference, $r(t) = \dfrac{t^k}{k!}$, it will be shown that the steady state error with respect to the reference is given by

$$e_{ss}^r = \lim_{s \to 0} \frac{(1 - T_r)}{s^k}.$$

Similarly, for a generic disturbance, $w(t) = \dfrac{t^k}{k!}$, it will be shown that the

Type of Controller	Optimum Gain
P	$K_P = 0.5K_u$
PI	$\begin{cases} K_P = 0.45K_u \\ K_I = 0.54K_u/T_u \end{cases}$
PID	$\begin{cases} K_P = 0.6K_u \\ K_I = 1.2K_u/T_u \\ K_D = 0.075K_uT_u \end{cases}$

Table 4.3 PID Controller Tuning: Stability Method

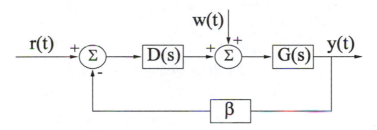

FIGURE 4.29
A General Control System

steady state error with respect to the disturbance is given by

$$e_{ss}^w = \lim_{s \to 0} \frac{-T_w}{s^k}.$$

The system type can then be defined in terms of these steady state errors e_{ss}^r, and e_{ss}^w. The general transfer function with respect to the reference input, $T_r(s)$, is obtained from Figure 4.29 by setting the disturbance to zero, i.e., $w(t) = 0$.

$$T_r(s) = \frac{D(s)G(s)}{1 + \beta D(s)G(s)}.$$

The general transfer function with respect to the disturbance, $T_w(s)$, is obtained from Figure 4.29 by setting the reference input to zero, i.e., $r(t) = 0$. Redrawing Figure 4.29 leads to the diagram in Figure 4.30, from which $T_w(s)$ is easily obtained.

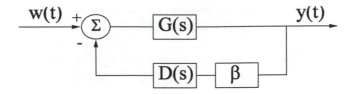

FIGURE 4.30
Obtaining the Transfer Function w.r.t the Disturbance

$$T_w(s) = \frac{G(s)}{1 + \beta D(s)G(s)}.$$

It is important to note that the same system characteristic equation is obtained from both $T_r(s)$ and $T_w(s)$,

$$1 + \beta D(s)G(s) = 0.$$

This is because a system only has one characteristic equation, which portrays the system dynamics.

The starting point in deriving the steady state error e_{ss}^r is considering the Laplace transform of the reference error,

$$
\begin{aligned}
E_r(s) &= R(s) - Y(s) \\
&= R(s)(1 - T_r) \\
e_{ss}^r &= \lim_{s \to 0} sE_r(s) \qquad\qquad \text{Final value theorem} \\
&= \lim_{s \to 0} sR(s)(1 - T_r) \\
&= \lim_{s \to 0} \frac{s(1 - T_r)}{s^{k+1}} \qquad \text{where } R(s) = \mathcal{L}\left[\frac{t^k}{k!}\right] = \frac{1}{s^{k+1}} \\
&= \lim_{s \to 0} \frac{(1 - T_r)}{s^k}.
\end{aligned}
$$

Similarly, the steady state error e_{ss}^w is derived by considering the Laplace transform of the error due to the disturbance (with no reference input).

$$E_w(s) = 0 - Y(s)$$

$$= -T_w W(s)$$

$$e_{ss}^w = \lim_{s \to 0} s E_w(s) \qquad \text{Final value theorem}$$

$$= -\lim_{s \to 0} s T_w W(s)$$

$$= \lim_{s \to 0} \frac{-s T_w}{s^{k+1}} \qquad \text{where } W(s) = \mathcal{L}\left[\frac{t^k}{k!}\right] = \frac{1}{s^{k+1}}$$

$$= \lim_{s \to 0} \frac{-T_w}{s^k}.$$

The system type is the degree (k) of the polynomial $\dfrac{t^k}{k!}$ that will make the steady state errors e_{ss}^r and e_{ss}^w nonzero constants. Thus, there are two types of system types: one with respect to reference and the other with respect to the disturbance.

Example 4.6 *Consider the closed-loop control system shown in the following diagram.*

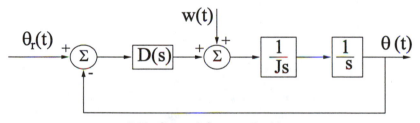

PID Control System Problem

(a) *Use proportional control* $D(s) = K_P$. *Does this controller provide additional damping?*

(b) *Use PD control,* $D(s) = K_P + K_D s$. *Determine the tracking and disturbance-rejection properties for step inputs on* $\theta_r(t)$ *and* $w(t)$.

(c) *Use PI control,* $D(s) = K_P + \dfrac{K_I}{s}$. *Discuss the effect of this controller on the stability of the system.*

(d) *Use PID control,* $D(s) = K_P + \dfrac{K_I}{s} + K_D s$. *Discuss the effect of this controller on the stability and steady state errors of the system. What are the two system types when the PID is employed?*

Solution 4.6 *The transfer function with respect to reference $T_r(s)$ is obtained from the block diagram by setting the disturbance to zero as follows:*

$$T_r(s) = \frac{\Theta(s)}{\Theta_r(s)} = \frac{\dfrac{D(s)}{Js^2}}{1 + \dfrac{D(s)}{Js^2}}$$

$$= \frac{D(s)}{D(s) + Js^2}.$$

The transfer function with respect to disturbance is obtained by setting the reference to zero, and then redrawing the diagram as shown in the following figure.

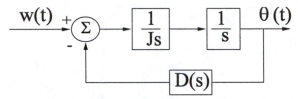

Obtaining the Transfer Function w.r.t. the Disturbance

$T_w(s)$ *is then obtained as follows:*

$$T_w(s) = \frac{\Theta(s)}{W(s)} = \frac{\dfrac{1}{Js^2}}{1 + \dfrac{D(s)}{Js^2}}$$

$$= \frac{1}{D(s) + Js^2}.$$

Thus, the general system characteristic equation for the system (it is the same from both $T_w(s)$ and $T_r(s)$) is given by

$$D(s) + Js^2 = 0.$$

a) Proportional (P) controller

$$D(s) = K_P.$$

The specific characteristic equation becomes

$$Js^2 + K_P = 0$$

$$s = \pm j\sqrt{\frac{K_P}{J}} \implies \xi = 0.$$

This means that the proportional controller does not provide any damping.

b) Proportional and Derivative (PD) Controller

$$D(s) = K_P + K_D s.$$

The transfer function with respect to reference is given by

$$T_r = \frac{D(s)}{D(s) + Js^2}$$

$$= \frac{[K_P + K_D s]}{Js^2 + [K_P + K_D s]}.$$

For tracking properties

$$e_{ss}^r = \lim_{s \to 0} \frac{(1 - T_r)}{s^k}$$

$$= \lim_{s \to 0} \frac{\left(1 - \dfrac{[K_P + K_D s]}{Js^2 + [K_P + K_D s]}\right)}{s^k}$$

$$= \lim_{s \to 0} \frac{\left(\dfrac{Js^2}{Js^2 + K_D s + K_P}\right)}{s^k}$$

$$= \lim_{s \to 0} \left(\frac{Js^{2-k}}{Js^2 + K_D s + K_P}\right). \qquad (4.45)$$

For a step reference input $\theta_r(t) = 1$, *which means* $\Theta_r(s) = \dfrac{1}{s} \Leftrightarrow k = 0$.
From Equation 4.45,

$$k = 0 \implies e_{ss}^r = \frac{0}{K_P} = 0.$$

This means there is a zero steady state error for a step input. For distur-bance rejection, the transfer function with respect to the disturbance is given by

$$T_w = \frac{1}{D(s) + Js^2}$$

$$= \frac{1}{Js^2 + [K_P + K_D s]}$$

The corresponding steady state error is given by

$$e_{ss}^w = \lim_{s \to 0} \frac{-T_w}{s^k}$$

$$= \lim_{s \to 0} \frac{-\dfrac{1}{Js^2 + [K_p + K_D s]}}{s^k}$$

$$= \lim_{s \to 0} \left(-\frac{s^{-k}}{Js^2 + [K_P + K_D s]} \right). \qquad (4.46)$$

For a step disturbance $w(t) = 1$, *which means* $\Omega_r(s) = \frac{1}{s} \Leftrightarrow k = 0$. *From Equation 4.46*

$$k = 0 \Longrightarrow e_{ss}^w = \frac{1}{K_P}.$$

This means there is always a steady state error due to a step disturbance. The error decreases as K_p *is increased.*

c) Proportional and Integral (PI) controller,

$$D(s) = K_P + \frac{K_I}{s}.$$

The characteristic equation takes the form

$$1 + \frac{D}{Js^2} = 0$$

$$\Longrightarrow 1 + \frac{K_P + \dfrac{K_I}{s}}{Js^2} = 0$$

$$\Longrightarrow Js^3 + K_P s + K_I = 0.$$

The coefficient of s^2 *is missing, which means (Routh-Hurwitz stability criterion) at least one pole is not in LHP, which implies that the PI is an unstable control strategy for this system.*

d) Proportional and Integral and Derivative (PID) controller

$$D(s) = K_P + \frac{K_I}{s} + K_D s.$$

This means that the characteristic equation of the system becomes

$$1 + \frac{1}{Js^2}[K_P + \frac{K_I}{s} + K_D s] = 0$$

$$\Longrightarrow Js^3 + K_D s^2 + K_P s + K_I = 0.$$

There is control over all the poles and the system can be made stable by choosing K_D, K_P, and K_I such that all the poles are strictly in the LHP. The transfer function with respect to the reference is given by

$$T_r = \frac{\dfrac{D(s)}{Js^2}}{1 + \dfrac{D(s)}{Js^2}}$$

$$= \frac{\dfrac{1}{Js^2}[K_P + \dfrac{K_I}{s} + K_D s]}{1 + \dfrac{1}{Js^2}[K_P + \dfrac{K_I}{s} + K_D s]}$$

$$= \frac{K_D s^2 + K_P s + K_I}{Js^3 + K_D s^2 + K_P s + K_I}.$$

The steady state error with respect to reference is given by

$$e_{ss}^r = \lim_{s \to 0} \frac{(1 - T_r)}{s^k}$$

$$= \lim_{s \to 0} \frac{(1 - T_r)}{s^0} \qquad \text{from } r(t) = 1 \Leftrightarrow R(s) = \frac{1}{s} \Longrightarrow k = 0$$

$$= \lim_{s \to 0} (1 - T_r)$$

$$= \lim_{s \to 0} \left(1 - \frac{K_D s^2 + K_P s + K_I}{Js^3 + K_D s^2 + K_P s + K_I}\right)$$

$$= 1 - \frac{K_I}{K_I}$$

$$= 0.$$

Hence, there is no steady state error with respect to a step input, which means with PID controller, the system can effectively track step inputs with no steady state errors. The transfer function with respect to the disturbance *is given by*

$$T_w = \frac{Y(s)}{W(s)} = \frac{G(s)}{1 + \beta D(s)G(s)}$$

$$= \frac{\dfrac{1}{Js^2}}{1 + \dfrac{D(s)}{Js^2}}$$

$$= \frac{\dfrac{1}{Js^2}}{1 + \dfrac{1}{Js^2}[K_P + \frac{K_I}{s} + K_D s]}$$

$$= \frac{s}{Js^3 + K_D s^2 + K_P s + K_I}.$$

The corresponding steady state error is given by

$$e_{ss}^w = \lim_{s \to 0} \frac{-T_w}{s^k}$$

$$= \lim_{s \to 0} \frac{-T_w}{s^0} \qquad from \ \ w(t) = 1 \Leftrightarrow W(s) = \frac{1}{s} \Longrightarrow k = 0$$

$$= \lim_{s \to 0} -T_w$$

$$= \lim_{s \to 0} -\frac{s}{Js^3 + K_D s^2 + K_P s + K_I}$$

$$= -\frac{0}{K_I}$$

$$= 0.$$

Hence, there is no steady state error with respect to a step disturbance, which means with PID controller the system effectively rejects step disturbances. In general, the system type with respect to the reference is the value of k such that

$$e_{ss}^r = \lim_{s \to 0} \frac{(1 - T_r)}{s^k} \equiv \ Nonzero \ constant.$$

For the specific unit feedback control system under discussion the expression *for e_{ss}^r can be simplified as follows:*

$$e_{ss}^r = \lim_{s \to 0} \frac{(1 - T_r)}{s^k}$$

$$= \lim_{s \to 0} \frac{\left(1 - \dfrac{K_D s^2 + K_P s + K_I}{J s^3 + K_D s^2 + K_P s + K_I}\right)}{s^k}$$

$$= \lim_{s \to 0} \frac{\left(\dfrac{J s^3}{J s^3 + K_D s^2 + K_P s + K_I}\right)}{s^k}$$

$$\implies e_{ss}^r = \lim_{s \to 0} \left(\frac{J s^{3-k}}{J s^3 + K_D s^2 + K_P s + K_I}\right). \tag{4.47}$$

In order to obtain the system type try increasing values of k starting from 0 in the expression in Equation 4.47 until a nonzero constant is obtained.

$$k = 0 \implies e_{ss}^r = \frac{0}{K_I} = 0$$

$$k = 1 \implies e_{ss}^r = \frac{0}{K_I} = 0$$

$$k = 2 \implies e_{ss}^r = \frac{0}{K_I} = 0$$

$$k = 3 \implies e_{ss}^r = \frac{J}{K_I} \quad \text{(a nonzero constant)}$$

$$\implies \text{system type with respect to reference is 3.}$$

This means with a PID controller this system can effectively track step, ramp, and parabolic inputs without any steady state errors. For cubic inputs there is a constant nonzero error and for higher-order inputs the system becomes unstable, i.e., the system is not able to track these higher-order inputs because there is an infinitely growing reference error.

In general, the system type with respect to the disturbance is the value of k such that

$$e_{ss}^w = \lim_{s \to 0} \frac{-T_w}{s^k} \equiv \text{Nonzero Constant.}$$

For the specific unit feedback control system under discussion the expression *for e_{ss}^w can be simplified as follows:*

$$e_{ss}^w = \lim_{s \to 0} \frac{-T_w}{s^k}$$

$$= -\lim_{s \to 0} \frac{\dfrac{s}{Js^3 + K_D s^2 + K_P s + K_I}}{s^k}$$

$$= -\lim_{s \to 0} \left(\frac{s^{1-k}}{Js^3 + K_D s^2 + K_P s + K_I} \right). \qquad (4.48)$$

In order to obtain the system type try increasing values of k starting from 0 in the expression in Equation 4.48 until a nonzero constant is obtained.

$$k = 0 \implies e_{ss}^w = -\frac{0}{K_I} = 0$$

$$k = 1 \implies e_{ss}^w = -\frac{1}{K_I} \implies \text{System type with respect to disturbance is 1.}$$

This means with a PID controller this system can effectively reject step disturbances without any steady state errors. For ramp inputs there is a constant nonzero error and for higher-order inputs the system becomes unstable, i.e., the system is not able to handle these higher-order disturbances because of infinitely growing errors.

4.12   Sensitivity

The sensitivity of a control system is concerned with how the controlled variable varies with respect to variations in system parameters. These parameters could be plant gains, controller gains or any other parameter on which the system depends. Although in the modeling carried out so far it has been assumed that these parameters are constant, in practical systems they could vary with changes in operating conditions such as temperature and pressure.

4.12.1   Definition of Sensitivity

Consider a system with a general (open- or closed-loop) transfer function $H(s)$, which is dependent on a parameter K. Suppose that a change in operating conditions causes the parameter K to drift from its original value to $K + \delta K$. This parameter variation in turn forces the transfer function to change from $H(s)$ to $H(s) + \delta H(s)$. The sensitivity S_K^H of the transfer function with respect to variation in the parameter K is defined as the ratio of $\dfrac{\delta H}{H}$ to $\dfrac{\delta K}{K}$ such that

$$S_K^H = \frac{\delta H}{H} \bigg/ \frac{\delta K}{K}$$

$$\approx \frac{dH}{H} \bigg/ \frac{dK}{K}$$

$$= \frac{K}{H} \frac{dH}{dK}.$$

4.12.2 Open- and Closed-Loop Sensitivity

Consider the general open and closed-loop transfer functions for a system with a proportional control K and plant gain $G(s)$

$$H_{ol}(s) = KG(s) \tag{4.49}$$

$$H_{cl}(s) = \frac{KG(s)}{1 + KG(s)}. \tag{4.50}$$

The open-loop sensitivity $S_K^{H_{ol}}$ with respect to parameter K (the gain) is obtained as follows:

$$S_K^{H_{ol}} = \frac{dH_{ol}}{H_{ol}} \bigg/ \frac{dK}{K}$$

$$= \frac{K}{H_{ol}} \frac{dH_{ol}}{dK}.$$

From Equation 4.49

$$\frac{dH_{ol}}{dK} = G(s).$$

Hence, the sensitivity function is given by

$$S_K^{H_{ol}} = \frac{K}{H_{ol}} \frac{dH_{ol}}{dK}$$

$$= \frac{K}{KG(s)} G(s)$$

$$= 1.$$

This means that

$$S_K^{H_{ol}} = \frac{\delta H_{ol}}{H_{ol}} \bigg/ \frac{\delta K}{K} = 1$$

$$\implies \frac{\delta H_{ol}}{H_{ol}} = \frac{\delta K}{K}, \tag{4.51}$$

which means that a 5% error in K would yield a 5% error in $H_{ol}(s)$, and hence, an error of 5% in the output (variable being controlled). Put differently, the system sensitivity with respect to proportional controller gain changes is 100% for open-loop control systems.

The closed-loop sensitivity $S_K^{H_{cl}}$ with respect to parameter K (the gain) is obtained as follows:

$$S_K^{H_{cl}} = \frac{dH_{cl}}{H_{cl}} \Big/ \frac{dK}{K}$$

$$= \frac{K}{Hcl} \frac{dH_{cl}}{dK}.$$

From Equation 4.50

$$\frac{dH_{cl}}{dK} = \frac{d}{dK} \left[\frac{KG(s)}{1 + KG(s)} \right]$$

$$= \frac{[1 + KG(s)]\, G(s) - KG(s)G(s)}{[1 + KG(s)]^2}$$

$$= \frac{G(s)}{[1 + KG(s)]^2}.$$

Hence the closed-loop sensitivity function is given by

$$S_K^{H_{cl}} = \frac{K}{H_{cl}} \frac{dH_{cl}}{dK}$$

$$= \frac{K}{\left[\dfrac{KG(s)}{1 + KG(s)} \right]} \frac{G(s)}{[1 + KG(s)]^2}$$

$$= \frac{1}{1 + KG(s)}.$$

This means that

$$S_K^{H_{cl}} = \frac{\delta H_{cl}}{H_{cl}} \Big/ \frac{\delta K}{K} = \frac{1}{1 + KG(s)}$$

$$\implies \frac{\delta H_{cl}}{H_{cl}} = \left(\frac{1}{1 + KG(s)} \right) \frac{\delta K}{K}.$$

Comparing this equation with open-loop sensitivity Equation 4.51, it is clear that the error in the controlled quantity is less sensitive to variations in the proportional controller gain K by a factor

$$1 + KG(s).$$

This is a major advantage of closed-loop systems over open-loop control systems. The term $1 + KG(s)$ is called the return difference of the feedback path.

Example 4.7 *A unity feedback control system has the following transfer function*

$$G(s) = \frac{K}{s(s+a)}$$

a) Compute the sensitivity of the closed-loop transfer function to changes in the parameter K.

b) Compute the sensitivity of the closed-loop transfer function to changes in the parameter a.

c) If the unity gain in the feedback changes to a value $\beta \neq 1$, compute the sensitivity of the closed-loop transfer function with respect to β.

Solution 4.7 *a) The transfer function of the system is given by*

$$H(s) = \frac{G(s)}{1 + G(s)} = \frac{\dfrac{K}{s(s+a)}}{1 + \dfrac{K}{s(s+a)}}$$

$$= \frac{K}{s^2 + as + K}.$$

From the general definition of sensitivity

$$S_K^H = \frac{dH}{H} \Big/ \frac{dK}{K}$$

$$= \frac{K}{H}\frac{dH}{dK}.$$

For the specific transfer function

$$\frac{dH}{dK} = \frac{(s^2 + as + K) - K}{(s^2 + as + K)^2}$$

$$= \frac{s^2 + as}{(s^2 + as + K)^2}.$$

Hence, the sensitivity function is given by

$$S_K^H = \frac{K}{H}\frac{dH}{dK}$$

$$= \frac{K(s^2 + as + K)}{K} \times \frac{s^2 + as}{(s^2 + as + K)^2}$$

$$= \frac{s^2 + as}{s^2 + as + K}.$$

Design and Analysis of Control Systems

b) From the general definition of sensitivity

$$S_a^H = \frac{dH}{H} / \frac{da}{a}$$

$$= \frac{a}{H} \frac{dH}{da}.$$

For the specific transfer function

$$\frac{dH}{da} = \frac{-sK}{(s^2 + as + K)^2}.$$

Hence, the sensitivity function is given by

$$S_a^H = \frac{a}{H} \frac{dH}{da}$$

$$= \frac{a(s^2 + as + K)}{K} \times \frac{-sK}{(s^2 + as + K)^2}$$

$$= \frac{-as}{(s^2 + as + K)}.$$

c) The transfer function of the system is given by

$$H(s) = \frac{G(s)}{1 + \beta G(s)}.$$

From the general definition of sensitivity

$$S_a^H = \frac{dH}{H} / \frac{d\beta}{\beta}$$

$$= \frac{\beta}{H} \frac{dH}{d\beta}.$$

For the specific transfer function

$$\frac{dH}{d\beta} = \frac{-G^2(s)}{[1 + \beta G(s)]^2}.$$

Hence the sensitivity function is given by

$$S_\beta^H = \frac{\beta}{H}\frac{dH}{d\beta}$$

$$= \frac{\beta(1+\beta G)}{G}\frac{-G^2}{(1+\beta G)^2}$$

$$= \frac{-\beta G}{1+\beta G} = \frac{\dfrac{-\beta K}{s(s+a)}}{1+\dfrac{\beta K}{s(s+a)}}$$

$$= \frac{-\beta K}{s(s+a)+\beta K}.$$

Example 4.8 *Consider the car's cruise control problem discussed in previous sections. The open-loop and closed-loop transfer functions were found to be*

$$H_{ol}(s) = KG(s) = \frac{bK}{ms+b} \tag{4.52}$$

$$H_{cl}(s) = \frac{KG(s)}{1+KG(s)} = \frac{\dfrac{bK}{ms+b}}{1+\dfrac{bK}{ms+b}} = \frac{bK}{ms+b+bK}. \tag{4.53}$$

Compare the sensitivity of the open-loop and closed-loop systems with respect to the controller gain K.

Solution 4.8 *Consider the open-loop and closed-loop transfer functions for the car's cruise control system. The open-loop sensitivity S_K^H with respect to parameter K (the gain) is obtained as follows:*

$$S_K^{H_{ol}} = \frac{dH_{ol}}{H_{ol}} \Big/ \frac{dK}{K}$$

$$= \frac{K}{H_{ol}}\frac{dH_{ol}}{dK}.$$

From Equation 4.52

$$\frac{dH_{ol}}{dK} = \frac{b}{ms+b}.$$

Hence the sensitivity function is given by

$$S_K^{H_{ol}} = \frac{K}{H_{ol}} \frac{dH_{ol}}{dK}$$

$$= \frac{K}{\left(\frac{bK}{ms+b}\right)} \frac{b}{ms+b}$$

$$= 1.$$

This means that

$$S_K^{H_{ol}} = \frac{\delta H_{ol}}{H_{ol}} \bigg/ \frac{\delta K}{K} = 1$$

$$\implies \frac{\delta H_{ol}}{H_{ol}} = \frac{\delta K}{K},$$

which means that a 5% error in K would yield a 5% error in $H_{ol}(s)$. This in turn means that an error of 5% in the controller gain K causes an error of 5% in the car speed (the variable or output being controlled).

The closed-loop sensitivity $S_K^{H_{cl}}$ with respect to parameter K (the gain) is obtained as follows:

$$S_K^{H_{cl}} = \frac{dH_{cl}}{H_{cl}} \bigg/ \frac{dK}{K}$$

$$= \frac{K}{Hcl} \frac{dH_{cl}}{dK}$$

From Equation 4.53

$$\frac{dH_{cl}}{dK} = \frac{d}{dK} \left(\frac{\dfrac{bK}{ms+b}}{1 + \dfrac{bK}{ms+b}} \right)$$

$$= \frac{\left(1 + \dfrac{bK}{ms+b}\right)\left(\dfrac{b}{ms+b}\right) - \dfrac{bK}{ms+b}\left(\dfrac{b}{ms+b}\right)}{\left(1 + \dfrac{bK}{ms+b}\right)^2}$$

$$= \frac{(ms+b+bK)b - bKb}{(ms+b+bK)^2}$$

$$= \frac{b(ms+b)}{(ms+b+bK)^2}.$$

Hence, the closed-loop sensitivity function is given by

$$S_K^{H_{cl}} = \frac{K}{H_{cl}} \frac{dH_{cl}}{dK}$$

$$= \frac{K}{\left(\dfrac{bK}{ms+b+bK}\right)} \times \frac{b(ms+b)}{(ms+b+bK)^2}$$

$$= \frac{ms+b}{ms+b+bK}$$

$$= \frac{1}{1 + \dfrac{bK}{ms+b}}.$$

This means that

$$S_K^{H_{ol}} = \frac{\delta H_{cl}}{H_{cl}} \bigg/ \frac{\delta K}{K} = \frac{1}{1 + \dfrac{bK}{ms+b}}$$

$$\implies \frac{\delta H_{cl}}{H_{cl}} = \left(\frac{1}{1 + \dfrac{bK}{ms+b}}\right) \frac{\delta K}{K}.$$

Hence, compared with the open-loop system, the error in the controlled quantity is less sensitive to variations in the plant gain K *by a factor*

$$1 + \frac{bK}{ms+b} \equiv 1 + KG(s).$$

If there is an error of 5% *in the controller gain* K *the error in the car speed will be*

$$\frac{5\%}{1 + \dfrac{bK}{ms+b}}.$$

Thus the percentage error will be far less than 5%, *hence this example illustrates a key advantage of closed-loop systems over open-loop systems.*

4.13   Stability

For a given control system, stability is usually the most important property to be determined. A stable system always gives responses appropriate to the stimulus. For linear time-invariant systems there are three ways of understanding and quantifying stability.

- Bounded Input Bounded Output (BIBO) stability.

- Asymptotic internal stability.

- Routh-Hurwitz stability.

4.13.1   Bounded Input-Bounded Output Stability

A system is said to be *bounded input-bounded output stable*, if every bounded input produces a bounded output. Consider a general system with an input $r(t)$, output $y(t)$, and impulse response $h(t)$. By using convolution

$$y(t) = \int_{-\infty}^{\infty} h(\tau)r(t-\tau)d\tau.$$

If the input $r(t)$ is bounded it means there exists a constant M such that

$$|r(t)| \leq M < \infty.$$

It follows that

$$|y| = |\int_{-\infty}^{\infty} h(\tau)r(t-\tau)d\tau|$$

$$\leq \int_{-\infty}^{\infty} |h(\tau)||r(t-\tau)|d\tau$$

$$\leq M \int_{-\infty}^{\infty} |h(\tau)|d\tau.$$

Therefore,the output is bounded if and only if

$$\int_{-\infty}^{\infty} |h(\tau)|d\tau$$

is bounded. Hence a system is said to be BIBO stable if and only if its impulse response $h(t)$ is such that

$$\int_{-\infty}^{\infty} |h(\tau)|d\tau \leq \infty.$$

4.13.2 Asymptotic Internal Stability

A general system transfer function is given by

$$H(s) = \frac{Y(s)}{U(s)}$$

$$= \frac{b_0 s^m + b_1 s^{m-1} + b_2 s^{m-2} + ... + b_m}{a_0 s^n + a_1 s^{n-1} + a_2 s^{n-2} + ... + a_n} \qquad (4.54)$$

$$= K \frac{\Pi_{j=1}^m (s - z_j)}{\Pi_{i=1}^n (s - p_i)}. \qquad (4.55)$$

The solution to the input-output (homogeneous) differential equation that corresponds to the characteristic equation from the transfer function represented in Equation 4.54 is given by

$$y(t) = \sum_{i=1}^n K_i e^{p_i t},$$

where $\{p_i\}$ are the roots of the characteristic equation and K_i depends on initial conditions. The system is stable if and only if $e^{p_i t}$ decays to zero for all poles $\{p_i\}$ as t is increased to ∞, i.e.,

$$e^{p_i t} \longrightarrow 0 \text{ as } t \longrightarrow \infty \quad \text{ for all } \{p_i\}.$$

This will happen if and only if all the poles $\{p_i\}$ are strictly in the LHP, that is,

$$\mathrm{Re}\{p_i\} < 0.$$

This is the *asymptotic internal stability criterion* and can be determined by computing location of the roots of the characteristic equation and checking whether the real parts of the roots are strictly less than zero.

4.13.3 Routh-Hurwitz Stability Criterion

It is not always easy to explicitly determine the roots of high-order polynomial functions (characteristic equations). The Routh-Hurwitz stability criterion allows the determination of stability without solving for the roots of the characteristic equation. This is achieved by analyzing the coefficients of the characteristic equation, and expressions derived from these coefficients. Consider the general characteristic equation

$$a_0 s^n + a_1 s^{n-1} + a_2 s^{n-2} + ... + a_n = 0.$$

From the definition of asymptotic stability, all roots of the characteristic equation must have negative real parts, which implies that a *necessary but not sufficient* condition for stability is that all the coefficients $\{a_i\}$ must be present and positive.

4.13.3.1   Summary of the Routh-Hurwitz Stability Procedure

- Inspect the characteristic equation, if any coefficient is missing (zero) or negative, then the system is unstable.

- If all the coefficients are present, then construct a triangular array that is a function of the coefficients $\{a_i\}$.

- For stability, all the elements in the first column of the array must be positive. This is the *necessary and sufficient* condition for stability.

4.13.3.2   The Array

Arrange the coefficients of the characteristic equation in two rows beginning with the first and second coefficients and followed by even-numbered and odd-numbered coefficients. Subsequent rows are then added. Routh-Hurwitz array

$$
\begin{array}{c|cccc}
s^n & 1 & a_2 & a_4 & \ldots \\
s^{n-1} & a_1 & a_3 & a_5 & \ldots \\
s^{n-2} & b_1 & b_2 & b_3 & \ldots \\
s^{n-3} & c_1 & c_2 & c_3 & \ldots \\
\ldots & \ldots & \ldots & \ldots \\
s^2 & * & * \\
s^1 & * \\
s^0 & *
\end{array}
$$

The terms are computed as follows:

$$b_1 = \frac{-\det \begin{bmatrix} 1 & a_2 \\ a_1 & a_3 \end{bmatrix}}{a_1} = \frac{a_1 a_2 - a_3}{a_1}$$

$$b_2 = \frac{-\det \begin{bmatrix} 1 & a_4 \\ a_1 & a_5 \end{bmatrix}}{a_1} = \frac{a_1 a_4 - a_5}{a_1}$$

$$c_1 = \frac{-\det \begin{bmatrix} a_1 & a_3 \\ b_1 & b_2 \end{bmatrix}}{b_1} = \frac{b_1 a_3 - a_1 b_2}{b_1}$$

$$c_2 = \frac{-\det \begin{bmatrix} a_1 & a_5 \\ b_1 & b_3 \end{bmatrix}}{b_1} = \frac{b_1 a_5 - a_1 b_3}{b_1}$$

If at least one element of the first column is not positive, this means that there are some roots in the RHP, which in turn means that the system is unstable. The number of roots in the RHP is equal to the number of sign changes. For example, if the first column consists of the following five elements:

$$2, \; -1, \; 7, \; 0.5, \; 13 \Longrightarrow 2 \;\; \text{sign changes} \Longrightarrow 2 \;\; \text{roots in the RHP.}$$

4.13.3.3 A Special Case: First Term in a Row is *Zero*

If the first term in one of the rows is zero replace the *zero* with ϵ and proceed with the Routh-Hurwitz stability procedure, and then apply the stability criterion as $\epsilon \longrightarrow 0$.

Example 4.9 *Consider the process plant control system shown below.*

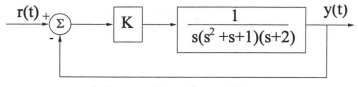

A Process Plant Control System

Find the range of K that will make the system stable.

Solution 4.9 *The closed-loop transfer function is obtained as follows:*

$$T(s) = \frac{KG(s)}{1 + KG(s)}$$

$$= \frac{\dfrac{K}{s(s^2 + s + 1)(s + 2)}}{1 + \dfrac{K}{s(s^2 + s + 1)(s + 2)}}$$

$$= \frac{K}{s(s^2 + s + 1)(s + 2) + K}$$

$$= \frac{K}{s^4 + 3s^3 + 3s^2 + 2s + K}.$$

Therefore, the characteristic equation is given by

$$s^4 + 3s^3 + 3s^2 + 2s + K = 0.$$

Using the Routh-Hurwitz procedure, an array of coefficients can be constructed.

$$
\begin{array}{cccc}
s^4 & 1 & 3 & K \\[2mm]
s^3 & 3 & 2 & 0 \\[2mm]
s^2 & \dfrac{7}{3} & K & \\[4mm]
s^1 & 2 - \dfrac{9K}{7} & & \\[4mm]
s^0 & K & &
\end{array}
$$

For stability, K must be positive, and all coefficients in the first column must be positive. Therefore,

$$K > 0 \quad and \quad \left(2 - \frac{9K}{7}\right) > 0$$

$$\implies 0 < K < \frac{14}{9}.$$

This is the range of K that will permit system stability. When $K = \dfrac{14}{9}$, the system becomes oscillatory with a constant amplitude. This is called marginal stability or indeterminate stability.

Example 4.10 *Consider the closed-loop control system shown below.*

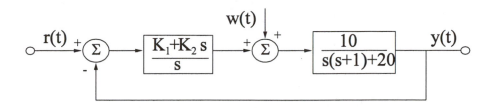

Closed Loop Control System

(a) *Determine the transfer function from r to y.*
(b) *Determine the transfer function from w to y.*
(c) *Find the range of (K_1, K_2) for which the system is stable.*
(d) *What is the system type with respect to r and w?*

Solution 4.10 *The solution is outlined below.*

a)

$$T_r(s) = \frac{Y(s)}{R(s)} = \frac{10(K_1 + K_2 s)}{s[s(s+1) + 20] + 10(K_1 + K_2 s)}.$$

b)

$$T_w(s) = \frac{Y(s)}{W(s)} = \frac{10s}{s[s(s+1) + 20] + 10(K_1 + K_2 s)}.$$

c) The characteristic equation is given by

$$s^3 + s^2 + (10K_2 + 20)s + 10K_1 = 0.$$

Routh-Hurwitz array

s^3	1	$10K_2 + 20$
s^2	1	$10K_1$
s^1	$10K_2 + 20 - 10K_1$	
s^0	$10K_1$	

For stability

$$10K_1 > 0 \Rightarrow K_1 > 0$$
$$10K_2 + 20 - 10K_1 > 0 \Rightarrow K_2 > K_1 - 2.$$

Therefore, the conditions for stability are given by

$$K_1 > 0 \text{ and } K_2 > K_1 - 2.$$

d) System type

   (i) The expression of the steady state error with respect to reference can be simplified as follows:

$$e_{ss}^r = \lim_{s \to 0} \frac{(1 - T_r)}{s^k}$$

$$= \lim_{s \to 0} \left[\frac{\left(1 - \dfrac{10(K_1 + K_2 s)}{s[s(s+1) + 20] + 10(K_1 + K_2 s)} \right)}{s^k} \right]$$

$$= \lim_{s \to 0} \left[\frac{\left(\dfrac{s[s(s+1) + 20]}{s[s(s+1) + 20] + 10(K_1 + K_2 s)} \right)}{s^k} \right]$$

$$= \lim_{s \to 0} \left[\frac{s^{1-k}[s(s+1) + 20]}{s[s(s+1) + 20] + 10(K_1 + K_2 s)} \right]. \tag{4.56}$$

In order to obtain the system type try increasing values of k starting from 0 in the expression in Equation 4.56 until a nonzero constant is obtained.

$$k = 0 \implies e_{ss}^r = \frac{0}{10K_1} = 0$$

$$k = 1 \implies e_{ss}^r = \frac{20}{10K_I} = \frac{2}{K_I} \quad \text{(a nonzero constant)}$$

$$\implies \text{System type with respect to reference is 1.}$$

(ii) The expression of the steady state error with respect to disturbance can be simplified as follows:

$$e_{ss}^w = \lim_{s \to 0} \frac{-T_w}{s^k}$$

$$= \lim_{s \to 0} \frac{-\left(\dfrac{10s}{s[s(s+1) + 20] + 10(K_1 + K_2 s)} \right)}{s^k}$$

$$= \lim_{s \to 0} \frac{-10s^{1-k}}{s[s(s+1) + 20] + 10(K_1 + K_2 s)}. \tag{4.57}$$

In order to obtain the system type try increasing values of k starting from 0 in the expression in Equation 4.57 until a nonzero constant is obtained.

$$k = 0 \Longrightarrow e^w_{ss} = -\frac{0}{10K_1} = 0$$

$$k = 1 \Longrightarrow e^w_{ss} = -\frac{10}{10K_I} = -\frac{1}{K_I} \quad (a\ nonzero\ constant)$$

$\Longrightarrow$ *System type with respect to disturbance is 1.*

Example 4.11 *The following figure shows the speed control of an assembly plant.*

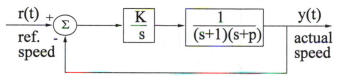

Speed Control of an Assembly Plant

Determine and plot the range of K and p that permits stable operation.

Solution 4.11 *The transfer function for the system is given by*

$$T(s) = \frac{K}{s^3 + (p+1)s^2 + ps + K}.$$

Stability condition is established by using the following Routh-Hurwitz array

s^3 1 p

s^2 $1+p$ K

s^1 b 0

s^0 K

From this array the conditions for stability are established as follows:

$$K > 0$$

$$(1+p) > 0 \Longrightarrow p > -1 \ but \ p > 0 \quad (coefficient\ of\ s)$$

$$\Longrightarrow p > 0$$

$$b = \frac{p^2 + p - K}{1+p} > 0$$

$$\Rightarrow K < \left(p^2 + p\right)$$

$$\Longrightarrow 0 < K < \left(p^2 + p\right) \ and \ p > 0$$

The curve of K vs. p where

$$K = p^2 + p,$$

is then plotted in Figure 4.31. For the system to be stable, the desired region is the shaded area, as illustrated in Figure 4.31

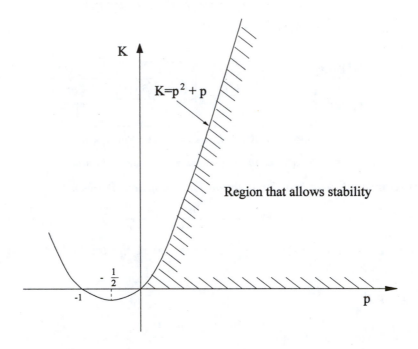

FIGURE 4.31
Stability Region for Assembly Plant

Example 4.12 *The feedback control system of a chemical plant is given below. How many of the system's poles are not in the LHP?*

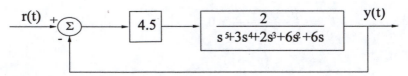

Chemical Plant Control System

Solution 4.12 *The system transfer function is given by*

$$T = \frac{KG(s)}{1 + KG(s)}$$

$$= \frac{4.5 \left(\dfrac{2}{s^5 + 3s^4 + 2s^3 + 6s^2 + 6s} \right)}{1 + 4.5 \left(\dfrac{2}{s^5 + 3s^4 + 2s^3 + 6s^2 + 6s} \right)}$$

$$= \frac{9}{s^5 + 3s^4 + 2s^3 + 6s^2 + 6s + 9}.$$

The next step is constructing the Routh-Hurwitz array.

$$
\begin{array}{c|ccc}
s^5 & 1 & 2 & 6 \\
\\
s^4 & 3 & 6 & 9 \\
\\
s^3 & \epsilon & 3 & 0 \\
\\
s^2 & \dfrac{6\epsilon - 9}{\epsilon} & 0 & 0 \\
\\
s^1 & 3 - \dfrac{\epsilon^2}{2\epsilon - 3} & 0 & 0 \\
\\
s^0 & 3 & 0 &
\end{array}
$$

The 0 *that is the first element of the* s^3 *row is replaced by* ϵ. *The stability criterion is applied by taking the limit as* $\epsilon \to 0$. *There are two sign changes in the first column of the array (from 0 to* $-\infty$ *and then from* $-\infty$ *to 3) which means that there are two poles which are not in the LHP.*

4.14   Problems

Problem 4.1 *Consider the following generic feedback control system.*

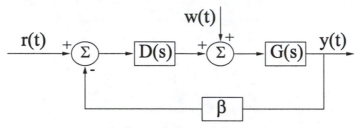

A General Control System

(a) *Find the transfer function from the reference $r(t)$ to the output $y(t)$ i.e. $T_r(s)$.*

(b) *Find the transfer function from the disturbance $w(t)$ to the output $y(t)$ i.e. $T_w(s)$.*

(c) *For a generic reference, $r(t) = \dfrac{t^k}{k!}$, show that the steady state error with respect to the reference is given by*

$$e^r_{ss} = \lim_{s \to 0} \frac{(1 - T_r)}{s^k}.$$

(d) *For a generic disturbance, $w(t) = \dfrac{t^k}{k!}$, show that the steady state error with respect to the disturbance is given by*

$$e^w_{ss} = \lim_{s \to 0} \frac{-T_w}{s^k}.$$

(e) *Explain the meaning of system type in terms of the steady state errors e^r_{ss}, and e^w_{ss}.*

Problem 4.2 *In the diagram for Problem 4.1, a PID controller is used in a unit feedback system such that*

$$D(s) = K_p + \frac{K_I}{s} + K_D s$$

$$G(s) = \frac{1}{Js^2}$$

$$r(t) = 1$$

$$w(t) = 1.$$

(a) What is the effect of the PID on the stability of this specific system?

(b) Find the steady state errors e_{ss}^r, e_{ss}^w for the system.

(c) What are the system types with respect to the reference and distur-bance, respectively?

Problem 4.3 *Design Project:*

The closed-loop control system of the DC motor can be derived and shown to be given by Figure 4.32. The terms $\Omega(s)$, $\Omega_r(s)$ are the Laplace trans-forms of the actual angular speed $w(t)$ and the desired angular speed $w_r(t)$, respectively. $E(s)$ and $U(s)$ are the Laplace transforms of the error and the control signals.

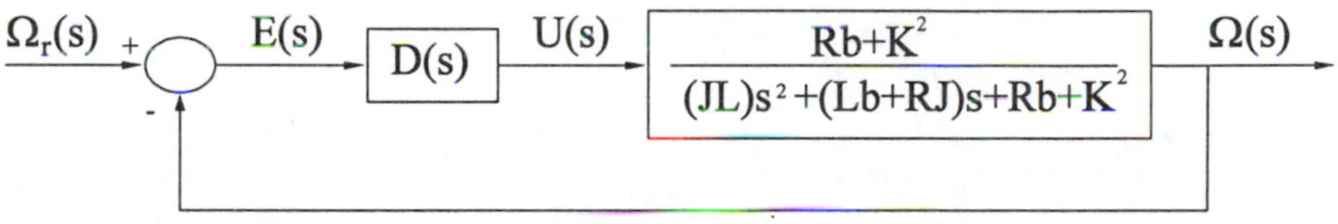

FIGURE 4.32
Closed-Loop Motor Speed Control (PID)

The controller $D(s)$ can take the form of any one of the four controllers P, PI, PD, and PID. A PID controller contains all three control components (proportional, derivative, and integral) and hence, its transfer function can be represented as follows

$$D(s) = K_P + \frac{K_I}{s} + K_D s$$
$$= \frac{K_D s^2 + K_P s + K_I}{s}.$$

Thus, the corresponding block diagram representation takes the form shown in Figure 4.33. This is the generic PID controller, and the other controllers P, PI, PD can be modeled in a similar fashion.

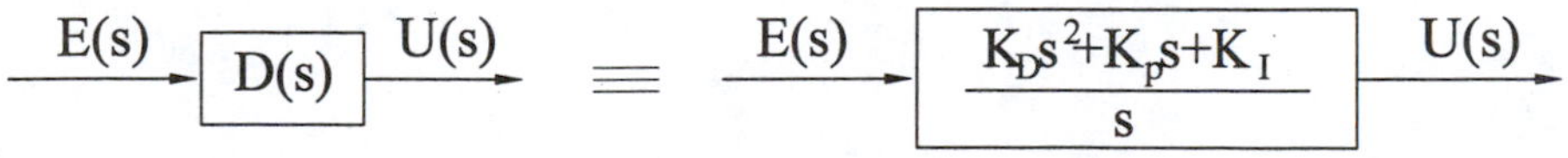

FIGURE 4.33
A PID Controller: Block Diagram Form

(a) Characteristics of PID Controllers

By employing MATLAB, implement four closed-loop control systems for the DC motor by using the four controllers (P, PI, PD, and PID) and the data provided. The objective is to satisfy the given design requirements.

From the results discuss the benefits and drawbacks of the controller components; Proportional (P), Integral (I), and Derivative (D). Also show and explain how the three components compensate for each other's drawbacks.

(b) Tuning of PID Controllers:

From the MATLAB implementation explain how you choose the gains (K_P, K_I, K_D) for the controllers (P, PI, PD and PID) ?

DATA

Assume the following values for the physical parameters, which were derived by experiment from an actual motor:

Moment of inertia of the rotor $J = 0.012 kg.m^2$

Damping ratio of the mechanical system $b = 0.105 Nms$

Electromotive force constant $K = 0.01 Nm/Amp$

Electric resistance $R = 1\Omega$

Electric inductance $L = 0.505H$

DESIGN REQUIREMENTS:

Desired angular speed $w_r(t) = 1 rad/\sec$

Settling time $t_s = 1 sec$,

Overshoot $M_p \leq 20\%$,

Steady state error $e_{ss}(t) \leq 5\%$

Problem 4.4 *Consider the control system shown in the following diagram.*

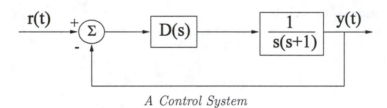

A Control System

The controller $D(s)$ is given by,

$$D(s) = K \frac{s + \alpha}{s^2 + \omega^2}.$$

(a) Prove that the system is capable of tracking a sinusoidal reference input, $r(t) = \sin(\omega t)$, with a zero steady state error.

(b) Find the range of K and α such that the closed-loop system remains stable. For the case where $\alpha = 1$, is the closed-loop system asymptotically stable for some set of values of K?

Problem 4.5 *Consider the helicopter altitude control problem shown in the following diagram,*

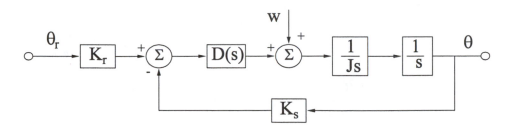

Helicopter Altitude Control Problem

where

$$J = helicopter\ inertia,$$
$$\theta_r(t) = reference\ helicopter\ attitude,$$
$$\theta(t) = actual\ helicopter\ attitude,$$
$$K_s = sensor\ gain,$$
$$K_r = reference\ scaling\ gain,$$
$$\omega(t) = disturbance\ torque.$$

(a) Use proportional control $D(s) = K_P$. Does this controller provide additional damping?

(b) Use PD control, $D(s) = K_P + K_D s$. Determine the tracking and disturbance-rejection properties (i.e., find e_{ss}^r and e_{ss}^w) for step inputs on $\theta_r(t)$ and $\omega(t)$.

(c) Use PI control, $D(s) = K_P + \dfrac{K_I}{s}$. Discuss the effect of this controller on the stability of the system.

(d) Use PID control, $D(s) = K_P + \dfrac{K_I}{s} + K_D s$. Discuss the effect of this controller on the stability and steady state errors of the system.

(e) What are the two system types (one with respect to the reference and the other with respect to the disturbance) when the PID controller is employed?

Problem 4.6 *A DC motor speed control is described by the differential equation*

$$\dot{y} + 60y = 600v_a - 1500\omega,$$

where $y(t)$ is the motor speed, $v_a(t)$ is the armature voltage, and $\omega(t)$ is the load torque. Assume the armature voltage is computed using the PI control law

$$K_P\left[r(t) - y(t)\right] + K_I \int_0^t \left[r(t) - y(t)\right] dt,$$

where $r(t)$ is the reference motor speed.

(a) Derive a block diagram representation of the system.

(b) (i) Compute the transfer function from $r(t)$ to $y(t)$ when $\omega(t) = 0$.

(ii) What is the steady state error due to a ramp reference motor speed, $r(t) = t$?

(iii) Deduce the system type with respect to the reference input $r(t)$.

(c) (i) Compute the transfer function from $\omega(t)$ to $y(t)$ when $r(t) = 0$.

(ii) What is the steady state error due to a disturbance input of the form $\omega(t) = t$?

(iii) Deduce the system type with respect to the disturbance $\omega(t)$.

(d) For the transfer function in (b) (i), compute the values for K_P and K_I so that the characteristic equation of the closed-loop system will have roots at $-60 \pm 60j$.

(e) Using the computed values of K_P and K_I, derive the system natural response due to the transfer function from $\omega(t)$ and $y(t)$.

(f) Verify the solutions to parts b (ii), c (ii), and (e) using MATLAB.

Problem 4.7 *Consider the following unit feedback control system.*

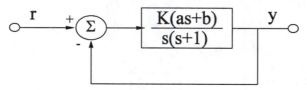

Control System Problem

(a) Discuss the effect of different values of (a, b, K) on the shape of the step response of the system.

(b) Determine the system type with respect to $r(t)$.

Problem 4.8 *Consider the following control system where the feedback gain β is subject to variations.*

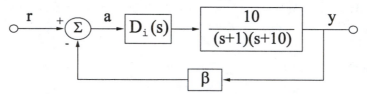

Control System with Variations in β

The objective is to design a controller for this system so that the output $y(t)$ accurately tracks the reference input $r(t)$.

(a) Let $\beta = 1$, and the following three options for the controller $D_i(s)$ are available:

$$D_1(s) = K, \ D_2(s) = \frac{K}{s}, \ D_3(s) = \frac{K}{s^2},$$

where K is a constant. Choose the controller (including a particular value for K) that will result in a type 1 system with a steady state error of less than $\frac{1}{10}$.

(b) Next, suppose that, due to harsh conditions, there is some attenuation in the feedback path that is best modeled by $\beta = 0.9$. Find the steady state error due to a ramp input for the choice of $D_i(s)$ in part (a).

(c) If $\beta = 0.9$, what is the system type for part (b)?

Problem 4.9 *A control system has the structure shown in the following diagram.*

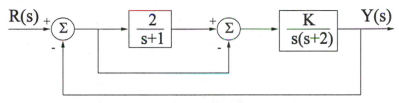

Control System

Determine the gain at which the system will become unstable. (*Ans: $0 < K < 1.5$*)

Problem 4.10 *Designers have developed a small, fast, vertical-takeoff fighter aircraft that is invisible to radar (stealth aircraft). This aircraft concept uses quickly turning jet nozzles to steer the airplane. The control system for the heading or direction control is shown in the following diagram.*

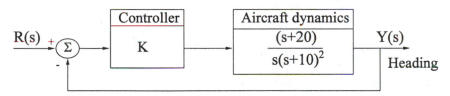

Vertical-Takeoff Aircraft

Determine the maximum gain of the system for stable operation.

Problem 4.11 *A system is represented by equation $\dot{\mathbf{x}} = \mathbf{A}\mathbf{x}$ where*

$$A = \begin{bmatrix} 0 & 1 & 0 \\ 0 & 0 & 1 \\ -1 & -c & -2 \end{bmatrix}$$

Find the range of c where the system is stable.

Problem 4.12 *Utilizing Routh-Hurwitz criterion, determine the stability of the following polynomials:*

   (a) $s^2 + 5s + 2$
   (b) $s^4 + s^3 + 3s^2 + 2s + K$
   (c) $s^5 + s^4 + 2s^3 + s + 5$
   (d) $s^5 + s^4 + 2s^3 + s^2 + s + K$

Problem 4.13 *Arc welding is one of the most important areas of application for industrial robots. In most manufacturing welding situations, uncertainties in dimensions of the part, geometry of the joint, and the welding process itself require the use of sensors for maintaining the weld quality. Several systems use a vision system to measure the geometry of the puddle of melted metal as shown in the following diagram, where the system uses a constant rate of feeding the wire to be melted.*

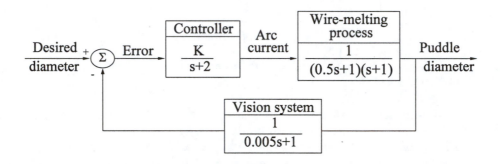

Arc Welding Control System

   (a) *Calculate the maximum value of K for the system that will result in a stable system.*
   (b) *For 1/2 of the maximum value found in part (a), determine the roots of the characteristic equation.*
   (c) *Estimate the overshoot of the system of part (b) when it is subjected to a step input.*

Problem 4.14 *A feedback control system is shown in the following diagram.*

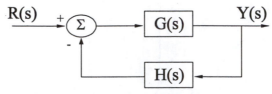

Feedback Control System

The process transfer function is

$$G(s) = \frac{K(s+40)}{s(s+10)},$$

and the feedback transfer function is $H(s) = 1/(s+20)$.

(a) *Determine the limiting value of gain* K *for a stable system.*

(b) *For the gain that results in marginal stability, determine the magnitude of the imaginary roots.*

(c) *Reduce the gain to 1/2 of the magnitude of the marginal value and determine the relative stability of the system (1) by shifting the axis and using the Routh-Hurwitz criterion and (2) by determining the root locations. Show that the roots are between* -1 *and* -2.

Problem 4.15 *A unity feedback control system is shown in the following diagram.*

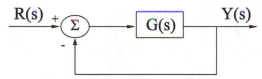

Unity Feedback Control System

Determine the relative stability of the system with the following transfer function by locating the complex roots in the s-plane.

(a)

$$G(s) = \frac{65 + 33s}{s^2(s+9)}$$

(b)

$$G(s) = \frac{24}{s(s^3 + 10s^2 + 35s + 50)}$$

(c)

$$G(s) = \frac{3(s+4)(s+8)}{s(s+5)^2}$$

Problem 4.16 *On July 16, 1993, the elevator in Yokohama's 70-story Landmark Tower, operating at a peak speed of 45 km/hr (28 mph), was inaugurated as the fastest super-fast elevator. To reach such a speed without leaving passengers' stomachs on the ground floor, the lift accelerates for longer periods, rather than more precipitously. Going up it reaches full speed only at the 27th floor; it begins to decelerate 15 floors later. The* *result is a peak acceleration similar to that of other skyscraper elevators-a bit less than a tenth of the force of gravity.*

Admirable ingenuity has gone into making this safe and comfortable. Special ceramic brakes had to be developed; iron ones would melt. Computer-controlled systems damp out vibrations. The lift has been streamlined to reduce wind noise as it hurtles up and down. One proposed system for the elevator's vertical position is shown in the following diagram.

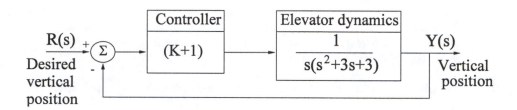

Elevator Vertical Position Control

Determine the range of K for a stable system.

Problem 4.17 *Consider the case of a navy pilot landing an aircraft on an aircraft carrier. The pilot has three basic tasks, the first of which is to guide the aircraft's approach to the ship along the extended centerline of the runway. The second task is maintaining the aircraft on the correct glideslope, the third task is that of maintaining the correct speed. A model of lateral position control system is shown in the following diagram.*

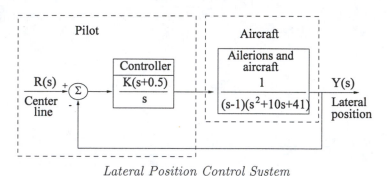

Lateral Position Control System

Determine the range of stability for $K \geq 0$.

Problem 4.18 *A chemical control system is shown in the following diagram. It is desired that the system be stable and the steady state error for a unit step input be less than or equal to 0.05 (5%).*

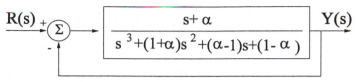

A Chemical Control System

(a) Determine the range of α that satisfies the error requirement.
(b) Determine the range of α that satisfies the stability required.
(c) Select an α that meets both requirements.

Problem 4.19 *The control of the spark ignition of an automotive engine requires constant performance over a wide range of parameters. The control system is shown in the following diagram, where a controller gain K is to be selected.*

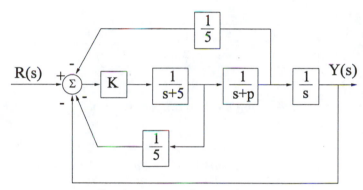

Spark Ignition Control System

The parameter p is equal to 2 for many autos but can equal *zero* for those with high performance. Select a gain K that will result in a stable system for both values of p.

Problem 4.20 *An automatically guided vehicle on Mars is represented by the following figure.*

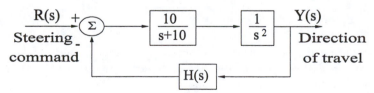

Mars Vehicle Control System

The system has a steerable wheel in both the front and back of the vehicle, and the design requires that $H(s) = Ks + 1$. Determine
(a) the value of K required for stability,

(b) the value of K when one root of the characteristic equation is equal to $s = -5$,

(c) the value of the two remaining roots for the gain in part (b).

(d) Find the response of the system to a step command for the gain selected in part (b).

Problem 4.21 *A traffic control signal is designed to control the distance between vehicles as shown in the following diagram*

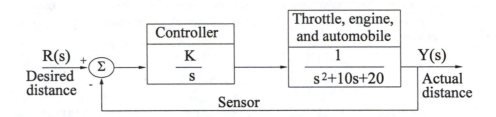

A Traffic Control Signal Control

(a) Determine the range of gain K for which the system is stable.

(b) If K_m is the maximum value of K so that the characteristic roots are on the $j\omega$-axis, then $K = K_m/N$ where $6 < N < 7$. We desire that the peak time be less than 2 seconds and the percentage overshoot be less than 18%. Determine an appropriate value of N.

Problem 4.22 *Consider a unity-feedback control system whose open-loop transfer function is*

$$G(s) = \frac{K}{s(Js + B)}$$

Discuss the effects that varying the values of K and B has on the steady state error in unit-ramp response. Sketch typical unit-ramp response curves for a small value, medium value, and large value of K.

Problem 4.23 *The following diagram shows three systems: System I is a positional servo system. System II is a positional servo system with PD control. System III is a positional servo system with velocity feedback.*

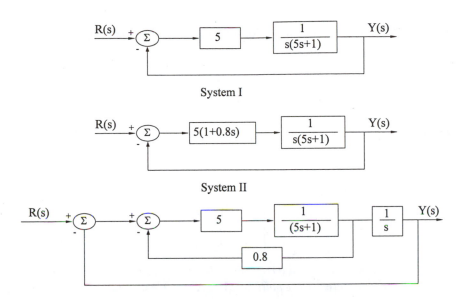

System I

System II

System III

Three Control Systems

Compare the unit-step, unit-impulse, and unit-ramp responses of the three systems. Which system is best with respect to speed of response and maximum overshoot in the step response?

Problem 4.24 *Consider the position control system shown in the following figure.*

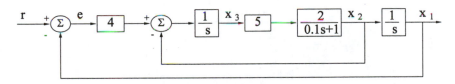

Control System: Multiple Loop Control

Write a MATLAB program to obtain a unit-step response and a unit-ramp response of the system. Plot curves $x_1(t)$ vs. t, $x_2(t)$ vs. t, and $e(t)$ vs. t [where $e(t) = r(t) - x_1(t)$] for both unit-step and unit-ramp response.

Problem 4.25 *Determine the range of K for stability of a unity-feedback control system whose open-loop transfer function is*

$$G(s) = \frac{K}{s(s+1)(s+2)}$$

Problem 4.26 *Consider the unity-feedback control system with the following open-loop transfer function:*

$$G(s) = \frac{10}{s(s-1)(2s+3)}$$

Is this system stable?

Problem 4.27 *Consider the system*

$$\dot{\mathbf{x}} = \mathbf{A}\mathbf{x}$$

where $\mathbf{A}$ *is given by*

$$\mathbf{A} = \begin{bmatrix} 0 & 1 & 0 \\ -b_3 & 0 & 1 \\ 0 & -b_2 & -b_1 \end{bmatrix}$$

($\mathbf{A}$ *is called the Schwarz matrix.) Show that the first column of the Routh's array of the characteristic equation* $|s\mathbf{I} - \mathbf{A}| = 0$ *consists of* $1, b_1, b_2,$ *and* b_1b_3.

Problem 4.28 *Without using the Routh-Hurwitz criterion, determine if the following systems are asymptotically stable, marginally stable, or unstable. In each case, the closed-loop transfer function is given.*
   (a)

$$M(s) = \frac{10(s+2)}{s^3 + 3s^2 + 5s}$$

   (b)

$$M(s) = \frac{K}{s^3 + 5s + 5}$$

   (c)

$$M(s) = \frac{100}{s^3 - 2s^2 + 3s + 10}$$

   (d)

$$M(s) = \frac{s-1}{(s+5)(s^2+2)}$$

Problem 4.29 *Using the Routh-Hurwitz criterion, determine the stability of the closed-loop system that has the following characteristic equations. Determine the number of roots of each equation that are in the right half s-plane and on the* $j\omega$-*axis.*
   (a) $s^3 + 25s^2 + 10s + 450 = 0$
   (b) $s^3 + 25s^2 + 10s + 50 = 0$
   (c) $s^3 + 25s^2 + 250s + 10 = 0$

Problem 4.30 *Given the system in state equation form,*

$$\frac{d\mathbf{x}(t)}{dt} = \mathbf{A}\mathbf{x}(t) + \mathbf{B}u(t)$$

where

(a) $\mathbf{A} = \begin{bmatrix} 1 & 0 & 0 \\ 0 & -3 & 0 \\ 0 & 0 & -2 \end{bmatrix}$ $\qquad \mathbf{B} = \begin{bmatrix} 1 \\ 0 \\ 1 \end{bmatrix}$

(b) $\mathbf{A} = \begin{bmatrix} 1 & 0 & 0 \\ 0 & -2 & 0 \\ 0 & 0 & 3 \end{bmatrix}$ $\qquad \mathbf{B} = \begin{bmatrix} 0 \\ 1 \\ 1 \end{bmatrix}$

Problem 4.31 *Robot control: Let us consider the control of a robot arm. It is predicted that there will be about 100,000 robots in service throughout the world by 2000. The robot is a six-legged micro robot system using highly flexible legs with high gain controllers that may become unstable and oscillate. Under this condition the characteristic polynomial is given by*

$$q(s) = s^5 + s^4 + 4s^3 + 24s^2 + 3s + 63$$

Problem 4.32 *Welding Control: Large welding robots are used in today's auto plants. The welding head is moved to different positions on the auto body, and rapid, accurate response is required. A block diagram of a welding head positioning system is shown in the following diagram.*

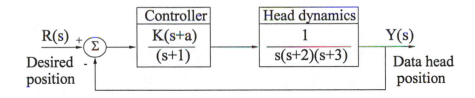

Welding Robot Control System

It is desired to determine the range of K and a for which the system is stable. The characteristic equation is

$$1 + G(s) = 1 + \frac{K(s+a)}{s(s+1)(s+2)(s+3)} = 0$$

Chapter 5

Root Locus Design Methods

5.1 Introduction

This chapter introduces and discusses the notion of root locus design, explains the procedure of creating root loci and outlines their uses. Definitions of the necessary terms are provided including a step-by-step guide to constructing a root locus, and details of how to design and evaluate controllers using the root locus method. Given a feedback control system, the root locus illustrates how the poles of the closed-loop system vary with system parameters, in particular the closed-loop gain. It is a method that shows how changes in the system's feedback characteristics and other parameters influence the pole locations. Root locus is a powerful graphic method for analysis and design of control systems.

Although the root locus method is commonly used to study the effect of control gain variations, it is also used to plot the roots of any polynomial expressed in the Evans root locus form. Most control systems work by regulating the system they are controlling around a desired operating point. In practice, control systems must have the ability not only to regulate around an operating point, but also to reject disturbances and to be robust to changes in their environment. The root locus method helps the designer of a control system to understand the stability and robustness properties of the controller at an operating point. Material in this chapter enables the reader to create a root locus and use the locus to understand the closed-loop system behavior given an open-loop system and a feedback controller. Case studies and examples that illustrate how to use the root locus for designing a control system are presented.

5.2   Root Locus

5.2.1   Background

Root locus is a powerful graphic method used in the analysis and design of control systems. Given a feedback control system, the root locus illustrates how the poles of the closed-loop system vary with system parameters in particular the closed-loop gain. It is a graph of the location of the roots as the system parameters vary in the s-plane. The study of control systems with respect to system parameters assumes importance in light of the following issues:

- How changes in the system's feedback characteristics and other parameters influence the pole locations

- Identifying the locations of the closed-loop pole in the s-plane as the parameter changes (this produces the root locus).

- Use of the root locus to design and analyze feedback control systems

- Use of computers (MATLAB) to generate root loci

- Root locus design when two or more parameters are varying, e.g., PID where there are three adjustable parameters

- Determination of control system stability from the intersection (or lack of intersection) of the root loci with the imaginary axis

- The addition of poles and zeros (compensation) to the open-loop transfer function in order to influence the root locus, thus satisfying design specifications (compensator design by the root locus method)

- Stability can be improved by addition of a zero and worsened by the addition of a pole

5.2.2   The Root Locus: Definition

Consider the block diagram of a general closed-loop system with a proportional controller gain K shown in Figure 5.1

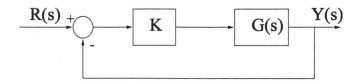

FIGURE 5.1
Closed Loop Unity Feedback Block Diagram

The closed-loop transfer function is given by

$$T(s) = \frac{Y(s)}{R(s)}$$

$$= \frac{KG(s)}{1 + KG(s)} \tag{5.1}$$

$$\equiv \frac{b(s)}{a(s)}, \tag{5.2}$$

where $a(s)$ and $b(s)$ are polynomials in s.

The characteristic equation of a system is based on the transfer function that models the system. There is only one characteristic equation for a given system. The characteristic equation is defined by equating the denominator of the transfer function to zero. Hence, for the system in Figure 5.1 the characteristic equation is given by $a(s) = 0$ and is thus obtained from

$$1 + KG(s) = 0. \tag{5.3}$$

The root locus is defined as the set of roots that satisfy this equation or as the path traced by the location of the roots of the characteristic equation (the poles of the closed-loop system) in the s-plane as the closed-loop control gain K is varied from zero to infinity, i.e., $K \geq 0$. Graphically, the locus is the set of paths in the complex plane traced by the closed-loop poles as the gain is varied. The characteristic equation defines where the poles will be located for any value of the control gain, K. In other words, it defines the characteristics of the system behavior for various values of controller gain. On the root locus, the characteristic equation is always satisfied.

In the context of root locus design methods, the control gain K is also called the root locus gain. As the gain is varied, the corresponding variations in the poles of the closed-loop system determine the root locus. As the gain increases from zero to infinity, the poles move from the open-loop poles along the locus toward open-loop zeros or toward infinity. The root locus gain, K, appears in both the numerator and the denominator of the closed-loop transfer function in Equation 5.1. The root locus is created using only the denominator (the characteristic equation) of the closed-loop transfer function.

The varying parameter does not have to be the controller gain, but rather any system parameter. Hence, in general, the root locus illustrates how the poles of the closed-loop system vary with changes in any system parameter. However, before the root locus is determined, the system characteristic equation must be written in the form of Equation 5.3. This form is called the Evans root locus form and it is generalized for any system parameter q by

$$1 + qP(s) = 0, \tag{5.4}$$

where $P(s)$ is a function of s. In this book, emphasis is placed on root loci with respect to the control gain K (also referred to as the root locus gain) in which case Equation 5.4 is reduced to Equation 5.3, that is, $q = K$ and $P(s) = G(s)$. However, all the principles and techniques that are discussed can be applied to root loci with respect to any varying system parameter by starting with Equation 5.4.

5.2.3   Magnitude and Angle Criteria

The magnitude and angle criteria are direct results of the definition of the root locus. They constitute another way of expressing the root locus requirements (magnitude and phase angle). The root locus characteristic Equation 5.3 can be expressed in terms of a magnitude and a phase angle as follows:

$$1 + KG(s) = 0$$

$$KG(s) = -1$$

$$|KG(s)| \angle KG(s) = -1 + j0.$$

The equation of the magnitude defines the magnitude criterion

$$|KG(s)| = 1. \tag{5.5}$$

Every point on the root locus must satisfy this criterion. The magnitude criterion is used to determine the locations of a set of roots in the s-plane for a given value of K.

The equation for the phase angle $\angle KG(s)$ defines the angle criterion

$$\angle KG(s) = \tan^{-1}\left(\frac{0}{-1}\right) \tag{5.6}$$

$$= -180° \pm 360k°, \tag{5.7}$$

where k is an integer. Note that $+180°$ could be used rather than $-180°$. The use of $-180°$ is just a convention and since $+180°$ and $-180°$ are the same angle, either angle produces the same result. Every point on the root locus must satisfy the angle criterion. This criterion is used to determine the departure angles for the parts of the root locus near the open-loop poles and the arrival angles for the parts of the root locus near the open-loop zeros.

When used together, the magnitude criterion and the angle criteria can be used to determine whether a point in the s-plane is on the root locus. On the root locus, the characteristic equation is always satisfied and hence both the magnitude and angle criteria are also always satisfied.

5.2.4 Breakpoint, Departure and Arrival Angles

The angle of departure is the angle at which the locus leaves a pole in the s-plane. The angle of arrival is the angle at which the locus arrives at a zero in the s-plane. By convention, both types of angles are measured relative to a ray starting at the origin and extending to the right along the real axis in the s-plane. Both arrival and departure angles are found using the angle criterion. When there are multiple poles or zeros at a point in the complex plane, the angles are evenly distributed about the point.

Breakpoints occur on the locus where two or more loci converge or diverge. Breakpoints often occur on the real axis, but they may appear anywhere in the s-plane. The loci that approach/diverge from a breakpoint do so at angles spaced equally about the breakpoint. The angles at which they arrive/leave are a function of the number of loci that approach/diverge from the breakpoint. Breakpoints indicate places on the locus where a multiple root exists for some value of the root locus gain. A breakpoint may have more than two loci leading to/from it. The breakpoint indicates a point where a third- or higher-order root exists for some value of K.

5.3 Constructing the Root Locus

This section outlines the steps to creating a root locus and illustrates the important properties of each step in the process. By the end of this section, the reader should be able to sketch a root locus given the open-loop poles and zeros of a system. Using these steps, the locus will be detailed enough to evaluate the stability and robustness properties of the closed-loop controller. In many cases, the designer of a control system needs a quick estimate of the behavior of the resulting closed-loop system and a root locus provides this kind of information.

5.3.1 Summary of the Root Locus Steps

The procedure of drawing a root locus can be summarized into eight steps:

Step 1 Express the characteristic equation in the Evans root locus form
Step 2 Obtain and draw the open-loop poles ($\times$) and zeros ($\circ$)
Step 3 Draw the part of the locus that lies on the real axis
Step 4 Locate the centroid and sketch the asymptotes (if any)
Step 5 Determine the breakpoint locations (if any)
Step 6 Determine the angles of arrival/departure
Step 7 Calculate the imaginary axis crossings (if any)
Step 8 Draw the rest of the locus by connecting the poles with the breakpoints, axis crossings, asymptotes, and arrival angles

It is important to note that one only has to draw the locus in the upper or lower half-plane since the root locus is always symmetric about the real axis.

5.3.2 Details of the Root Locus Steps

Each of the above steps is covered in detail in this section and the important properties of each step are illustrated. Using these steps, the locus will be detailed enough to evaluate the stability and robustness properties of the closed-loop controller.

Step 1 Evans Root Locus Form

Write the system characteristic equation in the Evans root locus form

$$1 + KG(s) = 0,$$

where K is the system parameter of interest and $G(s)$ is a function of s.

Step 2 Open-Loop Zeros and Poles

Locate the open-loop poles and zeros and denote them by $\times$ and $\circ$, respectively, on the s-plane. Factor $G(s)$ into poles and zeros and rewrite the characteristic equation as follows:

$$1 + KG(s) = 0$$

$$1 + K\frac{\Pi_{i=1}^{m}(s - z_i)}{\Pi_{j=1}^{n}(s - p_j)} = 0$$

$$\Pi_{j=1}^{n}(s - p_j) + K\Pi_{i=1}^{m}(s - z_i) = 0.$$

Since the locus represents the path of the roots (specifically, paths of the closed-loop poles) as the root locus gain is varied, the starting point is location of the roots when the gain of the closed-loop system is 0. Each locus starts at an open-loop pole and ends at a open-loop zero. If the system has more poles than zeros, then some of the loci end at zeros located infinitely from the poles. Draw the poles and zeros exactly as they appear in the open-loop system. Include all of the poles and zeros, i.e., poles and zeros of both the controller and the uncontrolled system. The poles will be the starting points of the loci, and the zeros will be the ending points.

When $K = 0$, then the roots of the characteristic equation give the poles of $G(s)$

When $K = \infty$, then the roots of the characteristic equation give the zeros of $G(s)$

The loci of characteristic equation roots begin at the poles and ends at the zeros of $G(s)$ as K increases from 0 to ∞. Note: most functions $G(s)$ have zeros that lie at ∞, because they have more poles than zeros i.e. $n > m$. (n is the number of poles and m is the number of zeros) $(n - m)$ branches of the root locus approaching the $(n - m)$ zeros at infinity

Step 3 Real Axis Crossings

Draw the part of the locus that lies on the real axis. Locate the segments of the real axis that are part of the root loci. The root locus on the real axis lies in a section to the left of an odd number of poles and zeros.

Many root loci have paths on the real axis. The real axis portion of the locus is determined by applying the following rule:

If an odd number of open-loop poles and open-loop zeros lie to the right of a point on the real axis, that point belongs to the root locus.

Note that the real axis section of the root locus is determined entirely by the number of open-loop poles and zeros and their relative locations. Since the final root locus is always symmetric about the real axis, the real axis part is fairly easy to carry out.

Start at positive infinity on the real axis. Move toward the origin until a pole or zero is encountered on the real axis. Draw a line from this pole/zero until the next pole or zero on the real axis is reached. If there are no more poles/zeros, the locus extends to negative infinity on the real axis. Otherwise, the locus starts again at the next pole/zero and continues to its successor, and so on.

If there are no poles or zeros on the real axis, then there will be no real axis component to the root locus. Some systems have more than one pole or zero at the same location (this indicates a double, triple, or even higher-order root to the characteristic equation). If there are an odd number of poles or zeros at the same location, the real axis part of the locus continues after the location of that pole/zero. If the number of poles/zeros at the location is even, the real axis part of the locus stops at that location. Pick any point on the real axis. If there is an odd number of roots to the right of that point, that point on the axis is a part of the locus. If there is a multiple root, then the real axis part depends on whether there is an even or odd number of roots at the same point.

Step 4 Centroid and Asymptotes

The asymptotes indicate where the poles will go as the gain approaches infinity. For systems with more poles than zeros, the number of asymptotes is equal to the number of poles minus the number of zeros. In some systems, there are no asymptotes; when the number of poles is equal to the number of zeros, then each locus is terminated at a zero rather than asymptotically to infinity. The asymptotes are symmetric about the real axis, and they stem from a point defined by the relative magnitudes of the open-loop roots. This point is called the centroid. Note that it is possible to draw a root locus for systems with more zeros than poles, but such systems do not represents physical systems. In these cases, some of the poles can be thought of as being located at infinity. First determine how many poles, n, and how many zeros, m, are in the system, then locate the centroid. The number of asymptotes is equal to the difference between the number of poles and the number of zeros. The location of the centroid σ on the real axis is given by:

$$\sigma = \frac{\sum_{i=1}^{n} \sigma_{p_i} - \sum_{j=1}^{m} \sigma_{z_j}}{n - m},$$

where p_i and z_j are the poles and zeros, respectively. Since p_i and z_j are symmetric about the real axis, their imaginary parts get cancelled out.

Once the centroid has been located, the next step is to draw the asymptotes at the proper angles. The asymptotes will leave the centroid and angles defined by

$$\pm 180° \left(\frac{2q + 1}{n - m} \right) \quad \text{where } q = 0, 1, 2, \dots (n - m - 1).$$

Step 5 Breakpoints

Breakpoints occur where two or more loci join, then diverge. Although they are most commonly encountered on the real axis, they can also occur elsewhere in the complex plane. Each breakpoint is a point where a double (or higher-order) root exists for some value of K. Mathematically, from the root locus equation

$$1 + KG(s) = 0$$
$$KG(s) = -1$$
$$K = \frac{-1}{G(s)}$$
$$= \frac{-1}{\left(\dfrac{b(s)}{a(s)}\right)}$$
$$= -\frac{a(s)}{b(s)},$$

where the transfer function $G(s)$ consists of a numerator, $b(s)$ and denominator, $a(s)$, then the breakpoints can be determined from the roots of

$$\frac{dK}{ds} = \frac{-\left[b(s)\dfrac{da(s)}{ds} - a(s)\dfrac{db(s)}{ds}\right]}{[b(s)]^2} = 0.$$

If K is real and positive at a value s that satisfies this equation, then the point is a breakpoint. There will always be an even number of loci around any breakpoint because for each locus that enters the breaking point, there must be one that leaves.

Perhaps the easiest way to find breakpoints is by trial and error. First, determine the characteristic equation of the system in terms of K. In the vicinity of the suspected breakpoint, substitute values for s in this equation. A breakpoint occurs when the characteristic equation is minimized. To calculate the breakpoint explicitly requires obtaining the derivative of the characteristic equation in terms of s and then equating to zero. The resulting equation is then solved for K and s.

Step 6 Angles of Arrival/Departure

The angle criterion determines which direction the roots move as the gain moves from zero (angles of departure, at open-loop poles) to infinity (angles of arrival, at open-loop zeros). An angle of departure/arrival is calculated at the complex open-loop poles and zeros.

Angle of departure.

At each complex pole, add up the angles from the zeros to the current pole, then subtract the angles from the other poles to the current pole. In mathematical terms, for a given pole, the angle of departure is

$$\theta_{dep} = 180 - \sum_{i=1}^{n} \theta_i - \sum_{j=1}^{m} \phi_j,$$

where θ_i is the angle between the *ith* pole and the given pole and ϕ_j is the angle between the *jth* zero and the given pole. These angles can be calculated using trigonometry.

Angle of arrival.

At each zero, add up the angles from the poles to the current zero, then subtract the angles from the other zeros to the current zero. In mathematical terms, for a given zero, the angle of arrival is

$$\theta_{arr} = 180 + \sum_{i=1}^{n} \theta_i - \sum_{j=1}^{m} \phi_j,$$

where θ_i is the angle between the *ith* pole the given zero, and ϕ_j is the angle between the *jth* zero and the given zero.

By convention, the arrival and departure angles are measured relative to the real axis, so that the positive real axis is 0.

Note that single poles and zeros on the real axis will always have arrival/departure angles equal to 0 or 180 degrees due to the symmetry of the complex conjugates.

Step 7 Axis Crossings

The points where the root locus intersects the imaginary axis indicate the values of K at which the closed-loop system is marginally stable. The closed-loop system will be unstable for any gain for which the locus is in the right half-plane of the complex plane.

If the root locus crosses the imaginary axis from left to right at a point where $K = K_0$ and then stays completely in the right half-plane, then the closed-loop system is unstable for all $K > K_0$. Therefore, knowing the value of K_0 is very useful.

Some systems are particularly complex when their locus dips back and forth across the imaginary axis. In these systems, increasing the root locus gain will cause the system to go unstable initially and then become stable again.

Not every locus will have imaginary axis crossings. First, determine if the locus will definitely cross the imaginary axis (for example, if there are more than two asymptotes), or if there is a good chance that the locus crosses the imaginary axis (for example, if there are poles or zeros close to the imaginary axis and/or the arrival/departure angles leads to the conclusion that the locus may cross the axis).

There are three ways to find the points where the locus intersects the imaginary axis:

(1) Trial and error (bracketing).

(2) The Routh-Hurwitz stability criterion.

(3) Solving for omega (ω_d) and K.

The method used depends on the accuracy required for the locations of the axis crossings.

Trial and error.

Start at the origin in the complex plane. Move up the imaginary axis in discrete steps and calculate the phase of the forward loop transfer function at each step. If the phase at the last point was less than 180 degrees and the phase at the current point is greater than 180 degrees (or vice versa) then an axis crossing lies between the two points. If the phase is equal to 180 degrees, then the point is on the locus and is an imaginary axis crossing point.

By bracketing regions on the imaginary axis, the axis crossings can be quickly determined. Rather than working up from the origin in regular steps, bracketing uses a binary search approach in which two points are tested, then another point is chosen based on whether there was a phase change between the two points. If there was a phase change, the third point is chosen between the two, if not, it is chosen outside the two.

Routh-Hurwitz Stability Criterion.

From the characteristic equation, the matrix of coefficients is created as is done when determining the stability of the system (as developed in Chapter 4). Then, from the matrix of coefficients, solve for K such that the stability criterion is met. Then solve for s to determine where on the imaginary axis the gain K is in effect. Note that this method can be very difficult to use, especially for systems with many poles and zeros.

Solving for ω_d and K.

Let $s = j\omega_d$ in the characteristic equation, equating both the real and imaginary parts to zero, then solve for ω_d and K. The values of ω_d are the frequencies at which the root loci cross the imaginary axis. The value of K is the root locus gain at which the crossing occurs.

Step 8 Sketch the Complete Root Locus

The complete root locus can be drawn by starting from the open-loop poles, connecting the real axis section, breakpoints, and axis crossings, then ending at either the open-loop zeros or along the asymptotes to infinity and beyond.

If the hand-drawn locus is not detailed enough to determine the behavior of your system, then MATLAB or some other computer tool can be used to calculate the locus exactly.

Now sketch in the rest of the locus. Use the asymptotes, arrival angles, departure angles, breakpoints, and axis crossings to guide the sketch. The final locus will include these points and will connect them smoothly. The shapes of the locus parts will depend on the proximity of the open-loop roots to each other.

In general, zeros tend to repel the locus, whereas poles tend to attract the locus. One locus segment tends to attract another locus segment until a breakpoint forms.

Typically, the only time needed to determine exactly the locus shape is when the locus is near the imaginary axis or in regions where a detailed understanding of the system behavior in the time domain is required. In these cases, if the previous steps did not yield locus details sufficiently accurate for the specific purposes, then use a computer tool to generate the locus exactly. These are some root locus plots for a variety of systems. They include the construction marks for arrival/departure angles, asymptotes, breakpoints, and axis crossings. Note that in some cases, a slight change in pole or zero coordinates can result in a markedly different locus. Note also, however, that such small changes to the roots will not change more general locus characteristics, such as the number of asymptotes.

5.3.3   Determining the Control Gain (Root Locus Gain)

The root locus shows graphically how the system roots will move as the root locus gain is changed. Often, however, one must determine the gain at critical points on the locus, such as points where the locus crosses the imaginary axis. The magnitude criterion is used to determine the value of the root locus gain K at any point on the root locus. The gain is calculated by multiplying the lengths of the distance between each pole to the point then dividing that by the product of the lengths of the distance between each zero and the point. Note that a linear change in position on the locus usually does not correspond to a linear change in the root locus gain.

5.3.4   Root Locus for Second-Order Systems

In Chapter 3, the s-plane or complex plane was introduced as a two-dimensional space defined by two orthogonal axes, the real number axis and the imaginary number axis. A complex pole (a complex number) has both a real component $(-\sigma)$ and an imaginary component (ω_d) such that

$$s \equiv -\sigma + j\omega_d.$$

As discussed in Chapter 3, any second-order system's characteristic equation can be represented by the general equation

$$s^2 + 2\xi\omega_n s + \omega_n^2 = 0.$$

The corresponding general roots of this equation are given by

$$s = \frac{-b \pm \sqrt{b^2 - 4ac}}{2a}$$

$$= \frac{-2\xi\omega_n \pm \sqrt{4\xi^2\omega_n^2 - 4\omega_n^2}}{2}$$

$$= -\xi\omega_n \pm \omega_n\sqrt{\xi^2 - 1}$$

$$= -\xi\omega_n \pm j\omega_n\sqrt{1 - \xi^2}$$

$$\equiv -\sigma \pm j\omega_d,$$

where

$$\sigma = \xi\omega_n \quad \text{and} \quad \omega_d = \omega_n\sqrt{1 - \xi^2}.$$

The following figure compares the generic root with a general point in the s-plane.

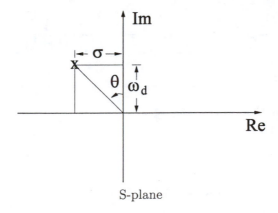

S-plane

From this figure it follows that

$$\tan\theta = \frac{\sigma}{\omega_d}$$

$$= \frac{\xi\omega_n}{\omega_n\sqrt{1-\xi^2}}$$

$$\tan\theta = \frac{\xi}{\sqrt{1-\xi^2}} = \frac{\sin\theta}{\cos\theta}$$

$$\Rightarrow \sin\theta = \xi \quad (\text{and} \quad \cos\theta = \sqrt{1-\xi^2})$$

$$\Rightarrow \theta = \sin^{-1}\xi.$$

Since the roots $(-\sigma \pm j\omega_d)$ and the control gain (root locus gain) K are related by the characteristic equation or the magnitude criteria

$$1 + KG(s) = 0$$

$$|KG(s)| = 1,$$

it follows that the gain K required to produce a particular damping ratio ξ or damping angle θ can be determined. Similarly, given the root locus gain (control gain) K, the system damping ratio ξ or damping angle θ can be obtained. These properties will be illustrated later in some examples.

5.4   Examples of Root Locus Design

Example 5.1 *Consider the feedback control system shown below where* $K \geq 0$.

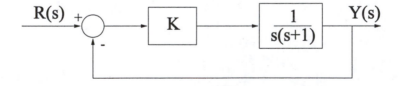

FIGURE 5.2
Root Locus Diagram

(a) *Find the root locus of the system with respect to the controller gain* K.
(b) *For what range of* K *is the system stable ?*

(c) Find the value of the gain K that will allow the system to have a root at

$$s = -\frac{1}{2} + j\frac{\sqrt{3}}{2}.$$

(d) Find the controller gain K that will be required to give the system a damping ratio of $\xi = 0.5$.

Solution 5.1 *(a) The closed-loop transfer function of the system is given by*

$$T = \frac{K\left[\dfrac{1}{s(s+1)}\right]}{1 + K\left[\dfrac{1}{s(s+1)}\right]}.$$

Hence, the characteristic equation (in Evans root locus form) is given by

$$1 + \frac{K}{s(s+1)} = 0$$

$$1 + \frac{K}{(s+0)(s+1)} = 0$$

The system has two poles one at $s = 0$, another at $s = -1$, and no zeros. Rewriting the equation gives

$$s(s+1) + K = 0$$

For $K = 0$, the roots of the characteristic equation are at 0, and -1. The two roots are given from the solution of the quadratic equation.

$$s^2 + s + K = 0$$

$$r_1, r_2 = -\frac{1}{2} \pm \frac{1}{2}\sqrt{1 - 4K}$$

The root locus is thus described by the following equations:

$$f(x) = \begin{cases} r_1, r_2 = -1 \quad and \quad 0 & for \quad K = 0 \\[2mm] r_1, r_2 = -\dfrac{1}{2} \pm \dfrac{1}{2}\sqrt{1 - 4K} & for \quad 0 \le K \le \dfrac{1}{4} \\[2mm] r_1, r_2 = -\dfrac{1}{2} \pm \dfrac{1}{2}j\sqrt{4K - 1} & for \quad K > \dfrac{1}{4}. \end{cases}$$

For $0 \leq K \leq \dfrac{1}{4}$ the roots are real (Note that K is not negative) and lie between -1 and 0. For $K = 0$, the roots are $-\dfrac{1}{2} \pm \dfrac{1}{2} \implies -1$ and 0, when $K > \dfrac{1}{4}$ the roots are complex. The two complex roots are of the form $-\dfrac{1}{2} \pm aj$, where a is a real positive constant. Thus, the root locus starts from the poles at 0 and -1 and move toward each other until both roots are equal to $-\dfrac{1}{2}$, then as K continues to increase the roots become complex and the locus moves from the point $(-\dfrac{1}{2}, 0)$ toward positive and negative infinity, parallel to the imaginary axis, describing the line $-\dfrac{1}{2} \pm aj$. The root locus is shown below

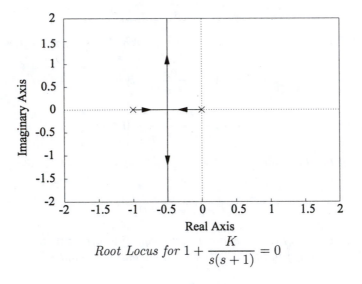

$$\text{Root Locus for } 1 + \frac{K}{s(s+1)} = 0$$

(b) In order to find the range of K for which the system is stable, the objective is to exclude any roots that are not in the LHP. Such a root is at $s = 0$ (where $K = 0$). Therefore, the range of K for which the system is stable is given by

$$K > 0.$$

(c) The value of the gain K that will allow the system to have a root at

$$s = -\frac{1}{2} + j\frac{\sqrt{3}}{2},$$

is determined by using the magnitude criterion.

$$|KG(s)| = 1$$

$$\left| \frac{K}{s(s+1)} \right| = 1$$

$$\frac{|K|}{|s| \, |s+1|} = 1$$

Substituting for the root and solving for K leads to

$$\implies \frac{K}{\left| -\frac{1}{2} + j\frac{\sqrt{3}}{2} \right| \left| -\frac{1}{2} + j\frac{\sqrt{3}}{2} + 1 \right|} = 1$$

$$\implies \frac{K}{\sqrt{\frac{1}{4} + \frac{3}{4}} \sqrt{\frac{1}{4} + \frac{3}{4}}} = 1$$

$$\implies \frac{K}{1 \times 1} = 1$$

$$\implies K = 1$$

If the root is at $s = -0.5$, K is similarly obtained from magnitude criterion

$$\left| \frac{K}{s(s+1)} \right| = 1$$

$$\left| \frac{K}{-\frac{1}{2}\left(-\frac{1}{2} + 1 \right)} \right| = 1$$

$$\left| \frac{K}{-\frac{1}{2}\left(\frac{1}{2} \right)} \right| = 1$$

$$K = \frac{1}{4}$$

If the root $s = 0$, K is similarly obtained from the magnitude criterion

$$\left| \frac{K}{0(0+1)} \right| = 1 \Rightarrow K = 0$$

(d) The general pole is of the form

$$s = -\sigma + j\omega_d,$$

where

$$\sigma = \frac{1}{2}$$

$$\text{and } \tan\theta = \frac{\sigma}{\omega_d} \implies \omega_d = \frac{\sigma}{\tan\theta}$$

If the damping ratio, $\xi = 0.5$, then the damping angle is given by

$$\theta = \sin^{-1}\xi = 30°$$

$$\implies \omega_d = \frac{\frac{1}{2}}{\tan 30°} = \frac{\sqrt{3}}{2}.$$

Therefore, the root for $\xi = 0.5$ is given by

$$s = -\frac{1}{2} + j\frac{\sqrt{3}}{2}.$$

Hence, the value of K that allows the system to have this root is then obtained by the magnitude criterion. In fact, this problem is now similar to (c). Substituting for the root in the expression for the magnitude criterion, and solving for K leads to

$$|KG(s)| = 1$$

$$\frac{K}{\left|-\frac{1}{2} + j\frac{\sqrt{3}}{2}\right|\left|-\frac{1}{2} + j\frac{\sqrt{3}}{2} + 1\right|} = 1$$

$$\frac{K}{\sqrt{\left(\frac{1}{4} + \frac{3}{4}\right)}\sqrt{\left(\frac{1}{4} + \frac{3}{4}\right)}} = 1$$

$$\frac{K}{1 \times 1} = 1$$

$$K = 1.$$

Example 5.2 *Finding root locus for the given block diagram.*

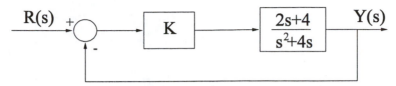

Feedback System

(a) *Find the root locus of the system with respect to the controller gain* K.

(b) *For what range of K is the system stable*

(c) *Find the value of the gain K that will allow the system to have a root at* $s = -1$.

Solution 5.2 *(a) The starting point is deriving the characteristic equation*

$$1 + KG = 0$$

$$1 + K\frac{2s+4}{s^2+4s} = 0 \Rightarrow 1 + \frac{2K(s+2)}{s^2+4s} = 0$$

The characteristic equation is

$$s(s+4) + 2K(s+2) = 0$$

$K = 0$ *gives poles at* $0, -4$ *and* $K = \infty$ *gives zeros at* -2.

$$s^2 + s(4+2K) + 4K = 0$$

$$s = -(2+K) \pm \sqrt{(2+K)^2 - 4K}$$

$$= -(2+K) \pm \sqrt{4 + K^2}$$

The expression under the square root, $(4 + K^2)$ *is always positive, and moreover*

$$(2+K) \geqq \sqrt{4 + K^2} \ for \ 0 \leq K \leq \infty$$

This implies that s is negative. As $K \to \infty$, $s_1 \to 2$ *and* $s_2 \to -2K$, *which means* $s \to \infty$. *The root locus (shown below) lies on the negative real axis between 0 and -2, -4 and* ∞ *as K goes from 0 to infinity.*

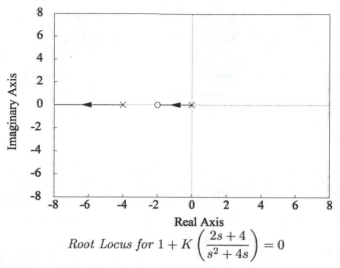

$$\text{Root Locus for } 1 + K\left(\frac{2s + 4}{s^2 + 4s}\right) = 0$$

(b) In order to find the range of K for which the system is stable, the objective is to exclude any roots that are not in the LHP. Such a root is at $s = 0$ (where $K = 0$). Therefore, the range of K for which the system is stable is given by

$$K > 0.$$

(c) Finding gain at specific root location say $s = -1$. Substituting $s = -1$ in the magnitude criterion

$$\frac{K\,|2s + 4|}{|s|\,|(s + 4)|} = 1$$

$$K = \frac{|-1|\,|-1 + 4|}{2\,|-1 + 2|} = \frac{3}{2}.$$

Example 5.3 *Consider the block diagram of a feedback control system shown below, where the controller gain is such that $K \geq 0$.*

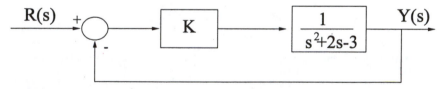

Root Locus System

(a) Find the system root locus with respect to the controller gain K.
(b) For what range of K is the system stable?
(c) Find the value of the gain K that will allow the system to have a root at $s = -1 + 3j$.

Solution 5.3 *(a) The closed-loop transfer function of the system is given by*

$$T = \frac{K\left[\dfrac{1}{s^2 + 2s - 3}\right]}{1 + K\left[\dfrac{1}{s^2 + 2s - 3}\right]}$$

Hence, the characteristic equation (characteristic) (in Evans root locus form) is given by

$$1 + \frac{K}{s^2 + 2s - 3} = 0$$
$$1 + \frac{K}{(s - 1)(s + 3)} = 0$$

The system has two poles, one at $s = -3$, another at $s = 1$, and no zeros. Rewriting the equation gives

$$s^2 + 2s + (K - 3) = 0$$

For $K = 0$, the roots of the characteristic equation are at -3, and 1. The two roots are given from the solution of the quadratic equation

$$s^2 + 2s + (K - 3) = 0$$
$$s = -1 \pm \sqrt{4 - K}.$$

For $0 \leq K \leq 4$ the roots are real (take note K is non negative) and lie between -3 and 1. For $K = 0$, the roots are $-1 \pm 2 \implies -3$ and 1, when $K > 4$ the roots are complex. The two complex roots are of the form $-1 \pm aj$, where a is a real positive constant. Thus, the root locus starts from the poles at 1 and -3 and move toward each other until both roots are equal to -1. Then as K continues to increase the roots become complex and the locus moves from point $(-1, 0)$ toward positive and negative infinity, parallel to the imaginary axis, describing $-1 \pm aj$. All points on the root locus must satisfy the magnitude and angle criteria. The root locus is shown in the following diagram.

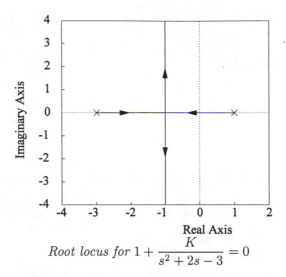

$$\textit{Root locus for } 1 + \frac{K}{s^2 + 2s - 3} = 0$$

(b) For the range of K that permits system stability, the objective is to exclude any roots that are·not in the LHP. Such roots are in the range $0 \le s \le 1$. The critical point is where the root locus crosses the imaginary axis into the RHP. Stability means the roots should be strictly in the LHP

$$-1 \pm \sqrt{4 - K} \;<\; 0$$

$$\Rightarrow \; 4 - K < 1$$

$$\Longrightarrow K > 3.$$

Therefore, the range of K for which the system is stable is given by

$$K > 3.$$

It is important to note that at $K = 3$ the system is marginally stable (it has a root 0 at the origin). The same result can be obtained by using the Routh Hurwitz criteria as shown below.

$$s^2 + 2s + (K - 3) = 0.$$

s^2	1	$K - 3$
s^1	2	0
s^0	$(K - 3)$	

$$(K - 3) > 0$$

$$K > 3.$$

The advantage of determining stability conditions from Routh-Hurwitz array as opposed to root locus plots is that it is not necessary know the location of the roots. This advantage is important in higher-order systems, where it is not easy to obtain the roots.

(c) The gain at the specific root location, $s = -1 + 3j$, is obtained by substituting this root in the expression for the magnitude criterion

$$\left| \frac{K}{s^2 + 2s - 3} \right| = 1$$

$$\frac{K}{|s^2 + 2s - 3|} = 1$$

$$\left| \frac{K}{(-1 + 3j)^2 + 2(-1 + 3j) - 3} \right| = 1$$

$$\left| \frac{K}{(1 - 6j - 9) + (-2 + 6j) - 3} \right| = 1$$

$$\left| \frac{K}{-13} \right| = 1$$

$$K = 13.$$

Example 5.4 *Consider the following block diagram where K has a fixed value (e.g., K=1) and the parameter of interest is c.*

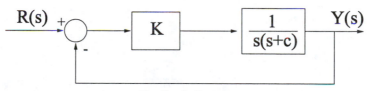

Feedback System

(a) Find the root locus of the system with respect to the parameter c.

(b) For what range of c is the system stable.

(c) Find the value of the gain c that will allow the system to have a root at $s = -\dfrac{1}{2}$.

Solution 5.4 *(a) The characteristic equation of the system is given by*

$$1 + KG(s) = 0$$

$$1 + \frac{K}{s(s + c)} = 0$$

$$s(s + c) + K = 0.$$

The Evans root locus form is obtained as follows:

$$s(s + c) + K = 0$$

$$s^2 + cs + K = 0$$

$$s^2 + K + cs = 0.$$

Dividing throughout by $(s^2 + K)$ gives the Evans root locus form

$$\frac{s^2 + K}{s^2 + K} + \frac{cs}{s^2 + K} = 0$$

$$1 + \frac{cs}{s^2 + K} = 0.$$

For $K = 1$, the Evans root locus form reduces to

$$1 + \frac{cs}{s^2 + 1} = 0.$$

The roots of the characteristic equation

$$s^2 + cs + 1 = 0,$$

are of the general form

$$s = -\frac{c}{2} \pm \frac{\sqrt{c^2 - 4}}{2}$$

$$\implies s = -\frac{c}{2} \pm j\frac{\sqrt{4 - c^2}}{2}$$

For $c = 0$ the roots are at $s = \pm j\sqrt{1}$. Using the general form of the roots, let the real part be $x = -\frac{c}{2}$, the imaginary part

$$y = \pm\frac{\sqrt{4 - c^2}}{2}$$

Consider the following function

$$x^2 + y^2 = \left(-\frac{c}{2}\right)^2 + \left(\pm\frac{\sqrt{4 - c^2}}{2}\right)^2$$

$$= \frac{c^2}{4} + \frac{4 - c^2}{4} = 1$$

$$= 1$$

This is the equation of a circle of radius 1 with its center at the origin. Since the parameter c is positive, the root locus will include only the LHP (or second and third quadrants of the circle) because x is always negative. The root locus is shown in the following figure.

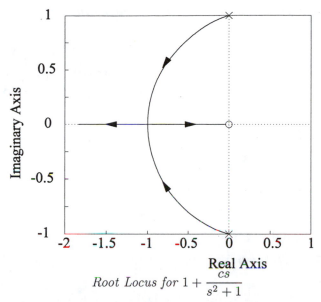

Root Locus for $1 + \dfrac{cs}{s^2 + 1}$

(b) In order to determine the range of c for which the system is stable, the objective is to exclude any roots that are not in the LHP. Such roots are at $s = \pm j1$ (where $c = 0$) and at $s = 0$ (where $c = \infty$). Therefore, the range of c for which the system is stable is given by

$$0 < c < \infty.$$

(c) The value of the parameter c that will allow the system to have a root at

$$s = -\frac{1}{2},$$

is obtained using the magnitude criterion.

$$|cP(s)| = 1$$

$$\left| \frac{cs}{s^2 + 1} \right| = 1$$

$$\frac{c\,|s|}{|s^2 + 1|} = 1$$

$$\implies \frac{c\left| -\dfrac{1}{2} \right|}{\left| \dfrac{1}{4} + 1 \right|} = 1$$

$$\implies c = \frac{5}{4} \times 2 = \frac{5}{2}.$$

Example 5.5 *The following figure shows the speed control system of an assembly plant.*

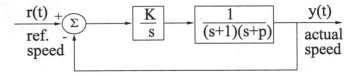

FIGURE 5.3
Control System for an Assembly Plant

(a) *Determine and plot the range of K and p that permits stable operation.*

(b) *In the assembly plant, if $p = 2$, sketch the root locus of the system as K varies from 0 to ∞*

Solution 5.5 *(a) The transfer function for the system is given by*

$$T(s) = \frac{K}{s^3 + (p+1)s^2 + ps + K}.$$

Stability condition is established by using the Routh-Hurwitz array

s^3	1	p
s^2	1+p	K
s^1	b	0
s^0	K	

where the conditions for stability are established as follows:

$$K > 0$$

$$(1 + p) > 0 \Longrightarrow p > -1 \ but \ p > 0 \qquad (coefficient \ of \ s)$$

$$\Longrightarrow p > 0$$

$$b = \frac{p^2 + p - K}{1 + p} > 0$$

$$\Rightarrow K < p^2 + p$$

$$\Longrightarrow 0 < K < p^2 + p \ \ and \ p > 0.$$

(b) *For $p = 2$ the characteristic equation, expressed in Evans root locus form, is given by*

$$1 + \frac{K}{s^3 + 3s^2 + 2s} = 0.$$

Using the eight root locus steps gives the locus shown below

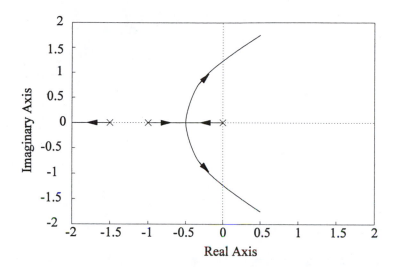

$$\text{Root Locus for } 1 + \frac{K}{s^3 + 3s^2 + 2s} = 0.$$

Example 5.6 *Consider a chemical processing plant represented by the following system,*

$$KG(s) = \frac{K}{s(s+1)(s+5)}.$$

(a) *Draw the root locus of the system.*
(b) *For what value of K are the roots on the imaginary axis?*

Solution 5.6 *(a) The starting point is expressing the transfer in the Evans root locus form.*

$$KG(s) = \frac{K}{s(s+1)(s+5)}$$

$$1 + KG(s) = 0$$

$$1 + \frac{K}{s(s+1)(s+5)} = 0$$

$$\frac{s(s+1)(s+5) + K}{s(s+1)(s+5)} = 0$$

The root locus is shown in the following diagram.

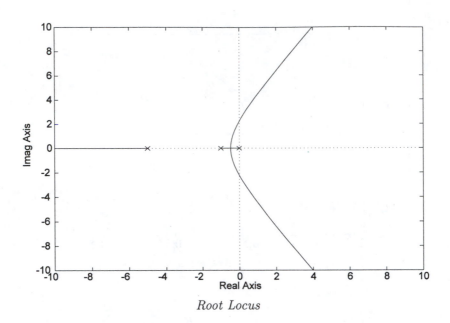

Root Locus

(b) The characteristic equation is given by

$$1 + KG(s) = 0$$

$$1 + \frac{K}{s(s+1)(s+5)} = 0$$

$$s(s+1)(s+5) + K = 0$$

$$s^3 + 6s^2 + 5s + K = 0$$

The Routh-Hurwitz array is obtained as

s^3	1	5
s^2	6	K
s^1	$\dfrac{30-K}{6}$	6
s^0	K	0

For stability, the coefficients in the first column should be greater than zero. On the imaginary axis the coefficients are equal to zero.

$$\implies \frac{30-K}{6} = 0 \ or \ K = 0$$

$$\implies K = 30 \ or \ 0.$$

Alternatively, the same results can be obtained by directly solving the char-

acteristic equation. This is done as follows:

$$s^3 + 6s^2 + 5s + K|_{s=j\omega} = 0$$
$$\implies (-6\omega^2 + K) + (-\omega^3 + 5\omega)j = 0$$
$$\implies \omega = \pm\sqrt{5}, 0 \quad K = 30, 30, 0$$
$$\implies K = 30 \; or \; 0$$

Example 5.7 *Draw root locus plots for the following systems*

(a) $\qquad KG(s) = \dfrac{K(s^2 + 2s + 12)}{s(s^2 + 2s + 10)}$

(b) $\qquad KG(s) = \dfrac{K(s+3)}{s(s+1)(s^2 + 4s + 5)}$

(c) $\qquad KG(s) = \dfrac{K(s+2)}{s^4}$

Solution 5.7 *(a)*

$$KG(s) = \frac{K(s^2 + 2s + 12)}{s(s^2 + 2s + 10)}$$
$$= \frac{K(s+1+j\sqrt{11})(s+1-j\sqrt{11})}{s(s+1+j3)(s+1-j3)}$$
$$n - m = 3 - 2 = 1$$
$$\phi_{asy} = 180$$

Angle of departure $\; : \; \phi = -18.4$

Angle of arrival $\; : \; \phi = 16.7°$

Imaginery axis crossing (jω crossing) : none (by Routh criterion)

Hence, the root locus is of the form shown in the next figure.

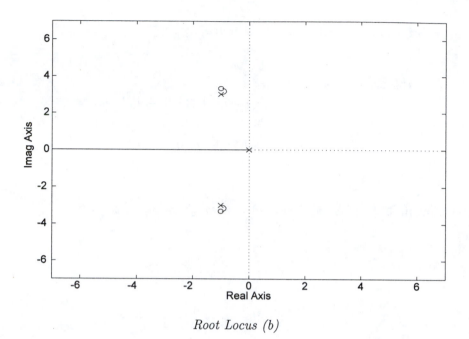

Root Locus (b)

(b)

$$KG(s) = \frac{K(s+3)}{s(s+1)(s^2+4s+5)}$$

$$KG(s) = \frac{K(s+3)}{s(s+1)(s+2+j)(s+2-j)}$$

$$n - m = 4 - 1 = 3$$

$$\phi_{asy} = \pm 60°, 180$$

$$\alpha = -\frac{2}{3}$$

Angle of departure from : $\phi = -153.4°$

Imaginary axis crossing (jω crossing) : $s = \pm j1.38$

Breakaway point : $s = -0.44$

Break-in point : $s = -3.65$

Hence, the root locus takes the form shown in the following diagram.

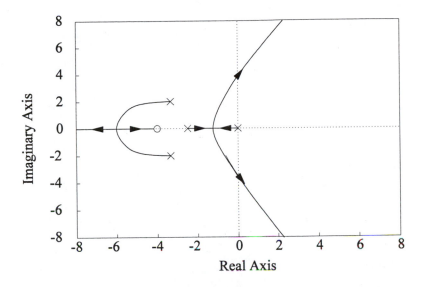

Root Locus (b)

(c)

$$KG(s) = \frac{K(s+2)}{s^4}$$

$$n - m = 4 - 1 = 3$$

$$\phi_{asy} = \pm 60°, 180$$

$$\alpha = \frac{2}{3}$$

Angle of departure from $s = 0 : \phi = \pm 45°, 135°$

Imaginary axis crossing (jω crossing) : $s = 0, K = 0$

Breakaway point : $s = 0$

Break-in point : $s = -\frac{8}{3}$

Thus, the root locus takes the structure shown in the next diagram.

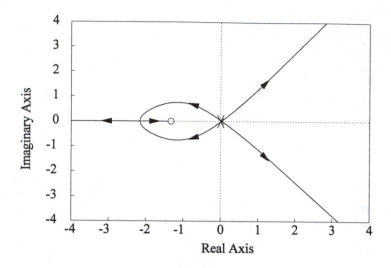

Root Locus (c)

Example 5.8 *Consider a modular robotic system that is described as follows,*

$$KG(s) = \frac{K(s+2)}{s(s-1)(s+6)^2}.$$

(a) Draw the root locus for the system.

(b) Verify that the root locus does not cross the imaginary axis by using the Routh-Hurwitz array.

Solution 5.8 *(a)*

$$KG(s) = \frac{K(s+2)}{s(s-1)(s+6)^2}$$

$$n - m = 4 - 1 = 3$$

$$\phi_{asy} = \pm 60°, 180$$

$$\alpha = -3$$

$$Breakaway\ point\ :\ s = 0.488$$

$$Imaginary\ axis\ crossing\ (j\omega\ crossing)\ :\ none$$

Hence, the root locus is of the form shown in the next diagram.

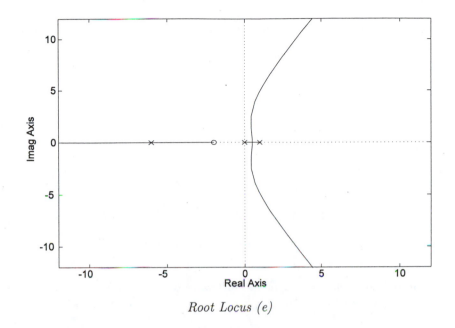

Root Locus (e)

(b) The characteristic equation is given by

$$1 + KG(s) = 0$$

$$1 + \frac{K(s+2)}{s(s-1)(s+6)^2} = 0$$

$$s(s-1)(s+6)^2 + K(s+2) = 0$$

$$s^4 + 11s^3 + 24s^2 + (K-36)s + 2K = 0$$

The Routh-Hurwitz array is obtained as

s^4	1	24	2K
s^3	11	K-36	
s^2	$\dfrac{300-K}{11}$	2K	
s^1	b_1	0	
s^0	2K		

For stability, all the coefficients in the first column must be greater than zero. At the point of crossing the imaginary axis the coefficient b_1 is zero.

$$b_1 = 0 \Rightarrow K^2 - 94K + 10800 = 0$$

$$K = \frac{-b \pm \sqrt{b^2 - 4ac}}{2a}$$

$$= \frac{94 \pm \sqrt{94^2 - 4 \times 10800}}{2}$$

$$= 47 \pm \sqrt{-8591}$$

$$= 47 \pm j92.69$$

$\Rightarrow K$ is not a real number since it has an imaginary component.

Hence, there is no imaginary axis $(s = j\omega)$ crossing for $K > 0$.

Problem 5.1 *Consider a dynamic system whose state-variable matrix equations are expressed as follows:*

$$\dot{\mathbf{x}} = \mathbf{A}\mathbf{x} + \mathbf{B}u$$
$$y = \mathbf{C}\mathbf{x} + \mathbf{D}u$$
$$u = r - y.$$

The matrices $\mathbf{A}$, $\mathbf{B}$, $\mathbf{C}$, *and* $\mathbf{D}$ *are given by*

$$\mathbf{A} = \begin{bmatrix} 0 & 1 & 0 \\ 0 & 0 & 1 \\ -160 & -56 & -14 \end{bmatrix}, \quad \mathbf{B} = \begin{bmatrix} 0 \\ 1 \\ -14 \end{bmatrix}$$

$$\mathbf{C} = \begin{bmatrix} 1 & 0 & 0 \end{bmatrix}, \quad \mathbf{D} = [0].$$

(a) Use MATLAB to find the transfer function of the system.
(b) Draw the system block diagram, indicating the matrices $\mathbf{A}$, $\mathbf{B}$, $\mathbf{C}$, *and* $\mathbf{D}$, *the vectors* $\dot{\mathbf{x}}$ *and* $\mathbf{x}$, *and the variables* $r(t)$, $u(t)$ *and* $y(t)$.
(c) Compare the root locus obtained by using the state-variable form and that obtained using the transfer function form.

Solution 5.9 *(a) The transfer function is obtained by using the following MATLAB command*

$$[\text{num}, \text{den}] = \text{ss2tf}(\mathbf{A}, \mathbf{B}, \mathbf{C}, \mathbf{D}).$$

$$\text{num} = \begin{bmatrix} 0 & 0 & 1 & 0 \end{bmatrix}$$
$$\text{den} = \begin{bmatrix} 1 & 14 & 56 & 160 \end{bmatrix}$$

Therefore,transfer function is given by

$$T(s) = \frac{s}{s^3 + 14s^2 + 56s + 160}$$

(b) The system block diagram is given by

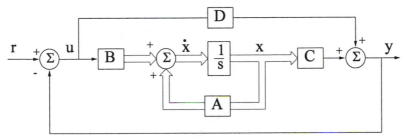

State Space System

(c) The root locus plot for this system can be obtained with MATLAB by use of the following command:

$$rlocus(A,B,C,D)$$

This command will produce the same root locus plot as can be obtained by use of the **rlocus(num,den)** *command, where* **num** *and* **den** *are obtained from*

$$[\text{num}, \text{den}] = \text{ss2tf}(A, B, C, D).$$

The root locus diagram is shown in the following diagram

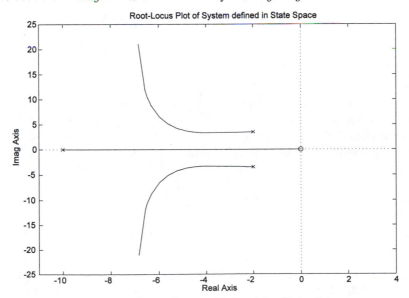

Root Locus for a System Defined in State Space.

5.5   Dynamic Compensation: Lead and Lag

The addition of poles and zeros (compensation) to the open-loop transfer function in order to reshape the root locus diagram to satisfy design specifications constitutes compensator design by the root locus method. For example, stability can be improved by the addition of a zero and worsened by the addition of a pole.

Lead compensation acts mainly to lower rise time and decrease the transient overshoot. Generally, lead compensation approximates derivative control. Lag compensation is usually used to improve the steady state accuracy of the system and it approximates integral control. Compensation with the transfer function of the form

$$D(s) = \frac{s+z}{s+p},$$

is called lead compensation is $z < p$ and lag compensation if $z > p$.

$$1 + KD(s)G(s) = 0$$

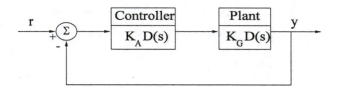

To explain the basic stabilizing effect of lead compensation of a system, a simplified $D(s) = s + z$ is considered. This is the same as the proportional-derivative (PI) control. This compensation is applied to the case of a second-order system with transfer function

$$KG(s) = \frac{K}{s(s+1)}.$$

The $G(s)$ has root locus as shown in the following figure and shown in the next figure is the root locus produced by $D(s)G(s)$ in the circle. The effect of zero is to move the root locus to the left more stable part of the s-plane. The root locus of the compensated case was produced by using $D(s) = s + 2$.

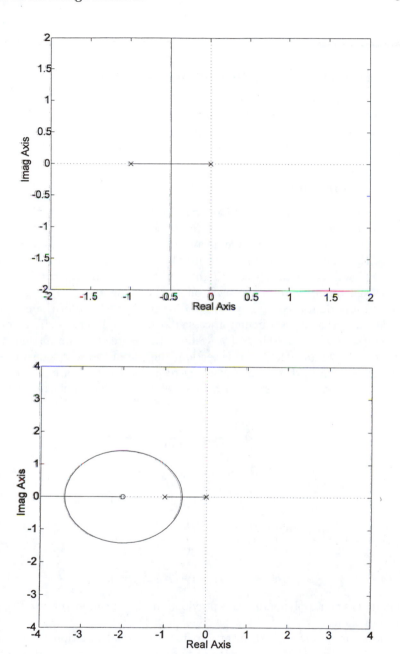

The trouble with choosing $D(s)$ based on only a zero is that the physical realization would contain a differentiator that would greatly amplify the inevitable high-frequency noise from the sensor signal.

5.6   Extensions of Root Locus Method

The root locus technique is a graphical scheme to show locations of possible roots of an equation as a real parameter varies. So far only polynomial equations for linear systems have been considered. Two additional types of systems are considered, systems with (i) time delay (ii) nonlinear elements.

5.6.1   Time Delay

Time delays often arise in control systems, both from delays in the process itself and from delays in the processing of sensed signals. Chemical plants often have processes with a time delay representing the time material takes to flow through the pipes. In measuring the altitude of a spacecraft en route to Mars, because of the speed of light, there is a significant time delay for the sensed quantity to arrive back on Earth. There is also a small time delay in any digital control system due to the cycle time of the computer and the fact that data is being processed at discrete intervals. Time delays always reduces the stability of a system; therefore, it is important to be able to analyze its effect. In this section, the use of the root locus for such analysis is presented, though frequency-response methods are easier.

As an example of a system with time delay, consider the system transfer function

$$G(s) = \frac{e^{-5s}}{(10s + 1)(60s + 1)}$$

where e^{-5s} term arises from the time delay. The root locus equations are

$$1 + KG(s) = 0$$

$$1 + K\frac{e^{-5s}}{(10s + 1)(60s + 1)} = 0$$

$$600s^2 + 70s + 1 + Ke^{-5s} = 0.$$

Since it is not a polynomial, the previous methods cannot be used to plot the root locus.

One approach is to find an approximate rational function for the non-rational function e^{-5s}. It consists of matching the series expansion of the transcendental function e^{-5s} with the series expansion of a rational function whose numerator is a polynomial of degree p and denominator is a polynomial of degree q. The result is called (p, q) Pade approximation to e^{-5s}. Initially, the approximant to e^{-s} is computed, and in the final result substitute $T_d s$ for s to allow for any delay. To illustrate the process begin with $(1,1)$ as an approximation. In this case, the objective is to select b_o, b_1, and a_o so that the error is given by

$$e^{-s} - \frac{b_o s + b_1}{a_o s + 1} = \varepsilon$$

is small. Expand the rational function using McLauren series and match as many terms of the initial terms as possible. The two series are

$$e^{-s} = 1 - s + \frac{s^2}{2!} - \frac{s^3}{3!} + \frac{s^4}{4!} - \ldots.$$

$$\frac{b_o s + b_1}{a_o s + 1} = b_1 + (b_o - a_o b_1)s - a_o(b_o - a_o b_1)s^2 + a_o^2(b_o - a_o b_1)s^3 + \ldots.$$

Matching the first four coefficients

$$b_1 = 1$$

$$(b_o - a_o b_1) = -1$$

$$-a_o(b_o - a_o b_1) = \frac{1}{2}$$

$$a_o^2 = -\frac{1}{6}.$$

Substituting $T_d s$ for s the resulting approximant is

$$e^{-T_d s} \cong \frac{1 - (T_d s/2)}{1 + (T_d s/2)}$$

If it is assumed that $p = q = 2$ then five parameters are obtained and a better match is possible. In this case $(2,2)$, the approximation is available as

$$e^{-T_d s} \cong \frac{1 - (T_d s/2) + (T_d s)^2/12}{1 + (T_d s/2) + (T_d s)^2/12}.$$

In some cases, a very crude approximation is acceptable, and the $(0,1)$ can be used, which is a first-order lag given by

$$e^{-T_d s} \cong \frac{1}{1 + T_d s}$$

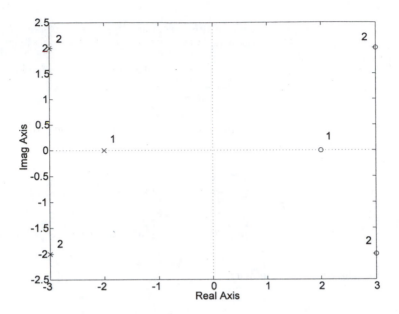

5.6.2 Nonlinear Systems

Real control systems are inherently nonlinear, and the linear analysis and design methods that have been described so far use linear approximations to the real models. There is one important category of nonlinear systems for which some significant analysis (and design) can be done. This comprises the systems in which the nonlinearity has no dynamics and is well approximated as a gain that varies as the size of its input signal varies. Some of such systems are shown in Figure 5.4.

5.7 Computer-Aided Determination of the Root Locus

In order to use the root locus as a design tool and to verify computer-generated loci, it is very important to be able to sketch root loci. The control engineer can then quickly predict, for design purposes, the effect of an added pole, or even several of them, or can quickly confirm computer output in a qualitative sense. For this reason, it is important to understand the guidelines for sketching the loci and be able to plot numerous example loci by hand. The computer can be used to determine accurate loci and to establish exact values of the parameters. It is especially useful in computing the closed-loop pole sensitivity to those parameters because their values may be known only to a certain tolerance at the time of the design and

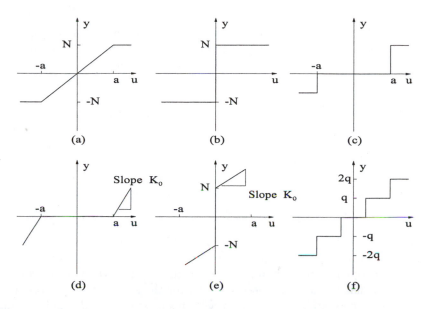

FIGURE 5.4
Examples of Nonlinear Elements

may be subject to drift over the life of the system. There are two basic approaches to machine computation of the root locus.

In the first approach the root locus problem is formulated as a polynomial in the form $a(s) + Kb(s) = 0$. For a sequence of values of K varying from near to zero to a large value, the polynomial for its n roots is solved by any of the many available numerical techniques. An advantage of this method is that it computes all the roots for each value of K, guaranteeing that a specific value, if required, is also included.

One of the disadvantages of the method is that the resulting root locations are very unevenly distributed in the s-plane. For example, near a point of multiple roots the sensitivity of the root locations to the K-value is very great, and the roots just *"fly through"* such points; the plots appear to be coarse, so it is easy to miss important features. (On the other hand, the method has the advantage that near a zero the root moves very slowly, since it takes an infinite value of the parameter to push the root all the way into the zero.

A second disadvantage of this method is that the equation must be a polynomial. In cases of time delay, it involves a transcendental equation, hence an approximation such as the Padé method must be used to reduce the given problem to an equivalent polynomial. Such approximations limit the range of values for which the results are accurate, and checking the accuracy is difficult unless a means is available to solve the true equation at critical points. A final disadvantage is that not many algorithms are able to solve polynomials easily at points of multiple roots. This problem is related to great sensitivity of the roots to the parameter at these points, as mentioned earlier. A method related to factoring is possible when a state-variable formulation is available.

The alternative to polynomial factoring is a method based on curve tracing. It hinges on the fact that a point on the positive root locus is a point where the phase of $G(s)$ is 180°. Thus, given a point on the locus at s_o with gain K_o, a circle of radius δ around s_o can be drawn and the circle searched for a new point where the angle condition is met and the new gain is larger than K_o. This method can be easily arranged to include a delay term such as $e^{-\lambda s}$; the resulting points will be spaced δ radians apart, a value that the designer can specify.

A disadvantage of this method is that only one branch of the locus is generated at a time (although computer logic can be easily set up through each of the open-loop poles to produce a complete graph). A second disadvantage is that the designer needs to monitor the selection of δ in order to ensure that the search for 180° converges on some points and avoid wasting too much time with a small value of δ at less critical points.

5.7.1   MATLAB

The issues involved in using MATLAB to determine the root locus of a dynamic system can be summarized as follows:

- Closed-loop poles

- Plotting the root locus of a transfer function

- Choosing a value of K from root locus

- Closed-loop response

- Key MATLAB commands used: *cloop, rlocfind, rlocus, sgrid, step*

Open-loop system
The following example illustrates the implementation for an open-loop system.

```
num=[1 7];
den=conv(conv([1 0],[1 5]),conv([1 15],[1 20]));
rlocus(num,den)
axis([-22 3 -15 15])
```

Choosing a value of K from the root locus

The plot above shows all possible closed-loop pole locations for a pure proportional controller. Obviously not all of those closed-loop poles will satisfy our design criteria. To determine what part of the locus is acceptable, the command *sgrid(ξ,ω_n)* can be used to plot lines of constant damping ratio and natural frequency. Its two arguments are the damping ratio ξ and natural frequency ω_n [these may be vectors if the objective is to look at a range of acceptable values]. In this problem, an overshoot less than 5% (which means a damping ratio ξ of greater than 0.7) and a rise time of 1 second (which means a natural frequency Wn greater than 1.8) are required. Enter in the MATLAB command window:

$\xi = 0.7;$
$\omega_n = 1.8;$
sgrid(ξ,ω_n)

From the plot above it can be seen that there is part of the root locus inside the desired region. So in this case only a proportional controller is required to move the poles to the desired region. The *rlocfind* command in MATLAB can be used to choose the desired poles on the locus:

$[kd, poles] = rlocfind(num, den)$

Closed-Loop Response

In order to find the step response, the closed-loop transfer function has to be known. This can be computed using the rules of block diagrams, or

MATLAB can be used to implement the closed-loop transfer function as follows:

[numcl, dencl] = cloop((kd)*num, den)

5.8   Problems

Problem 5.2 *Two systems have the following plant gains*

$$G(s) = \frac{3s - 4}{s^2 - 2s + 2}$$

$$G(s) = -\frac{3s - 4}{s^2 - 2s + 2}.$$

(a) Obtain the root loci for the two systems.

(b) Is there any value of controller gain K that will stabilize either of these plants?

Problem 5.3 *Consider a system with the following plant transfer function (where the controller gain $K = 1$).*

$$G(s) = \frac{s^2 + s + a}{s(s+1)(s^2 + s + 1.25)(s^2 + s + 4.25)}$$

(a) Plot the root locus for the plant as a varies from 1.25 to 4.25.

(b) Find the exact value of a for which the pair of poles which asymptotically approach zero changes.

Problem 5.4 *Determine the system type for following unity feedback systems whose forward path transfer functions are given below.*

$$(a) \quad G(s) = \frac{K}{(1+s)(1+10s)(1+20s)}$$

$$(b) \quad G(s) = \frac{10(s+1)}{s^3(s^2 + 5s + 5)}$$

$$(c) \quad G(s) = \frac{100}{s^3(s+2)^2}$$

Problem 5.5 *Determine the step and ramp error constants of unity feedback systems whose forward path transfer functions are given by:*

$$(a) \quad G(s) = \frac{1000}{(1+0.1s)(1+10s)}$$

$$(b) \quad G(s) = \frac{K(1+2s)(1+4s)}{s^2(s^2 + s + 1)}$$

$$(c) \quad G(s) = \frac{100}{s(s^2 + 10s + 100)}$$

Problem 5.6 *The forward path transfer functions of unity feedback control systems are given below.*

$$(a) \quad G(s) = \frac{K(s+4)}{s(s^2 + 4s + 4)(s+5)(s+6)}$$

$$(b) \quad G(s) = \frac{K}{s(s+2)(s+5)(s+10)}$$

$$(c) \quad G(s) = \frac{K(s^2 + 2s + 10)}{s(s+5)(s+10)}$$

$$(d) \quad G(s) = \frac{K(s^2 + 4)}{(s+2)^2(s+5)(s+6)}$$

Construct the root loci for $K \geq 0$.

Problem 5.7 *The characteristic equation of the liquid level control system is written as*

$$0.006s(s + 12.5)(As + K) + 250N = 0.$$

(a) For $A = K = 5$, construct the root loci of the characteristic equation as N varies from 0 to ∞.

(b) For $N = 10$ and $K = 50$, construct the root loci of the characteristic equation for $A \geq 0$.

(c) For $A = 50$ and $N = 20$, construct root loci for $K \geq 0$.

Problem 5.8 *The block diagram of a control system with feedback is shown in following diagram.*

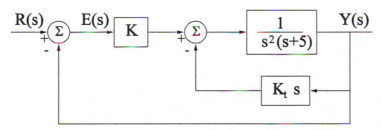

(a) Construct the root loci of the characteristic equation for $K \geq 0$ when $K_t = 0$.

(b) Set $K = 10$, construct the root loci of the characteristic equation for $K_t \geq 0$.

Chapter 6

Frequency-Response Design Methods

6.1 Introduction

In most of the work in previous chapters the input signals used were functions such as an impulse, a step, and a ramp function. In this chapter, the steady state response of a system to a sinusoidal input signal (sinusoid) is considered. It will be observed that the response of a linear time-invariant system to a sinusoidal input signal is an output sinusoidal signal at the same frequency as the input. However, the magnitude and phase of the output signal differ from those of the input sinusoidal signal, and the amount of difference is a function of the input frequency. Hence, the frequency response of a system is defined as the steady state response of the system to a sinusoidal input signal. The sinusoid is a unique input signal, and the resulting output signal for a linear system is sinusoidal in the steady state.

Consequently, the objective of the chapter is to investigate the steady state response of the system to sinusoidal inputs as the frequency varies. The design of feedback control systems in industry is probably accomplished using frequency-response methods more often than any other. Frequency-response design is popular primarily because it provides good designs in the face of uncertainty in the plant model. For example, for systems with poorly known or changing high-frequency resonances, their feedback compensation can be modified to alleviate the effects of those uncertainties. This modification is carried out more easily using frequency-response design than any other method. Another advantage of using frequency response is the ease with which experimental information can be used for design purposes. Raw measurements of the output amplitude and phase of a plant undergoing a sinusoidal input excitation are sufficient to design a suitable feedback control system. No intermediate processing of the data (such as finding poles and zeros or determining system matrices) is required to arrive at the system model. The wide availability of computers has rendered this advantage less important now than it was years ago; however, for relatively simple systems, frequency response is often still the most cost effective design method.

Hence, the frequency-response design method offers a practical and important alternative approach to the analysis and design of control systems. The frequency response of a system is defined as the steady state response of the system to a sinusoidal input signal. The sinusoid is a unique input signal, and the resulting output signal for a linear system, as well as signals throughout the system, is sinusoidal in the steady state; it differs from the input waveform only in amplitude and phase angle. The transfer function $G(s)$ is analyzed when $s = jw$ and methods for graphically displaying the complex number $G(jw)$ as ω varies are developed. The main techniques covered include Bode plots, polar plots and Nyquist plots. Time-domain performance measures are developed in terms of the frequency response and then applied in system performance evaluation.

6.1.1   Magnitude and Phase Angle

The frequency response of a system with a general transfer function $G(s)$ is defined as the steady state response of the system to a sinusoidal input signal $u(t)$. The sinusoid is a unique input signal with amplitude U_o and frequency ω, and the resulting output signal $y(t)$ is also sinusoidal and it differs from the input waveform only in amplitude and phase angle.

$$u(t) = U_o \sin \omega t \tag{6.1}$$

$$U(s) = \frac{U_o \omega}{s^2 + \omega^2}. \tag{6.2}$$

From the definition of a transfer function, the output signal $y(t)$ can be established as shown below

$$G(s) = \frac{Y(s)}{U(s)}$$

$$Y(s) = G(s)U(s)$$

$$= G(s)\frac{U_o \omega}{s^2 + \omega^2}$$

$$y(t) = \mathcal{L}^{-1}\left[G(s)\frac{U_o \omega}{s^2 + \omega^2}\right]$$

$$y(t) = AU_o \sin(\omega t + \phi), \tag{6.3}$$

where A and ϕ are the magnitude and phase angle of the transfer function $G(s)$. Thus, the sinusoidal output $y(t)$ differs from the input $u(t)$ in amplitude by a constant factor A, and in phase by angle ϕ. To establish A and

ϕ the substitution of $s = j\omega$ is effected in $G(s)$ such that

$$G(j\omega) = G(s)|_{s=j\omega}$$
$$= \operatorname{Re} G(j\omega) + j \operatorname{Im} G(j\omega)$$
$$= X(\omega) + jY(\omega)$$
$$= Ae^{j\phi}.$$

A is the magnitude of the transfer function $G(j\omega)$

$$A = |G(j\omega)|$$
$$= \sqrt{[\operatorname{Re} G(j\omega)]^2 + [\operatorname{Im} G(j\omega)]^2}$$
$$= \sqrt{X(\omega)^2 + Y(\omega)^2}.$$

ϕ is the phase angle of the function $G(j\omega)$

$$\phi = \arg G(j\omega)$$
$$= \angle G(j\omega)$$
$$= \tan^{-1} \left[\frac{\operatorname{Im} G(j\omega)}{\operatorname{Re} G(j\omega)} \right]$$
$$= \tan^{-1} \left[\frac{Y(\omega)}{X(\omega)} \right].$$

The term $\arg G(j\omega)$ is the shorthand form of the expression, *"the argument of $G(j\omega)$"* which means the phase angle of the complex function $G(j\omega)$.

Example 6.1 *Find the magnitude and phase angle of the transfer function*

$$G(s) = \frac{5}{3s + 4}.$$

Solution 6.1 *The solution proceeds as follows:*

$$G(jw) = \frac{5}{3(j\omega) + 4}$$
$$= \frac{5(4 - j3\omega)}{(4 + j3w)(4 - j3\omega)}$$
$$= \frac{20 - j15\omega}{(16 + 9\omega^2)}$$
$$= \frac{20}{(16 + 9\omega^2)} + \frac{-j15\omega}{(16 + 9\omega^2)}$$
$$\implies G(j\omega) = X(\omega) + jY(\omega)$$

Therefore, the magnitude is given by

$$A = |G(j\omega)|$$

$$= \sqrt{X^2(\omega) + Y^2(\omega)}$$

$$= \sqrt{\frac{400}{(16 + 9\omega^2)^2} + \frac{225\omega^2}{(16 + 9\omega^2)^2}}$$

$$= \frac{1}{(16 + 9\omega^2)}\sqrt{400 + 225\omega^2}$$

$$= \frac{5}{\sqrt{16 + 9\omega^2}}.$$

The phase angle is given by

$$\phi = \arg G(j\omega)$$

$$= \tan^{-1}\frac{Y(\omega)}{X(\omega)}$$

$$= \tan^{-1}\left(\frac{-15\omega}{20}\right)$$

$$= -\tan^{-1}\left(\frac{3\omega}{4}\right).$$

6.1.2   Combining Magnitudes and Phase Angles

Consider a transfer function that is composed of five transfer functions such that,

$$G(s) = \frac{G_1(s)G_2(s)}{G_3(s)G_4(s)G_5(s)}.$$

The overall magnitude and phase angle can be expressed in terms of the individual magnitudes and phase angles. The analysis is carried out as follows:

$$G(j\omega) = \frac{G_1(j\omega)G_2(j\omega)}{G_3(j\omega)G_4(j\omega)G_5(j\omega)}$$

$$= \frac{|G_1(j\omega)|e^{j\arg G_1(j\omega)}|G_2(j\omega)|e^{j\arg G_2(j\omega)}}{|G_3(j\omega)|e^{j\arg G_3(j\omega)}|G_4(j\omega)|e^{j\arg G_4(j\omega)}|G_5(j\omega)|e^{j\arg G_5(j\omega)}}$$

$$= \frac{A_1 e^{j\phi_1} A_2 e^{j\phi_2}}{A_3 e^{j\phi_3} A_4 e^{j\phi_4} A_5 e^{j\phi_5}}$$

$$= \frac{A_1 A_2}{A_3 A_4 A_5} e^{j(\phi_1 + \phi_2 - \phi_3 - \phi_4 - \phi_5)}$$

$$= A e^{j\phi}.$$

Therefore, the composite magnitude and phase angle are given by

$$A = \frac{|G_1(j\omega)||G_2(j\omega)|}{|G_3(j\omega)||G_4(j\omega)||G_5(j\omega)|} \tag{6.4}$$

$$\phi = \phi_1 + \phi_2 - \phi_3 - \phi_4 - \phi_5. \tag{6.5}$$

This means that the composite magnitude and phase angle can be obtained from the magnitudes and phase angles of the individual transfer functions that constitute the composite transfer function. Equations 6.4 and 6.5 can be generalized for a transfer function with m multiplier transfer functions and n divisor transfer functions such that

$$G(s) = \frac{\Pi_{j=1}^{m} G_j(s)}{\Pi_{i=1}^{n} G_i(s)}, \tag{6.6}$$

where the product symbol Π is defined by the expression

$$\Pi_{j=1}^{m} G_j(s) = G_1(s) G_2(s) ... G_m(s).$$

The magnitude and phase angle are obtained by using the substitution $s = j\omega$ in Equation 6.6.

$$G(s) = \frac{\Pi_{j=1}^{m} G_j(j\omega)}{\Pi_{i=1}^{n} G_i(j\omega)}$$

$$\implies A = \frac{\Pi_{j=1}^{m} |G_j(j\omega)|}{\Pi_{i=1}^{n} |G_i(j\omega)|} \tag{6.7}$$

$$\Longrightarrow \phi = \sum_{j=1}^{m} \phi_j - \sum_{i=1}^{n} \phi_i. \qquad (6.8)$$

These formulae in Equations 6.7 and 6.8 allow the computation of the magnitude and phase angle of a transfer function without first expressing the transfer function in terms of a real part and an imaginary part as done in Example 6.1. Applying Equations 6.7 and 6.8 in Example 6.1 simplifies the solution,

$$G(jw) = \frac{5}{3(jw) + 4} = \frac{G_1(jw)}{G_2(jw)}$$

$$A = \frac{|G_1(jw)|}{|G_2(jw)|} = \frac{|5|}{|j3w + 4|}$$

$$= \frac{5}{\sqrt{16 + 9w^2}}$$

$$\phi = \arg G_1(jw) - \arg G_2(jw)$$

$$= \phi_1 - \phi_2$$

$$= \tan^{-1}\left(\frac{0}{5}\right) - \tan^{-1}\left(\frac{3w}{4}\right)$$

$$= -\tan^{-1}\left(\frac{3w}{4}\right).$$

Example 6.2 *Consider the RC filter circuit shown in Figure 6.1.*

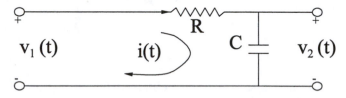

FIGURE 6.1
A Simple RC Filter

  Find the magnitude and phase angle of the circuit's transfer function, which is given by

$$G(s) = \frac{V_2(s)}{V_1(s)}.$$

Solution 6.2 *By using the KCL and KVL, expressions for the input and output voltages are determined.*

$$v_1(t) = i(t)R + \int \frac{i(t)}{C}dt$$

$$v_2(t) = \int \frac{i(t)}{C}dt.$$

Taking Laplace transforms of the two equations leads to

$$V_1(s) = I(s)\left[R + \frac{1}{sC}\right]$$

$$V_2(s) = I(s)\left[\frac{1}{sC}\right].$$

Hence, the transfer function is given by

$$
\begin{aligned}
G(s) &= \frac{V_2(s)}{V_1(s)} \\[2mm]
&= \frac{I(s)\left[\dfrac{1}{sC}\right]}{I(s)\left[R + \dfrac{1}{sC}\right]} \\[4mm]
&= \frac{1}{1 + sRC} \\[2mm]
&\Longrightarrow G(jw) = \frac{1}{1 + jwRC}.
\end{aligned}
$$

The magnitude and phase angle are obtained using Equations 6.7 and 6.8,

$$A = \frac{|G_1(j\omega)|}{|G_2(j\omega)|} = \frac{|1|}{|1 + j\omega RC|}$$

$$= \frac{1}{\sqrt{1 + (\omega RC)^2}}$$

$$\phi = \phi_1 - \phi_2$$

$$= \tan^{-1}\left(\frac{0}{1}\right) - \tan^{-1}\left(\frac{\omega RC}{1}\right)$$

$$= -\tan^{-1}(\omega RC).$$

Example 6.3 *Find the magnitude and phase angle of the transfer function*

$$G(s) = \frac{s(2s+1)}{(s+1)(2s+3)(s^2 + 2s + 10)}.$$

Solution 6.3 *The magnitude and phase angles are computed by using Equations 6.7 and 6.8*

$$G(jw) = \frac{jw(2jw + 1)}{(jw + 1)(2jw + 3)((jw)^2 + 2jw + 10)}$$

$$\Rightarrow \frac{G_1(jw)G_2(jw)}{G_3(jw)G_4(jw)G_5(jw)}$$

$$A = \frac{|G_1(jw)||G_2(jw)|}{|G_3(jw)||G_4(jw)||G_5(jw)|}$$

$$= \frac{|jw||(2jw + 1)|}{|(jw + 1)||(2jw + 3)||((jw)^2 + 2jw + 10)|}$$

$$= \frac{w\sqrt{(4w^2 + 1)}}{\sqrt{(w^2 + 1)}\sqrt{(4w^2 + 9)}\sqrt{[4w^2 + (10 - w^2)^2]}}$$

$$\phi = \phi_1 + \phi_2 - \phi_3 - \phi_4 - \phi_5$$

$$= \tan^{-1}\left(\frac{w}{0}\right) + \tan^{-1}\left(\frac{2w}{1}\right) - \tan^{-1}\left(\frac{w}{1}\right) - \tan^{-1}\left(\frac{2w}{3}\right)$$

$$- \tan^{-1}\left(\frac{2w}{10 - w^2}\right).$$

6.2   Bode Plots: The Principles

In this section the principles behind Bode plots are introduced on the foundation of the previous discussion on polar plots. The advantages of Bode plots and the techniques used to draw them are then presented.

6.2.1   Background

It is standard practice to measure power gain in decibels, that is, units of one-tenth of a bel such that

$$|H(jw)|_{db} = 10 \log_{10}\left(\frac{P_2}{P_1}\right).$$

Power can be expressed in terms of voltage and resistance,

$$P = \frac{V^2}{R}$$

$$\implies |H(jw)|_{db} = 10 \log_{10} \left(\frac{V_2^2}{R} \Big/ \frac{V_1^2}{R} \right)$$

$$= 10 \log_{10} \left(\frac{V_2^2}{V_1^2} \right)$$

$$\implies |H(jw)|_{db} = 20 \log_{10} \left(\frac{V_2}{V_1} \right).$$

It is within this context that the Bode magnitude or logarithmic magnitude M is defined for any transfer function $G(s)$.

$$M = |G(jw)|_{db} = 20 \log_{10} |G(jw)|$$

$$= 20 \log_{10} A.$$

The Bode plots for a transfer function consist of two figures: A plot of the logarithmic magnitude M with respect to frequency ω, and the phase angle ϕ with respect to frequency ω. The horizontal scale ω is logarithmically calibrated. The logarithmic plots are used to simplify the determination of graphical portrayal of the system frequency response. It is important to note that logarithmic scales are used because they considerably simplify the construction, manipulation, and interpretation of the Bode plots. The set of Bode equations (M and ϕ) for the RC filter circuit in Example 2 are given by

$$M = 20 \log_{10} |G(jw)|$$

$$= 20 \log_{10} \frac{1}{\sqrt{1 + (wRC)^2}}$$

$$= 20 \log_{10} 1 - 20 \log_{10} \sqrt{1 + (wRC)^2}$$

$$= -20 \log_{10} \sqrt{1 + (wRC)^2} \tag{6.9}$$

$$\phi = -\tan^{-1}(wRC). \tag{6.10}$$

The next sections deal with techniques that are employed to draw Bode plots from equations such as Equations 6.9 and 6.10.

6.2.2 Advantages of Bode Plots

A Bode plot gives the magnitude and phase of a system in response to a sinusoidal input for a range of frequencies. Each point on the Bode plot represents the system response for a given frequency input. Such plots can be used to establish system stability (by finding stability margins), study speed of response and establish how different vibrations affect a control system.

- Bode plots of systems in series or parallel can be simply added to produce the composite Bode plots.

- For reciprocal factors, the magnitude and phase need only be changed in sign.

- The phase-gain relationship is given in terms of logarithms of phase and gain.

- Low- to high-frequency system behavior can be displayed on a single Bode plot.

- Gain crossover, phase crossover, gain margin, and phase margin are easily determined from the Bode plots.

- Bode plots can be determined experimentally without explicit knowledge of the transfer function.

- For design purposes, the effects of adding controllers and their parameters are easily visualized on the Bode plot.

- Dynamic compensator design can be based entirely on Bode plots.

- Bode plots can be used to stabilize systems by changing system gains.

- Bode plots can also be used to study the impact of non-frequency inputs.

It is essential for the control systems engineer to be able to hand-plot frequency responses. This technique enables the engineer to analyze simple systems and also to check computer results for more complicated examples. Approximations can be used to quickly sketch the frequency response and deduce stability as well as determine the form of the needed dynamic compensations. Hand plotting is also useful in the interpretation of experimentally generated frequency-response data.

6.2.3 Bode Plot Techniques

A general transfer function can be expressed in terms of a constant, zero factors and pole factors as follows:

$$G(s) = K \frac{\Pi_{j=1}^{m}(s + z_j)}{\Pi_{i=1}^{n}(s + p_i)}$$

$$G(jw) = K \frac{\Pi_{j=1}^{m}(jw + z_j)}{\Pi_{i=1}^{n}(jw + p_i)} = Ae^{j\phi}, \qquad (6.11)$$

where A is the magnitude given by

$$A = K \frac{\Pi_{j=1}^{m}|(jw + z_j)|}{\Pi_{i=1}^{n}|(jw + p_i)|},$$

and ϕ is the phase angle.

$$\phi = \sum_{j=1}^{m} \arg(jw + z_j) + \sum_{i=1}^{n} \arg \frac{1}{(jw + p_i)}.$$

Expressing the magnitude A in decibels gives the Bode magnitude M,

$$M = |G(jw)|_{db}$$

$$= 20 \log_{10} A$$

$$= 20 \log_{10} \left(K \frac{\Pi_{j=1}^{m}|(jw + z_j)|}{\Pi_{i=1}^{n}|(jw + p_i)|} \right)$$

$$= 20 \log_{10} K + \sum_{j=1}^{m} 20 \log_{10} |(jw + z_j)| + \sum_{i=1}^{n} 20 \log_{10} \frac{1}{|(jw + p_i)|}.$$

Hence, the magnitude in decibels and phase angle for any transfer function can be obtained as follows:

$$M = 20 \log_{10} K + \sum_{j=1}^{m} 20 \log_{10} |(jw + z_j)| + \sum_{i=1}^{n} 20 \log_{10} \frac{1}{|(jw + p_i)|} \quad (6.12)$$

$$\phi = \sum_{j=1}^{m} \arg(jw + z_j) + \sum_{i=1}^{n} \arg \frac{1}{(jw + p_i)}. \qquad (6.13)$$

From these two equations it can be deduced that any transfer function whose Bode plots are of interest can be first expressed in terms of a constant, zero factors, and pole factors. The Bode plots of the constant and the individual factors are then drawn separately and then simply added to produce the composite Bode plots (ϕ and M) for the entire transfer function. There are six unique forms that these zero and pole factors can take: a simple (first-order) zero factor, a simple (first-order) pole factor, an integrator factor, a derivative factor, a complex (second-order) pole factor and a complex (second-order) zero factor. The form of a simple zero factor (with a real zero at $s = -z_j$) is established as follows:

$$G(s) = (s + z_j)$$

$$= z_j(1 + \frac{s}{z_j})$$

$$\Longrightarrow z_j \text{ (a constant) and } \left(1 + \frac{s}{z_j}\right) \text{ (a simple zero factor)}.$$

Similarly, a simple pole factor (with a real pole at $s = -p_i$) takes the form

$$G(s) = \frac{1}{(s + p_i)}$$

$$= \frac{1}{p_i(1 + \frac{s}{p_i})}$$

$$\Longrightarrow \frac{1}{p_i} \text{ (a constant) and } \frac{1}{\left(1 + \frac{s}{p_i}\right)} \text{ (a simple pole factor)}.$$

The integrator factor (a pole at $s = 0$) is of the form $1/s$ and derivative factor (a zero at $s = 0$) takes the form of s. A complex (second-order) pole factor occurs when there are two complex poles, $-p_1$ and $-p_2$, (a conjugate pair such that $p_1 = a + jb$ and $p_2 = a - jb$). This can be formulated as follows:

$$G(s) = \frac{1}{(s + (a + jb))} \frac{1}{(s + (a - jb))}$$

$$= \frac{1}{s^2 + 2as + (a^2 + b^2)}$$

$$\Longrightarrow \frac{1}{s^2 + 2\xi w_n s + w_n^2} \quad \text{(expressing result in standard form)}$$

Therefore, $\quad G(s) = \dfrac{1}{w_n^2 \left[1 + 2\xi \left(\dfrac{s}{w_n} \right) + \left(\dfrac{s}{w_n} \right)^2 \right]}$ (6.14)

$$\Rightarrow \frac{1}{w_n^2} \text{ (a constant gain)} \quad \text{and}$$

$$\frac{1}{\left[1 + 2\xi \left(\dfrac{s}{w_n} \right) + \left(\dfrac{s}{w_n} \right)^2 \right]} \quad \text{(a complex pole factor)}.$$

A complex (second-order) zero factor occurs when there are two complex zeros $-z_1$ and $-z_2$, (a conjugate pair such that $z_1 = a + jb$ and $z_2 = a - jb$).

$$
\begin{aligned}
G(s) &= (s + (a + jb))(s + (a - jb)) \\
&= s^2 + 2as + (a^2 + b^2) \\
\Longrightarrow\ & s^2 + 2\xi w_n s + w_n^2 \\
&= w_n^2 \left[1 + 2\xi \left(\frac{s}{w_n} \right) + \left(\frac{s}{w_n} \right)^2 \right]
\end{aligned}
$$
(6.15)

$$\Rightarrow w_n^2 \quad \text{(a constant gain)} \quad \text{and}$$

$$\left[1 + 2\xi \left(\frac{s}{w_n} \right) + \left(\frac{s}{w_n} \right)^2 \right] \quad \text{(a complex zero factor)}.$$

These are the six unique zero and pole factors that a transfer function can be broken into. Hence, including the constant function, there are seven exhaustive elementary functions that one has to be able to draw in order to be in a position to establish the Bode plots of any transfer function. An advantage of Bode representation is that for reciprocal factors the log-magnitude and phase need only be changed in sign. Hence, log-magnitude and phase of the simple pole factor are the negatives of those of simple zero factor, the log-magnitude and phase for the integrator factor are the negatives of those of the derivative factor, and the log-magnitude and phase of the complex pole factor are negatives of those of the complex zero factor. Hence, a careful study and construction of Bode plots for the following four classes of factors is sufficient and exhaustive.

6.2.3.1 Four Classes of Basic Factors

- Constant Factors

$$G(s) = K \text{ and } G(s) = \frac{1}{K}.$$

- Simple (First-Order) Zero and Pole Factors

$$G(s) = 1 + \frac{s}{z_j} \text{ and } G(s) = \frac{1}{\left(1 + \dfrac{s}{p_i}\right)}.$$

- Integrator and Derivative Factors

$$G(s) = \frac{1}{s} \text{ and } G(s) = s.$$

- Complex (Second-Order) Pole and Zero Factors

$$G(s) = \frac{1}{\left[1 + 2\xi\left(\dfrac{s}{w_n}\right) + \left(\dfrac{s}{w_n}\right)^2\right]} \text{ and } G(s) = \left[1 + 2\xi\left(\frac{s}{w_n}\right) + \left(\frac{s}{w_n}\right)^2\right].$$

The approach is to take any given transfer function and express it in terms of the basic factors. The next step is to draw the Bode plots corresponding to these functions using the method of asymptotes, then add them up to get the Bode plot for the whole system.

6.3   Constant Factors (Gain)

The simplest factor is a positive constant function (greater or less than one) depicted by

$$G(s) = K \text{ or } G(s) = \frac{1}{K}$$

$$\implies G(jw) = K \text{ or } G(jw) = \frac{1}{K}.$$

6.3.1   Magnitude

The magnitude is expressed by

$$M = 20\log_{10}|G(jw)|$$
$$= 20\log_{10}|K|.$$

Similarly for the reciprocal $\dfrac{1}{K}$,

$$M = 20 \log_{10} |G(jw)|$$

$$= 20 \log_{10} |\frac{1}{K}|$$

$$= -20 \log_{10} |K|.$$

For example,

$$G(s) = 10$$

$$G(jw) = 10$$

$$M = 20 \log_{10} 10 = 20$$

$$G(s) = \frac{1}{10}$$

$$M = 20 \log_{10} \left(\frac{1}{10} \right)$$

$$= 0 - 20 \log 10 = -20$$

6.3.2   Phase Angle

The phase angles of constant factors K and $\dfrac{1}{K}$ are obtained as follows:

$$\phi = \arg K$$

$$= \tan^{-1} \left(\frac{0}{K} \right)$$

$$= 0$$

$$\phi = \arg \frac{1}{K}$$

$$= \tan^{-1} \left(\frac{0}{1} \right) - \tan^{-1} \left(\frac{0}{K} \right)$$

$$= 0.$$

The magnitudes and phase angles can then be plotted as shown in Figure 6.2 and Figure 6.3. If K is negative, the bode magnitude is the same as that for positive K, but the phase angle is $-180°$. It is important to note that $-180°$ and $180°$ represent the same angle. The choice of -180 is simply a convention adopted in most literature and computer packages such as MATLAB.

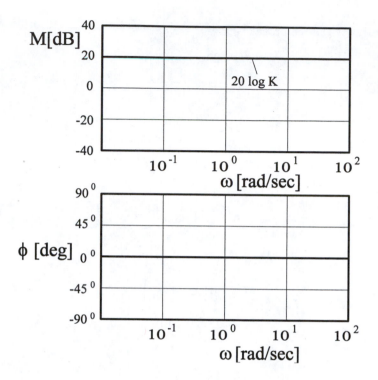

FIGURE 6.2
Bode Plots for a Constant Factor $G(s) = K = 10$

6.4   A Simple Zero Factor

In the simple zero factor form established above, let the zero at $s = -z_j$ be represented by $s = -w_c$, where w_c is called the cut-off, breakaway or corner frequency. This is the frequency at which two Bode asymptotes meet, and its importance will become clear as the Bode plots are constructed.

$$G(s) = 1 + \frac{s}{\omega_c}$$

$$\Rightarrow G(jw) = 1 + \frac{jw}{\omega_c}.$$

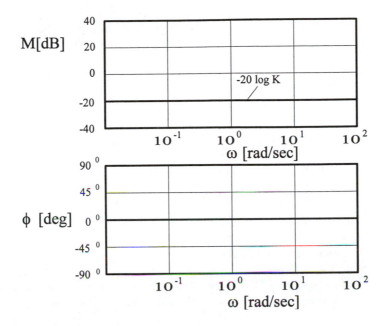

FIGURE 6.3

Bode Plots for a Constant Factor $G(s) = \dfrac{1}{K} = \dfrac{1}{10}$

6.4.1 Magnitude

$$M = 20\log_{10}\sqrt{\left(1+\left(\frac{\omega}{\omega_c}\right)^2\right)}.$$

For low frequencies, that is, $w \ll w_c$, it follows that

$$M = 20\log_{10}\sqrt{\left(1+\left(\frac{\omega}{\omega_c}\right)^2\right)}$$

$$\implies M \approx 20\log_{10}\sqrt{1}$$

$$\implies M \approx 0.$$

Hence, the asymptote for $w \ll w_c$ is $M = 0$. For high frequencies, that is, $w \gg w_c$, the expression for the magnitude can be simplified.

$$M = 20 \log_{10} \sqrt{\left(1 + \left(\frac{\omega}{\omega_c}\right)^2\right)}$$

$$\Longrightarrow M \approx 20 \log_{10} \left(\frac{\omega}{\omega_c}\right) \qquad \left(\text{because } \left(\frac{\omega}{\omega_c}\right)^2 \gg 1 \text{ for } \omega \gg \omega_c\right)$$

$$\Longrightarrow M \approx 20 \log_{10} \omega - 20 \log_{10} \omega_c \approx 20 \log_{10} \omega$$

$$\Longrightarrow \text{gradient of } M = 20 db/decade \text{ and } M = 0 \text{ at } \omega = \omega_c.$$

Hence, the asymptote for $w \gg w_c$ is $M = 20 \log_{10} w$. At the cut-off frequency ($w = w_c$) the two asymptotes ($w \ll w_c$ and $w \gg w_c$) meet with the value of the magnitude being $M = 0$. With the two asymptotes ($w \ll w_c$ and $w \gg w_c$) established including the value of M at the cut-off frequency ($w = w_c$), an *asymptote* Bode plot of M can be drawn. From the asymptote curve, the actual plot can then be sketched, as illustrated in Figure 6.4.

6.4.2   Phase Angle

$$\phi = \tan^{-1} \left(\frac{w}{w_c} / 1\right)$$

$$= \tan^{-1} \left(\frac{w}{w_c}\right). \tag{6.16}$$

Evaluating this expression at limit frequencies leads to

$$w = 0 \Longrightarrow \phi = 0 \tag{6.17}$$

$$w \longrightarrow \infty \Longrightarrow \phi \longrightarrow 90° \tag{6.18}$$

At the cut-off frequency, the phase angle is

$$\phi = \tan^{-1} \left(\frac{w_c}{w_c}\right)$$

$$= \tan^{-1} 1$$

$$= 45°.$$

With the limits in Equation 6.17 and 6.18 and the cut-off phase angle (45°), the phase angle plot can be sketched as shown in Figure 6.4 using the method of asymptotes. For any simple zero factor the only parameter that needs to be specified is the breakaway frequency (e.g., $w_c = 10$).

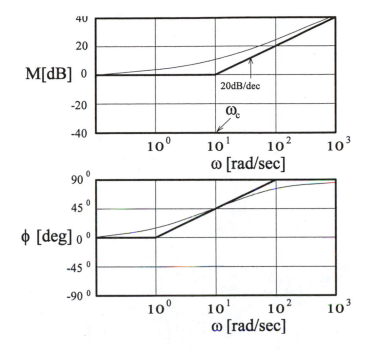

FIGURE 6.4

Bode Plots for a Simple Zero Factor $G(s) = 1 + \dfrac{s}{w_c}$

6.5 A Simple Pole Factor

In the simple pole factor form established above let the pole at $s = -p_i$ be represented by $s = -w_c$.

$$G(s) = \frac{1}{1 + \dfrac{s}{w_c}}$$

$$\Rightarrow G(jw) = \frac{1}{1 + \dfrac{jw}{w_c}} \cdot$$

6.5.1　Magnitude

$$M = 20 \log_{10} \frac{1}{\sqrt{\left(1 + \left(\dfrac{w}{w_c}\right)^2\right)}}$$

$$= -20 \log_{10} \sqrt{\left(1 + \left(\dfrac{w}{w_c}\right)^2\right)}$$

$$\equiv - \text{ the magnitude for the zero factor } \left(1 + \frac{s}{w_c}\right).$$

For low frequencies, that is, $w \ll w_c$

$$M = -20 \log_{10} \sqrt{\left(1 + \left(\dfrac{w}{w_c}\right)^2\right)}$$

$$\Longrightarrow M \approx -20 \log_{10} \sqrt{1}$$

$$\Longrightarrow M \approx 0.$$

Hence, the asymptote for $w \ll w_c$ is $M = 0$. For high frequencies $w \gg w_c$

$$M = -20 \log_{10} \sqrt{\left(1 + \left(\dfrac{w}{w_c}\right)^2\right)}$$

$$\Longrightarrow M \approx -20 \log_{10} \left(\frac{w}{w_c}\right) \qquad \left(\text{because } \left(\frac{w}{w_c}\right)^2 \gg 1 \text{ for } w \gg w_c\right)$$

$$\Longrightarrow M \approx -20 \log_{10} w + 20 \log_{10} w_c \approx -20 \log_{10} w \qquad (6.19)$$

$$\Longrightarrow \text{gradient of } M = -20 db/decade \text{ and } M = 0 \text{ at } w = w_c.$$

Hence, the asymptote for $w \gg w_c$ is $M = -20 \log_{10} w$. At the breakaway frequency ($w = w_c$) the two asymptotes ($w \ll w_c$ and $w \gg w_c$) meet with the value of the magnitude being ($M = 0$). With the two asymptotes ($w \ll w_c$ and $w \gg w_c$) established including the value of M at the cut-off frequency, an *asymptote* Bode plot of M can be drawn. From the asymptote curve the actual plot can then be sketched as illustrated in Figure 6.5.

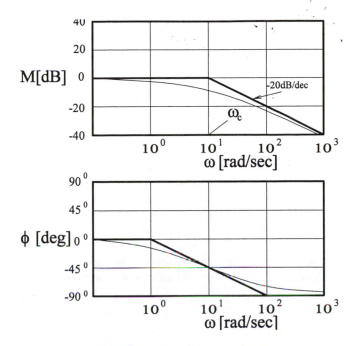

FIGURE 6.5

Bode Plots for a Simple Pole Factor $G(s) = 1/\left(1 + \dfrac{s}{w_c}\right)$

6.5.2 Phase Angle

$$\phi = \tan^{-1}(0/1) - \tan^{-1}\left(\frac{w}{w_c}/1\right)$$

$$\Longrightarrow \phi = -\tan^{-1}\left(\frac{w}{w_c}\right) \tag{6.20}$$

$$\Longrightarrow \phi = -\text{ Phase angle for }\left(1 + \frac{s}{w_c}\right).$$

Taking limits in this expression leads to

$$w = 0 \Longrightarrow \phi = 0 \tag{6.21}$$

$$w \longrightarrow \infty \Longrightarrow \phi \longrightarrow -90°. \tag{6.22}$$

At the cut-off frequency the phase angle is

$$\phi = -\tan^{-1}\left(\frac{w_c}{w_c}\right)$$

$$= -\tan^{-1} 1$$

$$= -45°.$$

With the limits in Equation 6.21 and 6.22 and the breakaway phase angle
(−45°), the phase angle plot can be sketched as shown in Figure 6.5. Note
that for any simple pole factor, the only parameter that needs to be specified
is the breakaway frequency (e.g., $w_c = 10$). The Bode plots in Figure 6.5
are simply the negatives of those for the simple zero factor shown in Figure
6.4, thus confirming the Bode property for reciprocal factors. A practical
example of a system containing a simple pole factor is the RC filter circuit
shown in Example 6.2. Its Bode plots will be similar to those in Figure 6.5
with a cut-off frequency $w_c = \dfrac{1}{RC}$.

6.6   An Integrator Factor

6.6.1   Magnitude

$$G(s) = \frac{1}{s}$$

$$\Rightarrow G(jw) = \frac{1}{jw}.$$

$$M = 20\log_{10}\frac{1}{w}$$

$$= -20\log_{10} w$$

$$\Longrightarrow \text{gradient of } M = -20db/decade.$$

To plot this straight line one point needs to be located, for example, $w =
1 \Longrightarrow M = 0$. With one point and a gradient the magnitude plot can be
drawn as shown in Figure 6.6.

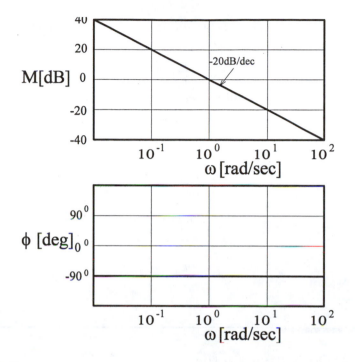

FIGURE 6.6

Bode Plots for an Integrator Factor $G(s) = \dfrac{1}{s}$

6.6.2   Phase Angle

$$
\begin{aligned}
\phi &= \tan^{-1}(0/1) - \tan^{-1}(w/0) \\
&= -\tan^{-1}\infty \\
&\Longrightarrow \phi = -90°.
\end{aligned}
\tag{6.23}
$$

The Bode plot for the integrator can thus be drawn as shown in Figure 6.6. There is no need for asymptotes. For a cascade of integrators such that

$$
G(s) = \frac{1}{s^k},
$$

the magnitude and phase are obtained as in the case of a single integrator.

$$
\begin{aligned}
M &= 20\log_{10}\left|\frac{1}{(jw)^k}\right| \\
&= -20k\log_{10} w \\
&\Longrightarrow \text{gradient of } M = -20k \ db/decade
\end{aligned}
$$

$$
\begin{aligned}
\phi &= \arg\frac{1}{(jw)^k} \\
&= \tan^{-1}(0/1) - k\tan^{-1}(w/0) \\
&= -k\tan^{-1}\infty \\
&\Longrightarrow \phi = -90k^\circ.
\end{aligned}
$$

6.7   A Derivative Factor

The magnitude and phase for the derivative factor are obtained in the same way as for the integrator factor and they are in fact the negatives of the magnitude and phase for the integral factor, respectively.

6.7.1   Magnitude

$$
G(s) = s
$$
$$
\Rightarrow G(jw) = jw
$$

$$
\begin{aligned}
M &= 20\log_{10} w \\
&\Longrightarrow \text{gradient of } M = 20db/decade.
\end{aligned}
$$

In order to plot this straight line one point has to be located, for example, $w = 1 \Longrightarrow M = 0$. With one point and a gradient, the magnitude plot can be drawn as shown in Figure 6.7.

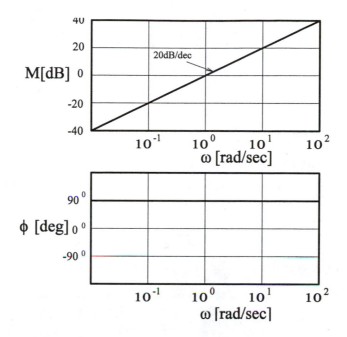

FIGURE 6.7
Bode Plots for a Derivative Factor $G(s) = s$

6.7.2   Phase Angle

$$\phi = \tan^{-1}(w/0)$$
$$= \tan^{-1}\infty$$
$$\implies \phi = 90° \tag{6.24}$$

The Bode plot for the derivative can thus be drawn as shown in Figure 6.7. There is no need for asymptotes and the plots are the negatives of the integrator plots. For a cascade of derivative factors such that

$$G(s) = s^k,$$

the magnitude and phase are obtained as in the case of a single derivative factor.

$$M = 20\log_{10} w^k$$
$$= 20k\log_{10} w$$
$$\implies \text{gradient of } M = 20k \; db/decade$$

$$
\begin{aligned}
\phi &= k\tan^{-1}(w/0) \\
&= k\tan^{-1}\infty \\
\Longrightarrow \phi &= 90k°.
\end{aligned}
$$

6.8   A Complex Pole Factor

As has already been shown, the complex pole factor is established by considering the general transfer function of a standard second-order system,

$$
H(s) = \frac{b(s)}{s^2 + 2\xi w_n s + w_n^2}
$$

such that the complex pole factor is given by

$$
G(s) = \frac{1}{\left[1 + 2\xi\left(\dfrac{s}{w_n}\right) + \left(\dfrac{s}{w_n}\right)^2\right]}
$$

$$
\Longrightarrow G(jw) = \frac{1}{\left[1 + 2\xi\left(\dfrac{jw}{w_n}\right) + \left(\dfrac{jw}{w_n}\right)^2\right]}. \tag{6.25}
$$

6.8.1   Magnitude

$$
M = 20\log_{10}\left|\frac{1}{\left[1 + 2\xi\left(\dfrac{jw}{w_n}\right) + \left(\dfrac{jw}{w_n}\right)^2\right]}\right|
$$

$$
= -20\log_{10}\left|\left[1 + 2\xi\left(\dfrac{jw}{w_n}\right) + \left(\dfrac{jw}{w_n}\right)^2\right]\right|
$$

$$
= -20\log_{10}\sqrt{\left(1 - \dfrac{w^2}{w_n^2}\right)^2 + \left(2\xi\dfrac{w}{w_n}\right)^2} \tag{6.26}
$$

For low frequencies, i.e., $w \ll w_n$ the magnitude becomes

$$
M \approx -20\log_{10}\sqrt{1}
$$

$$
\approx 0
$$

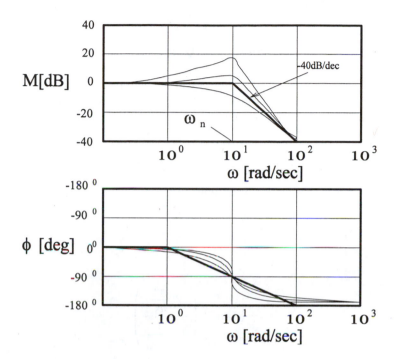

FIGURE 6.8
A Complex Pole Factor with Various Values of ξ (0.1, 0.3, 1.0)

Hence, the asymptote for $w \ll w_n$ is $M = 0$. For high frequencies $w \gg w_n$,

$$M = -20 \log_{10} \sqrt{\left(1 - \frac{w^2}{w_n^2}\right)^2 + \left(2\xi \frac{w}{w_n}\right)^2}$$

$$\Longrightarrow M \approx -40 \log_{10}\left(\frac{w}{w_n}\right)$$

$$\Longrightarrow M \approx -40 \log_{10} w + 40 \log_{10} w_n \tag{6.27}$$

$$\Longrightarrow \text{gradient of } M = -40db/decade \text{ and } M = 0 \text{ at } w = w_c.$$

Hence, the asymptote for $w \gg w_n$ is $M = -40 \log_{10} w$. At the breakaway frequency ($w = w_n$) the two asymptotes ($w \ll w_n$ and $w \gg w_n$) meet with the value of the magnitude being ($M = 0$). The two asymptotes ($w \ll w_n$ and $w \gg w_n$) are independent of the value of the damping ratio ξ. In the actual Bode plot derived from the asymptotes, near the frequency ($w = w_n$) a resonant peak M_{peak} occurs, as can be expected from Equation 6.25. The resonant peak is of the form

$$M_{peak} = -20 \log_{10} 2\xi.$$

This clearly shows that the damping ratio determines the magnitude of the resonant peak. There are large peaks for small values of ξ. The magnitude Bode plots are shown in Figure 6.8 for various values of ξ. For example for $\xi = 0.1$ the magnitude plot has a resonant peak at

$$M_{peak} = -20 \log_{10} (2 \times 0.1)$$

$$= 13.98.$$

6.8.2   Phase Angle

$$\phi = -\tan^{-1} \left(\frac{2\xi \dfrac{w}{w_n}}{1 - \dfrac{w^2}{w_n^2}} \right). \tag{6.28}$$

This means the phase angle is a function of both w and ξ. Analyzing this expression leads to

$$w = 0 \implies \phi = 0°$$

$$w = w_n \implies \phi = -90°$$

$$w = \infty \implies \phi = -180°.$$

The phase angle curve is skew symmetric about the inflection point, where $\phi = -90°$, with $0°$ and $-180°$ as asymptotes as shown in Figure 6.8. This figure shows the phase Bode plots for various values of ξ.

6.9   A Complex Zero Factor

The Bode plots for a complex zero factor are obtained by simply reversing the sign of the magnitude M and phase angle ϕ for the complex pole factor. If necessary, they can also be derived from first principles and analyzed just as done for the complex pole factor.

6.9.1 Magnitude

$$M = 20 \log_{10} \sqrt{\left(1 - \frac{w^2}{w_n^2}\right)^2 + \left(2\xi \frac{w}{w_n}\right)^2}. \tag{6.29}$$

6.9.2 Phase Angle

$$\phi = \tan^{-1} \left(\frac{2\xi \dfrac{w}{w_n}}{1 - \dfrac{w^2}{w_n^2}} \right). \tag{6.30}$$

6.10 Drawing Bode Plots

Given a transfer function, express and break it up into the basic functions. Identify the cut-off frequencies w_c or w_n and draw the basic Bode plots corresponding to these functions using the method of asymptotes, then add them up to get the Bode plot for the whole system.

Example 6.4 *Consider the following system transfer function,*

$$G(s) = \frac{2s + 2}{s^2 + 10s}.$$

(a) Show that the function can be broken into constant, zero, pole, and integral factors.
(b) Obtain the overall system Bode plots.

Solution 6.4 *(a) The transfer function can be broken into basic transfer functions as follows:*

$$G(s) = \frac{2s + 2}{s^2 + 10s}$$

$$= \frac{\frac{1}{5}(1 + s)}{s(1 + \frac{s}{10})}$$

$$\Rightarrow \frac{1}{5}, \quad (1 + s), \quad \frac{1}{(1 + \frac{s}{10})}, \quad \frac{1}{s}.$$

The cut-off frequency for the zero function is given by $w_c = 1$ and that for the pole function by $w_c = 10$.

(b) The four basic plots are then separately drawn as discussed above, and then added together to produce the composite Bode plots. The results are shown in the following diagrams:

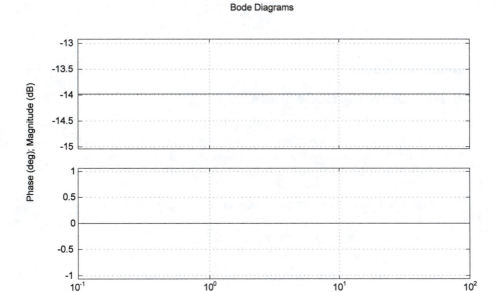

Bode Diagrams

Bode Plots for the Constant Factor $1/2$

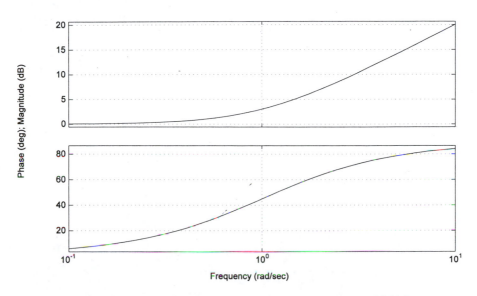

Bode Plots for the Zero Function $(1 + s)$

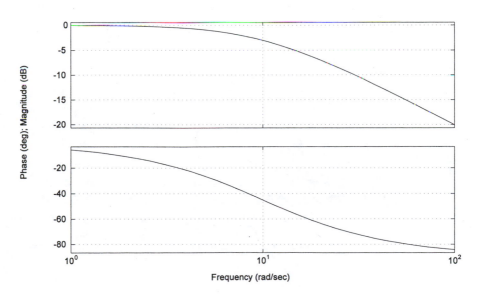

Bode Plots for the Pole Function $1/(1 + s/10)$

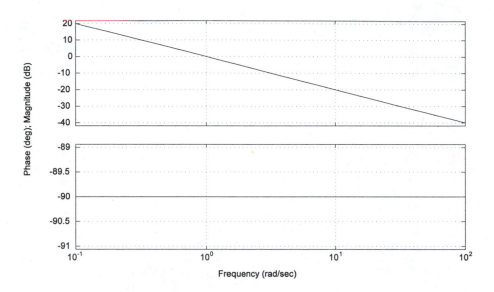

Bode Plots for the Integrator Factor 1/s

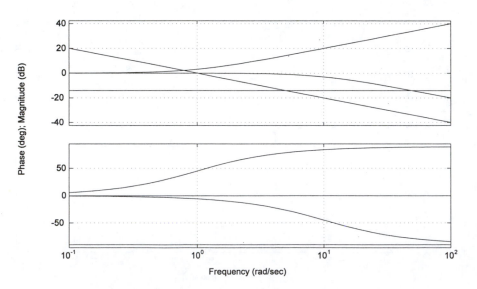

Summing up the Bode Plots of the Simple Factors

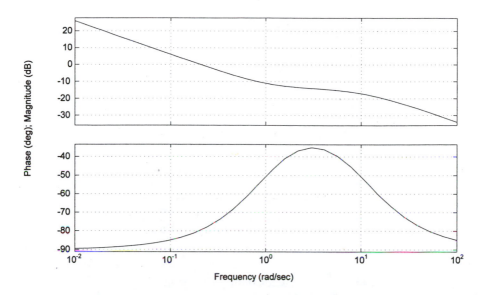

Bode Plots for the Entire Transfer Function $(2s + 2)/\{s(s + 10)\}$

Example 6.5 *Consider the following transfer function,*

$$G(s) = \frac{1}{s^2 + 10s + 100}.$$

(a) Express the transfer function in terms of a constant factor and a complex pole factor.

(b) Draw the Bode plots for the two factors.

(c) Derive the overall system Bode plots.

Solution 6.5 *(a) The starting point is expressing the transfer function in* *terms of basic factors. This is achieved as follows:*

$$G(s) \;=\; \frac{1}{s^2 + 10s + 100}$$

$$=\; \frac{1}{100 \left[1 + \dfrac{s}{10} + \left(\dfrac{s}{10} \right)^2 \right]}$$

$$G(jw) \;=\; \frac{1}{100 \left[1 + \dfrac{jw}{10} + \left(\dfrac{jw}{10} \right)^2 \right]}$$

$$\Longrightarrow\; \frac{1}{w_n^2 \left[1 + 2\xi \left(\dfrac{jw}{w_n} \right) + \left(\dfrac{jw}{w_n} \right)^2 \right]}.$$

$$\Longrightarrow \frac{1}{100} \qquad \textit{(a constant gain)} \;\; \textit{and}$$

$$\frac{1}{\left[1 + \dfrac{jw}{10} + \left(\dfrac{jw}{10} \right)^2 \right]} \qquad \textit{(a complex pole factor)}$$

where $\;\; w_n \;=\; 10 \;\; and \;\; \xi = 0.5$

$$\Longrightarrow M_{peak} = -20 \log_{10} 2\xi = 0.$$

(b), (c) The two functions, a constant gain, and a complex pole factor are then sketched and added together. The results are shown in the following diagrams:

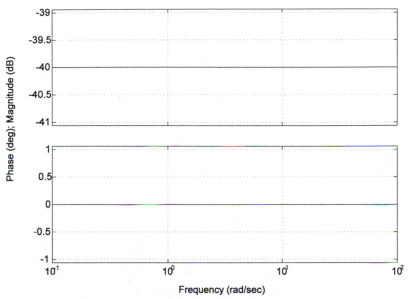

Bode Plots for the Constant Factor $1/100$

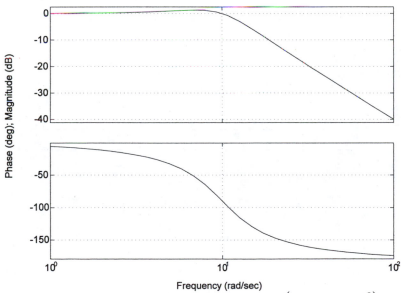

Bode Plots for the Complex Pole Factor $1/\left(1 + \frac{s}{10} + \left(\frac{s}{10}\right)^2\right)$

Bode Diagrams

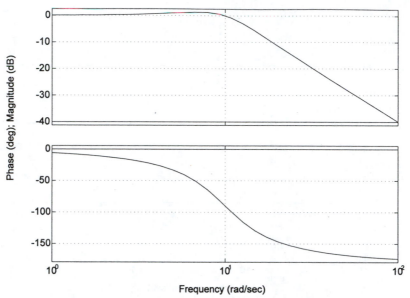

Summing up the Bode Plots of the Simple Factors

Bode Diagrams

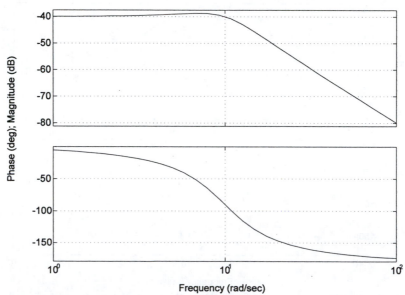

Bode Plots for the Entire Transfer Function $1/(s^2 + 10s + 100)$

Example 6.6 *(a) Show that the following transfer function can be broken into zero, pole, integrator, and complex factors.*

$$G(s) = \frac{1000(1+s)}{s(s+2)(s^2+8s+64)}.$$

(b) Sketch the Bode plots of these basic factors.
(c) Deduce the overall Bode plots for the system.

Solution 6.6 *(a) The transfer function can be expressed in terms of basic factors as follows:*

$$G(s) = \frac{1000(1+s)}{s(s+2)(s^2+8s+64)}$$

$$= \frac{\dfrac{1000}{2 \times 64}(1+s)}{s(1+\dfrac{s}{2})[1+\dfrac{s}{8}+(\dfrac{s}{8})^2]}$$

$$= \frac{\dfrac{125}{16}(1+s)}{s(1+\dfrac{s}{2})[1+\dfrac{s}{8}+(\dfrac{s}{8})^2]}$$

$$\Rightarrow \frac{125}{16}, \quad (1+s), \quad \frac{1}{(1+\dfrac{s}{2})}, \quad \frac{1}{s}, \quad \frac{1}{1+\dfrac{s}{8}+(\dfrac{s}{8})^2}.$$

Thus there are five basic plots. The cut-off frequency for the zero factor is given $w_c = 1$ and that for the pole factor by $w_c = 2$. For the complex pole, the natural frequency w_n and the damping ratio ξ are found by comparing it to the standard form, i.e.

$$\frac{1}{1+\dfrac{s}{8}+\left(\dfrac{s}{8}\right)^2}$$

$$= \frac{1}{1+\dfrac{jw}{8}+\left(\dfrac{jw}{8}\right)^2}$$

$$\Longrightarrow \frac{1}{\left[1+2\xi\left(\dfrac{jw}{w_n}\right)+\left(\dfrac{jw}{w_n}\right)^2\right]}$$

$$\Longrightarrow w_n = 8 \ and \ \xi = 0.5$$

$$\Longrightarrow M_{peak} = -20\log_{10} 2\xi = 0.$$

(b), (c) The five basic plots can now be drawn separately and then added to produce the composite Bode plots. The results are shown in the following diagrams:

Bode Diagrams

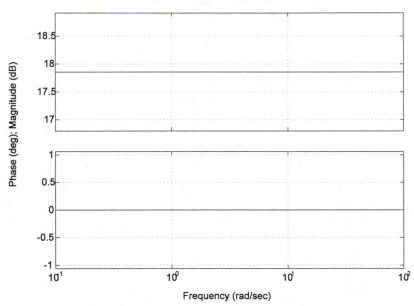

Bode Plots for the Constant Factor $125/16$

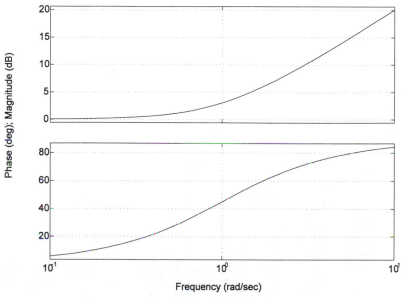

Bode Plots for the Zero Factor $(1 + s)$

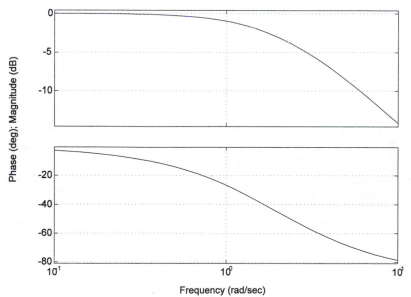

Bode Plots for the Pole Factor $1/(1 + s/2)$

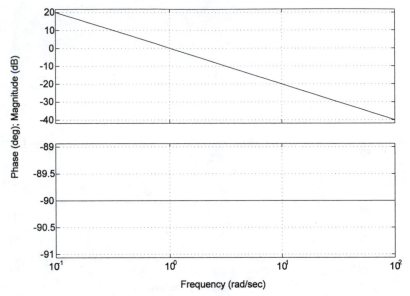

Bode Plots for the Integrator Factor $1/s$

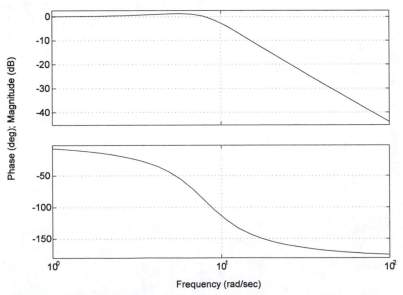

Bode Plots for the Complex Pole Factor $1/\{1 + s/8 + (s/8)^2\}$

Bode Diagrams

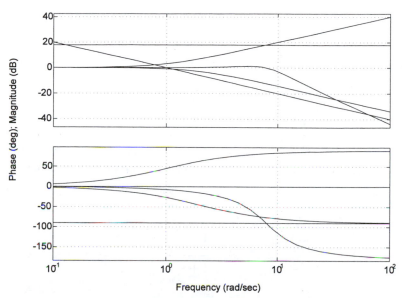

Summing up the Bode Plots of the Simple Factors

Bode Diagrams

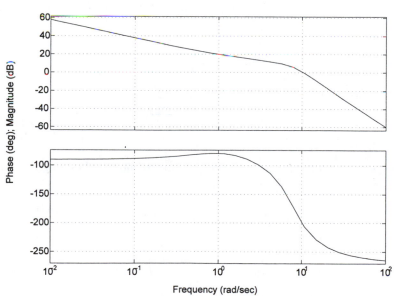

Bode Plots for the Entire Transfer Function
$1000(s + 1)/\{s(s + 2)(s^2 + 8s + 64)\}$

Solution 6.7 *These examples illustrate the techniques of hand drawing Bode plots.*

6.11   Nonminimum Phase Systems

A nonminimum phase system has at least one zero in the right-hand plane (RHP). Such systems arise when the system includes a nonminimum phase element or when there is an unstable minor loop. Consider two transfer functions that have the same structure except that one system has a zero at $\dfrac{1}{w_z}$ while the other has a zero at $-\dfrac{1}{w_z}$ such that

$$G_1(s) = \frac{1 + \dfrac{s}{w_z}}{1 + \dfrac{s}{w_p}}, \qquad G_2(s) = \frac{1 - \dfrac{s}{w_z}}{1 + \dfrac{s}{w_p}}, \qquad 0 \le \frac{s}{w_z} \le \frac{s}{w_p}.$$

6.11.1   Magnitude and Phase

The two systems have the same magnitude ($M_1 = M_2$) but different phase angles (ϕ_1 and ϕ_2). The two complex quantities $G_1(jw)$ and $G_2(jw)$ differ by a factor

$$
\begin{aligned}
G(jw) &= \frac{G_2(jw)}{G_1(jw)} \\[2mm]
&= \frac{1 - \dfrac{jw}{w_z}}{1 + \dfrac{jw}{w_z}}.
\end{aligned}
$$

The magnitude of this factor is always unity while its phase angle is given by

$$
\begin{aligned}
\phi &= \phi_2 - \phi_1 \\
&= \arg G(jw) \\
&= \tan^{-1}\left(-\frac{jw}{w_z}\right) - \tan^{-1}\left(\frac{jw}{w_z}\right) \\
&= -2\tan^{-1}\left(\frac{jw}{w_z}\right) \\
\Longrightarrow\ & \phi \text{ varies from } 0 \text{ to } 180° \text{ as } w \text{ varies from } 0 \text{ to } \infty.
\end{aligned}
$$

This is the difference between the phase angles ϕ_1 and ϕ_2. The Bode plots for the two systems $G_1(s)$ and $G_2(s)$ are given in Figure 6.9, where $\dfrac{1}{w_z} = \dfrac{1}{100}$ and $\dfrac{1}{w_p} = \dfrac{1}{10}$.

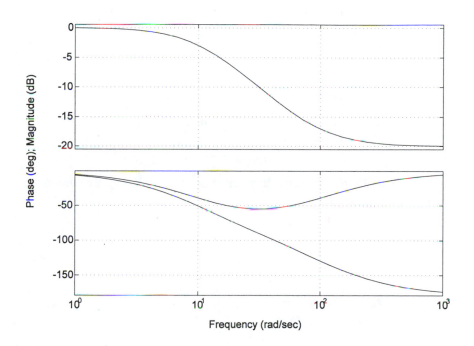

Bode Diagrams

FIGURE 6.9
The Effect of a Non-Minimum Phase System: Bode Plots of $G_1(s)$ and $G_2(s)$

Nonminimum phase systems arise in two different ways. One way is simply when a system includes a nonminimum phase element or elements. The other situation may arise in the case where a minor loop is unstable. Nonminimum phase systems are slow in response because of their faulty behavior at the start of response. In most practical systems excessive phase lag should be carefully avoided. In particular in system design where speed of response is of primary importance nonminimum phase elements should be avoided.

6.12   Time Delay (Transport Lag)

The time delay function or transport lag is a type of nonminimum phase system that has an excessive phase lag with no attenuation at high frequencies. Such transport delays exist in thermal, hydraulic, and pneumatic systems. For example, when a tap is turned on at one end of a long hose-pipe, it takes some time before the water appears at the far end. Also, when the feed to a conveyor belt is changed, it takes some time for the change to be observed at the other end. A pure time delay function can be modeled by considering a general system $f(t)$ delayed by time T, such that the system is represented by $f(t - T)$. Taking Laplace transforms of this function leads to

$$\mathcal{L}\, f(t - T) = e^{-sT} F(s).$$

Hence, the Laplace transform of a pure time delay is given by

$$G(s) = e^{-sT}.$$

In order to deal with the time delay function, the Pade approximation was required in root locus analysis,

$$G(s) = e^{-sT}$$

$$= \frac{e^{-s\frac{T}{2}}}{e^{s\frac{T}{2}}} \simeq \frac{1 - \frac{sT}{2}}{1 + \frac{sT}{2}}. \tag{6.31}$$

In frequency-response methods (Bode and Nyquist) an exact analysis of the delay function is possible. The frequency response of the delay is given by the magnitude and phase of $G(jw)$.

$$G(jw) = e^{-sT}\big|_{s=jw} = e^{-jwT}$$

$$= \cos jwT - \sin jwT \quad \text{(from Euler's equation)}.$$

6.12.1   Magnitude

The magnitude is expressed by

$$M = 20 \log_{10} |G(jw)|$$

$$= 20 \log_{10} |\cos^2 (jwT) + \sin^2 (jwT)|$$

$$= 20 \log 1$$

$$= 0.$$

6.12.2 Phase Angle

The phase angle of a time delay function is obtained as follows:

$$\phi = \arg G(jw)$$

$$= \tan^{-1}\left(\frac{-\sin jwT}{\cos jwT}\right)$$

$$= -wT \quad \text{(radians)}$$

$$= -57.3wT \quad \text{(degrees)}.$$

This means the phase angle varies linearly with the frequency w. The phase angle ($\phi = -wT$) is shown in Figure 6.10. It is important to note that the curve for the phase angle is not a linear curve because it is being plotted with respect to $\log w$.

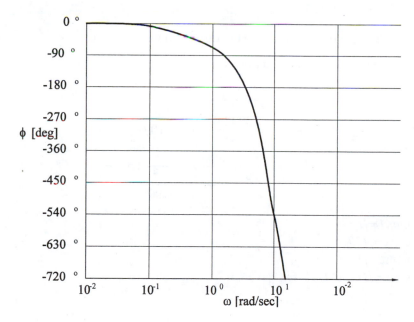

FIGURE 6.10
Phase Angle for a Pure Time Delay Function $G(s) = e^{-sT}$

6.13   Bode Plots Using MATLAB

Bode plots can be drawn using MATLAB software. The key command is *"bode(.)"* and it is applied to the system transfer function model or the state-variable matrix model. A few sample algorithms are presented here to illustrate the procedure involved.

6.13.1   Sample Algorithms

6.13.1.1   A Single Plot

Consider the transfer function

$$G(s) = \frac{2s^2 + 3s}{s^3 + 12s^2 + 7s + 11}.$$

num=[2 3 0];
den=[1 12 7 11];
bode(num,den)

6.13.1.2   Several plots on the same curve

Consider the transfer function

$$G(s) = \frac{1}{100 \left[1 + \dfrac{s}{10} + \left(\dfrac{s}{10} \right)^2 \right] s}$$

$$\implies \frac{1}{100}, \quad \frac{1}{1 + \dfrac{s}{10} + \dfrac{s^2}{100}} \quad \text{and} \quad \frac{1}{s}.$$

num1=[1];
den1=[100];
num2=[1];
den2=[1/100 1/10 1];
num3=[1];
den3=[1 0];
sys1=tf(num1,den1);
sys2=tf(num2,den2);
sys3=tf(num3,den3);
bode(sys1,sys2,sys3)

6.13.1.3 System in State Space

Consider a system with the following state-variable matrices

$$A = \begin{bmatrix} 0 & 1 \\ -24.5 & -4.5 \end{bmatrix}, \quad B = \begin{bmatrix} 0 \\ 24.5 \end{bmatrix}, \quad C = \begin{bmatrix} 1 & 0 \end{bmatrix}, \quad D = \begin{bmatrix} 0 \end{bmatrix}.$$

A=[0 1;-24.5 -4.5];

B=[0;24.5];

C=[1 0];

D=[0];

bode(A,B,C,D)

6.14 System Models From Experiment Frequency Data

Bode plots can be used to derive system models or verify models (obtained by other methods) by using information from experimental frequency data. The model obtained is in the form of a transfer function deduced directly from the frequency response. The issues and methods involved in the derivation of dynamic models by using experimental data were covered in Chapter 2. Sometimes the only practical or feasible way to obtain a model is in the form of an approximate one from frequency-response data. It is relatively easy to obtain the frequency response of a system experimentally. The procedure is to introduce a sinusoidal input and then measure the gain (logarithm of the amplitude ratio of output to input) and the phase difference between output and input. The curves plotted from this data then constitute the system model. Using the methods given in previous sections, one can derive the model directly from this information. Model verification using experimental data is accomplished by extracting an approximate transfer function from the plots by fitting straight lines to the data and estimating breakpoints (cut-off frequency), that is, finding the poles and zeros. General plots of complex pole factors with different damping ratios (Figure 6.8) are used to estimate the damping ratios of complex factors from the frequency overshoot. The model obtained from the data is then compared with the model derived by other methods for purpose of model validation.

6.15 Nyquist Analysis

For most systems, an increasing gain eventually causes instability. However, this relationship between gain and stability margins is not always valid. Occasionally the relationship reverses itself; that is, the system (e.g., an amplifier) becomes unstable when the gain is decreased. These conflicting observations constitute the central motivation behind Nyquist analysis, which is based on a result from complex variables theory known as the argument principle. Nyquist analysis is a graphical procedure carried out in the frequency domain in order to determine absolute and relative stability of closed-loop control systems. Information about stability is available directly from a graph of the sinusoidal open-loop transfer function $G(s)$, once the feedback system has been put into canonical form.

6.15.1 Advantages of Nyquist Method

There are several reasons for which the Nyquist method can be chosen to determine information about system stability.

- Routh-Hurwitz, BIBO, and asymptotic stability methods are often inadequate because, with few exceptions, they can only be used for determining absolute stability. Nyquist analysis is used for the determination of both absolute and relative stability, and for the evaluation of the closed-loop frequency response.

- The Nyquist method can handle time delay functions, i.e., terms like e^{-Ts}, in the characteristic equation without requiring the use of approximations, whereas the Routh-Hurwitz method is only strictly applicable to systems whose characteristic equation is a finite polynomial in s. In order to deal with time delay functions it is necessary to employ approximations for the function e^{-Ts} when using the Routh-Hurwitz method. Hence, the Nyquist method yields exact results for both absolute and relative stability of the systems.

- Nyquist techniques are also useful for obtaining transfer functions of components or systems from experimental frequency-response data. The polar plot may be directly graphed from sinusoidal steady state measurements on the components making up the open-loop transfer function.

- Nyquist methods are very useful in the determination of system stability properties when transfer functions of loop components are not available in analytic forms, or when physical systems are to be tested and evaluated experimentally.

6.16   Polar Plots

As developed in sections on Bode plots, the transfer function $G(s)$ can be represented in the frequency domain as a sinusoidal transfer function by substituting jw for s in $G(s)$ such that

$$G(jw) = G(s)|_{s=jw}$$
$$= \operatorname{Re} G(jw) + j \operatorname{Im} G(jw)$$
$$= X(w) + jY(w)$$
$$= Ae^{j\phi},$$

where A is the magnitude of the transfer function $G(jw)$

$$A = |G(jw)|$$
$$= \sqrt{X(w)^2 + Y(w)^2},$$

and ϕ is the phase angle of the function $G(jw)$

$$\phi = \arg G(jw)$$
$$= \tan^{-1} \frac{Y(w)}{X(w)}.$$

The resulting form $G(jw) = Ae^{j\phi}$ is a complex function of the single variable w and can be plotted in two dimensions with w as a parameter. The complex function $G(jw)$ can be written in the following three equivalent forms

$$G(jw) = A \arg G(jw) \qquad \text{(Polar Form)}$$

$$G(jw) = A(\cos \phi + j \sin \phi) \qquad \text{(Euler form)}$$

$$G(jw) = X(w) + jY(w) \qquad \text{(Complex Form)}$$

A polar plot of $G(jw)$ is a graph of $Y(w)$ vs. $X(w)$ in the finite portion of the $G(jw)$-plane for $-\infty \le w \le \infty$. The polar plot of $G(jw)$ can also be generated on a polar coordinate system. The magnitude and phase angle pair denoted by (M, ϕ) are plotted with w varying between $-\infty \le w \le \infty$. However, for practical systems negative frequency is not defined and hence the range of interest is reduced to $0 \leq w \leq \infty$. The locus of $G(jw)$ is identical on either the complex plane or polar coordinate system. The choice of coordinates depends on which form of the equations above is easier to use. Experimental data is usually expressed in terms M and ϕ, in this case, polar coordinates are the natural choice. The construction of polar plots proceeds in the same way as for the Bode plots by considering the four classes of factors.

6.17   Four Classes of Basic Factors

As was the case with Bode plots the four classes are: (1) constant factors, (2) simple zero and pole factors, (3) integrator and derivative factors, and (4) complex pole and zero factors.

- Constant Factors

$$G(s) = K .$$

A is the magnitude of the transfer function $G(jw)$

$$A = |G(jw)|$$
$$= |K|.$$

ϕ is the phase angle of the function $G(jw)$

$$\phi = \arg G(jw)$$
$$= \tan^{-1}\left(\frac{0}{K}\right)$$
$$= 0.$$

The polar plot is a point on the real axis at a distance $|K|$ from the origin.

- Simple (First-Order) Zero and Pole Factors

$$G(s) = 1 + \frac{s}{w_c} \text{ and } G(s) = \frac{1}{\left(1 + \dfrac{s}{w_c}\right)}.$$

Consider the simple pole

$$G(s) \;=\; \frac{1}{(1 + \dfrac{s}{w_c})}$$

$$\implies G(jw) = \frac{1}{(1 + \dfrac{jw}{w_c})}$$

The magnitude and phase angle are thus given by

$$A = \frac{1}{\sqrt{1 + \left(\dfrac{w}{w_c}\right)^2}}$$

$$\phi = -\tan^{-1}\left(\frac{w}{w_c}\right).$$

To draw the polar plots, find the values of the magnitude and phase angle, i.e., the polar pair (A, ϕ), for various values of w. In particular, consider $w = 0, w = w_c$ and $w = \infty$.

$$w = 0 \implies (1, 0)$$

$$w = w_c \implies (1/\sqrt{2}, -45)$$

$$w = \infty \implies (0, -90).$$

The resulting locus for $0 \le w \le \infty$ is the lower semicircle of a circle of radius 0.5 and Cartesian center $(0, 0.5)$ as shown in Figure 6.11. The arrows on the curve show the direction of increasing frequency w. The graph for the frequency range $-\infty \le w \le 0$ is the mirror image of the lower semicircle about its diameter, i.e., the upper semicircle.

6.17.0.1 The RC Filter Circuit

The RC filter circuit is a good example of a simple pole factor. Its transfer function was derived in the sections on Bode plots. The polar plot of the circuit can be drawn in the same way as for the generic simple pole, with cut-off frequency as $w_c = \dfrac{1}{RC}$.

$$G(s) \;=\; \frac{V_2(s)}{V_1(s)} = \frac{1}{1 + sRC}$$

$$\implies G(jw) = \frac{1}{1 + jwRC}.$$

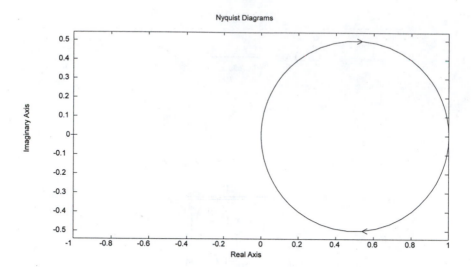

FIGURE 6.11

Polar Plot for $G(s) = \dfrac{1}{1 + \dfrac{s}{w_c}}$

The magnitude and phase angle are given by

$$A = \frac{1}{\sqrt{1 + (wRC)^2}}$$

$$\phi = -\tan^{-1} wRC.$$

In order to draw the polar plots, the values of A and ϕ for various values of w are obtained, in particular, the following frequencies are considered; $w = 0, w = w_c, w = \infty$. As was the case in the generic simple pole, the polar plot for the frequency range $0 \le w \le \infty$ is the lower semicircle of a circle of radius 0.5 and center $(0, 0.5)$ as shown in Figure 6.12, where the arrows on the curve indicate the direction of increasing frequency w.

The polar plot of the simple zero factor can be drawn in the same way as has been done for the simple pole factor.

$$G(s) = 1 + \frac{s}{w_c}$$

$$\implies G(jw) = 1 + \frac{jw}{w_c}$$

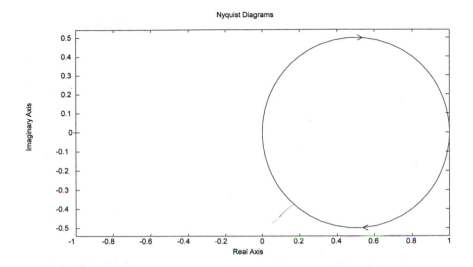

FIGURE 6.12
Polar Plot for $G(s) = \frac{1}{1+sRC}$

Magnitude and phase angle are given by

$$A = \sqrt{1 + \left(\frac{w}{w_c}\right)^2}$$

$$\phi = \tan^{-1}\left(\frac{w}{w_c}\right).$$

The polar plot is drawn by first determining the magnitude and phase angle at different frequencies, in particular,

$$w = 0 \implies (1, 0)$$
$$w = w_c = 1 \implies (\sqrt{2}, 45)$$
$$w = \infty \implies (\infty, 90°).$$

This means that the polar plot is simply the upper half of the straight line graph passing through Cartesian point $(1, 0)$ in the complex plane and parallel to the imaginary axis. Thus the polar plot of the simple zero factor has an appearance completely different from that of the simple pole factor.

- Integrator and Derivative Factors

$$G(s) = \frac{1}{s} \text{ and } G(s) = s.$$

Consider the integrator factor first. The magnitude A is obtained as follows:

$$G(s) = \frac{1}{s}$$

$$\Longrightarrow A = |\frac{1}{jw}| = \frac{1}{w}.$$

The phase angle ϕ is given by

$$\phi = \arg G(jw)$$

$$= 0 - \tan^{-1}\left(\frac{w}{0}\right)$$

$$= -90°.$$

This means that the polar plot of the integrator factor is the negative imaginary axis. For the derivative factor, the same analysis can be carried out and the magnitude A is obtained as follows:

$$G(s) = s$$

$$\Longrightarrow A = |jw| = w$$

The phase angle ϕ is given by

$$\phi = \arg G(jw)$$

$$= \tan^{-1}\left(\frac{w}{0}\right)$$

$$= 90°.$$

This means that the polar plot of the derivative factor is the positive imaginary axis.

- Complex (Second-Order) Pole and Zero Factors

$$G(s) = \frac{1}{\left[1 + 2\xi\left(\dfrac{s}{w_n}\right) + \left(\dfrac{s}{w_n}\right)^2\right]} \quad \text{and} \quad G(s) = \left[1 + 2\xi\left(\dfrac{s}{w_n}\right) + \left(\dfrac{s}{w_n}\right)^2\right].$$

The low and high frequency portion of the polar plots for the complex pole factor are determined by using expressions of its magnitude and phase angle. The magnitude A of the transfer function is obtained by using $s = j\omega$ in $G(s)$. Thus,

$$A = |G(j\omega)|$$

$$= \left| \frac{1}{\left[1 + 2\xi \left(\dfrac{j\omega}{\omega_n} \right) + \left(\dfrac{j\omega}{\omega_n} \right)^2 \right]} \right|$$

$$= \frac{1}{\sqrt{\left(1 - \dfrac{\omega^2}{\omega_n^2} \right)^2 + \left(2\xi \dfrac{\omega}{\omega_n} \right)^2}}.$$

The phase angle ϕ takes the form

$$\phi = \arg G(j\omega)$$

$$= -\tan^{-1} \left(\frac{2\xi \dfrac{\omega}{\omega_n}}{1 - \dfrac{\omega^2}{\omega_n^2}} \right).$$

The polar plot is drawn by first evaluating the magnitude and phase angle, i.e., the polar pair (A, ϕ), at different frequencies.

$$\omega = 0 \Longrightarrow (1, 0)$$

$$\omega = \omega_n \Longrightarrow \left(\frac{1}{2\xi}, -90° \right)$$

$$\omega = \infty \Longrightarrow (0, -180°).$$

The polar plot starts at the polar point $(1, 0)$ and ends at $(0, -180°)$ as w increases from 0 to ∞. Thus, the high-frequency portion is tangent to the negative real axis. The values of $G(jw)$ in the frequency range of interest can be calculated directly or by use of the Bode diagram. Similar analysis can be conducted for the complex zero factor.

6.17.1 Properties of Polar Plots

- The polar plot for the complex function

$$G(jw) + a,$$

where a is any complex constant, is identical to the plot for $G(jw)$ with the origin of coordinates shifted to the point $-a$.

- The polar plot of the transfer function of a time-invariant, constant-coefficient, linear system exhibits conjugate symmetry, i.e., the graph for $-\infty \leq w \leq 0$ is the mirror image about the horizontal axis of the graph for $0 \leq w \leq \infty$.

- The polar plot can be constructed directly from Bode plot. Values of magnitude and phase angle at various frequencies w on the Bode plot represent points along the locus of the polar plot.

- Constant increments of frequency are not generally separated by equal intervals along the polar plot.

6.17.2   Nyquist Path

Nyquist analysis is based on the argument principle from the mathematics field known as Complex variables. This principle states that the Nyquist Path is a closed contour in the s-plane that completely encloses the entire right half of the s-plane (RHP). So that Nyquist Path should not pass through any poles of $G(s)$, small semicircles along the imaginary axis or at the origin of $G(s)$ are required in the path if $G(s)$ has poles on the Imaginary axis $(s = jw)$ or at the origin $(s = 0)$. The radii ℓ of these small circles are interpreted as approaching zero in the limit. In order to enclose the RHP at infinity, a large semicircular path is drawn in the RHP and the radius R of this semicircle is taken as infinite in the limit. It is apparent that every pole and zero of $G(s)$ in the RHP is enclosed by the Nyquist Path when it is mapped into the $G(s)$-plane.

6.17.3   Nyquist Diagram

The Nyquist Diagram is an extension of the polar plot. It is a mapping of the entire Nyquist Path into the $G(s)$ and it is constructed using mapping properties. The Nyquist diagram is the polar coordinate representation of $G(jw)$ in the $G(s)$-plane. It is a mapping of the positive imaginary axis of the s-plane into the $G(s)$-plane for a close mapping in the $G(s)$-plane, the negative imaginary axis of the s-plane is also mapped. The conjugate complex property of imaginary numbers ensures $G(jw)$ is reflected in the real axis. Substitute jw for s in the transfer function and evaluate the resulting complex number.

6.17.4   Plotting and Analyzing the Nyquist Plot

- Check $G(s)$ for poles on the imaginary axis (jw) and at the origin.

- Sketch the image of path in the $G(s)$ plane.

- Draw the mirror image about the real axis of the sketch from 2.

- Use equation to plot the image of path. This path at infinity usually plots into a point in the $G(s)$.

- Employ equation to plot the image of path.

- Connect all curves drawn in the previous steps.

- Evaluate the number of clockwise encirclements of -1, and call that number N.

- Determine the number of unstable (RHP) poles of $G(s)$, and call that number P.

- Calculate the number of unstable closed-loop roots, Z; $\; Z = N + P$.

- For stability, $Z = 0$, which means no closed-loop poles in the RHP.

- The Nyquist plot will always be symmetric with respect to the real axis.

6.17.5   Nyquist Stability Criterion

The Nyquist stability criterion relates the open-loop frequency response to the number of closed-loop poles of the system in the RHP. The study of the Nyquist criterion will allow the determination of stability from the frequency response of a complex system, for example, with one or more resonances, where the magnitude curve crosses 1 several times and/or the phase crosses 180° several times. It is also very useful in dealing with open-loop, unstable systems, nonminimum-phase systems and systems with pure delays (transportation lags). A closed-loop control system is absolutely stable if all the roots of the characteristic equation have negative real parts. Put differently, the poles of the closed-loop transfer function, or the zeros of the denominator

$$1 + KG(s)$$

of the closed-loop transfer function, must lie in the left-half plane (LHP). The Nyquist stability criterion establishes the number of zeros of $1 + KG(s)$ in the RHP directly from the Nyquist stability plot of KG(s).

6.17.5.1   Statement of Nyquist Stability Criterion

Let P be the number of poles of $KG(s)$ in the RHP, and let N be the number of CW encirclements of the $(-1, 0)$ point, i.e., $KG(s) = -1$ in the $KG(s)$-plane. The closed-loop control system whose open-loop transfer function is KG(s) is stable if and only if

$$N = -P \leq 0.$$

If $N > 0$ (i.e., the system is unstable), the number of zeros Z of $1 + KG(s)$ in the RHP is determined by

$$Z = N + P.$$

If $N \leq 0$, the $(-1,0)$ is not enclosed by the Nyquist stability plot. Therefore, $N \leq 0$ (stable system) if the region to the right of the contour in
the prescribed direction does not include the $(-1,0)$ point. Shading of this
region helps significantly in determining whether $N \leq 0$; that is, if and only
if, the $(-1,0)$ point does not lie in the shaded region.

Example 6.7 *Consider the system described by*

$$KG(s) = \frac{1}{s(s+1)}.$$

*Check the Nyquist stability plot. The region to the right of the contour
has been shaded. Clearly the $(-1,0)$ point is not in the shaded area, which
means $N \leq 0$. The poles of $KG(s)$ are at $s = 0$ and $s = -1$, neither of
which are in the in the RHP, $P = 0$*

$$N = -P = 0$$

6.17.6   Nyquist Diagrams Using MATLAB

Use of MATLAB
numG=[1];
denG=[1 2 1];
Nyquist(numG,denG)

6.17.6.1   System in state space

A=[0 1; -25   -4];
B=[0; 25];
C=[1 0];
D=[0];
Nyquist(A,B,C,D)

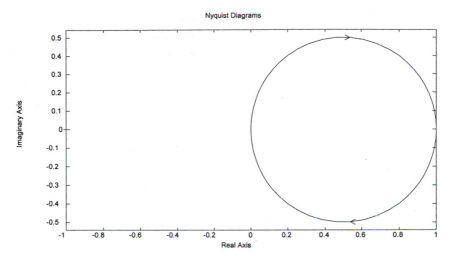

Nyquist Diagram for $KG(s) = \dfrac{1}{s+1}$

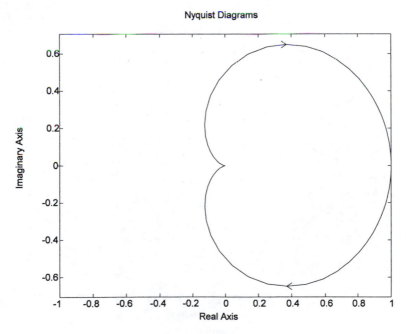

Nyquist Plot for $KG(s) = \dfrac{1}{(s+1)^2}$

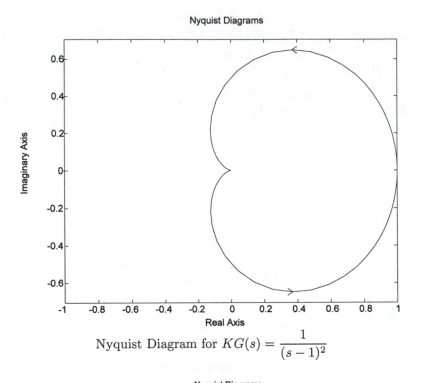

Nyquist Diagram for $KG(s) = \dfrac{1}{(s-1)^2}$

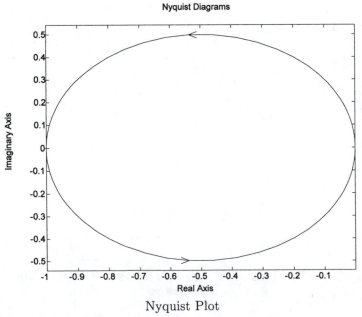

Nyquist Plot

These are the Nyquist plots obtained using MATLAB.

6.18 Argument and Rouché's Theorem

In this section, Cauchy's residue theorem is used to derive two theoretical results that have important practical applications to Nyquist analysis. These results pertain to functions whose isolated singularities are poles.

Definition 6.1 *A function f is said to be meromorphic in a domain D if at every point of D it is either analytic or has a pole.*

Theorem 6.1 *(Argument principle) If f is analytic and nonzero at each point of a simple closed positively oriented contour C and is meromorphic inside C, then*

$$\frac{1}{2\pi i} \int_C \frac{f'(z)}{f(z)} dz = N_o(f) - N_p(f)$$

where $N_o(f)$ and $N_p(f)$ are, respectively, the number of zeros and poles of f inside C (multiplicity included).

Theorem 6.2 *(Rouché's theorem) If f and g are each functions which are analytic inside and on a simple closed contour C and if the strict inequality*

$$|f(z) - g(z)| < |f(z)|$$

holds at each point on C, then f and g must have the same total number of zeros (counting multiplicity) inside C.

6.19 Examples of Nyquist Analysis

In this section detailed examples of Nyquist plots are developed and analyzed.

Example 6.8

$$GH(s) = \frac{1}{(s + p_1)(s + p_2)}; \quad where \ \ p_1, \ p_2 > 0$$

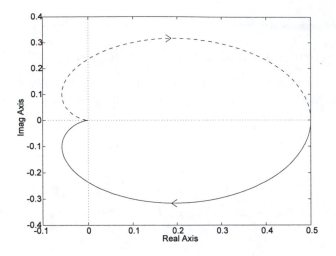

Nyquist Plot

$$ab \rightarrow s = j\omega \ 0 < \omega < \infty$$

$$GH(s) = \frac{1}{(s + p_1)(s + p_2)}$$

$$GH(jw) = \frac{1}{(j\omega + p_1)(j\omega + p_2)}$$

$$= \frac{1}{\sqrt{p_1^2 + \omega^2}\sqrt{p_2^2 + \omega^2}} \angle - \tan^{-1}\frac{\omega}{p_1} - \tan^{-1}\frac{\omega}{p_2}$$

$$\omega \rightarrow 0 \ GH = \frac{1}{p_1 p_2} \angle 0^\circ$$

$$\omega \rightarrow \infty \ GH = \frac{1}{\infty} \angle - 90^\circ - 90^\circ \Rightarrow 0 \angle - 180^\circ$$

$$|GH| \ ranges \ from \ \frac{1}{p_1 p_2} \ to \ 0$$

$$\angle GH \ from \ \angle 0^\circ \ to \ \angle - 180^\circ$$

Angle from first quarter to second quarter $\Rightarrow$ *Crossing the imaginary axis*

$$bc \Rightarrow s = \lim_{R\to\infty} Re^{j\theta}$$

$$GH = \frac{1}{(Re^{j\theta} + p_1)(Re^{j\theta} + p_2)}$$

$$= \frac{1}{R^2 e^{2j\theta} + (p_1 + p_2)\,Re^{j\theta} + p_1 p_2}\angle - \theta$$

$$= 0\angle - \theta^\circ$$

$$N = Z - P$$

$$0 = Z - 0$$

$$Z = 0 \to stable$$

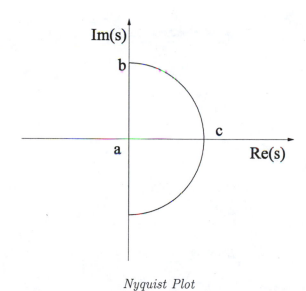

Nyquist Plot

Example 6.9

$$GH(s) = \frac{1}{s}$$

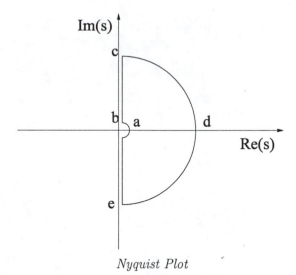

Nyquist Plot

$\overline{bc}$

$$s = j\omega \quad 0<\omega<\infty$$

$$GH = \frac{1}{j\omega} = \frac{1}{\omega}\angle - 90^{o}$$

$$\omega \to 0 \quad \infty\angle - 90^{o}$$

$$\omega \to \infty \quad 0\angle - 90^{o}$$

$\overline{cde}$

$$s = \lim_{R\to\infty} Re^{j\theta}$$

$$GH = \frac{1}{Re^{j\theta}} = \frac{1}{R}\angle - \theta$$

$$R \to \infty \quad GH \to 0\angle - \theta$$

$\overline{fab}$

$$s = \lim_{\rho\to 0} \rho e^{j\theta} \quad -90^{o} < \theta < 90^{o}$$

$$= \frac{1}{\rho e^{j\theta}} = \frac{1}{\rho}\angle - \theta$$

$$GH \ as \ \rho \to 0 = \infty\angle - \theta$$

$$\infty\angle 90^{o} \ to \ \infty\angle - 90^{o}$$

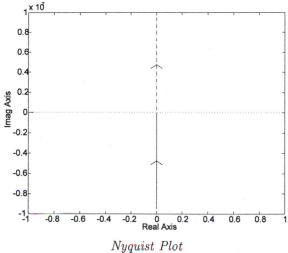

Nyquist Plot

Example 6.10

$$GH(s) = \frac{1}{s(s+p_1)(s+p_2)} \qquad p_1, p_2 > 0$$

The system has pole at zero. For the path $\overline{bc}$

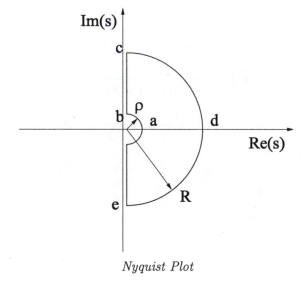

Nyquist Plot

$\overline{bc}$

$$s = j\omega \qquad o < \omega < \infty$$

$$GH(jw) = \frac{1}{j\omega(j\omega + p_1)(j\omega + p_2)}$$

$$= \frac{1}{\omega\sqrt{\omega^2 + p_1}\sqrt{\omega^2 + p_2}} \angle -90 - \tan^{-1}\frac{\omega}{p_1} - \tan^{-1}\frac{\omega}{p_2}$$

$$as \; \omega \to 0 \quad GH \to \frac{1}{0}\angle -90^\circ \to \infty\angle -90^\circ$$

$$as \; \omega \to \infty \quad GH \to \frac{1}{\infty}\angle -90^\circ - 90^\circ - 90^\circ = 0\angle -270^\circ = 0\angle 90^\circ$$

$\overline{cde}$

$$s = \lim_{R \to \infty} Re^{j\theta} \qquad 90^\circ < \theta < -90^\circ$$

$$\frac{1}{Re^{j\theta}(Re^{j\theta} + p_1)(Re^{j\theta} + p_2)}$$

$$as \; R \to \infty \quad 0\angle -\theta$$

$$0\angle -90^\circ \; to \; 0\angle 90^\circ$$

$\overline{fab}$

$$s = \rho e^{j\theta} \qquad -90^\circ \leq \theta < 90^\circ$$

$$GH = \frac{1}{\rho e^{j\theta}(\rho e^{j\theta} + p_1)(\rho e^{j\theta} + p_2)}$$

$$= \frac{1}{\rho e^{j\theta} p_1 p_2} = \frac{1}{\rho p_1 p_2}\angle -\theta$$

$$\infty\angle -\theta$$

$$\infty\angle 90^\circ \to \infty\angle -90^\circ$$

Example 6.11

$$GH = \frac{1}{s^2}$$

The system has pole at zero. For path $\overline{ad}$

$$s = j\omega \qquad 0 < \omega < \infty$$

$$GH = \frac{1}{(j\omega)^2} = \frac{1}{\omega.\omega}\angle - 180^o$$

$$\omega \rightarrow 0 \qquad GH \rightarrow \frac{1}{0}\angle - 180^o = \infty\angle - 180^o$$

$$\omega \rightarrow \infty \qquad GH \rightarrow \frac{1}{\infty}\angle - 180^o = 0\angle - 180^o$$

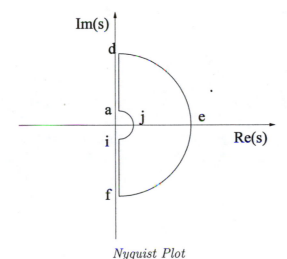

Nyquist Plot

To check take $\omega = 1$

$$GH = \frac{1}{1}\angle - 180^o$$

$\overline{def}$

$$s = \lim_{R \rightarrow \infty} Re^{j\theta} \qquad 90^o \le \theta \le -90^o$$

$$GH(jw) = \frac{1}{(Re^{j\theta})(Re^{j\theta})}$$

$$= \frac{1}{R^2 e^{2j\theta}} = \frac{1}{R^2}\angle - 2\theta$$

as $R \rightarrow \infty = 0\angle - 2\theta$

$$0\angle - 180^o \rightarrow 0\angle 180^o$$

$\overline{ija}$

$$s = \lim_{\rho \to 0} \rho e^{j\theta} \qquad -90° \leq \theta \leq 90°$$

$$GH = \frac{1}{(\rho e^{j\theta})^2} = \frac{1}{\rho^2 e^{2j\theta}} = \frac{1}{\rho^2}\angle - 2\theta$$

$$as \; \rho \to 0 = \infty\angle 180° \to \infty\angle - 180°$$

Note: there should be one semicircle since it is Type 1.

$$N = Z + P$$
$$Z = N - P$$
$$= 1 - 0$$
$$Z = 1 \qquad \Rightarrow unstable$$

Example 6.12

$$GH(s) = \frac{K}{s(s-1)}$$

The system has a pole at zero. For path $\overline{de}$

$$G(j\omega) = \frac{K}{(j\omega)(j\omega - 1)}$$

$$= \frac{K}{\omega.\sqrt{\omega^2 + 1}}\angle - 90° - \tan^{-1}\omega$$

$$\omega \to 0 \quad G(j\omega) = \frac{1}{0}\angle - 90°$$

$$\omega \to \infty \quad G(j\omega) = \frac{1}{\infty}\angle - 90° - 90°$$

$$= 0\angle - 180°$$

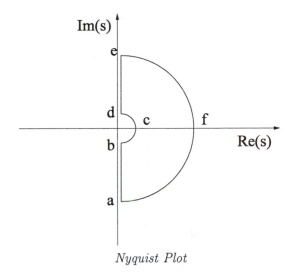

Nyquist Plot

To check put $\omega = 1$

$$\frac{K}{1\sqrt{1+1}}\angle -90^{\circ} - \tan^{-1} 1$$

$\overline{efa}$

$$s = \lim_{R \to \infty} \mathrm{Re}^{j\theta} \qquad 90^{\circ} \le \theta \le -90^{\circ}$$

$$GH = \frac{K}{\mathrm{Re}^{j\theta}(\mathrm{Re}^{j\theta}-1)} = \frac{K}{R}\angle -2\theta$$

as $R \to \infty \qquad 0\angle -2\theta = 0\angle 180^{\circ}$

$\overline{bcd}$

$$s = \lim_{\rho \to 0} \rho e^{j\theta} \qquad -90^{\circ} \le \theta \le 90^{\circ}$$

$$GH = \frac{K}{\rho e^{j\theta}(\rho e^{j\theta}-1)} = \frac{K}{\rho}\angle -2\theta$$

as $\rho \to 0 = \infty\angle -2\theta$

$$N = Z - P$$

$$1 = Z + 0 \qquad unstable$$

Example 6.13

$$KG(s) = \frac{K(s+2)}{(s+10)}$$

Does not have a pole at zero, so for path $\overline{ab}$

$$s = j\omega \qquad 0 < \omega < \infty$$

$$G(j\omega) = \frac{j\omega + 2}{(j\omega + 10)} = \frac{\sqrt{\omega^2 + 4}}{\sqrt{\omega^2 + 100}} \angle \tan^{-1}\frac{\omega}{2} - \tan^{-1}\frac{\omega}{10}$$

$$\omega \to 0 \qquad G(j\omega) = \sqrt{\frac{4}{100}} \angle 0 - 0 = \frac{2}{10}\angle 0^\circ$$

$$\omega \to \infty \qquad G(j\omega) = 1\angle 90^\circ - 90^\circ = 1\angle 0^\circ$$

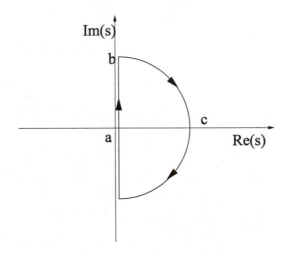

for $\overline{bcd}$

$$s = \lim_{R \to \infty} Re^{j\theta}$$

$$as \; R \to \infty \qquad \frac{Re^{j\theta} + 2}{Re^{j\theta} + 10} = 1\angle 0^\circ$$

$$Z = N + P$$

$$P \to 0 \quad no \; poles \; on \; RHP$$

$$N \to 0 \quad no \; circle \; for \; -1$$

$$Z = 0 \Rightarrow \quad stable \; for \; any \; K > 0$$

The root locus is also shown in the following diagram.

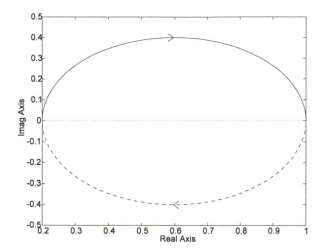

These examples illustrate Nyquist analysis.

6.20   Stability margins

Quite a significant class of control systems become unstable if the gain increases past a critical point. The gain margin (GM) and the phase margin (PM) are two quantities that measure the stability margin of a control system, and they are directly related to the stability criterion.

6.20.1   Gain Margin (GM)

The gain margin is the factor by which the gain is less than the neutral stability value. It can be read directly from the Bode plot by measuring the vertical distance between $|KG(jw)|$ curve and the $|KG(jw)| = 1$ line at the frequency where $\arg G(jw) = 180°$. This means the GM is the factor by which the gain K can be raised before instability arises and the system is unstable if

$$|GM| < 1 \quad \text{or} \quad \log_{10} |GM| < 0dB.$$

The GM can also be determined from the root locus with respect to K by noting two values of K; at the point where the locus crosses the imaginary (jw) axis (K_{jw}), and at the nominal closed-loop poles (K_{nom}). The gain margin is the ratio of these gains such that

$$GM = \frac{K_{jw}}{K_{nom}}$$

6.20.2   Phase Margin (PM)

The phase margin is the amount by which the phase of $G(jw)$ exceeds $-180°$ when $|KG(jw)| = 1$. This is an alternate way of measuring the degree to which the stability conditions are satisfied. A positive PM is necessary for stability. Together, the two stability margins GM and PM determine how far the complex quantity $G(jw)$ passes from the -1 point, which is another way of stating the neutral-stability point. These margins can also be defined from Nyquist plots as measures of how close the Nyquist plot comes to encircling the point -1. The GM indicates how much the gain can be raised before instability occurs and the PM is the difference between the phase of $G(jw)$ and 180° when of $KG(jw)$ crosses the circle $|KG(jw)| = 1$. A stable system, i.e., one with no Nyquist encirclements, has a positive value of PM. It is however, easier to determine stability margins from Bode plots than from Nyquist plots. The crossover frequency refers to the frequency at which the gain is unity, i.e.,

$$K = 1 \ \text{ or } \ \log_{10} K = 0 \ dB$$

One of the benefits of frequency-response design is the ease with which the effects of the gain changes ΔK.

6.20.3   Relative Stability

The relative stability of a feedback control system is easily determined from the polar plot or Nyquist stability plot. The phase crossover frequency w_c is that frequency at which the phase angle of $KG(s)$ is -180, i.e., the frequency at which the polar plot crosses the negative real axis. The gain margin is given by

$$\text{gain margin} = \frac{1}{|KG(jw_c)|}.$$

The notion of relative stability is central in analyzing control systems.

6.21   Gain and Phase Relationship

One of Bode's important contributions is the theorem that states that for any stable minimum-phase system (one with no RHP zeros or poles), the phase of $G(jw)$ is uniquely related to the magnitude of $G(jw)$. For the Bode plots defined in terms of log scale (i.e., $\log_{10}|G(jw)|$ vs. $\log_{10} w$ such that the slope is kdB) the theorem is fairly simplified

$$\arg G(jw) \approx k \times 90°$$

When $|KG(jw)| = 1$,

$$\arg G(jw) \approx -90° \quad \text{if} \quad k = -1$$
$$\arg G(jw) \approx -180° \quad \text{if} \quad k = -2.$$

Stability is achieved if

$$\arg G(jw) > -180° \quad \text{for } PM > 0$$

This theorem is formally stated as follows:

$$argG(jw) = \frac{1}{\pi} \int_{-\infty}^{+\infty} \left(\frac{dM}{du} \right) W(u) du$$

where

$$M = \log_{10} |G(jw)|$$

$$u = \log_{10} \left(\frac{w}{w_c} \right) \quad \text{(normalized frequency)}$$

$$W(u) = \log_{10}(\coth |u|/2) \quad \text{(weighting function)}.$$

6.22   Compensation

Compensation is when dynamic elements are typically added to feedback control systems to improve their stability and error characteristics. Basic types of feedback: proportional (P), derivative (D), and integral (I). There are two kinds of dynamic compensation: the lead network, which approximates proportional-derivative (PD) feedback, and the lag network, which approximates proportional-integral (PI) control. In this section, compensation is discussed in terms of frequency-response characteristics. In many cases, the compensation will be implemented in a microprocessor. Techniques for converting the continuous compensation D(s) into a form that can be coded in the computer will be discussed under Digital Control Systems. The frequency response stability analysis to this point has considered the closed-loop system to have the characteristic equation

$$1 + KG(s) = 0.$$

With the introduction of compensation, the closed-loop characteristic equation becomes

$$1 + D(s)G(s) = 0,$$

where $D(s)$ is a PI, PD or PID controller. The previous discussion pertaining to the frequency response of $KG(s)$ applies directly to the compensated case if the frequency response of $D(s)G(s)$ is analyzed.

6.22.1   PD Compensation

The starting point is compensation design using the frequency response with PD control. The compensator transfer function is given by

$$D(s) = K_P + K_D s.$$

6.22.2   Lead Compensation

In order to alleviate the high-frequency amplification of the PD compensation, a first-order pole is added in the denominator at frequencies higher than the breakpoint of the PD compensator.

$$D(s) = \frac{K_P + K_D s}{1 + \dfrac{s}{w_p}}.$$

6.22.3   PI Compensation

In many problems it is important to keep the bandwidth low and also to reduce the steady state error. For this purpose, a proportional-integral (PI) or lag compensator is useful.

$$D(s) = K_P + \frac{1}{K_I s}.$$

6.22.4   Lag Compensation

Lag compensation approximates PI control. For frequency-response design, it is more convenient to write the transfer function of the lag compensation in the form

$$D(s) = \frac{K_P s + \dfrac{1}{K_I}}{s + \dfrac{1}{w_p}}.$$

6.22.5 PID Compensation (Lead-Lag Compensator)

Sometimes it is effective to use both lead and lag compensation. By combining the derivative and integral feedback, PID is established,

$$D(s) = K_P + K_D s + \frac{1}{K_I s}.$$

This compensation is roughly equivalent to combining lead and lag compensators in the same design, and so is sometimes referred to as a lead-lag compensator. Hence, it can provide simultaneous improvement in transient and steady state responses. Although lead-lag compensation approximates PID control, the two approaches involve two completely different strategies.

6.22.6 Summary of Compensation Characteristics

- PD control adds phase lead at all frequencies above the breakpoint. If there is no change in gain on the low-frequency asymptote, PD compensation will increase the crossover frequency and the speed of response. The increase in magnitude of the frequency response at the higher frequencies will increase the system's sensitivity to noise.

- Lead compensation adds phase lead at a frequency band between the two breakpoints, which are usually selected to bracket the crossover frequency. If there is no change in gain on the low frequency asymptote, lead compensation will increase both the crossover frequency and the speed of response over the uncompensated system. If K is selected so that the low-frequency magnitude is unchanged, then the steady state errors of the system will increase.

- PI control increases the frequency-response magnitude at frequencies below the breakpoint, thereby decreasing steady state errors. It also contributes phase lag below the breakpoint, which must be kept at a low enough frequency to avoid degrading the stability excessively.

- Lag compensation increases the frequency response magnitude at frequencies below the two breakpoints, thereby decreasing steady state errors. Alternatively, with suitable adjustments in K, lag compensation can be used to decrease the frequency-response magnitude at frequencies above the two breakpoints so that w_c yields an acceptable phase margin. Lag compensation also contributes phase lag between the two breakpoints, which must be kept at frequencies low enough to keep the phase decrease from degrading the PM excessively.

6.23   Problems

Problem 6.1 *Draw Bode, Nyquist plots for a sixth-order Pade approximation to a pure delay of 1 second.*

Problem 6.2 *Sketch the polar plot of the frequency response for the following transfer functions:*

$$(a) \qquad KG(s) = \frac{1}{(1 + 0.5s)(1 + 2s)}$$

$$(b) \qquad KG(s) = \frac{(1 + 0.5s)}{s^2}$$

$$(c) \qquad KG(s) = \frac{s + 10}{s^2 + 6s + 10}$$

$$(d) \qquad KG(s) = \frac{30(s + 8)}{s(s + 2)(s + 4)}$$

Problem 6.3 *A rejection network that can be utilized instead of the twin-T network is the bridged-T network shown below.*

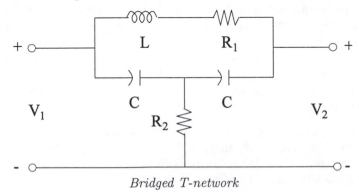

Bridged T-network

The transfer function of this network is given by

$$G(s) = \frac{s^2 + \omega_n^2}{s^2 + 2(\omega_n s/Q) + \omega_n^2},$$

where $\omega_n^2 = \dfrac{2}{LC}$, and $Q = \dfrac{\omega_n L}{R_1}$, and R_2 is adjusted so that $R_2 = \dfrac{(\omega_n L)^2}{4R_1}$.
Show that the expression given for $G(s)$ is correct.

Problem 6.4 *A control system for controlling the pressure in a closed chamber is shown below, together with its flow graph model.*

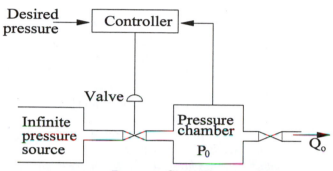

Pressure Controller

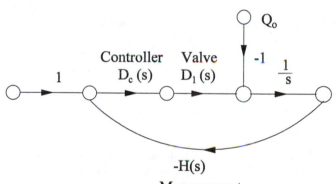

Flow Graph Model

The transfer function for the measuring element is

$$G(s) = \frac{100}{s^2 + 15s + 100},$$

and the transfer function for the valve is

$$D_1(s) = \frac{1}{(0.1s+1)(s/15+1)}.$$

The controller function is given by

$$D_c(s) = s + 1.$$

Obtain the frequency response characteristics for the loop transfer function

$$D_c(s)D_1(s)G(s)\left[\frac{1}{s}\right].$$

Problem 6.5 *Autonomous, self-driven vehicles can be used in warehouses, airports, and many other applications. These vehicles follow a wire embedded in the floor and adjust the steerable front wheel in order to maintain proper direction. The sensing coils, mounted on the front wheel assembly, detect an error in the direction of travel and adjust the steering. The overall control system open-loop transfer function is*

$$DG(s) = \frac{K}{s(s+\pi)^2} = \frac{K_v}{s(s/\pi+1)^2}.$$

It is desired to have a bandwidth of the closed-loop system exceed $2\pi rad/s$.
(a) Set $K_v = 2\pi$ and sketch the Bode diagram.
(b) Using the Bode diagram, obtain the logarithmic-magnitude vs. phase angle curve.

Problem 6.6 *(a) Calculate the magnitude and phase of*

$$G(s) = \frac{1}{s+1},$$

for $\omega = 0.1,\ 0.2,\ 0.5,\ 1,\ 2,\ 5,$ and $10 rad/sec$.
(b) Sketch the asymptotes for $G(s)$, and compare these with the computed results from part (a).

Problem 6.7 *Draw the Bode plots for each of the following systems. Compare the sketches with the plots obtained using MATLAB.*

$$(a) \quad G(s) = \frac{1}{(s+1)^2(s^2+s+4)}$$

$$(b) \quad G(S) = \frac{s}{(s+1)(s+10)(s^2+5s+2500)}$$

$$(c) \quad G(s) = \frac{4s(s+10)}{(s+50)(4s^2+5s+4)}$$

$$(d) \qquad G(s) = \frac{10(s+4)}{s(s+1)(s^2+2s+5)}$$

$$(e) \qquad G(s) = \frac{1000(s+1)}{s(s+2)(s^2+8s+64)}$$

$$(f) \qquad G(s) = \frac{(s+5)(s+3)}{s(s+1)(s^2+s+4)}$$

$$(g) \qquad G(s) = \frac{4000}{s(s+40)}$$

$$(h) \qquad G(s) = \frac{100}{s(1+0.1s)(1+0.5s)}$$

$$(i) \qquad G(s) = \frac{1}{s(1+s)(1+0.02s)}$$

Problem 6.8 *A certain system is represented by the asymptotic Bode diagram shown below. Find and sketch the response of this system to a unit step input (assuming zero initial conditions).*

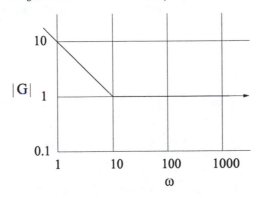

Problem 6.9 *Prove that the magnitude slope of −1 in the Bode plot corresponds to −20dB per decade.*

(a) Sketch the polar plot for an open-loop system with transfer function $1/s^2$, that is, sketch

$$\left. \frac{1}{s^2} \right|_{s=C_1},$$

where C_1 is a contour enclosing the entire RHP. (Hint: assume C_1 takes a small detour around the poles at $s = 0$).

(b) Repeat part (a) for an open-loop system whose transfer function is

$$G(s) = \frac{1}{s^2 + \omega_o^2}.$$

Problem 6.10 *Draw a Nyquist diagram for each of the following systems, and compare the result with that obtained using the MATLAB command Nyquist.*

(a) $KG(s) = \dfrac{K(s+2)}{s+10}$

(b) $KG(s) = \dfrac{K}{(s+10)(s+2)^2}$

(c) $KG(s) = \dfrac{K(s+10)(s+1)}{(s+100)(s+2)^3}$

Using the plots, estimate the range of K for which each system is stable, and qualitatively verify the result using root locus plot (generated by hand or using MATLAB).

Problem 6.11 *Draw a Nyquist diagram for each of the following systems, and compare the result with that obtained using the MATLAB command Nyquist.*

(a) $KG(s) = \dfrac{K(s+1)}{s+2}$

(b) $KG(s) = \dfrac{K}{(s+1)(s+2)^2}$

(c) $KG(s) = \dfrac{K(s+4)(s+1)}{s(s+2)^4}$

Problem 6.12 *The forward-path transfer function of a unity feedback control system is*

$$G(s) = \frac{K}{s(s+6.54)}$$

Analytically, find the resonant peak M_{peak}, resonant frequency ω_r, and bandwidth BW of the closed-loop system for the following values of K:
    (a) $K = 5$ (b) $K = 21.39$ (c) $K = 100$.

Problem 6.13 *Use MATLAB to find the resonant peak, M_{peak}, resonant frequency ω_r, and bandwidth BW of the following unity feedback control* systems. *Make sure that the systems are stable.*

$$\text{(a)} \quad G(s) = \frac{5}{s(1 + 0.5s)(1 + 0.1s)}$$

$$\text{(b)} \quad G(s) = \frac{10(s + 1)}{s(s + 2)(s + 10)}$$

$$\text{(c)} \quad G(s) = \frac{0.5}{s(s^2 + s + 1)}$$

$$\text{(d)} \quad G(s) = \frac{100e^{-s}}{s(s^2 + 5s + 5)}$$

Chapter 7

State-Space Design Methods

7.1 Introduction

In Chapter 2, it was shown that by using state-variable matrix modeling, a dynamic system described by higher-order differential equations can be expressed in terms of simple first-order differential equations. Such state-variable models form the basis of state-space design and analysis. These are methods that use state variables, i.e., the analysis and the design are carried out in the state space. The techniques involve designing dynamic compensation by working directly with the state-variable descriptions of dynamic systems. State space methods are simpler because they deal directly with the system states that are in simple first-order differential equations. Further advantages of state space methods include the ability to study more general models, facilitating the use of ideas of geometry in differential equations, providing connections between internal and external descriptions, the ability to handle multi-input multi-output (MIMO) systems, and easy implementation using software such as MATLAB.

The use of the state space approach has often been referred to as *modern* control system design as opposed to *classical* control system design (transfer-function based approaches, root locus and frequency-response methods). This chapter discusses state-space methods of analysis and design for a broad range of dynamic systems. The role of the transfer function, the different state-space canonical forms, and the state-transition matrix are introduced. The theoretical basics and design issues involved in the state-space subjects of observability, controllability, similarity transformation, full-state feedback, optimal control, estimator design, and compensator design are then discussed and illustrated. A number of design examples are provided including discussions about their implementation in MATLAB.

7.2   The Block Diagram and the Transfer Function

7.2.1   State-Space Description and the Block Diagram

The state-space description of a linear system was previously discussed under the modeling of dynamic systems (Chapter 2). It involves four system matrices (or vectors): the input control signal vector (or scalar) $\mathbf{u}$, the state vector $\mathbf{x}$ as well as its time derivative $\dot{\mathbf{x}}$, and the output vector $\mathbf{y}$. The general form is given by

$$\dot{\mathbf{x}} = \mathbf{A}\mathbf{x} + \mathbf{B}\mathbf{u} \tag{7.1}$$

$$\mathbf{y} = \mathbf{C}\mathbf{x} + \mathbf{D}\mathbf{u}. \tag{7.2}$$

The system state $\mathbf{x}$ is an n-vector where n is the number of the states in that system and $\mathbf{u}$ is an m-vector of control inputs. The output of the system is represented by an l-vector $\mathbf{y}$. The system matrices $\mathbf{A}$, $\mathbf{B}$, $\mathbf{C}$ and $\mathbf{D}$ are known, respectively, as the plant (or generally as the system) matrix, input matrix, output matrix and the feed-through matrix. The plant matrix $\mathbf{A}$ is an $n \times n$ square matrix while the input matrix $\mathbf{B}$ is a $n \times m$ matrix. Matrices $\mathbf{C}$ and $\mathbf{D}$ have dimensions $l \times n$ and $l \times m$ respectively. At present, only linear time-invariant (LTI) systems are considered, hence, these system matrices are constant. The corresponding block diagram for this system is shown in Figure 7.1

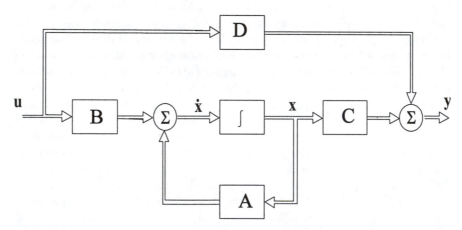

FIGURE 7.1
The Block Diagram of a System in State Variable Form

As illustrated in Figure 7.1, the physical significance of the different matrices can easily be seen where the feed-through matrix $\mathbf{D}$ feeds part of the input control **u** signal forward to the output signal **y**, the output matrix **C** transforms the system states **x** to output signals **y** and the input matrix **B** transforms the input control signal **u** to the derivative of the system states **ẋ**. Finally the plant **A**, (or system matrix) that represents the plant dynamics transforms the state vector **x** to its time derivative **ẋ**. Figure 7.1 does not include the reference input that is necessary in the control system. Such control systems in which there are no reference inputs $r(t)$ are known as *regulators*.

Normally the state, control input, and the output vector are functions of time. Therefore, the state-space presentation for a *single input single output* (SISO) system can be given as

$$\dot{\mathbf{x}}(t) = \mathbf{A}\mathbf{x}(t) + \mathbf{B}u(t) \tag{7.3}$$

$$y(t) = \mathbf{C}\mathbf{x}(t) + \mathbf{D}u(t), \tag{7.4}$$

where in this case (SISO) the input and output are scalar quantities $u(t)$ and $y(t)$, respectively. The input matrix **B** and the feedforward matrix **D** become column matrix or n-vector and a scalar constant, respectively, while the output matrix **C** is a row matrix or the transpose of an n-vector. In MATLAB the system matrices **A**, **B**, **C**, **D**, can be obtained from the system transfer function by the command "tf2ss". However, as it will be discussed later, MATLAB returns system matrices in one of the canonical forms that might be different from the results one could get by normal transformation of the transfer function to state-variable form. The MATLAB syntax for "tf2ss" is

$$[\mathsf{A}, \mathsf{B}, \mathsf{C}, \mathsf{D}] = \mathsf{tf2ss}(\mathsf{num}, \mathsf{den}),$$

where num is a row matrix of the coefficients of the numerator of the transfer function in descending powers of s. For MIMO systems, it will have as many rows as the number of outputs in the output vector y, therefore, for SISO system it will be a row vector whose elements are the coefficients of the descending powers of s in the numerator of $G(s)$. Vector den contains the coefficients of the denominator of the transfer function also in descending powers of s.

Example 7.1 *The dynamics of a hypothetical system can be presented by a third-order linear ordinary differential equation as*

$$\frac{d^3x}{dt^3} + 4\frac{d^2x}{dt^2} + 3\frac{dx}{dt} + 5x - 9 = 0.$$

If the system output is linearly dependent on the parameter x such that $y = x$, transform the system differential equation into the corresponding state-variable form. Do the same problem using MATLAB. (Hint: Assume a control step input of strength 9).

Solution 7.1 *Letting $x_1 = x$, $x_2 = \dot{x}$, $x_3 = \ddot{x}$, it follows that*

$$\dot{x}_1 = x_2$$

$$\dot{x}_2 = x_3$$

$$\dot{x}_3 = \frac{d^3 x}{dt^3} = -5x_1 - 3x_2 - 4x_3 + 9$$

$$y = x_1.$$

In state-variable form this becomes

$$\begin{bmatrix} \dot{x}_1 \\ \dot{x}_2 \\ \dot{x}_3 \end{bmatrix} = \begin{bmatrix} 0 & 1 & 0 \\ 0 & 0 & 1 \\ -5 & -3 & -4 \end{bmatrix} \begin{bmatrix} x_1 \\ x_2 \\ x_3 \end{bmatrix} + \begin{bmatrix} 0 \\ 0 \\ 1 \end{bmatrix} 9$$

$$y = \begin{bmatrix} 1 & 0 & 0 \end{bmatrix} \begin{bmatrix} x_1 \\ x_2 \\ x_3 \end{bmatrix} + \begin{bmatrix} \frac{1}{3} \end{bmatrix} 9.$$

Therefore, the system matrices become

$$A = \begin{bmatrix} 0 & 1 & 0 \\ 0 & 0 & 1 \\ -5 & -3 & -4 \end{bmatrix}, \quad B = \begin{bmatrix} 0 \\ 0 \\ 1 \end{bmatrix}, \quad C = \begin{bmatrix} 1 & 0 & 0 \end{bmatrix}, \quad D = [0].$$

To get the state-space matrices using MATLAB, there is need to know the system transfer function. The transfer function of this system for a step input of strength 9 can be derived (using methods discussed in previous chapters) and shown to be

$$T(s) = \frac{s^2}{s^3 + 4s^2 + 3s + 5}.$$

As such, the MATLAB code for this problem becomes
        den=[1 4 3 5];
        num=[1 0 0];
        [A,B,C,D]=tf2ss(num,den).
    The resulting system matrices are

$$A = \begin{bmatrix} -4 & -3 & -5 \\ 1 & 0 & 0 \\ 0 & 1 & 0 \end{bmatrix}, \quad B = \begin{bmatrix} 1 \\ 0 \\ 0 \end{bmatrix}, \quad C = \begin{bmatrix} 1 & 0 & 0 \end{bmatrix}, \quad D = [0].$$

Although structurally there is a difference between the MATLAB results and the results presented above, these two are representing the same information. The structural difference arises because MATLAB presents the state vector starting with the higher-order element of the state in the form

$$\mathbf{x} = \begin{bmatrix} x_3 \\ x_2 \\ x_1 \end{bmatrix}$$

whereas the results of the computation are presented starting with the lower-order element of the state vector resulting in

$$\mathbf{x} = \begin{bmatrix} x_1 \\ x_2 \\ x_3 \end{bmatrix}.$$

Since these differences do occur, the reader is reminded to exercise caution in arranging the state vector when using MATLAB in cases like the one discussed above.

7.2.2 Transfer Function, Poles, and Zeros

Given a state description for a SISO system, there may be an interest in the system transfer function. This can be achieved by taking the Laplace transform for the state-variable equations which results in two equations with three variables $\underline{x}(s)$, $y(s)$ and $u(s)$. Eliminating $\underline{x}(s)$ and combining the two equations give one equation in $y(s)$ and $u(s)$. The transfer function is thus extracted as the function that maps $u(s)$ to $y(s)$ as shown below.

Consider the SISO system given in Equations 7.3 and 7.4. The Laplace transform of the whole equation is given as

$$s\mathbf{x}(s) - \mathbf{x}(t_o) = \mathbf{A}\mathbf{x}(s) + \mathbf{B}\ \mathbf{u}(s) \tag{7.5}$$

$$\mathbf{y}(s) = \mathbf{C}\mathbf{x}(s) + \mathbf{D}u(s). \tag{7.6}$$

The first of these equations can be written as

$$(s\mathbf{I} - \mathbf{A})\mathbf{x}(s) = \mathbf{B}u(s) + \mathbf{x}(t_o)$$

while the second can also be rewritten as

$$\mathbf{C}^{-1}[\mathbf{y}(s) - \mathbf{D}u(s)] = \mathbf{x}(s)$$

Combining the two equations by elimination of $\mathbf{x}(s)$ yields

$$(s\mathbf{I} - \mathbf{A})\mathbf{C}^{-1}[y(s) - Du(s)] = \mathbf{B}u(s) + \mathbf{x}(t_o)$$

so that on isolation of $y(s)$ from $u(s)$ it follows that

$$(s\mathbf{I} - \mathbf{A})\mathbf{C}^{-1}y(s) = [\mathbf{B} + (s\mathbf{I} - \mathbf{A})\mathbf{C}^{-1}\mathbf{D}]u(s) + \mathbf{x}(t_o).$$

Since the transfer function does not depend on the initial condition $\mathbf{x}(t_o)$, one can easily extract the transfer function by setting $\mathbf{x}(t_o) = 0$ so that

$$T(s) = \frac{y(s)}{u(s)} = \mathbf{C}(s\mathbf{I} - \mathbf{A})^{-1}\mathbf{B} + \mathbf{D}. \tag{7.7}$$

The matrix $(s\mathbf{I}-\mathbf{A})^{-1}$ has special importance in state-space analysis and is known as the system resolvent matrix $\Phi(s)$. It will be shown later that the resolvent matrix is a Laplace transform of the transition matrix $\Phi(t)$, which represents the natural dynamics of the system.

The transfer function is given by

$$\Phi(s) = (s\mathbf{I} - \mathbf{A})^{-1} = \frac{\text{adj}(s\mathbf{I} - \mathbf{A})}{\det|s\mathbf{I} - \mathbf{A}|}, \tag{7.8}$$

where $adj(s\mathbf{I} - \mathbf{A})$ refers to the *adjoint (or adjugate)* of matrix $(s\mathbf{I} - \mathbf{A})$, which means the transpose of the matrix of the cofactors of $(s\mathbf{I}-\mathbf{A})$. Hence, the transfer function can be written as

$$T(s) = \frac{y(s)}{u(s)} = \frac{\mathbf{C}[\text{adj}(s\mathbf{I} - \mathbf{A})]\mathbf{B} + \mathbf{D}|s\mathbf{I} - \mathbf{A}|}{|s\mathbf{I} - \mathbf{A}|}. \tag{7.9}$$

From this transfer function (Equation 7.9), the characteristic equation is obtained by setting the denominator equal to zero, i.e.,

$$|s\mathbf{I} - \mathbf{A}| = 0. \tag{7.10}$$

As can be seen, the characteristic equation of the system equals the algebraic characteristic equation of the matrix $\mathbf{A}$ whose roots are the eigenvalues of $\mathbf{A}$. Therefore, since the poles of a system are the roots of its characteristic equation, it follows that the poles of the system in question will be just the eigenvalues of the plant (system) matrix. The zeros of the system are determined from the numerator of the transfer function. Since for a SISO system $\mathbf{B}$ is a column matrix while $\mathbf{C}$ is a row matrix and $\mathbf{D}$ is a scalar, it can be seen that the numerator represents a scalar polynomial in s, the roots of which are the zeros of the system in question. Thus to get the zeros of the system, the solution of the following equation must be determined,

$$\mathbf{C}[\text{adj}(s\mathbf{I} - \mathbf{A})]\mathbf{B} + \mathbf{D}|s\mathbf{I} - \mathbf{A}| = 0. \qquad (7.11)$$

In MATLAB the transfer function of a system can be derived from its state variable presentation by the command "ss2tf". This returns the two vectors that contains the coefficients of the numerator and denominator of the transfer function respectively. Since the command has been designed to handle both MIMO and SISO systems, it becomes necessary to specify the input for which the transfer function is required. However, if the input is not specified, MATLAB sets the default input. The syntax is

$$[\text{num}, \text{den}] = \text{ss2tf}(\mathsf{A}, \mathsf{B}, \mathsf{C}, \mathsf{D}, \text{iu}),$$

where A,B,C and D are the system matrices and iu is an integer between 1 and n specifying the input for which the transfer function is required. The poles and zeros of the system can also be determined in MATLAB by using the command "ss2zp". This returns a matrix of the zeros z_i of the transfer functions where each column represents the zeros corresponding to one input y_i and two vectors one of which is a vector containing the poles p_i of the system and the other is a vector that contains the gains k_i of each numerator transfer function. For SISO systems however, the zeros will be returned in a column vector instead of a matrix. The transfer function in terms and zeros and poles is given by

$$T(s) = k \frac{(s - z_1)(s - z_2) \cdots (s - z_{n-1})(s - z_n)}{(s - p_1)(s - p_2) \cdots (s - p_{m-1})(s - p_m)}. \qquad (7.12)$$

The syntax for the command "ss2zp" is

$$[\mathsf{z}, \mathsf{p}, \mathsf{k}] = \text{ss2zp}(\mathsf{A}, \mathsf{B}, \mathsf{C}, \mathsf{D}, \text{iu}),$$

where iu represents the single input in the system for which the poles and zeros are sought. In general, the use of this command is the same as the "ss2tf" discussed previously.

Example 7.2 *For the system whose state variable description has the following matrices; determine the system transfer function, characteristic equation, poles and zeros. Comment on the system stability. Use MATLAB to do the same problem.*

$$A = \begin{bmatrix} 1 & 2 & 6 \\ 2 & 8 & 4 \\ 6 & 4 & 4 \end{bmatrix}, \qquad B = \begin{bmatrix} 1 \\ 3 \\ 0 \end{bmatrix}, \qquad C = \begin{bmatrix} 0 \\ 0 \\ 1 \end{bmatrix}^T, \qquad D = [0].$$

Solution 7.2 *The system transfer function is given by Equation 7.9 as*

$$T(s) = \frac{Y(s)}{U(s)} = \frac{\mathbf{C}[adj(s\mathbf{I} - \mathbf{A})]\mathbf{B} + \mathbf{D}|s\mathbf{I} - \mathbf{A}|}{|s\mathbf{I} - \mathbf{A}|}.$$

In this case it becomes

$$T(s) = \frac{y(s)}{u(s)} = \frac{\mathbf{C}[adj(s\mathbf{I} - \mathbf{A})]\mathbf{B}}{|s\mathbf{I} - \mathbf{A}|}$$

$$(s\mathbf{I} - \mathbf{A}) = \begin{bmatrix} s-1 & -2 & -6 \\ -2 & s-8 & -4 \\ -6 & -4 & s-4 \end{bmatrix}.$$

Therefore,

$$adj(s\mathbf{I} - \mathbf{A}) = \begin{bmatrix} (s-8)(s-4) - 16 & 2(s-4) - 24 & 8 + 6(s-8) \\ 2(s-4) - 24 & (s-1)(s-4) - 36 & 4(s-1) - 12 \\ 8 + 6(s-8) & 4(s-1) - 12 & (s-1)(s-8) - 4 \end{bmatrix}.$$

The numerator of the transfer function that gives the system zeros is

$$\mathbf{C}[adj(s\mathbf{I} - \mathbf{A})]\mathbf{B} = \begin{bmatrix} 0 & 0 & 1 \end{bmatrix} \begin{bmatrix} s^2 - 12s + 16 & 2s + 16 & 6s - 40 \\ 2s + 16 & s^2 - 5s - 32 & 4s + 8 \\ 6s - 40 & 4s + 8 & s^2 - 9s + 4 \end{bmatrix} \begin{bmatrix} 1 \\ 3 \\ 0 \end{bmatrix}$$

$$= 18.0s - 16.0$$

Therefore, the system zero is at $s = 0.8889$.
The characteristic equation is given by

$$|s\mathbf{I} - \mathbf{A}| = 0$$

$$\begin{vmatrix} s-1 & -2 & -6 \\ -2 & s-8 & -4 \\ -6 & -4 & s-4 \end{vmatrix} = 0$$

$$(s-4)[(s-1)(s-8) + 56] = 0,$$

which in expanded form gives the characteristic equation of the system as

$$s^3 - 13s^2 + 100s - 256 = 0.$$

Therefore, the system transfer function becomes

$$T(s) = \frac{18s - 16}{s^3 - 13s^2 + 100s - 256}.$$

The roots of the characteristic equation, which are also the eigenvalues of A, are the poles of the system. The simplified characteristic equation of the system is

$$(s-4)\left[s^2 - 9s + 64\right] = 0,$$

which gives the poles at $s = 4$, -3.7614, *and* $s = 12.7614$. *Since all the system poles except one are positive, the system is not stable.*

In MATLAB, *this problem could be solved giving the same results as follows:*

```
A=[1 2 6;2 8 4;6 4 4];
B=[1 3 0];
B=B'
C=[0 1];
D=0;
[z,p,k]=ss2zp(A,B,C,D,1)
```

7.3 System Response: The State-Transition Matrix

For linear time-invariant (LTI) systems, the state-space description given by Equation 7.4

$$\dot{\mathbf{x}}(t) = \mathbf{A}\ \mathbf{x}(t) + \mathbf{B}\ \mathbf{u}(t),$$

is a system of first-order linear ordinary differential equations in the system states. In this equation, the control input is regarded as a forcing function. The time response of such systems (without feed-through) depends on the output matrix $\mathbf{C}$ and the system state vector $\underline{\mathbf{x}}(t)$ such that

$$\mathbf{y}(t)\ =\mathbf{C}\ \mathbf{x}(t). \tag{7.13}$$

Since for LTI (linear time-invariant) systems, $\mathbf{C}$ is constant so the system response depends entirely on the state vector. Therefore, in the analysis of such system time response, one is interested in getting the time function of the state vector. There are several analytical techniques for determining the system time response, but in this chapter, focus will be placed on two methods only as described in the following sections.

7.3.1 Direct Solution of the Differential Equation

A system of equations in state-space form has an analytical solution analogous to that of a first-order differential equation of the form

$$\frac{dx(t)}{dt} = ax(t) +\ bu(t). \tag{7.14}$$

This equation can be rewritten as

$$\frac{dx(t)}{dt} - ax(t) = bu(t),$$

where $u(t)$ is known as the forcing function. The description matrices replace the coefficients in the solution form.

The solution for such an equation has two parts; the homogeneous part that assumes that the forcing function $u(t)$ is zero, and the nonhomogeneous part that takes into account the effect of the forcing function. The general solution is obtained by using the superposition principle to sum up the two parts of the solution and the particular solution is then obtained by taking into consideration the given initial conditions. The general form of the homogeneous part is

$$x_{\text{hom}} = Ce^{at}. \tag{7.15}$$

The nonhomogeneous part can be obtained by the method of variation of parameters where

$$x_{\text{nonhom}} = C(t)e^{at} \tag{7.16}$$

so that

$$\frac{dx(t)}{dt} = C'(t)e^{at} + aC(t)e^{at},$$

which gives

$$C(t) = \int_0^t e^{-a\tau}bu(\tau)d\tau. \tag{7.17}$$

The independent time variable t in the integral of Equation 7.17 has been changed to τ in order to distinguish it from the t in the upper limit of the integration. Combining these results gives the nonhomogeneous solution as

$$x_{\text{nonhom}} = e^{at}\int_0^t e^{-a\tau}bu(\tau)d\tau$$

$$= \int_0^t e^{-a(t-\tau)}bu(\tau)d\tau. \tag{7.18}$$

The general solution to the given first-order equation, which is a sum of homogeneous and nonhomogeneous parts of the solution, becomes

$$x(t) = Ce^{at} + \int_0^t e^{-a(t-\tau)}bu(\tau)d\tau. \qquad (7.19)$$

If the initial condition at $t = 0$ is $x(t_o)$, then the constant C is found to be $C = x(t_o)$. This gives the particular solution as

$$x(t) = x(t_o)e^{at} + \int_0^t e^{-a(t-\tau)}bu(\tau)d\tau. \qquad (7.20)$$

The state-space solution also can be obtained to be similar to Equation 7.20. Before proceeding, it is worth recalling that the exponential e^{at} can be expressed in power series as

$$e^{at} = 1 + at + \frac{1}{2!}(at)^2 + \frac{1}{3!}(at)^3 + \frac{1}{4!}(at)^4 + \frac{1}{5!}(at)^5 + \ldots\ldots \qquad (7.21)$$

Analogous to the scalar expansion is the matrix exponential $\mathbf{e}^{\mathbf{A}t}$, which is defined as

$$\mathbf{e}^{\mathbf{A}t} = \mathbf{I} + \mathbf{A}t + \frac{1}{2!}(\mathbf{A}t)^2 + \frac{1}{3!}(\mathbf{A}t)^3 + \frac{1}{4!}(\mathbf{A}t)^4 + \frac{1}{5!}(\mathbf{A}t)^5 + \ldots, \qquad (7.22)$$

where $\mathbf{A}$ is a square matrix and $\mathbf{I}$ is an identity matrix. With this information in mind, attention is paid to the state Equation 7.4

$$\dot{\mathbf{x}}(t) = \mathbf{A}\ \mathbf{x}(t) + \mathbf{B}\ u(t)$$

whose homogeneous form is

$$\frac{d\mathbf{x}(t)}{dt} = \mathbf{A}\ \mathbf{x}(t) \qquad (7.23)$$

or

$$d\mathbf{x}(t) = \mathbf{A}\ \mathbf{x}(t)\ dt,$$

which gives

$$\mathbf{x}(t)_{\text{hom}} = \mathbf{x}(t_o) + \mathbf{A}\int_0^t \mathbf{x}(\tau)d\tau. \qquad (7.24)$$

Since $\underline{x}(t)$ is on both sides of Equation 7.24, the equation is revolving about $\underline{x}(t)$ (disregard the subscript $_{\text{hom}}$) resulting in an infinite power series as follows:

$$\mathbf{x}(t)_{\text{hom}} = \mathbf{x}(t_o) + \mathbf{A} \int_0^t \mathbf{x}(\tau)d\tau$$

$$= \mathbf{x}(t_o) + \mathbf{A} \int_0^t \left(\mathbf{x}(t_o) + \mathbf{A} \int_0^t \mathbf{x}(\tau)d\tau \right) d\tau$$

$$= \mathbf{x}(t_o) + \mathbf{A} \int_0^t \left(\mathbf{x}(t_o) + \mathbf{A} \int_0^t \left[\mathbf{x}(t_o) + \mathbf{A} \int_0^t \mathbf{x}(\tau)d\tau \right] d\tau \ldots \right) d\tau.$$

Expansion of this series and simplification gives

$$\mathbf{x}(t)_{\text{hom}} = \left[\mathbf{I} + \mathbf{A}t + \frac{1}{2!}(\mathbf{A}t)^2 + \frac{1}{3!}(\mathbf{A}t)^3 + \frac{1}{4!}(\mathbf{A}t)^4 + \ldots\ldots \right] \mathbf{x}(t_o). \tag{7.25}$$

As has been shown above, the bracketed power series represents a matrix exponential $\mathbf{e}^{\mathbf{A}t}$, therefore, the homogeneous solution of this state equation can be written as

$$\mathbf{x}(t)_{\text{hom}} = \mathbf{e}^{\mathbf{A}t}\mathbf{x}(t_o), \tag{7.26}$$

which is analogous to the scalar homogeneous solution shown before.

The matrix exponential $\mathbf{e}^{\mathbf{A}t}$ is very important in control theory and is given a special name, the fundamental matrix, or, more commonly, the state-transition matrix expressed by

$$\Phi(t) = \mathbf{e}^{\mathbf{A}t}. \tag{7.27}$$

Physically, the state-transition matrix transforms the state from the given initial condition at t_o to another condition at t in the absence of the control effort. It represents the natural dynamics or free response of the system excited by initial conditions only. Normally it is presented with the two time limits as $\Phi(t, t_o)$, which indicates that it is transforming the system state at t_o to another state at t, however, at this time it is sufficient to show it with only one time element as $\Phi(t)$.

The nonhomogeneous part is again obtained using the method of variation of parameters by letting

$$\mathbf{x}(t)_{\text{nonhom}} = \mathbf{e}^{\mathbf{A}t}\underline{\mathbf{C}}(t) \tag{7.28}$$

so that

$$\dot{\underline{x}}(t)_{\text{nonhom}} = \mathbf{A}\mathbf{e}^{\mathbf{A}t}\underline{\mathbf{C}}(t) + \mathbf{e}^{\mathbf{A}t}\underline{\mathbf{C}}'(t),$$

which, when used in the nonhomogeneous equation, gives the coefficient of the nonhomogeneous solution as

$$\underline{\mathbf{C}}'(t) = \mathbf{e}^{-\mathbf{A}t}\mathbf{B}u(t)$$

or

$$\underline{\mathbf{C}}(t) = \int_0^t \mathbf{e}^{-\mathbf{A}\tau}\mathbf{B}u(\tau). \tag{7.29}$$

Hence, the nonhomogeneous solution becomes

$$\mathbf{x}(t)_{\text{nonhom}} = \mathbf{e}^{\mathbf{A}t}\int_0^t \mathbf{e}^{-\mathbf{A}\tau}\mathbf{B}u(\tau)d\tau$$

$$= \int_0^t \mathbf{e}^{\mathbf{A}(t-\tau)}\mathbf{B}u(\tau)d\tau. \tag{7.30}$$

As discussed earlier, the general solution $\underline{x}(t)$ is a linear combination of the nonhomogeneous and the homogeneous solutions, thus

$$\mathbf{x}(t) = \mathbf{e}^{\mathbf{A}t}\mathbf{x}(t_o) + \int_0^t \mathbf{e}^{\mathbf{A}(t-\tau)}\mathbf{B}u(\tau)d\tau \tag{7.31}$$

or by replacing $\mathbf{e}^{\mathbf{A}t}$ with $\Phi(t)$ Equation 7.31 becomes

$$\mathbf{x}(t) = \Phi(t)\mathbf{x}(t_o) + \int_0^t \Phi(t-\tau)\mathbf{B}u(\tau)d\tau. \tag{7.32}$$

The similarity with the scalar equation and this matrix solution can be seen here.

Since Equation 7.22 for the state transition (fundamental matrix) is a sum of an infinite series, one can use it only by truncation after a number of terms to obtain just an approximation of the transition matrix. The exact computation of $e^{\mathbf{A}t}$ can be time consuming, particularly if the dimension of A is very large. One of the simplest ways of computing $e^{\mathbf{A}t}$ is by diagonalization of matrix $\mathbf{A}$ so that $e^{\mathbf{A}t}$ will also be diagonal. It is known from linear algebra that if matrix $\mathbf{A}$ has real distinct eigenvalues, then there exists a transformation matrix $\mathbf{P}$ such that

$$\mathbf{P}^{-1}\mathbf{A}\mathbf{P} = \Lambda. \tag{7.33}$$

Λ is called the modal matrix and satisfies the equation

$$\Lambda = diag[\lambda_1(A),\ \lambda_2(A),\ \lambda_3(A),\ \cdots\ \lambda_n(A)] \tag{7.34}$$

where $\lambda_i(A)\quad i = 1, 2 \cdots n$ are the eigenvalues of $\mathbf{A}$. Therefore,

$$\mathbf{A} = \mathbf{P}\Lambda\mathbf{P}^{-1} \tag{7.35}$$

so that

$$e^{\mathbf{A}t} = e^{\mathbf{P}\Lambda\mathbf{P}^{-1}t}, \tag{7.36}$$

which by the definition in Equation 7.22, becomes

$$e^{\mathbf{A}t} = \mathbf{I} + \mathbf{P}\Lambda\mathbf{P}^{-1}t + \frac{1}{2!}(\mathbf{P}\Lambda\mathbf{P}^{-1})^2t^2 + \cdots + \frac{1}{n!}(\mathbf{P}\Lambda\mathbf{P}^{-1})^nt^n + \cdots$$

$$= \sum_{n=0}^{\infty} \frac{1}{n!}(\mathbf{P}\Lambda\mathbf{P}^{-1})^nt^n. \tag{7.37}$$

But since

$$(\mathbf{P}\Lambda\mathbf{P}^{-1})^n = (\mathbf{P}\Lambda\mathbf{P}^{-1})(\mathbf{P}\Lambda\mathbf{P}^{-1})(\mathbf{P}\Lambda\mathbf{P}^{-1})\cdots(\mathbf{P}\Lambda\mathbf{P}^{-1})$$

$$= \mathbf{P}\Lambda^n\mathbf{P}^{-1}$$

then

$$e^{\mathbf{A}t} = \mathbf{P}\left(\sum_{n=0}^{\infty}\frac{1}{n!}\Lambda^nt^n\right)\mathbf{P}^{-1} \tag{7.38}$$

where

$$\sum_{n=0}^{\infty}\frac{1}{n!}\Lambda^nt^n = e^{\Lambda t}. \tag{7.39}$$

This means

$$e^{\mathbf{A}t} = \mathbf{P}e^{\Lambda t}\mathbf{P}^{-1}. \tag{7.40}$$

Now, if Λ is a diagonal matrix, it can be shown that $e^{\Lambda t}$ is also a diagonal matrix

$$e^{\Lambda t} = diag\left[e^{\lambda_1(A)t},\ e^{\lambda_2(A)t},\ e^{\lambda_3(A)t}\cdots e^{\lambda_n(A)t}\right] \tag{7.41}$$

Example 7.3 *Use the diagonalization method to determine the transition matrix for the system matrix* $\boldsymbol{A}$ *of*

$$A = \begin{bmatrix} 6 & -2 & -7 \\ -2 & -3 & 2 \\ 1 & -2 & -2 \end{bmatrix}$$

Solution 7.3 *The eigenvalues for this matrix are* $-1, -3, 5$, *with corresponding eigenvectors*

$$\mathbf{v}_{\lambda=-1} = \begin{bmatrix} 1 \\ 0 \\ 1 \end{bmatrix}, \qquad \mathbf{v}_{\lambda=-3} = \begin{bmatrix} 1 \\ 1 \\ 1 \end{bmatrix}, \qquad \mathbf{v}_{\lambda=5} = \begin{bmatrix} -5 \\ 1 \\ -1 \end{bmatrix}$$

therefore, the diagonal matrix Λ *is given by*

$$\Lambda = \begin{bmatrix} -1 & 0 & 0 \\ 0 & -3 & 0 \\ 0 & 0 & 5 \end{bmatrix},$$

and the corresponding modal matrix $\mathbf{P}$ *is given by*

$$\mathbf{P} = \begin{bmatrix} 1 & 1 & -5 \\ 0 & 1 & 1 \\ 1 & 1 & -1 \end{bmatrix}.$$

By using Equation 7.40 it follows that

$$e^{\mathbf{A}t} = \begin{bmatrix} 1 & 1 & -5 \\ 0 & 1 & 1 \\ 1 & 1 & -1 \end{bmatrix} \begin{bmatrix} e^{-t} & 0 & 0 \\ 0 & e^{-3t} & 0 \\ 0 & 0 & e^{5t} \end{bmatrix} \begin{bmatrix} 1 & 1 & -5 \\ 0 & 1 & 1 \\ 1 & 1 & -1 \end{bmatrix}^{-1}$$

$$= \begin{bmatrix} -\frac{1}{4}\left(2e^{-6t} - e^{-8t} - 5\right)e^{5t} & -e^{-t} + e^{-3t} & \frac{1}{4}\left(6e^{-6t} - e^{-8t} - 5\right)e^{5t} \\ \frac{1}{4}e^{-3t} - \frac{1}{4}e^{5t} & e^{-3t} & -\frac{1}{4}e^{-3t} + \frac{1}{4}e^{5t} \\ -\frac{1}{4}\left(2e^{-6t} - e^{-8t} - 1\right)e^{5t} & -e^{-t} + e^{-3t} & \frac{1}{4}\left(6e^{-6t} - e^{-8t} - 1\right)e^{5t} \end{bmatrix}$$

Due to the need to compute eigenvalues and eigenvectors, as well as the modal matrix and its inverse, the method described in the foregoing paragraph turns out to be inefficient also for large systems.

The alternative approach is by using the Cayley-Hamilton theorem, which requires that every matrix satisfy its characteristic equation. That is, if the characteristic equation of an $n \times n$ matrix $\mathbf{A}$ is

$$\psi(\lambda) = \lambda^n + a_n\lambda^{n-1} + a_{n-1}\lambda^{n-2} + \cdots\cdots + a_3\lambda^2 + a_2\lambda + a_1$$

then

$$\psi(\mathbf{A}) = \mathbf{A}^n + a_n\mathbf{A}^{n-1} + a_{n-1}\mathbf{A}^{n-2} + \cdots\cdots + a_3\mathbf{A}^2 + a_2\mathbf{A} + a_1\mathbf{I}. \quad (7.42)$$

This theorem sets the maximum degree of a matrix polynomial, which is expected. A direct extension of this theorem gives the best alternative to the computation of $e^{\mathbf{A}t}$. This comes from the well known result of polynomials, which states that any polynomial $P(x)$ can be expressed as a product of two polynomials $f(x)$, $q(x)$ and a remainder $r(x)$.

$$P(x) = f(x)q(x) + r(x) \tag{7.43}$$

where the degree of $r(x)$ is such that

$$\deg\left[r\left(x\right)\right] \leq \deg\left[f(x)\right] - 1.$$

The polynomial $q(x)$ is known as the quotient while the $r(x)$ is the remainder. This method is known in mathematics as the quotient-remainder presentation of polynomials.

If the eigenvalues of $\mathbf{A}$ are λ_1, λ_2, λ_3, $\cdots$, λ_n and its characteristic function is $\psi(\lambda)$, then any other polynomial in λ, say $f(\lambda)$, can be expressed in terms of $\psi(\lambda)$ as

$$f(\lambda) = \psi(\lambda)q(\lambda) + r(\lambda), \tag{7.44}$$

where $\deg\left[r(\lambda)\right] \leq n - 1$. Extending this equation to matrix polynomial $f(\mathbf{A})$ gives

$$f(\mathbf{A}) = \psi(\mathbf{A})q(\mathbf{A}) + r(\mathbf{A}).$$

However, since $\psi(\mathbf{A}) = 0$, then

$$f(\mathbf{A}) = r(\mathbf{A}).$$

This indicates that any matrix polynomial $f(\mathbf{A})$ can be presented as a polynomial of, at most, order $n - 1$

$$f(\mathbf{A}) = \beta_0\mathbf{I} + \beta_1\mathbf{A} + \beta_2\mathbf{A}^2 + \beta_3\mathbf{A}^3 + \cdots + \beta_{n-1}\mathbf{A}^{n-1}$$

where the coefficients β_i must satisfy

$$r(\lambda_i) = \beta_0 + \beta_1\lambda_i + \beta_2\lambda_i^2 + \beta_3\lambda_i^3 + \cdots + \beta_{n-1}\lambda_i^{n-1}$$

for each of the eigenvalues of λ_i of $\mathbf{A}$.

Extension of this result to matrix exponential reduces the infinite series to a finite matrix polynomial

$$e^{\mathbf{A}t} = \alpha_0\mathbf{I} + \alpha_1\mathbf{A}t + \alpha_2\mathbf{A}^2t^2 + \alpha_3\mathbf{A}^3t^3 + \cdots + \alpha_{n-1}\mathbf{A}^{n-1}t^{n-1} \tag{7.45}$$

where the coefficients α_i satisfy

$$r(\lambda_i) = \alpha_0 + \alpha_1\lambda_i + \alpha_2\lambda_i^2 + \alpha_3\lambda_i^3 + \cdots + \alpha_{n-1}\lambda_i^{n-1}.$$

If λ_i is an eigenvalue of $\mathbf{A}$, then

$$e^{\lambda_i} = r(\lambda_i)$$

and for cases where λ_i has multiplicity k, $k > 1$, the following equations apply

$$e^{\lambda_i} = \left. \frac{dr(\lambda)}{d\lambda} \right|_{\lambda=\lambda_i}$$

$$e^{\lambda_i} = \left. \frac{d^2 r(\lambda)}{d\lambda^2} \right|_{\lambda=\lambda_i}$$

$$\ldots \ldots$$

$$e^{\lambda_i} = \left. \frac{d^{k-1} r(\lambda)}{d\lambda^{k-1}} \right|_{\lambda=\lambda_i}.$$

This method is known as the Cayley Hamilton remainder technique.

Example 7.4 *Use the Cayley-Hamilton remainder technique to determine* e^{At} *if matrix $\mathbf{A}$ is given as*

$$A = \begin{bmatrix} 0 & 1 & 0 \\ 0 & -2 & -5 \\ 0 & 1 & 2 \end{bmatrix}.$$

Solution 7.4 *This is a 3×3 matrix, therefore, $n = 3$, and, according to Equation 7.45, it can be expressed as*

$$e^{\mathbf{A}t} = \alpha_0 \mathbf{I} + \alpha_1 \mathbf{A}t + \alpha_2 \mathbf{A}^2 t^2$$

$$= \begin{bmatrix} \alpha_0 & t\alpha_1 - 2t^2\alpha_2 & -5t^2\alpha_2 \\ 0 & \alpha_0 - 2t\alpha_1 - t^2\alpha_2 & -5t\alpha_1 \\ 0 & t\alpha_1 & \alpha_0 + 2t\alpha_1 - t^2\alpha_2 \end{bmatrix},$$

and the coefficients must satisfy

$$r(\lambda) = e^\lambda = \alpha_0 + \alpha_1 \lambda + \alpha_2 \lambda^2.$$

For each eigenvalue λ of A. The eigenvalues of $\mathbf{A}t$ can be determined and are $\lambda_1 = 0$, $\lambda_2 = it$, and $\lambda_3 = -it$ where $i = \sqrt{-1}$, therefore,

$$e^0 = \alpha_0 + \alpha_1(0) + \alpha_2(0)^2$$
$$e^{it} = \alpha_0 + \alpha_1(it) + \alpha_2(it)^2$$
$$e^{-it} = \alpha_0 + \alpha_1(-it) + \alpha_2(-it)^2$$

which gives the solution as

$$\alpha_0 = 1, \quad \alpha_1 = \frac{\sin t}{t}, \quad \alpha_2 = \frac{1 - \cos t}{t^2}.$$

Substituting these values in the matrix for $e^{\mathbf{A}t}$ gives

$$e^{\mathbf{A}t} = \begin{bmatrix} 1 & \sin t - 2 + 2\cos t & -5 + 5\cos t \\ 0 & -2\sin t + \cos t & -5\sin t \\ 0 & \sin t & 2\sin t + \cos t \end{bmatrix}.$$

7.3.2 System Response by Laplace Transform Method

The approach discussed in the preceding section is somewhat mechanical and is in some respect termed as the classical approach. The approach that is direct and simpler is that of using Laplace transforms. The Laplace transform of the system Equation 7.4 gives

$$s\mathbf{x}(s) - \mathbf{x}(t_o) = \mathbf{A}\mathbf{x}(s) + \mathbf{B}u(s). \tag{7.46}$$

This can be rearranged as

$$(s\mathbf{I} - \mathbf{A})\mathbf{x}(s) = \mathbf{x}(t_o) + \mathbf{B}u(s)$$

or

$$\mathbf{x}(s) = (s\mathbf{I} - \mathbf{A})^{-1}\mathbf{x}(t_o) + (s\mathbf{I} - \mathbf{A})^{-1}\mathbf{B}u(s). \tag{7.47}$$

The matrix $(s\mathbf{I} - \mathbf{A})^{-1}$ is the system resolvent matrix defined in Equation 7.8, which is a Laplace transform of the state-transition matrix $\Phi(t)$,

$$\Phi(s) = (s\mathbf{I} - \mathbf{A})^{-1}.$$

Therefore, the inverse Laplace transform of resolvent matrix gives the state transition equation as

$$\Phi(t) = \mathcal{L}^{-1}[(s\mathbf{I} - \mathbf{A})^{-1}]. \tag{7.48}$$

With this definition, the inverse Laplace transform of Equation 7.47 becomes

$$\mathbf{x}(t) = \Phi(t)\mathbf{x}(t_o) + \mathcal{L}^{-1}[(s\mathbf{I} - \mathbf{A})^{-1}\mathbf{B}u(s)]. \qquad (7.49)$$

By the application of the convolution theorem on $\mathcal{L}^{-1}[(s\mathbf{I} - \mathbf{A})^{-1}\mathbf{B}u(s)]$ results in the complete solution of the equation as

$$\mathbf{x}(t) = \Phi(t)\mathbf{x}(t_o) + \int_0^t \Phi(t - \tau)\mathbf{B}u(\tau)d\tau. \qquad (7.50)$$

It can easily be seen how simple the Laplace transform approach is. Since the convolution theorem applies equally well to the two components of the convolution

$$\mathcal{L}^{-1}[(s\mathbf{I} - \mathbf{A})^{-1}\mathbf{B}u(s)] = \int_0^t \Phi(t - \tau)\mathbf{B}u(\tau)d\tau = \int_0^t \Phi(\tau)\mathbf{B}u(t - \tau)d\tau.$$
$$(7.51)$$

The choice of which term between $\Phi(t)$ and $u(t)$ should be subjected to a time delay in the integration depends on the overall computation advantage offered.

Example 7.5 *For the system described by the state-space equation*

$$\begin{bmatrix} \dot{x}_1(t) \\ \dot{x}_2(t) \\ \dot{x}_3(t) \end{bmatrix} = \begin{bmatrix} 0 & 1 & 0 \\ 0 & -2 & -5 \\ 0 & 1 & 2 \end{bmatrix} \begin{bmatrix} x_1(t) \\ x_2(t) \\ x_3(t) \end{bmatrix} + \begin{bmatrix} 0 \\ 1 \\ 0 \end{bmatrix} u(t)$$

determine its resolvent matrix and the state-transition matrix using the Laplace transform. Compare the result with that obtained by using the Cayley Hamilton remainder technique. If the initial state is

$$x_1(0) = 2, \quad x_2(0) = 0, \quad x_3(0) = 2,$$

determine the time response of the states to a step function $u(t) = 2$, and if the output equation is

$$y(t) = \begin{bmatrix} 1 & 1 & 0 \end{bmatrix} \begin{bmatrix} x_1(t) \\ x_2(t) \\ x_3(t) \end{bmatrix},$$

determine the output $y(t)$ under these conditions.

Solution 7.5 *The resolvent matrix was defined in Equation 7.8 as*

$$\Phi(s) = (s\mathbf{I} - \mathbf{A})^{-1}.$$

Therefore, for this system,

$$\Phi(s) = \begin{bmatrix} s & -1 & 0 \\ 0 & s+2 & 5 \\ 0 & -1 & s-2 \end{bmatrix}^{-1}$$

$$= \begin{bmatrix} \dfrac{1}{s} & \dfrac{s-2}{s(s^2+1)} & -\dfrac{5}{s(s^2+1)} \\ 0 & \dfrac{s-2}{s^2+1} & -\dfrac{5}{s^2+1} \\ 0 & \dfrac{1}{s^2+1} & \dfrac{s+2}{s^2+1} \end{bmatrix}.$$

From the resolvent matrix, the state-transition matrix can be determined using equation 7.48 as

$$\Phi(t) = \mathcal{L}^{-1}\left[(s\mathbf{I} - \mathbf{A})^{-1}\right]$$

$$= \begin{bmatrix} \mathcal{L}^{-1}\left[\frac{1}{s}\right] & \mathcal{L}^{-1}\left[\dfrac{s-2}{s(s^2+1)}\right] & \mathcal{L}^{-1}\left[-\dfrac{5}{s(s^2+1)}\right] \\ 0 & \mathcal{L}^{-1}\left[\dfrac{s-2}{s^2+1}\right] & \mathcal{L}^{-1}\left[-\dfrac{5}{s^2+1}\right] \\ 0 & \mathcal{L}^{-1}\left[\dfrac{1}{s^2+1}\right] & \mathcal{L}^{-1}\left[\dfrac{s+2}{s^2+1}\right] \end{bmatrix},$$

where

$$\mathcal{L}^{-1}\left[\frac{1}{s}\right] = 1$$

$$\mathcal{L}^{-1}\left[\frac{s-2}{s(s^2+1)}\right] = \mathcal{L}^{-1}\left[-\frac{2}{s} + \frac{2s}{s^2+1} + \frac{1}{s^2+1}\right] = -2 + 2\cos t + \sin t$$

$$\mathcal{L}^{-1}\left\{-\frac{5}{s(s^2+1)}\right\} = \mathcal{L}^{-1}\left[-\frac{5}{s} + 5\frac{s}{s^2+1}\right] = -5 + 5\cos t$$

$$\mathcal{L}^{-1}\left[\frac{s-2}{s^2+1}\right] = \mathcal{L}^{-1}\left[\frac{s}{s^2+1} - \frac{2}{s^2+1}\right] = \cos t - 2\sin t$$

$$\mathcal{L}^{-1}\left[-\frac{5}{s^2+1}\right] = -5\sin t$$

$$\mathcal{L}^{-1}\left[\frac{1}{s^2+1}\right] = \sin t$$

$$\mathcal{L}^{-1}\left[\frac{s+2}{s^2+1}\right] = \mathcal{L}^{-1}\left[\frac{s}{s^2+1} + \frac{2}{s^2+1}\right] = \cos t + 2\sin t.$$

Therefore, the state-transition matrix becomes

$$
e^{\mathbf{A}t} = \begin{bmatrix} 1 & -2 + 2\cos t + \sin t & -5 + 5\cos t \\ 0 & \cos t - 2\sin t & -5\sin t \\ 0 & \sin t & \cos t + 2\sin t \end{bmatrix}.
$$

As can be seen, this state-transition matrix is the same as that which was obtained in the previous example. The free response of the system (homogeneous solution) for the given initial condition is given as

$$
\mathbf{x}(t) = e^{\mathbf{A}t}\mathbf{x}(0).
$$

This means

$$
\begin{bmatrix} x_1(t) \\ x_2(t) \\ x_3(t) \end{bmatrix}_{hom} = \begin{bmatrix} 1 & -2 + 2\cos t + \sin t & -5 + 5\cos t \\ 0 & \cos t - 2\sin t & -5\sin t \\ 0 & \sin t & \cos t + 2\sin t \end{bmatrix} \begin{bmatrix} 2 \\ 0 \\ 2 \end{bmatrix}
$$

$$
= \begin{bmatrix} -8 + 10\cos t \\ -10\sin t \\ 2\cos t + 4\sin t \end{bmatrix},
$$

and the nonhomogeneous solution is given by

$$
\mathbf{x}(t)_{inh} = \mathcal{L}^{-1}[(s\mathbf{I} - \mathbf{A})^{-1}\mathbf{B}u(s)].
$$

where for the step input $u(t) = 2$ it follows that $u(s) = \dfrac{2}{s}$.

Therefore, the following expression is obtained,

$$(s\mathbf{I} - \mathbf{A})^{-1}\mathbf{B}u(s) = \begin{bmatrix} \dfrac{1}{s} & \dfrac{s-2}{s(s^2+1)} & -\dfrac{5}{s(s^2+1)} \\[3mm] 0 & \dfrac{s-2}{s^2+1} & -\dfrac{5}{s^2+1} \\[3mm] 0 & \dfrac{1}{s^2+1} & \dfrac{s+2}{s^2+1} \end{bmatrix} \begin{bmatrix} 0 \\ 1 \\ 0 \end{bmatrix} \left(\dfrac{2}{s}\right)$$

$$= \begin{bmatrix} \dfrac{2}{s^2}\dfrac{s-2}{s^2+1} \\[3mm] 2\dfrac{s-2}{s(s^2+1)} \\[3mm] \dfrac{2}{s(s^2+1)} \end{bmatrix}.$$

Thus, the time response becomes

$$\begin{bmatrix} x_1(t) \\ x_2(t) \\ x_3(t) \end{bmatrix}_{nonhom} = \begin{bmatrix} \mathcal{L}^{-1}\left[\dfrac{2}{s^2}\dfrac{s-2}{s^2+1}\right] \\[3mm] \mathcal{L}^{-1}\left[2\dfrac{s-2}{s(s^2+1)}\right] \\[3mm] \mathcal{L}^{-1}\left[\dfrac{2}{s(s^2+1)}\right] \end{bmatrix}$$

$$= \begin{bmatrix} 2 - 4t - 2\cos t + 2\sin t \\ -4 + 2\sin t + 4\cos t \\ 2 - 2\cos t \end{bmatrix}.$$

Therefore, the time response

$$\mathbf{x}(t) = \mathbf{x}(t)_{hom} + \mathbf{x}(t)_{nonhom}$$

becomes

$$\begin{bmatrix} x_1(t) \\ x_2(t) \\ x_3(t) \end{bmatrix} = \begin{bmatrix} -8 + 10\cos t \\ -10\sin t \\ 2\cos t + 4\sin t \end{bmatrix} + \begin{bmatrix} 2 - 4t - 2\cos t + 2\sin t \\ -4 + 2\sin t + 4\cos t \\ 2 - 2\cos t \end{bmatrix}$$

$$= \begin{bmatrix} -6 + 8\cos t - 4t + 2\sin t \\ -8\sin t - 4 + 4\cos t \\ 4\sin t + 2 \end{bmatrix}.$$

The output y(t) is measured through the output matrix as

$$y(t) = \begin{bmatrix} 1 & 1 & 0 \end{bmatrix} \begin{bmatrix} x_1(t) \\ x_2(t) \\ x_3(t) \end{bmatrix}$$

$$= \begin{bmatrix} 1 & 1 & 0 \end{bmatrix} \begin{bmatrix} -6 + 8\cos t - 4t + 2\sin t \\ -8\sin t - 4 + 4\cos t \\ 4\sin t + 2 \end{bmatrix}$$

$$= -10 + 12\cos t - 4t - 6\sin t.$$

7.4 System Controllability and Observability

7.4.1 Controllability

For a given system presentation using matrices $\mathbf{A}$ and $\mathbf{B}$, the effect of the control input to the system states is given by its controllability property. The system configuration $(\mathbf{A}, \mathbf{B})$ is said to be controllable if every unconstrained control input $u(t)$ can affect the system state $\underline{x}(t_o)$ so as to place it at another location $\underline{x}(t_1)$. Otherwise, if the matrices $\mathbf{A}$ and $\mathbf{B}$ are such that the control input $u(t)$ has no effect on the system states $\underline{x}(t)$, then this system is said to be uncontrollable. Since, as it will be seen later, the same input-output relationship of the system can be shown to have many different presentations in state-space form, the system controllability turns out to be a property of the system presentation (configuration) rather than a property of the system itself. Controllability is one of the most important properties that must be ensured before any attempt in controller design is done.

System controllability is associated with the controllability grammian matrix $\mathbf{M}$ defined as

$$\mathbf{M} = \int_o^t \Phi(\tau)\mathbf{B}\mathbf{B}^T\Phi^T(\tau)d\tau. \tag{7.52}$$

The system is said to be controllable if this matrix $\mathbf{M}$ is positive definite and non-singular. For the development of complete controllability of a linear time-invariant system, use is made of the nonhomogeneous part of its time response which must drive the system response state $\underline{x}(t)$ to zero from some initial state $\underline{x}(0)$. Now, as has been shown in Equation 7.50, the time response of the system is

$$\underline{x}(t) = \underline{\Phi}(t)\underline{x}(0) + \int_0^t \Phi(t - \tau)\mathbf{B}u(t)d\tau,$$

where the nonhomogeneous part is given in Equation 7.30 as

$$\underline{x}(t)_{\text{inh}} = \int_0^t \underline{\Phi}(\tau)\mathbf{B}u(t - \tau)d\tau.$$

If the system time response is to be zero by the presence of the nonhomogeneous response, then

$$
\begin{aligned}
0 &= \underline{\Phi}(t)\underline{x}(0) + \int_0^t \Phi(t - \tau)\mathbf{B}u(t)d\tau \\
&= e^{\mathbf{A}t}\underline{x}(0) + \int_0^t e^{\mathbf{A}(t-\tau)}\mathbf{B}u(t)d\tau \\
&= e^{\mathbf{A}t}[\underline{x}(0) + \int_0^t e^{-\mathbf{A}\tau}\mathbf{B}u(t)d\tau],
\end{aligned}
\tag{7.53}
$$

establishment of the controllability property is done through expressing the initial state as a function of the control input $u(t)$. From Equation 7.53 it follows that

$$-\underline{x}(0) = \int_0^t e^{-\mathbf{A}\tau}\mathbf{B}u(t)d\tau. \tag{7.54}$$

The matrix exponential $e^{-\mathbf{A}\tau}$ can be expressed in an infinite power series form as

$$e^{-\mathbf{A}\tau} = \mathbf{I} - \mathbf{A}\tau + \frac{1}{2}\mathbf{A}^2\tau^2 - \frac{1}{3!}\mathbf{A}^3\tau^3 + \frac{1}{4!}\mathbf{A}^4\tau^4 + \ldots\ldots$$

However, using Cayley-Hamilton theorem, it can also be expressed as a matrix polynomial with at most $(n - 1)$ power terms so that

$$e^{-\mathbf{A}\tau} = r_o\mathbf{I} + r_1\mathbf{A} + r_2\mathbf{A}^2 + r_3\mathbf{A}^3 + r_4\mathbf{A}^4 + \cdots + r_{n-1}\mathbf{A}^{n-1},$$

where r_i are scalar functions of τ and eigenvalues of $\mathbf{A}$. If this expansion is used in the above expression for $-\underline{x}(0)$ it yields

$$-\mathbf{x}(0) = \int_0^t [r_o\mathbf{I} + r_1\mathbf{A} + r_2\mathbf{A}^2 + r_3\mathbf{A}^3 + r_4\mathbf{A}^4 + \cdots + r_{n-1}\mathbf{A}^{n-1}]\mathbf{B}u(t)d\tau$$

$$= [\mathbf{B} \; \vdots \mathbf{AB} \vdots \mathbf{A}^2\mathbf{B} \; \vdots \mathbf{A}^3\mathbf{B} \; \vdots \mathbf{A}^4\mathbf{B} \; \vdots \cdots \vdots \mathbf{A}^{n-1}\mathbf{B} \;] \int_0^t \begin{bmatrix} v_0(\tau) \\ v_1(\tau) \\ v_2(\tau) \\ v_3(\tau) \\ \vdots \\ v_{n-1}(\tau) \end{bmatrix} d\tau$$

$$= \mathcal{C} \int \underline{\mathbf{V}}(\tau)d\tau,$$

where $\underline{\mathbf{V}}(\tau)$ is a vector of $v_i(\tau) = \mathrm{r}_i(\tau)u(\tau)$. The matrix $\mathcal{C}$ is an n x mn matrix called the system controllability matrix given by

$$\mathcal{C} = [\mathbf{B} \; \vdots \mathbf{AB} \; \vdots \mathbf{A}^2\mathbf{B} \; \vdots \mathbf{A}^3\mathbf{B} \; \vdots \mathbf{A}^4\mathbf{B} \; \vdots \cdots \vdots \mathbf{A}^{n-1}\mathbf{B}]. \qquad (7.55)$$

For the system to be controllable, this matrix must be a full row rank matrix, however for SISO systems, it is sufficient to observe that the matrix $\mathcal{C}$ is a square n x n and for it to be full row rank it must also be non-singular with a non-zero determinant such that

$$\det |\mathcal{C}| \neq 0. \qquad (7.56)$$

Essentially, the controllability is an indication that there are no pole-zero cancellations in the transfer function. Another test for the system controllability may be by direct observation in the signal flow chart. For a controllable system configuration, its control signal $u(t)$ is linked to all the system states $\underline{\mathbf{x}}(\mathbf{t})$ in the signal flow chart as shown in Figure 7.2.

In MATLAB the system controllability matrix can be constructed by command "ctrb(A,B)" which returns the controllability matrix $\mathcal{C}$. Also the system controllability can be established by examination of its controllability grammian as seen before. This is accomplished in MATLAB by the command "gram(A,B)" which constructs the controllability grammian $\mathbf{M}$ of Equation 7.52. The syntax for the gram command is

$$M = \mathrm{gram}(A, B). \qquad (7.57)$$

Example 7.6 *The input and the plant matrices for a certain system are*

$$\mathbf{A} = \begin{bmatrix} 2 & 3 & 0 \\ 3 & 5 & 5 \\ 4 & 3 & 6 \end{bmatrix} \qquad \mathbf{B} = \begin{bmatrix} 3 \\ 2 \\ 5 \end{bmatrix}$$

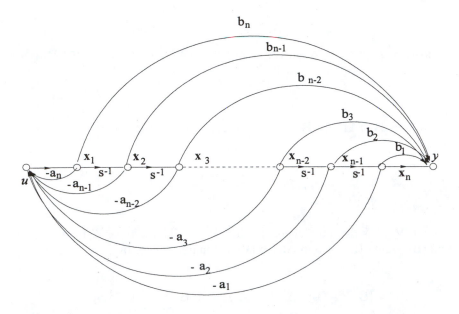

FIGURE 7.2
The Signal Flow Graph of a Controllable System

Establish whether the system is controllable. Confirm the results using MATLAB.

Solution 7.6 *The system controllability matrix (given in Equation 7.55) is given by*

$$\mathcal{C} = [\, B \quad \vdots AB \quad \vdots A^2 B \,]$$

where

$$\mathbf{B} = \begin{bmatrix} 3 \\ 2 \\ 5 \end{bmatrix}$$

$$\mathbf{AB} = \begin{bmatrix} 2 & 3 & 0 \\ 3 & 5 & 5 \\ 4 & 3 & 6 \end{bmatrix} \begin{bmatrix} 3 \\ 2 \\ 5 \end{bmatrix}$$

$$= \begin{bmatrix} 12.0 \\ 44.0 \\ 48.0 \end{bmatrix}.$$

$$\mathbf{A}^2\mathbf{B} = \mathbf{A}(\mathbf{AB}) = \begin{bmatrix} 2 & 3 & 0 \\ 3 & 5 & 5 \\ 4 & 3 & 6 \end{bmatrix} \begin{bmatrix} 12.0 \\ 44.0 \\ 48.0 \end{bmatrix}$$

$$= \begin{bmatrix} 156.0 \\ 496.0 \\ 468.0 \end{bmatrix}$$

Therefore, the controllability matrix $\mathcal{C}$ *becomes*

$$\mathcal{C} = \begin{bmatrix} 3 & 12 & 156 \\ 2 & 44 & 496 \\ 5 & 48 & 468 \end{bmatrix},$$

from which it follows that it has full rank 3 and is non-singular with determinant $|\mathcal{C}| = -10\,464$*. Therefore, the system is controllable.*

In MATLAB the controllability matrix could be formed through the following code with the same results

```
A=[2 3 0;3 5 5;4 3 6];
B=[3;2;5];
Co=ctrb(A,B)
result=det(Co).
```

Alternatively, one could establish the controllability grammian and find out if it has full rank by checking whether it has a determinant.

```
A=[2 3 0;3 5 5;4 3 6];
B=[3;2;5];
M=gram(A,B)
result=det(M).
```

Example 7.7 *Consider a system*

$$\dot{\mathbf{x}}(t) = \mathbf{A}\mathbf{x}(t) + \mathbf{B}\mathbf{u}(t)$$

If the matrices **A** *and* **B** *are known to be*

$$\mathbf{A} = \begin{bmatrix} -2 & 0 & 0 \\ 0 & 0 & 1 \\ 0 & -3 & -3 \end{bmatrix} \qquad \mathbf{B} = \begin{bmatrix} 0 \\ 1 \\ 1 \end{bmatrix},$$

determine the controllability matrix and grammian and establish whether this system is controllable. The results of the computation can be confirmed by using MATLAB.

Solution 7.7 *For the given* 3×3 *matrix system, the controllability matrix is given by*

$$\mathcal{C} = \begin{bmatrix} \mathbf{B} \vdots \mathbf{AB} \vdots \mathbf{A}^2\mathbf{B} \end{bmatrix}.$$

Using the same procedures as in the previous example, the controllability matrix is obtained as

$$\mathcal{C} = \begin{bmatrix} 0 & 0 & 0 \\ 1 & 1 & -6 \\ 1 & -6 & 15 \end{bmatrix}.$$

Clearly, it follows that the first row of the controllability matrix contains zeros only, which is an indication that the matrix is singular, and hence the system is uncontrollable. It can be seen that the determinant of $\mathcal{C}$ is zero.

To establish the controllability grammian the state-transition matrix $\Phi(t)$ for the system is required, which can be calculated using any of the methods discussed before. Unfortunately, the transition matrix for this problem has elements made of very long exponential terms that are not shown here. However, multiplication of this transition matrix with $\mathbf{B}$ gives

$$\Phi(t)\mathbf{B} = \begin{bmatrix} 0 \\ \frac{1}{2}e^{-\frac{3}{2}t-\frac{1}{2}it\sqrt{3}} + \frac{5}{6}ie^{-\frac{3}{2}t-\frac{1}{2}it\sqrt{3}}\sqrt{3} - \frac{5}{6}ie^{-\frac{3}{2}t+\frac{1}{2}it\sqrt{3}}\sqrt{3} + \frac{1}{2}e^{-\frac{3}{2}t+\frac{1}{2}it\sqrt{3}} \\ \frac{1}{2}e^{-\frac{3}{2}t-\frac{1}{2}it\sqrt{3}} - \frac{3}{2}ie^{-\frac{3}{2}t-\frac{1}{2}it\sqrt{3}}\sqrt{3} + \frac{3}{2}ie^{-\frac{3}{2}t+\frac{1}{2}it\sqrt{3}}\sqrt{3} + \frac{1}{2}e^{-\frac{3}{2}t+\frac{1}{2}it\sqrt{3}} \end{bmatrix}$$

Without going further (of course because of the length of the result), one could see right from here that due to the zero element in the first entry, then matrix $\Phi(t)\mathbf{B}\mathbf{B}^T\Phi(t)^T$ will have the following form

$$\Phi(t)\mathbf{B}\mathbf{B}^T\Phi(t)^T = \begin{bmatrix} 0 & 0 & 0 \\ 0 & * & * \\ 0 & * & * \end{bmatrix},$$

where $(*)$ indicates the non-zero entries. This indicates that the controllability grammian is singular and hence, the system is uncontrollable. (The complete evaluation of the controllability grammian and verification of the results is left to the reader as a special exercise)

7.4.2   Observability

System observability is another property of interest in the state-space control design. Like controllability, this property has nothing to do with the system as a whole but rather with the system presentation. The system is said to be observable if the output has all the components of the state. In this case, it becomes possible to estimate the system states from the output measurements. For linear time-invariant systems, the observability property is governed by the observability grammian matrix $\mathbf{N}$. It must be non-singular, i.e., it should be invertible. To get a clear picture of this situation, consider a system whose state-space output equation in the absence of feed-through is

$$y(t) = \mathbf{C}\mathbf{x}(t), \tag{7.58}$$

where the homogeneous response of this system is

$$\mathbf{x}(t) = \Phi(t)\mathbf{x}(t_o). \tag{7.59}$$

These two equations can be combined to give

$$y(t) = \mathbf{C}\Phi(t)\mathbf{x}(t_o). \tag{7.60}$$

Now if $\mathbf{C}\Phi(t)$ were invertible, one could easily get $\mathbf{x}(t_o)$ from the output and hence the system states in Equation 7.59 so that the system would be observable. However, this is not normally the case, as matrix $\mathbf{C}\Phi(t)$ is not necessarily a square matrix, e.g., for a SISO system it is a row matrix. Therefore, the procedure to determine the system observability goes on by some mathematical manipulation of pre-multiplying both sides of Equation 7.60 by $\Phi^T(t)\mathbf{C}^T$ and integrating as

$$\int_0^t \Phi^T(\tau)\mathbf{C}^T y(\tau)d\tau = \int_0^t \Phi^T(\tau)\mathbf{C}^T \mathbf{C}\Phi(t)d\tau\,\mathbf{x}(t_o),$$

which gives the observability grammian as

$$\mathbf{N}(t) = \int_0^t \Phi^T(\tau)\mathbf{C}^T \mathbf{C}\Phi(t)d\tau \tag{7.61}$$

so that

$$\int_0^t \Phi^T(\tau)\mathbf{C}^T y(\tau)d\tau = \mathbf{N}(t)\,\mathbf{x}(t_o). \tag{7.62}$$

Therefore, $\mathbf{x}(t_o)$ can be determined from the output measurements $y(t)$ as

$$\mathbf{x}(t_o) = \mathbf{N}^{-1}(t)\int_0^t \Phi^T(\tau)\mathbf{C}^T y(\tau)d\tau, \tag{7.63}$$

which requires that the controllability grammian $\mathbf{N}(t)$ must be invertible and hence non-singular.

For LTI systems, which are the main subject of this discussion, the most direct and sufficient test for system observability is given by the observability matrix. The observability matrix can best be derived using MIMO models. However, since the concern has been about SISO systems, the following approach is adopted which leads to same conclusion. It has been shown that in Equation 7.60

$$y(t) = \mathbf{C}\Phi(t)\mathbf{x}(t_o) \tag{7.64}$$

where

$$\Phi(t) = e^{\mathbf{A}t} = \mathbf{I} + \mathbf{A}t + \frac{1}{2}\mathbf{A}^2 t^2 + \frac{1}{3!}\mathbf{A}^3 t^3 + \frac{1}{4!}\mathbf{A}^4 t^4 + \cdots$$

which according to Cayley-Hamilton theorem, can be shown to be

$$\Phi(t) = e^{\mathbf{A}t} = r_o \mathbf{I} + r_1 \mathbf{A} + r_2 \mathbf{A}^2 + r_3 \mathbf{A}^3 + r_4 \mathbf{A}^4 + \cdots + r_{n-1}\mathbf{A}^{n-1}$$

where r_i are some coefficients that are functions of time. When used in the output Equation 7.64 it gives

$$y(t) = \mathbf{C}\left(r_o \mathbf{I} + r_1 \mathbf{A} + r_2 \mathbf{A}^2 + r_3 \mathbf{A}^3 + r_4 \mathbf{A}^4 + \cdots + r_{n-1}\mathbf{A}^{n-1}\right)\mathbf{x}(t_o).$$

This is can be expressed as

$$y(t) = \sum_{i=0}^{n-1} r_i \mathbf{C}\mathbf{A}^i \mathbf{x}(t_o).$$

The row vector $\sum r_i \mathbf{C}\mathbf{A}^i$ can be written as

$$\sum_{i=0}^{i=n-1} r_i \mathbf{C}\mathbf{A}^i = \begin{bmatrix} r_o & r_1 & r_2 & r_3 & \cdots\cdots & r_{n-1} \end{bmatrix} \begin{bmatrix} C \\ CA \\ CA^2 \\ CA^3 \\ \vdots \\ CA^{n-1} \end{bmatrix}$$

$$= \mathbf{r}\mathcal{O}$$

so that

$$y = \mathbf{r}\mathcal{O}\mathbf{x}(t_o),$$

where $\mathbf{r}$ and $\mathbf{x}(t_o)$ are n-row and n-column vectors respectively while $\mathcal{O}$ is a $n \times n$ matrix called the observability matrix of the system in Equations 7.3 and 7.4. Normally for MIMO systems $\mathcal{O}$ is an $n \times n$ matrix.

$$\mathcal{O} = \begin{bmatrix} C \\ CA \\ CA^2 \\ CA^3 \\ \vdots \\ CA^{n-1} \end{bmatrix}. \tag{7.65}$$

The system is said to be observable if the observability matrix has a column rank n, and for linear SISO systems, it suffices to observe that this observability matrix will be non-singular and hence

$$\det |\mathcal{O}| \neq 0. \tag{7.66}$$

In the corresponding signal flow chart, the output needs to be linked to all the system states as shown in Figure 7.3

In MATLAB, the observability matrix of a system is established by command "obsv(A,C)" where C is the output matrix. Its syntax is

$$Ob = obsv(A, C).$$

Another command to establish observability of the system is from its grammian as discussed before. In MATLAB, the observability grammian is established by the command "gram(A',C')." Note that, although the same gram command is used for establishing the controllability grammian, the difference in the result lies in the input arguments. While, on testing the system controllability, the input arguments were the system matrix $\mathbf{A}$ as well as the input matrix $\mathbf{B}$, when this command is used for establishing the observability grammian, the input arguments become the transpose of the system matrix as well as the transpose of the output matrix, i.e., $\mathbf{A}^T$ and $\mathbf{C}^T$ respectively. In this way, it returns the system observability grammian given in Equation 7.61. The syntax for the gram command is

$$N = gram(A', C')$$

Example 7.8 *If the output matrix C in the previous example is*

$$C = [1 \ \ 2 \ \ 1],$$

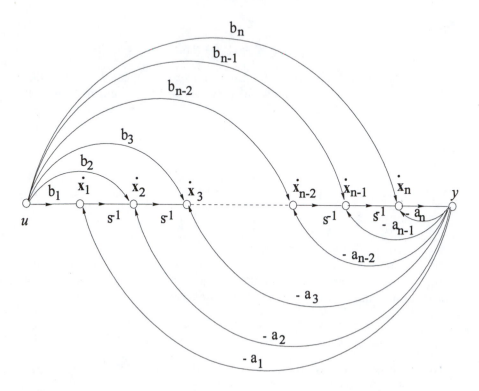

FIGURE 7.3
The Signal Flow Chart for an Observable System

establish the observability of this system. Remember that

$$A = \begin{bmatrix} 2 & 3 & 0 \\ 3 & 5 & 5 \\ 4 & 3 & 6 \end{bmatrix}, \qquad B = \begin{bmatrix} 3 \\ 2 \\ 5 \end{bmatrix}.$$

Solution 7.8 *The system observability matrix was given in Equation 7.65, which for the system in question becomes*

$$\mathcal{O} = \begin{bmatrix} C \\ CA \\ CA^2 \end{bmatrix}$$

Now given that

$$C = \begin{bmatrix} 1 & 2 & 1 \end{bmatrix}$$

then

$$CA = \begin{bmatrix} 1\ 2\ 1 \end{bmatrix} \begin{bmatrix} 2\ 3\ 0 \\ 3\ 5\ 5 \\ 4\ 3\ 6 \end{bmatrix}$$

$$= \begin{bmatrix} 12\ 16\ 16 \end{bmatrix}$$

and

$$CA^2 = (CA)A = \begin{bmatrix} 12\ 16\ 16 \end{bmatrix} \begin{bmatrix} 2\ 3\ 0 \\ 3\ 5\ 5 \\ 4\ 3\ 6 \end{bmatrix}$$

$$= \begin{bmatrix} 136\ 164\ 176 \end{bmatrix}.$$

Therefore, the observability matrix becomes

$$\mathcal{O} = \begin{bmatrix} 1 & 2 & 1 \\ 12 & 16 & 16 \\ 136 & 164 & 176 \end{bmatrix},$$

which has full rank 3 and hence is non-singular with determinant $|\mathcal{O}|=112$. *Therefore, the system is observable. The corresponding MATLAB code is*

    A=[2 3 0;3 5 5;4 3 6];
    B=[1;2;1];
    Oo=obsv(A,C)
    solution=det(Oo);

Alternatively, one could establish the observability grammian as

    N=gram(A',C')
    solution=det(N);

Example 7.9 *Examine the observability of a system whose plant and output matrices* **A** *and* **C** *are*

$$\mathbf{A} = \begin{bmatrix} 3\ 4 \\ 1\ 3 \end{bmatrix} \qquad \mathbf{C} = \begin{bmatrix} 1\ 2 \end{bmatrix}.$$

Also determine the system observability grammian and establish the observability by examining this grammian

Solution 7.9 *For this* 2×2 *system, the observability matrix is constructed as follows*

$$\mathcal{O} = \begin{bmatrix} \mathbf{C} \\ \mathbf{CA} \end{bmatrix}$$

$$= \begin{bmatrix} 1\ 2 \\ 5\ 10 \end{bmatrix}.$$

The determinant is found to be $[10 - 5(2)] = 0$. *Clearly, this system observability matrix is singular and hence, the system is unobservable.*

To get the observability grammian, one needs to know its state-transition matrix, which can be determined using either of the methods discussed earlier and is

$$\Phi(t) = \begin{bmatrix} \frac{4}{5}e^t + \frac{1}{5}e^{5t} & -\frac{2}{5}e^t + \frac{2}{5}e^{5t} \\ -\frac{2}{5}e^t + \frac{2}{5}e^{5t} & \frac{1}{5}e^t + \frac{4}{5}e^{5t} \end{bmatrix}$$

therefore, for a given $\mathbf{C}$

$$\Phi(t)^T \mathbf{C}^T = \begin{bmatrix} e^{5t} \\ 2e^{5t} \end{bmatrix}$$

so that

$$\Phi(t)^T \mathbf{C}^T \mathbf{C} \Phi(t) = \begin{bmatrix} e^{10t} & 2e^{10t} \\ 2e^{10t} & 4e^{10t} \end{bmatrix}$$

$$= \begin{bmatrix} 1 & 2 \\ 2 & 4 \end{bmatrix} e^{10t}.$$

Thus, the observability grammian becomes

$$N(t) = \int_0^t \Phi(\tau)^T \mathbf{C}^T \mathbf{C} \Phi(\tau) d\tau$$

$$= \begin{bmatrix} 1 & 2 \\ 2 & 4 \end{bmatrix} \int_0^t e^{10\tau} d\tau$$

$$= \begin{bmatrix} \frac{1}{10} & \frac{2}{10} \\ \frac{2}{10} & \frac{4}{10} \end{bmatrix} \left(e^{10t} - 1 \right).$$

Again, the observability grammian of this system is singular with determinant 0. This conforms with the results in the previous part, which indicates the system is unobservable.

7.5 Similarity Transformations: Canonical Forms

7.5.1 Similarity Transformation

The state-space description of a system can be expressed using another variable without losing the system input-output relationship. This transformation involves redefinition of the state variables, and the system matrices leaving the control as well as the output signal unaltered. It can be easily observed that, since this transformation involves the system matrices and states only, preserving the input-output relationship of the system, the system properties such as the transfer function, poles, zeros, and the characteristic equation remain unchanged. Such transformation is known as the similarity transformation and requires pre-definition of the state transformation matrix $\mathbf{P}$, which transforms any other state variable $\mathbf{z}(t)$ to $\mathbf{x}(t)$

$$\mathbf{x}(t) = \mathbf{P}\underline{\mathbf{z}}(t), \tag{7.67}$$

so that

$$\dot{\mathbf{x}}(t) = \mathbf{P}\dot{\underline{\mathbf{z}}}(t).$$

Under similarity transformation, the transformed dynamic state Equations 7.3 and 7.4 become

$$\mathbf{P}\dot{\underline{\mathbf{z}}}(t) = \mathbf{AP}\underline{\mathbf{z}}(t) + \mathbf{B}u(t) \tag{7.68}$$
$$y(t) = \mathbf{CP}\underline{\mathbf{z}}(t) + \mathbf{D}u(t), \tag{7.69}$$

which can be rearranged as

$$\dot{\underline{\mathbf{z}}}(t) = \mathbf{P}^{-1}\mathbf{AP}\underline{\mathbf{z}}(t) + \mathbf{P}^{-1}\mathbf{B}u(t) \tag{7.70}$$
$$y(t) = \mathbf{CP}\underline{\mathbf{z}}(t) + \mathbf{D}u(t). \tag{7.71}$$

This state-space dynamic equation can be simplified by redefining the transformed system matrices as

$$\mathbf{A}^* = \mathbf{P}^{-1}\mathbf{AP} \tag{7.72}$$
$$\mathbf{B}^* = \mathbf{P}^{-1}\mathbf{B} \tag{7.73}$$
$$\mathbf{C}^* = \mathbf{CP} \tag{7.74}$$
$$\mathbf{D}^* = \mathbf{D} \tag{7.75}$$

and hence giving a system of the following form

$$\dot{\underline{z}}(t) = \mathbf{A}^*\underline{z}(t) + \mathbf{B}^*\mathbf{u}(t) \qquad (7.76)$$

$$y(t) = \mathbf{C}^*\underline{z}(t) + \mathbf{D}^*\mathbf{u}(t). \qquad (7.77)$$

As stated before, since the dynamics of the system are preserved by similarity transformation, the eigenvalues, eigenvectors, poles, zeros, and the system characteristic equation are all preserved. Using such transformation, several forms of state descriptions of the system can be defined, known as canonical form, as given in the next sections. Knowing the transformation matrix P, the similarity transformation can be carried out in MATLAB using the command "ss2ss" and its syntax is

$$[\mathsf{At}, \mathsf{Bt}, \mathsf{Ct}, \mathsf{Dt}] = \mathsf{ss2ss}(\mathsf{A}, \mathsf{B}, \mathsf{C}, \mathsf{D}, \mathsf{P}).$$

However, for some of the special canonical forms discussed in the following sections where the transformation matrices are well established, special commands besides the "ss2ss" do exist, as will be seen next. However, for some other forms, it remains for the user to determine the transformation matrix $\mathbf{P}$ in advance and use the "ss2ss" command.

7.5.2   Controllable Canonical Form

Controllable canonical form is a system description in which all states can be modified by the control input. This form is very convenient in the design of the system controllers. For any given system presentation $\mathbf{A}$, $\mathbf{B}$, $\mathbf{C}$ and $\mathbf{D}$, the similarity transformation matrix $\mathbf{P}$, which puts the system in controllable canonical form, is given as a product of the controllability matrix $\mathcal{C}$

$$\mathcal{C} = [\mathbf{B} \ \vdots \mathbf{AB} \ \vdots \mathbf{A}^2\mathbf{B} \ \vdots \mathbf{A}^3\mathbf{B} \ \vdots \mathbf{A}^4\mathbf{B} \ \vdots \cdots \vdots \mathbf{A}^{n-1}\mathbf{B} \ ],$$

and the triangular matrix $\mathbf{Q}$ of coefficients of the characteristic equation.

$$|s\mathbf{I} - \mathbf{A}| = s^n + a_n s^{n-1} + a_{n-1} s^{n-2} + \cdots\cdots a_3 s^2 + a_2 s + a_1 = 0,$$

where Q is an upper triangular matrix given by

$$\mathbf{Q} = \begin{bmatrix} a_2 & a_3 & a_4 & \cdots & a_{n-1} & a_n & 1 \\ a_3 & a_4 & \cdots & a_{n-1} & a_n & 1 & 0 \\ a_4 & \vdots & a_{n-1} & a_n & 1 & 0 & 0 \\ \vdots & a_{n-1} & a_n & 1 & 0 & 0 & \vdots \\ a_{n-1} & a_n & 1 & 0 & 0 & \vdots & 0 \\ a_n & 1 & 0 & 0 & \cdots & 0 & 0 \\ 1 & 0 & 0 & \cdots & 0 & 0 & 0 \end{bmatrix} \tag{7.78}$$

so that

$$\mathbf{P} = \mathcal{C}\mathbf{Q}. \tag{7.79}$$

The transformed system will then be in controllable canonical form in which

$$\mathbf{A}^* = \mathbf{P}^{-1}\mathbf{A}\mathbf{P} = \begin{bmatrix} 0 & 1 & 0 & 0 & \vdots & 0 & 0 \\ \dot{0} & 0 & 1 & 0 & \vdots & 0 & 0 \\ 0 & 0 & 0 & 1 & \vdots & 0 & 0 \\ \cdots & \cdots & \cdots & \cdots & \ddots & \cdots & \cdots \\ 0 & 0 & 0 & 0 & \vdots & 1 & 0 \\ 0 & 0 & 0 & 0 & \vdots & 0 & 1 \\ -a_1 & -a_2 & -a_3 & -a_4 & \vdots & -a_{n-1} & -a_n \end{bmatrix} \tag{7.80}$$

Other system matrices will be transformed according to Equations 7.72 through to Equation 7.75, and the states are transformed according to Equation 7.67. Important to note is that for SISO systems the input matrix $\mathbf{B}^*$ in controllable canonical form gets the following form.

$$\mathbf{B}^* = \begin{bmatrix} 0 \\ 0 \\ 0 \\ \cdots \\ 1 \end{bmatrix} \tag{7.81}$$

while the output matrix $\mathbf{C}^*$ becomes

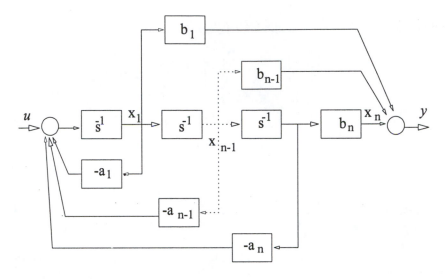

FIGURE 7.4
System in Controllable Canonical Form

$$\mathbf{C}^* = \begin{bmatrix} b_1 & b_1 & \cdots & b_{n-2} & b_{n-1} \end{bmatrix}. \qquad (7.82)$$

The signal flow chart and the block diagram corresponding to the controllable canonical form are given in Figure 7.2 and Figure 7.4 respectively. It can easily be seen from these figures that the control signal $u(t)$ has some effect to each of the system states $\dot{x}_i$ and hence, the meaning of controllability.

In MATLAB, there is no direct command for transformation of a system to controllable canonical form as discussed here. However, a number of tools are available in MATLAB and their combination can lead to system controllable canonical form. This includes the matrix manipulation power of MATLAB as well as other computational tools. The other tools available include the "canon" command. This command in companion form returns the system matrix into companion (observable) canonical form whose transpose is the controllable canonical form. Also the "ctrbf" command returns a decomposition of the system matrices into the controllable and uncontrollable subspaces. it is highly recommended to consult the MATLAB users manual on the use of the "canon" and the "ctrbf" commands for the system transformation to controllable canonical form.

Example 7.10 *The coefficient matrices of the state equations in the dy-*

namic system given by Equation 7.4 are

$$\mathbf{A} = \begin{bmatrix} 1\ 2\ 2 \\ 1\ 4\ 3 \\ 3\ 1\ 3 \end{bmatrix} \qquad \mathbf{B} = \begin{bmatrix} 1 \\ 0 \\ 1 \end{bmatrix} \qquad \mathbf{C} = \begin{bmatrix} 2\ 1\ 1 \end{bmatrix}.$$

Transform the system to state controllable canonical form.

Solution 7.10 *The characteristic equation of* $\mathbf{A}$ *is*

$$|sI - A| = \begin{vmatrix} s-1 & -2 & -2 \\ -1 & s-4 & -3 \\ -3 & -1 & s-3 \end{vmatrix}$$

$$= s^3 - 8s^2 + 8s + 1.$$

Thus the coefficients of the characteristic equation are $a_1 = 1$, $a_2 = 8$, *and* $a_3 = -8$. *The matrix of the coefficients is*

$$\mathbf{Q} = \begin{bmatrix} 8 & -8 & 1 \\ -8 & 1 & 0 \\ 1 & 0 & 0 \end{bmatrix}.$$

The controllability matrix of the system is

$$\mathcal{C} = \begin{bmatrix} B \vdots AB \vdots A^2B \end{bmatrix}$$

$$= \begin{bmatrix} 1\ 3\ 23 \\ 0\ 4\ 37 \\ 1\ 6\ 31 \end{bmatrix}.$$

Therefore, from Equation 7.79 the transformation matrix $\mathbf{P}$ *becomes*

$$\mathbf{P} = \mathcal{C}\mathbf{Q}$$

$$= \begin{bmatrix} 1\ 3\ 23 \\ 0\ 4\ 37 \\ 1\ 6\ 31 \end{bmatrix} \begin{bmatrix} 8 & -8 & 1 \\ -8 & 1 & 0 \\ 1 & 0 & 0 \end{bmatrix}$$

$$= \begin{bmatrix} 7.0 & -5.0 & 1.0 \\ 5.0 & 4.0 & 0 \\ -9.0 & -2.0 & 1.0 \end{bmatrix}.$$

Thus, from Equations 7.72 through to 7.75 the transformed matrices $\mathbf{A}^*$, $\mathbf{B}^*$ *and* $\mathbf{C}^*$ *can be obtained for the controllable canonical form. This is left as an exercise for the reader.*

7.5.3   Observable Canonical Form

This is a form of system presentation in which the system is completely observable as defined in the preceding sections. The similarity transformation matrix that brings the system in observable canonical form is given as the inverse of the product of the characteristic coefficients matrix $\mathbf{Q}$ and the observability matrix $\mathcal{O}$, both of which have been defined in the preceding sections with

$$\mathcal{O} = \begin{bmatrix} \mathbf{C} \\ \mathbf{CA} \\ \mathbf{CA}^2 \\ \mathbf{CA}^3 \\ \vdots \\ \mathbf{CA}^{(n-2)} \\ \mathbf{CA}^{(n-1)} \end{bmatrix},$$

and

$$\mathbf{Q} = \begin{bmatrix} a_2 & a_3 & a_4 & \cdots & a_{n-1} & a_n & 1 \\ a_3 & a_4 & \cdots & a_{n-1} & a_n & 1 & 0 \\ a_4 & \vdots & a_{n-1} & a_n & 1 & 0 & 0 \\ \vdots & a_{n-1} & a_n & 1 & 0 & 0 & \vdots \\ a_{n-1} & a_n & 1 & 0 & 0 & \vdots & 0 \\ a_n & 1 & 0 & 0 & \cdots & 0 & 0 \\ 1 & 0 & 0 & \cdots & 0 & 0 & 0 \end{bmatrix}.$$

The observable canonical form transformation matrix is then given as

$$\mathbf{P} = (\mathbf{Q}\mathcal{O})^{-1}. \tag{7.83}$$

Hence, the transformed system can be deduced using the previous transformation formulae. Of interest can be the plant matrix $\mathbf{A}^* = \mathbf{P}^{-1}\mathbf{AP}$, which takes the following general form

$$\mathbf{A}^* = \mathbf{P}^{-1}\mathbf{A}\mathbf{P} = \begin{bmatrix} 0 & 0 & 0 & 0 & \vdots & 0 & -a_1 \\ 1 & 0 & 0 & 0 & \vdots & 0 & -a_2 \\ 0 & 1 & 0 & 0 & \vdots & 0 & -a_3 \\ \cdots & \cdots & \cdots & \cdots & \ddots & \cdots & \cdots \\ 0 & 0 & 1 & 0 & \vdots & 0 & -a_{n-2} \\ 0 & 0 & 0 & 1 & \vdots & 0 & -a_{n-1} \\ 0 & 0 & 0 & 0 & \vdots & 1 & -a_n \end{bmatrix}. \tag{7.84}$$

The input matrix becomes

$$\mathbf{B}^* = \mathbf{P}^{-1}\mathbf{B} = \begin{bmatrix} b_1 \\ b_2 \\ b_3 \\ \vdots \\ b_n \end{bmatrix}, \tag{7.85}$$

and the output matrix becomes

$$C = \begin{bmatrix} 0 & 0 & \cdots & 1 \end{bmatrix}. \tag{7.86}$$

The corresponding signal flow chart and block diagram for observable canonical form are given in Figure 7.3 and Figure 7.5 respectively. Note that in these diagrams, the output $y(t)$ is connected to each of the system states, which makes it possible to estimate the system states from the output measurements

Example 7.11 *Transform the system*

$$\underline{\dot{x}} = \mathbf{A}\mathbf{x} + \mathbf{B}u$$
$$y = \mathbf{C}\mathbf{x}$$

into observable canonical form, given that the system matrices are

$$\mathbf{A} = \begin{bmatrix} 1 & 2 & 2 \\ 1 & 4 & 3 \\ 3 & 1 & 3 \end{bmatrix} \qquad \mathbf{B} = \begin{bmatrix} 1 \\ 0 \\ 1 \end{bmatrix} \qquad \mathbf{C} = \begin{bmatrix} 2 & 1 & 1 \end{bmatrix}.$$

Solution 7.11 *The observability matrix for this system is*

$$\mathcal{O} = \begin{bmatrix} 2 & 1 & 1 \\ 6 & 9 & 10 \\ 45 & 58 & 69 \end{bmatrix}.$$

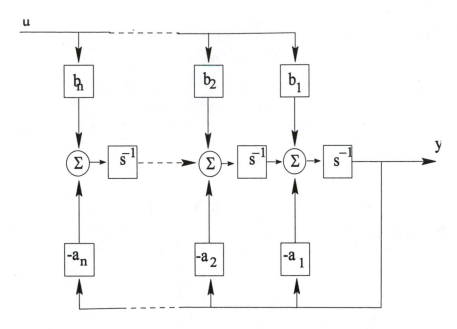

FIGURE 7.5
System in Observable Canonical Form

From the previous example, for this system

$$Q = \begin{bmatrix} 8 & -8 & 1 \\ -8 & 1 & 0 \\ 1 & 0 & 0 \end{bmatrix}.$$

Thus the observable transformation matrix $\mathbf{P}$ *is*

$$\mathbf{P} = (\mathbf{Q}\mathcal{O})^{-1}$$

$$= \begin{bmatrix} \dfrac{1}{61} & -\dfrac{3}{61} & \dfrac{9}{61} \\[2mm] -\dfrac{14}{61} & -\dfrac{19}{61} & -\dfrac{4}{61} \\[2mm] \dfrac{12}{61} & \dfrac{25}{61} & \dfrac{47}{61} \end{bmatrix}.$$

Thus, the observable canonical form becomes

$$\mathbf{A}^* = \mathbf{P}^{-1}\mathbf{A}\mathbf{P}$$

$$= \begin{bmatrix} 0 & 0 & -1 \\ 1 & 0 & -8 \\ 0 & 1 & 8 \end{bmatrix}$$

$$\mathbf{B}^* = \begin{bmatrix} 10 \\ -8 \\ 3 \end{bmatrix}$$

$$\mathbf{C}^* = \begin{bmatrix} 0 & 0 & 1 \end{bmatrix}.$$

Example 7.12 *In this example, the objective is to demonstrate the effect of trying to transform a system to observable canonical form. Consider the system*

$$\mathbf{A} = \begin{bmatrix} 3 & 4 \\ 1 & 2 \end{bmatrix} \qquad \mathbf{C} = \begin{bmatrix} 1 \\ 2 \end{bmatrix},$$

which has been identified before as unobservable.

Solution 7.12 *The characteristic equation of A is*

$$\psi(s) = s^2 - 6s + 5$$

so that

$$\mathbf{Q} = \begin{bmatrix} -6 & 1 \\ 1 & 0 \end{bmatrix}$$

The observability has been determined before and is given by

$$\mathcal{O} = \begin{bmatrix} 1 & 2 \\ 5 & 10 \end{bmatrix},$$

therefore, the observable transformation matrix becomes

$$\mathbf{P} = (\mathbf{Q}\mathcal{O})^{-1}$$

$$= \left(\begin{bmatrix} -6 & 1 \\ 1 & 0 \end{bmatrix} \begin{bmatrix} 1 & 2 \\ 5 & 10 \end{bmatrix} \right)^{-1}$$

$$= \begin{bmatrix} -1 & -2 \\ 1 & 2 \end{bmatrix}^{-1}.$$

From here it is clear that the matrix $\mathbf{Q}\mathcal{O}$ is singular, therefore, it is not invertible, in which case no transformation matrix $\mathbf{P}$ for observable canonical form can be formed. This is the reason that the system is said to be unobservable.

7.5.4   Diagonal Canonical Form

Diagonal canonical form, also known as modal canonical form, plays an important role in the analysis of MIMO systems, where decoupling of the states becomes very crucial. Under diagonal canonical forms, the transformed plant matrix is a diagonal one, with the diagonal elements corresponding to the eigenvalues of the plant matrix. It can be seen that if the plant matrix has n distinct eigenvalues λ_i ($i = 1, 2 \ .. \ n$) so that each eigenvalue has a distinct eigenvector $\underline{v}_i$, then the transformation matrix $\mathbf{P}$ is the augmented matrix of the distinct eigenvectors, also known as the modal matrix, for that particular system.

$$\mathbf{P} = [\ \underline{v}_1 \ \vdots \underline{v}_2 \ \ \vdots \underline{v}_3 \ \ \vdots \cdots \vdots \underline{v}_n \ \ ]. \tag{7.87}$$

If, however, the system is already in the controllable canonical form with known eigenvalues λ_i ($i = 1, 2 \ .. \ n$), the transformation matrix $\mathbf{P}$ can be derived to be the Vandermonde matrix, which is formed by powers of the eigenvalues.

$$\mathbf{P} = \begin{bmatrix} 1 & 1 & 1 & \cdots & 1 \\ \lambda_1 & \lambda_2 & \lambda_3 & \cdots & \lambda_n \\ \lambda_1^2 & \lambda_2^2 & \lambda_3^2 & \cdots & \lambda_n^2 \\ \lambda_1^3 & \lambda_2^3 & \lambda_3^3 & \cdots & \lambda_n^3 \\ \cdots & \cdots & \cdots & \cdots & \cdots \\ \lambda_1^{n-1} & \lambda_2^{n-1} & \lambda_3^{n-1} & \cdots & \lambda_n^{n-1} \end{bmatrix} \tag{7.88}$$

so that the transformed system takes on matrices as given in equations.

Example 7.13 *The following system*

$$\dot{\mathbf{x}}(t) = \mathbf{A}\mathbf{x}(t) + \mathbf{B}\mathbf{u}(t)$$

the plant matrix $\mathbf{A}$ is given as

$$\mathbf{A} = \begin{bmatrix} 0 & 1 & 0 \\ 0 & 0 & 1 \\ -6 & -11 & -6 \end{bmatrix}.$$

Find the transformation matrix $\mathbf{P}$, which will transform it to diagonal canonical form. If the input matrix is

$$\mathbf{B} = \begin{bmatrix} 0 \\ 0 \\ 1 \end{bmatrix},$$

find the corresponding input matrix in diagonal canonical form.

Solution 7.13 *The given system matrix is already in controllable canonical form, therefore, to transform it to diagonal canonical form, the transformation matrix must be the Vandermonde matrix, which requires prior knowledge of the system eigenvalues. For the system matrix A, the eigenvalues are computed from its characteristic equation*

$$|s\mathbf{I} - \mathbf{A}| = \begin{vmatrix} s & -1 & 0 \\ 0 & s & -1 \\ 6 & 11 & s+6 \end{vmatrix}$$

$$= s^3 + 6s^2 + 11s + 6$$

$$= (s+3)(s+2)(s+1).$$

Therefore, the eigenvalues are $\lambda_1 = -1, \lambda_2 = -2, \lambda_3 = -3$, so, according to equation 7.88, the Vandermonde matrix is

$$\mathbf{P} = \begin{bmatrix} 1 & 1 & 1 \\ -1 & -2 & -3 \\ 1 & 4 & 9 \end{bmatrix}.$$

This transforms the system to diagonal canonical form according to Equations 7.72 through to 7.75 as

$$\mathbf{A}^* = \mathbf{P}^{-1}\mathbf{A}\mathbf{P}$$

$$= \begin{bmatrix} 1 & 1 & 1 \\ -1 & -2 & -3 \\ 1 & 4 & 9 \end{bmatrix}^{-1} \begin{bmatrix} 0 & 1 & 0 \\ 0 & 0 & 1 \\ -6 & -11 & -6 \end{bmatrix} \begin{bmatrix} 1 & 1 & 1 \\ -1 & -2 & -3 \\ 1 & 4 & 9 \end{bmatrix}$$

$$= \begin{bmatrix} -1 & 0 & 0 \\ 0 & -2 & 0 \\ 0 & 0 & -3 \end{bmatrix}.$$

Also the input matrix **B** *transforms to*

$$\mathbf{B}^* = \mathbf{P}^{-1}\mathbf{B}$$

$$= \begin{bmatrix} 1 & 1 & 1 \\ -1 & -2 & -3 \\ 1 & 4 & 9 \end{bmatrix}^{-1} \begin{bmatrix} 0 \\ 0 \\ 1 \end{bmatrix}$$

$$= \begin{bmatrix} \frac{1}{2} \\ -1 \\ \frac{1}{2} \end{bmatrix}.$$

7.5.5   Jordan Canonical Form

This is an alternative to the diagonal canonical form for unsymmetrical systems with multiple order eigenvalues for which diagonal transformation is not possible. The Jordan canonical form (JCF) is a block diagonal matrix in which each $n \times n$ diagonal block matrix corresponds to n-multiplicity eigenvalues. The main diagonal elements of the JCF are the eigenvalues (with their multiplicities) and for the repeated eigenvalues the entry above the main diagonal is a 1 and the rest are 0s. A typical structure of the Jordan canonical form is

$$
\begin{bmatrix}
\lambda_1 & 1 & . & \vdots & . & \vdots & . & . & \vdots & . & . & 0 \\
0 & \lambda_1 & 1 & \vdots & 0 & \vdots & & . & \vdots & . & . & . \\
0 & 0 & \lambda_1 & \vdots & & \vdots & 0 & & \vdots & . & . & . \\
\cdots & \cdots & \cdots & \vdots & \cdots & \vdots & \cdots & \cdots & \vdots & \cdots & \cdots & \cdots \\
. & 0 & . & \vdots & \lambda_2 & \vdots & & & \vdots & . & . & . \\
\cdots & \cdots & \cdots & \vdots & \cdots & \vdots & \cdots & \cdots & \vdots & \cdots & \cdots & \cdots \\
. & . & & \vdots & & \vdots & \lambda_3 & 1 & \vdots & . & 0 & . \\
. & . & . & \vdots & 0 & \vdots & 0 & \lambda_3 & \vdots & . & & . \\
\cdots & \cdots & \cdots & \vdots & \cdots & \vdots & \cdots & \cdots & \vdots & \cdots & \cdots & \cdots \\
. & . & . & \vdots & . & \vdots & & . & \vdots & . & . & . \\
. & . & . & \vdots & . & \vdots & 0 & \vdots & . & \lambda_{n-1} & 0 \\
0 & . & . & \vdots & . & \vdots & & . & \vdots & . & 0 & \lambda_n
\end{bmatrix} . \qquad (7.89)
$$

The transformation to Jordan canonical form is done by a matrix that is formed by augmentation of the system of generalized eigenvectors: $\mathbf{p}_1$, $\mathbf{p}_2$, $\mathbf{p}_3$...... $\mathbf{p}_n$ so that

$$
T = \begin{bmatrix} \mathbf{p}_1 & \mathbf{p}_2 & \mathbf{p}_3 & \cdots & \mathbf{p}_n \end{bmatrix} . \qquad (7.90)
$$

Since the idea of generalized eigenvectors can be confusing, it is worth recalling some ideas from linear algebra here. It is known that if $\mathbf{A}$ is n x n matrix with n eigenvalues λ_i, $i = 1, 2, 3 \dots n$, its non-zero eigenvectors $\mathbf{p}_i$ must satisfy the equation

$$(\lambda_i \mathbf{I} - \mathbf{A})\mathbf{p}_i = 0. \tag{7.91}$$

This results in n-eigenvectors, each corresponding to one eigenvalue. If, however, $\mathbf{A}$ has multiple order eigenvalues and is non-symmetric, not all the eigenvectors will be found by Equation 7.91 above. This equation will result in eigenvectors that are less in numbers than n, which is required for the matrix of order n like $\mathbf{A}$. The remaining eigenvectors corresponding to each λ_j of m-order multiplicity are found by using the $m - 1$ generalized eigenvector equations

$$(\lambda_j \mathbf{I} - \mathbf{A})\mathbf{p}_{j+q} = \mathbf{p}_{j+q-1} \qquad q = 1, \ 2, \ \dots \ (m-1). \tag{7.92}$$

For $q = 1$, the normal Equation 7.91 holds.

7.6 Transfer Function Decomposition

When a transfer function is given in the classical rational form, one may wish to decompose it to one of the state-space canonical forms without having to go back to the time domain. Three decomposition approaches are possible, all of which require the preparation of the signal flow charts depending on how the transfer function is given. These are direct decomposition, cascade decomposition, and parallel decomposition. This section will present the principles of these decomposition methods.

7.6.1 Direct Decomposition

Direct decomposition takes advantage of the fact that the coefficients of the numerator and denominator of the transfer function can by inspection be used to form the observable or controllable canonical form. The method constructs the signal flow chart from the transfer function from which the appropriate canonical form can be extracted. The method for constructing the signal flow chart from the transfer function for the controllable canonical form differs from that of the observable canonical form. The two methods are discussed here.

7.6.1.1 Direct Decomposition to Controllable Canonical Form

For a transfer function

$$G(s) = \frac{Y(s)}{U(s)} = \frac{N(s)}{D(s)}$$

this goes through four stages as:

(1) Express $G(s)$ in negative powers of s by multiplying numerator and denominator by s^{-n}

$$\frac{Y(s)}{U(s)} = \frac{N(s)s^{-n}}{D(s)s^{-n}}$$

(2) Multiply $N(s)$ and $D(s)$ by a dummy variable $x(s)$ representing the state variables.

$$\frac{Y(s)}{U(s)} = \frac{N(s)s^{-n}x(s)}{D(s)s^{-n}x(s)}$$

(3) Equate the numerator and denominator of both sides thereby forming two equations.

$$Y(s)D(s)s^{-n}x(s) = U(s)N(s)s^{-n}x(s)$$

(4) Construct the signal flow chart using the two equations by forming states using

$$x_i = s^{-i}x(s) \text{ where } i = 1, 2, .., n$$

7.6.1.2 Direct Decomposition to Observable Canonical Form

This goes through three stages as:

(1) Express G(s) in negative powers of s.

$$\frac{Y(s)}{U(s)} = \frac{N(s)s^{-n}}{D(s)s^{-n}}$$

(2) Cross-multiply $N(s)$ and $D(s)$ by $u(s)$ and $y(s)$ respectively, and express $y(s)$ in terms of ascending negative powers of s, $y(s)$ itself and $u(s)$.

(3) Draw the corresponding signal flow chart.

Cascade decomposition is a special form of direct decomposition that applies to a transfer function presented in the pole-zero form. A separate signal flow chart is made for each factor of the p-z transfer function and then connected in cascade to complete the signal flow chart for the whole system. Parallel decomposition applies when the transfer function is expressed as a sum of its partial fractions. As such, it can be applied to any system for which the denominator of the transfer function has been factored. A separate signal flow graph is made for each of the partial fractions. The overall signal flow graph for the system is a parallel combination of the individual signal flow graphs for the partial fractions.

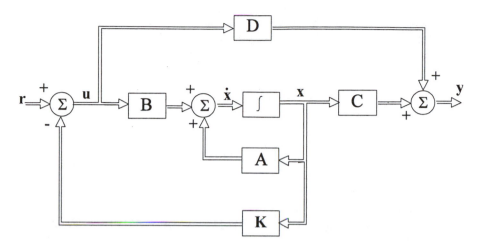

FIGURE 7.6
Block Diagram for The State Feedback Control

7.7   Full State Feedback Control

State feedback control has been widely applied in most control systems
not only because of its simplicity, but also its flexibility in handling system dynamics for disturbance rejection and stability characteristics. Under
state feedback control, the control action is achieved by feeding back a linear combination of the system states to the reference input $r(t)$ through a
feedback matrix $\mathbf{K}$, as shown in Fig 7.6 to produce the control input $u(t)$.
In this way, the control signal $u(t)$ becomes

$$u(t) = -\mathbf{K}\mathbf{x}(t) + r(t). \tag{7.93}$$

For the case of regulators where $r(t) = 0$, the control signal will be

$$u(t) = -\mathbf{K}\mathbf{x}(t). \tag{7.94}$$

Full state feedback control scheme assumes that all the system states $x_i(t)$
in the state vector $\underline{\mathbf{x}}(t)$ are available for feedback. However, in real practice,
it may not be possible to have all the system states because of either the
large quantity or the cost of sensors that would be needed to measure all
the system states. Some of these states might require costly and specialized
sensors. Even when all states are available for measurement and there is
still' the problem of measurement noise. Furthermore, in some cases, it is

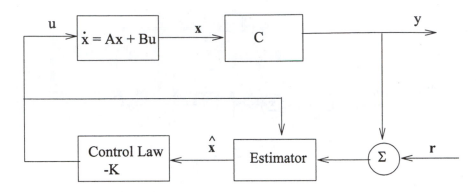

FIGURE 7.7
Combined Estimator and Control Law in Full State Feedback Control

desirable to do state transformations to enable design insights, but the new states may have no physical meaning and hence they cannot be measured. All these issues lead to the idea of state estimation in which the whole state vector can be estimated using data which is obtained from measurements of only a few states of the state vector or more usually from the output. As such, the control design is normally carried out in two stages.

(a) Control law design, which assumes that all states of the system are available for feedback

(b) Observer (or estimator) design, which estimates the entire state vector from measurements of the portion of the state vector from

$$y(t) = \mathbf{C}\mathbf{x}(t) + D(u).$$

Together, the estimator and the control provide the full state feedback control and the structure for such control is shown in Figure 7.7.

This section addresses the first stage of the full state feedback control design with the assumption that the entire state vector is available for feedback. Estimator design will be discussed in later sections. With the control input $u(t)$ given by Equation 7.93, the state equation of the system in the closed-loop becomes

$$\dot{\mathbf{x}}(t) = \mathbf{A}\mathbf{x}(t) + \mathbf{B}[-\mathbf{K}\mathbf{x}(t) + r(t)]$$
$$= (\mathbf{A} - \mathbf{B}\mathbf{K})\mathbf{x}(t) + \mathbf{B}r(t). \tag{7.95}$$

If the system is completely controllable, then the feedback matrix $\mathbf{K}$ exists for which the characteristic equation of the closed-loop system becomes

$$|s\mathbf{I} - \mathbf{A} + \mathbf{BK}| = 0. \tag{7.96}$$

(recall Equation 7.10) The task of the control design becomes that of determining the elements of matrix $\mathbf{K}$. The most popular design technique for determination of $\mathbf{K}$ for SISO systems is the pole placement method, which is discussed in the next section. It should be noted that for SISO systems $\mathbf{B}$ is a column vector while $\mathbf{K}$ is a row vector, therefore, the product $\mathbf{BK}$ is an outer product of vectors $\mathbf{B}$ and $\mathbf{K}$, which result in a matrix so that matrix algebra rules are still observed in Equations 7.95 and 7.96.

7.7.1 Pole Placement Design Method

Pole placement is a method that seeks to place the poles of the closed-loop system at some predetermined locations. Although this method has some drawbacks in handling complex systems, it is still fairly sufficient for most small control systems and it gives the best introduction to the design of complex systems. The basic concept behind the method is to get $\mathbf{K}$, which will satisfy the closed-loop transfer function in Equation 7.96 at desired pole locations s_i, $i = 1, 2, \cdots n$. Implementation of the method will be described here, through the following illustrative example in which a regulator is assumed, i.e., no reference input. (The reference input will be added after some discussion on the state estimators).

Suppose the system

$$\underline{\dot{\mathbf{x}}}(t) = \mathbf{A}\mathbf{x}(t) + \mathbf{B}u(t)$$

is to be controlled by full state feedback such that

$$u(t) = -\mathbf{K}\mathbf{x}(t),$$

where the closed-loop poles are placed at locations $p_1, p_2, p_3, \ldots p_n$. This means that the required closed-loop transfer function of the controlled system is given by

$$\psi(s) = (s - p_1)(s - p_2)(s - p_3) \cdots (s - p_n) = 0, \tag{7.97}$$

which can be expanded as

$$\psi(s) = s^n + q_n s^{n-1} + q_{n-1} s^{n-2} \cdots + q_3 s^2 + q_2 s + q_1 = 0. \tag{7.98}$$

Let the system matrix A and the input matrix B, respectively, be

$$\mathbf{A} = \begin{bmatrix} a_{11} & a_{12} & \cdots & a_{1n} \\ a_{21} & a_{22} & \cdots & a_{2n} \\ a_{31} & a_{31} & \cdots & a_{3n} \\ \vdots & \vdots & \cdots & \vdots \\ a_{n1} & a_{n2} & \cdots & a_{nn} \end{bmatrix}$$

$$\mathbf{B} = \begin{bmatrix} b_1 \\ b_2 \\ b_3 \\ \vdots \\ b_n \end{bmatrix}.$$

Therefore, if the feedback matrix K is

$$\mathbf{K} = \begin{bmatrix} k_1 & k_2 & k_3 & \cdots & k_n \end{bmatrix},$$

then the closed-loop system has the following system matrix

$$\mathbf{A} - \mathbf{BK} = \begin{bmatrix} a_{11} - b_1 k_1 & a_{12} - b_1 k_2 & \cdots & a_{1n} - b_1 k_n \\ a_{21} - b_2 k_1 & a_{22} - b_2 k_2 & \cdots & a_{2n} - b_2 k_n \\ a_{21} - b_2 k_1 & a_{32} - b_3 k_2 & \cdots & a_{3n} - b_3 k_n \\ \cdots & \cdots & \cdots & \cdots \\ a_{n1} - b_n k_1 & a_{n2} - b_n k_2 & \cdots & a_{nn} - b_n k_n \end{bmatrix}$$

whose characteristic function is

$$\psi(s) = \mid s\mathbf{I} - \mathbf{A} + \mathbf{BK} \mid$$

$$= \begin{vmatrix} s - a_{11} + b_1 k_1 & -a_{12} + b_1 k_2 & \cdots & -a_{1n} + b_1 k_n \\ -a_{21} + b_2 k_1 & s - a_{22} + b_2 k_2 & \cdots & -a_{2n} + b_2 k_n \\ -a_{21} + b_2 k_1 & -a_{32} + b_3 k_2 & \cdots & -a_{3n} + b_3 k_n \\ \cdots & \cdots & \cdots & \cdots \\ -a_{n1} + b_n k_1 & -a_{n2} + b_n k_2 & \cdots & s - a_{nn} + b_n k_n \end{vmatrix} \qquad (7.99)$$

$$= 0.$$

Comparison of this characteristic equation and the demanded one in 7.98 can lead to the determination of the values of k_i and hence matrix $\mathbf{K}$. However, as it can be seen, the algebra behind such a problem is very cumbersome and might in some cases be insoluble. On the other hand, however, if system $(\mathbf{A}, \mathbf{B})$ is controllable, the closed-loop system can be expressed in its controllable canonical form as

$$\mathbf{A}^* - \mathbf{B}^* \mathbf{K}^* = \begin{bmatrix} a_{11}^* - b_1^* k_1^* & a_{12}^* - b_1^* k_2^* & \cdots & a_{1n}^* - b_1^* k_n^* \\ a_{21}^* - b_2^* k_1^* & a_{22}^* - b_2^* k_2^* & \cdots & a_{2n}^* - b_2^* k_n^* \\ a_{21}^* - b_2^* k_1^* & a_{32}^* - b_3^* k_2^* & \cdots & a_{3n}^* - b_3^* k_n^* \\ \cdots & \cdots & \cdots & \cdots \\ a_{n1}^* - b_n^* k_1^* & a_{n2}^* - b_n^* k_2^* & \cdots & a_{nn}^* - b_n^* k_n^* \end{bmatrix}.$$

Matrices $\mathbf{A}^*$ and $\mathbf{B}^*$ are calculated using Equations 7.72 and 7.73. For SISO systems, they have been found to be as given in Equations 7.80 and 7.81. In this case, the closed-loop transfer function becomes

$$| s\mathbf{I} - \mathbf{A}^* + \mathbf{B}^* \mathbf{K}^* | = \begin{vmatrix} s & -1 & \cdots & 0 \\ 0 & s & \cdots & 0 \\ 0 & 0 & \cdots & 0 \\ \cdots & \cdots & \cdots & \cdots \\ -a_1 - k_1^* & -a_2 - k_2^* & \cdots & s - a_n - k_n^* \end{vmatrix}$$

whose expansion can easily be determined to be

$$\psi(s) = s^n + (a_n + k_n^*)s^{n-1} + (a_{n-1} + k_{n-1}^*)s^{n-2} + \cdots$$
$$\cdots + (a_3 + k_3^*)s^2 + (a_2 + k_2^*)s + (a_1 + k_1^*). \tag{7.100}$$

Comparison of this equation with the demanded one in 7.98 shows that

$$a_i + k_i^* = q_i \qquad i = 1, 2, 3 \cdots n,$$

from which the elements of the feedback matrix can be computed as

$$k_i^* = q_i - a_i \qquad i = 1, 2, 3 \cdots n$$

or in vector form

$$\mathbf{K}^* = \mathbf{q} - \mathbf{a} \tag{7.101}$$

where

$$\mathbf{q} = \begin{bmatrix} q_1 & q_2 & \cdots & q_n \end{bmatrix} \tag{7.102}$$
$$\mathbf{a} = \begin{bmatrix} a_1 & a_2 & \cdots & a_n \end{bmatrix}. \tag{7.103}$$

It is emphasized again that this procedure applies only for SISO systems in controllable canonical form, and that the order of the elements in vectors **a, q** and **K*** are as shown above. Improper order of the elements will give wrong results. This matrix **K*** is the feedback gain for the system in controllable canonical form, i.e., the control effort is such that

$$u(t) = -\mathbf{K}^*\mathbf{x}^*(t), \tag{7.104}$$

where from Equation 7.67

$$\mathbf{x}^*(t) = \mathbf{P}^{-1}\mathbf{x}(t).$$

Therefore, for the original system (not in control canonical form) this control effort becomes

$$u(t) = -\mathbf{K}^*\mathbf{P}^{-1}\mathbf{x}(t) \tag{7.105}$$

so that the corresponding feedback gain matrix **K** is

$$\mathbf{K} = \mathbf{K}^*\mathbf{P}^{-1}. \tag{7.106}$$

In MATLAB, pole placement design is accomplished by using command "place(A,B,P)", which computes the state feedback matrix **K** such that the eigenvalues of $\mathbf{A} - \mathbf{BK}$ are those specified in vector **P**. Whenever used, the matrix algebra rules must be observed as to the dimensions of **A, B** and **P** so that equation $\mathbf{A} - \mathbf{BK}$ remains valid. This means that **P** and **B** must be an n-dimensional vector where **A** is an $n \times n$ square matrix. If complex eigenvalues are required, they must appear in consecutive complex conjugate pairs in **P**. The "place" syntax is:

$$\mathsf{K} = \mathsf{place}(\mathsf{A}, \mathsf{B}, \mathsf{P}).$$

Example 7.14 *For system matrices given as*

$$A = \begin{bmatrix} 2 & 1 & 1 \\ 2 & 3 & 4 \\ -1 & -1 & -2 \end{bmatrix} \qquad B = \begin{bmatrix} 1 \\ 2 \\ 1 \end{bmatrix},$$

design a state feedback controller K to place the poles of the system at $s = -2, -5, -6$ *. Confirm the results using MATLAB.*

Solution 7.14 *The given system is not in controllable canonical form. Transforming it into a controllable canonical form using methods discussed previously the transformation matrix* **P** *is needed. With this transformation matrix, the whole exercise of transforming the system can be skipped,* *as will be seen in this example. First, get the characteristic equation of the uncontrolled system.*

$$|s\mathbf{I} - \mathbf{A}| = 0$$
$$\implies s^3 - 3s^2 - s + 3 = 0,$$

*from which the vector **a** of the coefficients is found to be*

$$\mathbf{a} = \begin{bmatrix} 3 & -1 & -3 \end{bmatrix}.$$

Now the controlled system needs to have the poles at -2, -5 and -6; thus the characteristic equation of the closed-loop system becomes

$$(s+2)(s+5)(s+6) = 0,$$

which expands to

$$s^3 + 13s^2 + 52s + 60 = 0.$$

*Therefore, the vector **q** of the coefficients of the controlled system becomes*

$$\mathbf{q} = \begin{bmatrix} 60 & 52 & 13 \end{bmatrix}$$

*which, according to Equation 7.101, gives the gain matrix **K** for the system in canonical form as*

$$\mathbf{K}^* = \mathbf{q} - \mathbf{a}.$$

From these results it follows that

$$\mathbf{K}^{*T} = \begin{bmatrix} 60 \\ 52 \\ 13 \end{bmatrix} - \begin{bmatrix} 3 \\ -1 \\ -3 \end{bmatrix}$$

$$\mathbf{K}^* = \begin{bmatrix} 57 & 53 & 16 \end{bmatrix}.$$

*This corresponds to the canonical gain matrix. In order to convert it into the gain matrix **K** of the original state-space system, the transformation matrix **P** is needed where*

$$\mathbf{P} = \mathcal{C}\mathbf{Q}$$

so that K can be calculated by using Equation 7.106. Now, from the characteristic equation of the uncontrolled system, the coefficients matrix is

$$\mathbf{Q} = \begin{bmatrix} -1 & -3 & 1 \\ -3 & 1 & 0 \\ 1 & 0 & 0 \end{bmatrix}$$

and the controllability matrix of the system is

$$\mathcal{C} = [\, \mathbf{B} \mid \mathbf{AB} \mid \mathbf{A}^2\mathbf{B} \,]$$

$$= \begin{bmatrix} 1 & 5 & 17 \\ 2 & 12 & 26 \\ 1 & -5 & -7 \end{bmatrix}.$$

Therefore, the transformation matrix P is

$$\mathbf{P} = \begin{bmatrix} 1 & 5 & 17 \\ 2 & 12 & 26 \\ 1 & -5 & -7 \end{bmatrix} \begin{bmatrix} -1 & -3 & 1 \\ -3 & 1 & 0 \\ 1 & 0 & 0 \end{bmatrix}$$

$$= \begin{bmatrix} 1 & 2 & 1 \\ -12 & 6 & 2 \\ 7 & -8 & 1 \end{bmatrix},$$

and the inverse is

$$\mathbf{P}^{-1} = \begin{bmatrix} \dfrac{11}{64} & -\dfrac{5}{64} & -\dfrac{1}{64} \\[2ex] \dfrac{13}{64} & -\dfrac{3}{64} & -\dfrac{7}{64} \\[2ex] \dfrac{27}{64} & \dfrac{11}{64} & \dfrac{15}{64} \end{bmatrix}.$$

Therefore, the required feedback gain is

$$\mathbf{K} = \begin{bmatrix} 57 & 53 & 16 \end{bmatrix} \begin{bmatrix} \dfrac{11}{64} & -\dfrac{5}{64} & -\dfrac{1}{64} \\[2ex] \dfrac{13}{64} & -\dfrac{3}{64} & -\dfrac{7}{64} \\[2ex] \dfrac{27}{64} & \dfrac{11}{64} & \dfrac{15}{64} \end{bmatrix},$$

which gives

$$\mathbf{K} = \begin{bmatrix} 27.313 & -4.1875 & -2.9375 \end{bmatrix}.$$

Note that the system matrix $\boldsymbol{A}$ *in control canonical form becomes*

$$\mathbf{A}^* = \mathbf{P}^{-1}\mathbf{AP}$$

$$= \begin{bmatrix} 0 & 1 & 0 \\ 0 & 0 & 1 \\ -3 & 1 & 3 \end{bmatrix}.$$

In MATLAB this problem could be solved as follows:
```
A = [1    2    1; 2    3    4; -1    -1    -2];
B = [1; 2; 1];
P = [-2    -5    -6];
K = place(A, B, P).
```

7.7.2   Pole Placement Using Ackermann's Formula

The pole placement method described in the previous section is cumbersome in that it requires the system to be converted to controllable canonical form and the Gain matrix $\mathbf{K}^*$ is computed element by element. Ackermann's formula offers an elegant way of determining the feedback matrix $\mathbf{K}$ for pole placement using information from the required closed-loop characteristic equation without converting the system to controllable canonical form. It gives a direct relationship between the closed-loop characteristic equation

$$\psi(s) = s^n + q_n s^{n-1} + q_{n-1} s^{n-2} + q_{n-2} s^{n-3} + \cdots + q_3 s^2 + q_2 s + q_1 = 0, \tag{7.107}$$

and the system matrix A (not in controllable canonical form) so that by using the earlier relationship between k_i, a_i and q_i matrix K is obtained directly where

$$\mathbf{K}^* = \mathbf{q} - \mathbf{a}. \tag{7.108}$$

Derivation of Ackermann's formula originates from Cayley-Hamilton's theorem, which states that every square matrix satisfies its characteristic equation. For matrix $\mathbf{A}$, whose characteristic equation is given by

$$\psi(s) = s^n + a_n s^{n-1} + a_{n-1} s^{n-2} + a_{n-2} s^{n-3} + \cdots + a_3 s^2 + a_2 s + a_1 = 0, \tag{7.109}$$

then according to Cayley-Hamilton's theorem

$$\psi(\mathbf{A}) = \mathbf{A}^n + a_n \mathbf{A}^{n-1} + a_{n-1} \mathbf{A}^{n-2} + a_{n-2} \mathbf{A}^{n-3} + \cdots + a_3 \mathbf{A}^2 + a_2 \mathbf{A} + a_1 \mathbf{I}$$
$$= 0.$$

It is assumed that A is not in controllable canonical form, therefore, the first task would be to convert it to controllable canonical form $(\mathbf{A}^*, \mathbf{B}^*)$ by similarity transformations

$$\mathbf{A}^* = \mathbf{P}^{-1}\mathbf{A}\mathbf{P} \tag{7.110}$$

$$\mathbf{B}^* = \mathbf{P}^{-1}\mathbf{B}. \tag{7.111}$$

The state vector will also be transformed to

$$\mathbf{x}^* = \mathbf{P}^{-1}\mathbf{x}. \tag{7.112}$$

Therefore

$$\psi(\mathbf{A}^*) = \mathbf{A}^{*n} + a_n\mathbf{A}^{*n-1} + a_{n-1}\mathbf{A}^{*n-2} + \cdots + a_3\mathbf{A}^{*2} + a_2\mathbf{A}^* + (a_1\mathbf{I} \tag{7.113}$$
$$= 0.$$

Similarly, for the controlled closed-loop system whose characteristic equation is given in Equation 7.107 the theorem will be satisfied in this way

$$\psi(\mathbf{A}^* - \mathbf{B}^*\mathbf{K}) = (\mathbf{A}^* - \mathbf{B}^*\mathbf{K}^*)^n + q_n(\mathbf{A}^* - \mathbf{B}^*\mathbf{K}^*)^{n-1} + \cdots$$
$$+ q_3(\mathbf{A}^* - \mathbf{B}^*\mathbf{K}^*)^2 + q_2(\mathbf{A}^* - \mathbf{B}^*\mathbf{K}^*) + q_1\mathbf{I} \tag{7.114}$$
$$= 0.$$

However, since there is no *a priori* information about matrix $\mathbf{A}^* - \mathbf{B}^*\mathbf{K}^*$, this equation is not immediately useful. Instead, the method gets simplified by letting $\mathbf{B}^*\mathbf{K}^* = \mathbf{0}$, which results in

$$\psi(\mathbf{A}^*) = \mathbf{A}^{*n} + q_n\mathbf{A}^{*n-1} + q_{n-1}\mathbf{A}^{*n-2} + \cdots q_2\mathbf{A}^* + q_1\mathbf{I}. \tag{7.115}$$

This step is done for mathematical convenience; in fact Equation 7.115 is no longer equal to zero. From Equation 7.112 $\mathbf{A}^{*n}$ can be written as

$$\mathbf{A}^{*n} = -(a_n\mathbf{A}^{*n-1} + a_{n-1}\mathbf{A}^{*n-2} + \cdots + a_3\mathbf{A}^{*2} + a_2\mathbf{A}^* + a_1\mathbf{I}), \tag{7.116}$$

which, when used in Equation 7.115 above, gives

$$\psi(\mathbf{A}^*) = (q_n - a_n)\mathbf{A}^{*n-1} + (q_{n-1} - a_{n-1})\mathbf{A}^{*n-2} + \cdots$$
$$\cdots + (q_2 - a_2)\mathbf{A}^* + (q_1 - a_1)\mathbf{I}. \tag{7.117}$$

If $\mathbf{e}_1$ is a vector of the first column of an identity matrix i.e.

$$\mathbf{e_1} = \begin{bmatrix} 1\,0\,0\cdots 0 \end{bmatrix}^T,$$

then because of the structure of A^*

$$\mathbf{e}_1^T \mathbf{A} = \begin{bmatrix} 0\,1\,0\cdots 0 \end{bmatrix}^T = \mathbf{e}_2^T$$
$$\mathbf{e}_1^T \mathbf{A^2} = \mathbf{e}_2^T \mathbf{A} = \begin{bmatrix} 0\,0\,1\cdots 0 \end{bmatrix}^T = \mathbf{e}_3^T$$
$$\mathbf{e}_1^T \mathbf{A^3} = \mathbf{e}_3^T \mathbf{A} = e_4^T$$
$$\cdots$$
$$\mathbf{e}_1^T \mathbf{A^{n-1}} = \mathbf{e}_{n-1}^T \mathbf{A} = \mathbf{e}_n^T. \tag{7.118}$$

Using this result and by multiplying Equation 7.117 by $\mathbf{e}_1^T$ gives

$$\mathbf{e}_1^T \psi(\mathbf{A^*}) = (q_n - a_n)\mathbf{e}_1^T \mathbf{A}^{*n-1} + (q_{n-1} - a_{n-1})\mathbf{e}_1^T \mathbf{A}^{*n-2} + \cdots$$
$$(q_2 - a_2)\mathbf{e}_1^T \mathbf{A^*} + (q_1 - a_1)\mathbf{e}_1^T \mathbf{I}$$
$$= (q_n - a_n)\mathbf{e}_n^T + (q_{n-1} - a_{n-1})\mathbf{e}_{n-1}^T + \cdots$$
$$+ (q_2 - a_2)\mathbf{e}_2^T + (q_1 - a_1)\mathbf{e}_1^T$$
$$= \begin{bmatrix} (q_n - a_n) & (q_{n-1} - a_{n-1})\cdots\cdots & (q_2 - a_2) & (q_1 - a_1) \end{bmatrix}. \tag{7.119}$$

Since from the previous section it was shown that

$$\mathbf{q} - \mathbf{a} = \mathbf{K^*},$$

then

$$\mathbf{e}_1^T \psi(\mathbf{A^*}) = \begin{bmatrix} k_n\ k_{n-1}\ k_{n-2}\ \cdots\ k_1 \end{bmatrix} = \mathbf{K^*} \tag{7.120}$$

where $\mathbf{K^*}$ is defined for the system in controllable canonical form. Hence, the control input $u(t)$ becomes

$$\mathbf{x^*} = -\mathbf{K^* x^*}$$
$$= -\mathbf{K^* P^{-1} x}. \tag{7.121}$$

Therefore, the required system gain based on the configuration is now in canonical form,

$$\mathbf{K} = \mathbf{K}^* \mathbf{P}^{-1}$$
$$= \mathbf{e}_1^T \psi(\mathbf{A}^*) \mathbf{P}^{-1}. \tag{7.122}$$

Since $\mathbf{A}^* = \mathbf{P}^{-1} \mathbf{A} \mathbf{P}$ (a similarity transformation) then $\mathbf{A}^{*n} = \mathbf{P}^{-1} \mathbf{A}^n \mathbf{P}$ so that

$$\psi(\mathbf{A}^*) = \mathbf{P}^{-1} \psi(\mathbf{A}) \mathbf{P}. \tag{7.123}$$

Hence, Equation 7.122 becomes

$$\mathbf{K} = \mathbf{e}_1^T \mathbf{P}^{-1} \psi(\mathbf{A}). \tag{7.124}$$

From the relation that

$$\mathbf{P} = \mathcal{C} \mathbf{Q}$$

it follows that

$$\mathbf{P}^{-1} = (\mathcal{C} \mathbf{Q})^{-1} = Q^{-1} \mathcal{C}^{-1}, \tag{7.125}$$

which, when used in Equation 7.124 above, for $\mathbf{K}$ gives

$$\mathbf{K} = \mathbf{e}_1^T Q^{-1} \mathcal{C}^{-1} \psi(\mathbf{A}). \tag{7.126}$$

It is interesting to note that $\mathbf{e}_1^T Q^{-1} = \mathbf{e}_n^T$ (the reader may verify this), so that Ackermann's formula gets the following form

$$\mathbf{K} = \mathbf{e}_n^T \mathcal{C}^{-1} \psi(\mathbf{A}) \tag{7.127}$$

or by letting $\mathbf{e}_n^T \mathcal{C}^{-1} = \mathbf{b}^T$ this formula can be written as

$$\mathbf{K} = \mathbf{b}^T \psi(\mathbf{A}). \tag{7.128}$$

Though attractive, the application of Ackermann's formula is limited to SISO systems only, where the assumptions for $\mathbf{A}^*$ and $\mathbf{B}^*$ hold. MATLAB implements Ackermann's formula through command "acker(A,B,P)". This command is similar in all respects to the command "place" discussed in the previous section in application. However, it is known to be numerically unreliable particularly for problems of order greater than 10, or weakly controllable systems. As such, for problems of that kind, it should be avoided. Its syntax is

$$K = ac\,ker(A, B, P).$$

Example 7.15 *Consider the system described by*

$$A = \begin{bmatrix} 2 & 1 & 1 \\ 2 & 3 & 4 \\ -1 & -1 & -2 \end{bmatrix} \qquad B = \begin{bmatrix} 1 \\ 2 \\ 1 \end{bmatrix}.$$

Use Ackermann's formula to design a state feedback controller K so that the characteristic equation of the system becomes

$$s^3 + 13s^2 + 52s + 60 = 0.$$

Solution 7.15 *The controllability matrix for the given system is*

$$\mathcal{C} = [\, B \mid AB \mid A^2B \,]$$

$$= \begin{bmatrix} 1 & 5 & 17 \\ 2 & 12 & 26 \\ 1 & -5 & -7 \end{bmatrix},$$

and hence, the inverse is

$$\mathcal{C}^{-1} = \begin{bmatrix} -0.3594 & 0.3906 & 0.5781 \\ -0.3125 & 0.1875 & -0.0625 \\ 0.1719 & -0.0781 & -0.0156 \end{bmatrix}.$$

The Cayley-Hamilton matrix of $\mathbf{A}$ in the closed-loop system is given by

$$\psi(\mathbf{A}) = \mathbf{A}^3 + 13\mathbf{A}^2 + 52\mathbf{A} + 60\mathbf{I}$$

$$= \begin{bmatrix} 243 & 117 & 117 \\ 202 & 328 & 308 \\ -85 & -85 & -65 \end{bmatrix}.$$

Therefore, direct application of Ackermann's formula

$$\mathbf{K} = \mathbf{e}_n^T \mathcal{C}^{-1} \psi(\mathbf{A})$$

gives

$$\mathbf{K} = \begin{bmatrix} 0 & 0 & 1 \end{bmatrix} \begin{bmatrix} 1 & 5 & 17 \\ 2 & 12 & 26 \\ 1 & -5 & -7 \end{bmatrix}^{-1} \begin{bmatrix} 243 & 117 & 117 \\ 202 & 328 & 308 \\ -85 & -85 & -65 \end{bmatrix}$$

which gives the gain matrix as

$$\mathbf{K} = \begin{bmatrix} 27.313 & -4.1875 & -2.9375 \end{bmatrix}.$$

It will be realized that this is the same problem as in the previous example. In order to minimize the computing round-off error, the inverse of C was used in its raw form to show the exactness of the results for the two methods. However, using the computed inverse of C will give the same results with some rounding-off errors as shown below

$$\mathbf{K} = \begin{bmatrix} 0 & 0 & 1 \end{bmatrix} \begin{bmatrix} -0.3594 & 0.3906 & 0.5781 \\ -0.3125 & 0.1875 & -0.0625 \\ 0.1719 & -0.0781 & -0.0156 \end{bmatrix} \begin{bmatrix} 243 & 117 & 117 \\ 202 & 328 & 308 \\ -85 & -85 & -65 \end{bmatrix}$$

$$= \begin{bmatrix} 27.322 & -4.1785 & -2.9285 \end{bmatrix}.$$

In MATLAB the Ackermann's formula could be used to solve this problem as shown below giving the same results.

```
A = [1   2   1; 2   3   4; -1   -1   -2];
B = [1; 2; 1];
P = roots([1   13   52   60]);
K = acker(A, B, P)
```

7.8   Introduction to Optimal Control

The pole placement and the consequent Ackermann's formula apply to SISO systems only. For MIMO systems, where there are many conflicting elements to be selected under some constraints to put the system under optimal conditions. The controller design turns out to be an optimization problem in which some form of performance index or cost function has to be optimized. This is called optimal control. In its general form optimal control handles all states with appropriate weights. As has been stated before, some of the states may not be measurable and so some means of estimating them from noisy measurements is done. The performance index, which has to be optimized under optimal control, is defined as a combination of the estimated state vector, the control vector and/or the output vector. Depending on the form of the performance index, state vector and estimation policy as well as underlying assumptions, a number of optimal control strategies can be defined. In this section, the basic principles of the optimal control methods are discussed, but first, a brief overview of optimization theory is given.

7.8.1   Overview of Optimization Theory

7.8.1.1   The Optimization Problem and Optimality Conditions

An optimization problem is a mathematical problem that involves finding the best or optimal solution of a given problem under some constraints. The general structure of an optimization problem can be posed as follows:

"Minimize (or maximize) a function $f(\mathbf{x})$ subject to the condition that $\underline{\mathbf{x}} \in \Omega$"

In this optimization problem, the function $f:\mathbb{R}^n \Rightarrow \mathbb{R}$ to be optimized is a real valued function called the objective function or the performance (cost) index. The vector $\mathbf{x}$ is an n-vector of n-independent variables, i.e.,

$$\mathbf{x} = [x_1, \quad x_2, \quad \cdots \quad x_n]^T \in \mathbb{R}^n$$

whose elements are called decision variables. The set $\Omega \in \mathbb{R}^n$ is called the constraint set or feasible set that captures the optimization constraints. Often the constraints (equality and inequality) set take the form

$$\Omega = \{\mathbf{x}:h(\mathbf{x}) = \mathbf{0}; \quad g(\mathbf{x}) \leq \mathbf{0}\}, \tag{7.129}$$

where h and g are given vector functions. Depending on the presence or absence of the constraints, the optimization may be referred to as constrained or unconstrained optimization.

Under unconstrained optimization of a real valued function $f(\mathbf{x})$, the first and necessary condition for it to be optimal at point $\underline{\mathbf{x}}^*$ is that the gradient of $f(\mathbf{x}^*)$ must be zero,

$$\nabla f(\mathbf{x}^*) = \left[\frac{\partial f(\mathbf{x}^*)}{\partial x_i}\right]^T = 0. \tag{7.130}$$

When this condition is met, the point $\mathbf{x}^*$ is called a stationary point. The stationary point may be a maximum, minimum, or an inflection point. In most control problems, the interest is in minimization of the given objective function, therefore, the minimum point has to be sought. The minimum of $f(\mathbf{x})$ at $\mathbf{x}^*$ is found when the second derivative or the Hessian of f is positive definite.

$$\nabla^2 f(\mathbf{x}^*) = \left[\frac{\partial^2 f(\mathbf{x}^*)}{\partial x_i \partial x_j}\right]^T > 0. \tag{7.131}$$

Note that for maximization problems, the same conditions must be satisfied but with opposite polarity for the second condition. In fact, a maximization problem may be turned into a minimization problem by sign inversion.

Example 7.16 *Determine the vector* $\mathbf{x} = [x_1 \quad x_2]^T$, *which minimizes the function* $f(\mathbf{x}) = 2x_1^4 + x_2^2 - 4x_1x_2 + 4$ *and the minimal value of* $f(\mathbf{x})$.

Solution 7.16 *The first condition for a stationary point is* $\nabla f(\mathbf{x}) = 0$ *and for the given function it becomes*

$$\nabla f(\mathbf{x}) = \begin{bmatrix} \dfrac{\partial f(\mathbf{x})}{\partial x_1} \\[3mm] \dfrac{\partial f(\mathbf{x})}{\partial x_2} \end{bmatrix} = \begin{bmatrix} 8x_1^3 - 4x_2 \\[2mm] 2x_2 - 4x_1 \end{bmatrix}.$$

Applying the first condition for a stationary point gives two equations in x_1 *and* x_2 *as*

$$8x_1^3 - 4x_2 = 0$$
$$2x_2 - 4x_1 = 0,$$

whose simultaneous solution gives

$$\mathbf{x} = \begin{bmatrix} 0 \\ 0 \end{bmatrix}, \quad \begin{bmatrix} 1 \\ 2 \end{bmatrix}, \quad \begin{bmatrix} -1 \\ -2 \end{bmatrix}.$$

The condition for a minimum is $\nabla^2 f(\mathbf{x}) > 0$ *and for the given function*

$$\nabla^2 f(\mathbf{x}^*) = \begin{bmatrix} \dfrac{\partial^2 f(\mathbf{x}^*)}{\partial x_i \partial x_j} \end{bmatrix}$$

$$= \begin{bmatrix} \dfrac{\partial^2 f(\mathbf{x})}{\partial x_1^2} & \dfrac{\partial^2 f(\mathbf{x})}{\partial x_1 \partial x_2} \\[4mm] \dfrac{\partial^2 f(\mathbf{x})}{\partial x_1 \partial x_2} & \dfrac{\partial^2 f(\mathbf{x})}{\partial x_2^2} \end{bmatrix}$$

$$= \begin{bmatrix} 24x_1^2 & -4 \\ -4 & 2 \end{bmatrix},$$

and for this to be positive definite, all its eigenvalues must be positive. Thus, expressions for the eigenvalues have to be determined from the characteristic equation

$$|\lambda \mathbf{I} - \nabla^2 f(\mathbf{x})| = \begin{vmatrix} \lambda - 24x_1^2 & 4 \\ 4 & \lambda - 2 \end{vmatrix} = 0$$

$$= (\lambda - 24x_1^2)(\lambda - 2) - 16 = 0$$

$$= \lambda^2 - (2 + 24x_1^2)\lambda + (48x_1^2 - 16) = 0$$

which gives the eigenvalues as

$$\lambda_1, \lambda_2 = (1 + 12x_1^2) \pm \sqrt{(1 + 12x_1^2)^2 + (16 - 48x_1^2)},$$

the conditions for all eigenvalues to be positive are

$$(1 + 12x_1^2) > 0$$

$$(1 + 12x_1^2) > \sqrt{(1 + 12x_1^2)^2 + (16 - 48x_1^2)},$$

which gives the limiting values of x_1 as $x_1 \geq \frac{1}{\sqrt{3}}$ and $x_1 \leq -\frac{1}{\sqrt{3}}$. For the functional stationary values found above, the only points that satisfy these conditions are

$$\mathbf{x} = \begin{bmatrix} 1 \\ 2 \end{bmatrix} \quad and \quad \begin{bmatrix} -1 \\ -2 \end{bmatrix}.$$

Hence, these are the values of $\mathbf{x}$ that minimize $f(\mathbf{x})$. The functional value at these points is $f(\mathbf{x}) = 1$. Note that at $\underline{\mathbf{x}} = [0\ 0]^T$ the functional value is $f(\mathbf{x}) = 3$ and hence it is not a minimum.

7.8.1.2 Constrained Optimization: The Lagrangian and the Lagrange Multipliers

When the constraint set is defined, the problem becomes a constrained optimization problem. The constraints may be equality or inequality constraints. For the purpose of this chapter, only equality constraints are considered.

Suppose the objective (vector) function $\mathbf{f}(\mathbf{x})$ is to be minimized subject to constraints that $\mathbf{h}(\mathbf{x}) = 0$. The standard procedure of solving such an optimization problem is to combine both the objective function $\mathbf{f}(\mathbf{x})$ and the constraint equation $\mathbf{h}(\mathbf{x})$ using Lagrange multipliers $\underline{\lambda}$ into one equation known as the Lagrangian $\mathcal{L}(\mathbf{x}, \lambda)$ where

$$\mathcal{L}(\mathbf{x}, \lambda) = \mathbf{f}(\mathbf{x}) + \underline{\lambda}^T \mathbf{h}(\mathbf{x}) \tag{7.132}$$

and $\underline{\lambda} \in \mathbb{R}^n$ is a vector of Lagrange multipliers, which are to be determined on the course of getting the solution to the given optimization problem. Thus, the necessary conditions that minimize $\mathbf{f}(\mathbf{x})$ and yet satisfy the constraint equation $\mathbf{h}(\mathbf{x})$, are then contained in the Lagrangian $\mathcal{L}(\mathbf{x}, \lambda)$ and are

$$\frac{\partial \mathcal{L}(\mathbf{x}, \lambda)}{\partial \lambda} = \mathbf{h}(\mathbf{x}) = 0 \tag{7.133}$$

$$\frac{\partial \mathcal{L}(\mathbf{x}, \lambda)}{\partial \mathbf{x}} = \nabla \mathbf{f}(\mathbf{x}) + \lambda^T \nabla \mathbf{h}(\mathbf{x}) = \mathbf{0}. \tag{7.134}$$

Although in most applications the values of the Lagrange multipliers are not needed, they must be determined, however, as intermediate values that allow complete determination of the optimal quantities of interest, i.e., vector $\mathbf{x}^*$ and possibly the minimum value of the objective function $\mathbf{f}(\mathbf{x}^*)$. By introducing the Lagrange multipliers, the constrained problem of minimizing the objective function $\mathbf{f}(\mathbf{x})$ is reduced to an unconstrained problem of minimizing the Lagrangian $\mathcal{L}(\mathbf{x}, \boldsymbol{\lambda})$ without constraints.

7.8.1.3   Objective Functions

In control systems design, the objective function is normally chosen by the control designer. This might be to minimize the absolute error, mean square error or anything that fits into the problem at hand. The objective function can be linear, quadratic or polynomial. However, in most applications, the quadratic objective function is used, i.e.,

$$ f(\mathbf{x})_\mathbf{Q} = \frac{1}{2}\mathbf{x}^\mathbf{T}\mathbf{W}\mathbf{x}, \tag{7.135} $$

though in some cases, the linear function

$$ f(\mathbf{x})_\mathbf{L} = \underline{\mathbf{W}}\mathbf{x} \tag{7.136} $$

is also used subject to some constraints. In the quadratic objective function, the matrix $\mathbf{W}$, known as the weight matrix, is chosen to be symmetric and positive definite. It expresses the relative importance of the various decision variables $\underline{\mathbf{x}}$ in optimizing the problem.

7.8.2   The Basic Optimal Control Problem

In designing feedback optimal controllers, the basic problem faced is that of selecting the elements of the feedback matrix $\mathbf{K}$ that optimize the various criteria imposed on the system states $\underline{\mathbf{x}}(t)$ and the control signals $\underline{\mathbf{u}}(t)$. These criteria are altogether contained in a quadratic objective function that is normally expressed as

$$ f(\mathbf{x}) = \frac{1}{2}\mathbf{x}^\mathbf{T}(T)\mathbf{W_0}\mathbf{x}(T) + \frac{1}{2}\int_0^T \left[\mathbf{x}^\mathbf{T}(t)\mathbf{W_1}\mathbf{x}(t) + \mathbf{u}^\mathbf{T}(t)\mathbf{W_2}\mathbf{u}(t)\right] dt, \tag{7.137} $$

where the weight matrices $\mathbf{W}_1$ and $\mathbf{W}_2$ are positive definite expressing the relative importance of the different states and controls in the controlled system as a whole. The term $\frac{1}{2}\mathbf{x}^\mathbf{T}(T)\mathbf{W_0}\mathbf{x}(T)$ is a penalty term that dictates the final required final state $\mathbf{x}(T)$, in which the weight matrix $\mathbf{W}_0$ is also symmetric positive definite. Therefore, it is required to minimize this performance function subject to the conditions that

$$\dot{\mathbf{x}}(t) = \mathbf{A}\mathbf{x}(t) + \mathbf{B}\underline{\mathbf{u}}(t) \tag{7.138}$$

where

$$\underline{\mathbf{u}}(t) = -\mathbf{K}\mathbf{x}(t). \tag{7.139}$$

As can be seen, this is a constrained optimization problem with equality constraints that can then be solved using the method of Lagrange multipliers. Important here is how correctly the weight matrices in the objective function are defined. Since the constraint Equation 7.137 is time dependent, it must also be satisfied at all times, therefore, the corresponding Lagrangian becomes

$$\mathcal{L}(\mathbf{x},\ \boldsymbol{\lambda}) = \frac{1}{2}\mathbf{x^T}(T)\mathbf{W_0}\mathbf{x}(T) + \frac{1}{2}\int_0^T \mathbf{x}^T(t)\mathbf{W}_1\mathbf{x}(t) + \underline{\mathbf{u}}^T(t)\mathbf{W}_2\underline{\mathbf{u}}(t) +$$

$$\boldsymbol{\lambda}(t)\left[\frac{d\mathbf{x}(t)}{dt} - \mathbf{A}\mathbf{x}(t) - \mathbf{B}\underline{\mathbf{u}}(t)\right] dt \tag{7.140}$$

$$= \phi(x(T)) + \int_0^T \mathcal{H}(\mathbf{x},\mathbf{u},\boldsymbol{\lambda}) dt. \tag{7.141}$$

Since the final state is fixed, unconstrained minimization of this Lagrangian over the time interval $[0, T]$ can be seen as just the unconstrained minimization of the Hamiltonian $\mathcal{H}(\mathbf{x},\mathbf{u},\boldsymbol{\lambda})$.

The conditions for optimization are obtained when the partial derivative of the Hamiltonian with respect to each of its variables is equal to zero. That is

$$\frac{\partial \mathcal{H}(\mathbf{x},\underline{\mathbf{u}},\boldsymbol{\lambda})}{\partial \boldsymbol{\lambda}} = \frac{d\mathbf{x}(t)}{dt} - \mathbf{A}\mathbf{x}(t) - \mathbf{B}\underline{\mathbf{u}}(t) = 0 \tag{7.142}$$

$$\frac{\partial \mathcal{H}(\mathbf{x},\underline{\mathbf{u}},\boldsymbol{\lambda})}{\partial \mathbf{x}} = \mathbf{W}_1\mathbf{x}(t) + \mathbf{A}^T\boldsymbol{\lambda}(t) + \frac{d\boldsymbol{\lambda}(t)}{dt} = 0 \tag{7.143}$$

$$\frac{\partial \mathcal{H}(\mathbf{x},\underline{\mathbf{u}},\boldsymbol{\lambda})}{\partial \mathbf{x}} = \mathbf{W}_2\underline{\mathbf{u}}(t) + \mathbf{B}^T\boldsymbol{\lambda}(t) = 0. \tag{7.144}$$

Equation 7.142 is the original state equation that was introduced in the Hamiltonian as a constraint, and Equation 7.143 is an equation resembling the state equation but in $\boldsymbol{\lambda}(t)$. These equations are known respectively as the state and costate equations, while Equation 7.144 is the control equation. The solution of the control equation gives

$$\underline{u}(t) = -\mathbf{W}_2^{-1}\mathbf{B}^\mathbf{T}\boldsymbol{\lambda}(t). \qquad (7.145)$$

Comparing this equation and Equation 7.139, it is evident that if the Lagrange multiplier as $\boldsymbol{\lambda}(t)$ can be determined as a function of $\mathbf{x}(t)$ then the optimal feedback gain matrix $\mathbf{K}$ will be determined as a function of the weight matrix $\mathbf{W}_2$ and the input matrix $\mathbf{B}$. This can be done by combining the three equations as follows: First, the control equation and the state equation are combined so as to remain with only two equations, which can hence be solved, (though at a cost). The combination of the state equation and the control equation gives

$$0 = \dot{\mathbf{x}} - \mathbf{A}\mathbf{x}(t) + \mathbf{B}\mathbf{W}_2^{-1}\mathbf{B}^T\boldsymbol{\lambda}(t). \qquad (7.146)$$

Thus, Equation 7.146 and the costate equation can be presented in a matrix form known as the Hamiltonian system, as follows:

$$\begin{bmatrix} \dot{\mathbf{x}}(t) \\ \dot{\boldsymbol{\lambda}}(t) \end{bmatrix} = \begin{bmatrix} \mathbf{A} & -\mathbf{B}\mathbf{W}_2^{-1}\mathbf{B}^T \\ -\mathbf{W}_1 & -\mathbf{A}^T \end{bmatrix} \begin{bmatrix} \mathbf{x}(t) \\ \boldsymbol{\lambda}(t) \end{bmatrix}, \qquad (7.147)$$

where the coefficient matrix is known as the Hamiltonian matrix. To find the optimal control, the Hamiltonian system must be solved, taking into account the boundary conditions $\underline{x}(t)$ and $\boldsymbol{\lambda}(t)$ and inputs over the interval [0,T]. However, normally, $\boldsymbol{\lambda}(0)$ and $\boldsymbol{\lambda}(T)$ are unknown and this further complicates the problem. For the purpose of this introductory material, the free final state $\underline{x}(T)$ given an initial state $\underline{x}(0)$ will be assumed over this control interval [0,T]. Notice that this becomes a two-point boundary value problem that still remains difficult to solve. A popular method that has been very successful in handling this problem is the sweep method, which was suggested by Bryson and Ho in 1975. This method assumes that $\underline{x}(t)$ and $\boldsymbol{\lambda}(t)$ are linearly related such that

$$\boldsymbol{\lambda}(t) = \mathbf{S}(t)\mathbf{x}(t). \qquad (7.148)$$

This reduces the two-point boundary value problem in $\mathbf{x}(t)$ and $\boldsymbol{\lambda}(t)$ to a single point problem in $\mathbf{S}(t)$ which can thus be solved with ease. Before dwelling on the solution of the Hamiltonian system in Equation 7.147, consider the time derivative of Equation 7.148. This becomes

$$\dot{\boldsymbol{\lambda}}(t) = \dot{\mathbf{S}}(t)\mathbf{x}(t) + \mathbf{S}(t)\dot{\mathbf{x}}(t). \qquad (7.149)$$

replacing $\dot{\mathbf{x}}(t)$ by the state Equation 7.146 and using the relation in 7.148 gives

$$\dot{\boldsymbol{\lambda}}(t) = \dot{\mathbf{S}}(t)\mathbf{x}(t) + \mathbf{S}(t)\left[\mathbf{A}\mathbf{x} - \mathbf{B}\mathbf{W}_2^{-1}\mathbf{B}^T\mathbf{S}(t)\mathbf{x}(t)\right]. \qquad (7.150)$$

Also by using Equation 7.148 in the costate equation gives $\dot{\boldsymbol{\lambda}}(t)$ as

$$\dot{\boldsymbol{\lambda}}(t) = -\mathbf{W}_1\mathbf{x}(t) - \mathbf{A}^T\mathbf{S}(t)\mathbf{x}(t). \tag{7.151}$$

Now, equating Equations 7.150 and 7.151 gives

$$-\mathbf{W}_1\mathbf{x}(t) - \mathbf{A}^T\mathbf{S}(t)\mathbf{x}(t) = \dot{\mathbf{S}}(t)\mathbf{x}(t) + \mathbf{S}(t)\left[\mathbf{A}\mathbf{x}(t) - \mathbf{B}\mathbf{W}_2^{-1}\mathbf{B}^T\mathbf{S}(t)\mathbf{x}(t)\right], \tag{7.152}$$

so that on elimination of $\underline{\mathbf{x}}(t)$ and doing some rearrangement, this problem reduces to

$$-\dot{\mathbf{S}}(t) = \mathbf{A}^T\mathbf{S}(t) + \mathbf{S}(t)\mathbf{A} - \mathbf{B}\mathbf{W}_2^{-1}\mathbf{B}^T\mathbf{S}(t) + \mathbf{W}_1, \tag{7.153}$$

for all $t < T$. This is a matrix equation known as the Algebraic Riccati Equation (shortened ARE) which must be solved for $\mathbf{S}(t)$ with the final condition of $\mathbf{S}(T)$ so that $\boldsymbol{\lambda}(t)$ in Equation 7.148 can be determined and hence, the control input in Equation 7.145 becomes

$$\underline{\mathbf{u}}(t) = -\mathbf{W}_2^{-1}\mathbf{B}^T\mathbf{S}(t)\mathbf{x}(t), \tag{7.154}$$

which gives the feedback gain matrix $\mathbf{K}$ as

$$\mathbf{K} = \mathbf{W}_2^{-1}\mathbf{B^T}\mathbf{S}(t). \tag{7.155}$$

Normally, the Riccati equation in 7.153 is solved by the backward in time approach. The gain obtained by solving the ARE this way is normally time varying as it varies with $\mathbf{S}(t)$ even if the system is time-invariant. For most practical applications, it is not desired to have a time varying $\mathbf{K}$, therefore, a steady state solution of ARE for $\mathbf{S}(t)$ is required. This steady state solution of ARE is obtained from Equation 7.153 by making $\dot{\mathbf{S}}(t) = 0$ so that the steady state Riccati equation becomes

$$0 = \mathbf{A}^T\mathbf{S}(t) + \mathbf{S}(t)\mathbf{A} - \mathbf{B}\mathbf{W}_2^{-1}\mathbf{B}^T\mathbf{S}(t) + \mathbf{W}_1. \tag{7.156}$$

If the steady state conditions of $\mathbf{x}(T)$ or $\mathbf{S}(T)$ are known (which is normally the case), the common approach to solve the Riccati equation is by backward propagation in time approach from time T to t giving $\mathbf{x}(t)$ and $\mathbf{S}(t)$ respectively, and this is normally done off-line. However, in this section, it has been assumed that the initial condition of $\mathbf{x}(0)$ is known, therefore, it is going to be solved by forward approach, which seems to be simpler than the backward approach, though basically both approaches are similar. A brief discussion on the backward approach is given at the end of this section.

The analytical solution of ARE is determined from the Hamiltonian Equation 7.147 and the sweep Equation 7.148

$$\begin{bmatrix} \dot{\mathbf{x}}(t) \\ \dot{\boldsymbol{\lambda}}(t) \end{bmatrix} = \begin{bmatrix} \mathbf{A} & -\mathbf{B}\mathbf{W}_2^{-1}\mathbf{B}^T \\ -\mathbf{W}_1 & -\mathbf{A}^\mathbf{T} \end{bmatrix} \begin{bmatrix} \mathbf{x}(t) \\ \boldsymbol{\lambda}(t) \end{bmatrix}$$

$$\boldsymbol{\lambda}(t) = \mathbf{S}(t)\mathbf{x}(t).$$

This is a system of linear time-invariant ordinary deferential equations which, as has been stated before, can be solved either by forward or backward propagation in time using an appropriate transition matrix of the Hamiltonian matrix $\mathcal{H}$

$$\mathcal{H} = \begin{bmatrix} \mathbf{A} & -\mathbf{B}\mathbf{W}_2^{-1}\mathbf{B}^T \\ -\mathbf{W}_1 & -\mathbf{A}^\mathbf{T} \end{bmatrix}. \tag{7.157}$$

The solution, $\mathbf{S}(t)$, of the Riccati equation can then be determined analytically in terms of the eigenvalues and eigenvectors of this Hamiltonian matrix. Due to the special structure of this matrix, if $\mathcal{H}$ has distinct eigenvalues, then it can be written as a diagonally structured matrix such that its eigenvalues are $\pm\lambda_i$, i.e., if λ_i is an eigenvalue of $\mathcal{H}$ then so is $-\lambda_i$. Consequently, a left modal transformation matrix $\mathbf{E}$ of $\mathcal{H}$ can be defined as

$$\mathbf{E} = \begin{bmatrix} \mathbf{E}_{11} & \mathbf{E}_{12} \\ \mathbf{E}_{21} & \mathbf{E}_{22} \end{bmatrix} \tag{7.158}$$

such that

$$\mathcal{H} = \mathbf{E}\hat{\mathbf{\Lambda}}\mathbf{E}^{-1}, \tag{7.159}$$

where $\hat{\Lambda}$ is a diagonal matrix of the eigenvalues of $\mathcal{H}$ arranged so that fast eigenvalues are in the upper left block matrix $-\Lambda$

$$\hat{\Lambda} = \begin{bmatrix} -\Lambda & \mathbf{0} \\ \mathbf{0} & \Lambda \end{bmatrix}. \tag{7.160}$$

Associated with the right modal transformation matrix is the right modal transformation matrix $\mathbf{D}^T$ defined as

$$\mathbf{D}^T = \mathbf{E}^{-1}, \tag{7.161}$$

which can be used in Equation 7.159 to give

$$\mathcal{H} = \mathbf{E}\hat{\mathbf{\Lambda}}\mathbf{D}^T.$$

In this case, $\mathbf{D}$ is defined as

$$\mathbf{D^T} = \begin{bmatrix} \mathbf{D}_{11}^T & \mathbf{D}_{21}^T \\ \mathbf{D}_{12}^T & \mathbf{D}_{22}^T \end{bmatrix}.$$

The right modal transformation matrix $\mathbf{D}$ is related with the left modal transformation matrix $\mathbf{E}$ as

$$\mathbf{D^T} = \begin{bmatrix} \mathbf{E}_{22}^T & -\mathbf{E}_{12}^T \\ -\mathbf{E}_{21}^T & \mathbf{E}_{11}^T \end{bmatrix}.$$

Therefore, the Hamiltonian system can then be written as

$$\begin{bmatrix} \dot{\mathbf{x}}(t) \\ \dot{\boldsymbol{\lambda}}(t) \end{bmatrix} = \begin{bmatrix} \mathbf{E}_{11} & \mathbf{E}_{12} \\ \mathbf{E}_{21} & \mathbf{E}_{22} \end{bmatrix} \begin{bmatrix} -\Lambda & 0 \\ 0 & \Lambda \end{bmatrix} \begin{bmatrix} \mathbf{D}_{11}^T & \mathbf{D}_{21}^T \\ \mathbf{D}_{12}^T & \mathbf{D}_{22}^T \end{bmatrix} \begin{bmatrix} \mathbf{x}(t) \\ \boldsymbol{\lambda}(t) \end{bmatrix}.$$

Defining the normal state vector $\mathbf{q}(t) = [\mathbf{q}_1(t) \quad \mathbf{q}_2(t)]^T$ as

$$\mathbf{q}(t) = \begin{bmatrix} \mathbf{q}_1(t) \\ \mathbf{q}_2(t) \end{bmatrix} = \begin{bmatrix} \mathbf{D}_{11}^T & \mathbf{D}_{21}^T \\ \mathbf{D}_{12}^T & \mathbf{D}_{22}^T \end{bmatrix} \begin{bmatrix} \mathbf{x}(t) \\ \boldsymbol{\lambda}(t) \end{bmatrix} \tag{7.162}$$

or equivalently

$$\begin{bmatrix} \mathbf{x}(t) \\ \boldsymbol{\lambda}(t) \end{bmatrix} = \begin{bmatrix} \mathbf{E}_{11} & \mathbf{E}_{12} \\ \mathbf{E}_{21} & \mathbf{E}_{22} \end{bmatrix} \begin{bmatrix} \mathbf{q}_1(t) \\ \mathbf{q}_2(t) \end{bmatrix}. \tag{7.163}$$

Some mathematical operations show that this normal state vector must satisfy the state dynamics

$$\begin{bmatrix} \dot{\mathbf{q}}_1(t) \\ \dot{\mathbf{q}}_2(t) \end{bmatrix} = \begin{bmatrix} -\Lambda & 0 \\ 0 & \Lambda \end{bmatrix} \begin{bmatrix} \mathbf{q}_1(t) \\ \mathbf{q}_2(t) \end{bmatrix}$$

whose solution is

$$\begin{bmatrix} \mathbf{q}_1(t) \\ \mathbf{q}_2(t) \end{bmatrix} = \begin{bmatrix} e^{-\Lambda t} & 0 \\ 0 & e^{\Lambda t} \end{bmatrix} \begin{bmatrix} \mathbf{q}_1(0) \\ \mathbf{q}_2(0) \end{bmatrix}, \tag{7.164}$$

where $\mathbf{q}(0)$ is the initial value of the normal state at $t = 0$. This shows that the normal state $\mathbf{q}_1(t)$ is completely stable while $\mathbf{q}_2(t)$ is completely unstable. If the state vector at $t = 0$ is $\mathbf{x}(0)$ then the initial normal states can be determined as

$$
\begin{bmatrix} \mathbf{q}_1(0) \\ \mathbf{q}_2(0) \end{bmatrix} = \begin{bmatrix} \mathbf{D}_{11}^T & \mathbf{D}_{21}^T \\ \\ \mathbf{D}_{12}^T & \mathbf{D}_{22}^T \end{bmatrix} \begin{bmatrix} \mathbf{x}(0) \\ \boldsymbol{\lambda}(0) \end{bmatrix}, \tag{7.165}
$$

which together with Equation 7.164 gives

$$
\begin{bmatrix} \mathbf{q}_1(t) \\ \mathbf{q}_2(t) \end{bmatrix} = \begin{bmatrix} e^{-\Lambda t} & \mathbf{0} \\ \mathbf{0} & e^{\Lambda t} \end{bmatrix} \begin{bmatrix} \mathbf{D}_{11}^T & \mathbf{D}_{21}^T \\ \\ \mathbf{D}_{12}^T & \mathbf{D}_{22}^T \end{bmatrix} \begin{bmatrix} \mathbf{x}(0) \\ \boldsymbol{\lambda}(0) \end{bmatrix}.
$$

On employing the sweep method it follows that

$$
\boldsymbol{\lambda}(0) = \mathbf{S}(0)\mathbf{x}(0),
$$

so that

$$
\begin{bmatrix} \mathbf{q}_1(t) \\ \mathbf{q}_2(t) \end{bmatrix} = \begin{bmatrix} e^{-\Lambda t} & \mathbf{0} \\ \mathbf{0} & e^{\Lambda t} \end{bmatrix} \begin{bmatrix} \left(\mathbf{D}_{11}^T + \mathbf{D}_{21}^T \mathbf{S}(0) \right) \mathbf{x}(0) \\ \\ \left(\mathbf{D}_{12}^T + \mathbf{D}_{22}^T \mathbf{S}(0) \right) \mathbf{x}(0) \end{bmatrix}
$$

$$
= \begin{bmatrix} e^{-\Lambda t} \left(\mathbf{D}_{11}^T + \mathbf{D}_{21}^T \mathbf{S}(0) \right) \mathbf{x}(0) \\ \\ e^{\Lambda t} \left(\mathbf{D}_{12}^T + \mathbf{D}_{22}^T \mathbf{S}(0) \right) \mathbf{x}(0) \end{bmatrix}.
$$

Using this in Equation 7.163 gives

$$
\begin{bmatrix} \mathbf{x}(t) \\ \boldsymbol{\lambda}(t) \end{bmatrix} = \begin{bmatrix} \mathbf{E}_{11} & \mathbf{E}_{12} \\ \mathbf{E}_{21} & \mathbf{E}_{22} \end{bmatrix} \begin{bmatrix} e^{-\Lambda t} \left(\mathbf{D}_{11}^T + \mathbf{D}_{21}^T \mathbf{S}(0) \right) \mathbf{x}(0) \\ \\ e^{\Lambda t} \left(\mathbf{D}_{12}^T + \mathbf{D}_{22}^T \mathbf{S}(0) \right) \mathbf{x}(0) \end{bmatrix}.
$$

Now, since $\mathbf{x}(t)$ must have a stable response, it must thus be generated by a stable normal state $\mathbf{q}_1(t)$, where

$$
\mathbf{q}_1(t) = e^{-\Lambda t} \left(\mathbf{D}_{12}^T + \mathbf{D}_{22}^T \mathbf{S}(0) \right) \mathbf{x}(0). \tag{7.166}
$$

This calls for

$$
\mathbf{E}_{12} e^{\Lambda t} \left(\mathbf{D}_{12}^T + \mathbf{D}_{22}^T \mathbf{S}(0) \right) \mathbf{x}(0) = 0.
$$

Since $\mathbf{E}_{12}$ is not necessarily a null matrix, this means

$$
\mathbf{D}_{12}^T + \mathbf{D}_{22}^T \mathbf{S}(0) = 0
$$

which gives

$$\mathbf{S}(0) = -\mathbf{D}_{22}^{-T}\mathbf{D}_{12}^{T}.$$

Since $\mathbf{D}_{22}^{-T} = \mathbf{E}_{11}^{-T}$, $\mathbf{D}_{12}^{T} = -\mathbf{E}_{21}^{T}$ then

$$\mathbf{S}(0) = \mathbf{E}_{11}^{-T}\mathbf{E}_{21}^{T}.$$

Knowing $\mathbf{S}(0)$ together with $\mathbf{x}(0)$, it becomes possible to solve by the linear system by forward propagation until the steady state is reached.

Alternatively, the solution of the Riccati equation can be determined from the boundary condition $\mathbf{S}(\mathbf{T})$ by backward approach if the steady state boundary condition of the Riccati equation $\mathbf{S}(\mathbf{T})$ is known. In this case, the Riccati solution can be determined by using Equation 7.166 with some slight modifications to be in the backward in time propagation where the steady state normal state of the stabilizing solution becomes

$$\mathbf{q}_1(T) = -(\mathbf{E}_{22} - \mathbf{S}(T)\mathbf{E}_{12})^{-1}(\mathbf{E}_{21} - \mathbf{S}(T)\mathbf{E}_{11}).$$

Since by backward method

$$\mathbf{q}_1(t) = e^{-\Lambda(T-t)}\mathbf{q}_1(t)e^{-\Lambda(T-t)}$$

then the analytical solution to the equation becomes

$$\mathbf{S}(t) = (\mathbf{E}_{21} + \mathbf{E}_{22}\mathbf{q}_1(t))(\mathbf{E}_{11} + \mathbf{E}_{12}\mathbf{q}_1(t))^{-1}.$$

Currently, almost every Computer Aided Control System Design (CACSD) packages include routines that solve both the steady state and the differential algebraic Riccati equation (ARE). In MATLAB, the routines ARE and RIC are used to solve this equation.

7.9   Estimator Design

So far, in the discussion on full state feedback control, it has been assumed that all states will be available for feedback. In practice, this is not always the case not only because of the cost of sensors that would be required to measure all the states, but also the fact that some of the states are physically inaccessible. The common approach used by most control systems is to take measurements of the outputs that are assumed to have sufficient information about the dynamics of a system. Based on these measurements, the entire state vector can be estimated. It is this estimated state vector $\hat{\mathbf{x}}$ that is fed back in the control loop to achieve the required control action

$$u(t) = -\mathbf{K}\hat{\mathbf{x}}(t). \tag{7.167}$$

The unit that does this state estimation is called the observer or simply the estimator. In this section, the general principles of state estimation are presented. Two methods of state estimation are available for this task. These are the full-order estimators and the reduced-order estimator. The full-order estimator reconstructs the entire state vector from a set of measurements while the reduced-order estimator reconstructs only those states that are not contained in the measurement. Both have their advantages and disadvantages, as will be seen. In this context the terms observer and estimator can be used interchangeably.

7.9.1 Full State Estimator

Consider an *idealized noiseless* LTI system (in this case a general MIMO system may be assumed which will assist further in the discussion of reduced-order estimators in the next section) whose dynamics may be presented as

$$\dot{\mathbf{x}}(t) = \mathbf{A}\mathbf{x}(t) + \mathbf{B}\mathbf{u}(t) \tag{7.168}$$

where the output is measured using

$$\mathbf{y}(t) = \mathbf{C}\mathbf{x}(t) + \mathbf{D}\mathbf{u}(t). \tag{7.169}$$

Here the description *"idealized noiseless"* has been stressed because real systems will have additive noise in both the measurement and the state equation. There is need to estimate the vector $\hat{\mathbf{x}}$ from the measurements $\mathbf{y}$ so that the error between these estimates and the true state is minimal. There may be several ways of defining the minimal error, such as minimum square error or minimum absolute error and many others, some of which will be discussed later. If the estimation error is defined as

$$\tilde{\mathbf{x}}(t) = \mathbf{x}(t) - \hat{\mathbf{x}}(t) \tag{7.170}$$

and since the estimate $\hat{\mathbf{x}}(t)$ must satisfy Equation 7.168, i.e.,

$$\dot{\hat{\mathbf{x}}}(t) = \mathbf{A}\hat{\mathbf{x}}(t) + \mathbf{B}\mathbf{u}(t) \tag{7.171}$$

thus, this estimation error vector $\tilde{\mathbf{x}}(t)$ will satisfy

$$\begin{aligned}
\dot{\tilde{\mathbf{x}}}(t) &= \dot{\mathbf{x}}(t) - \dot{\hat{\mathbf{x}}}(t) \\
&= \mathbf{A}\mathbf{x}(t) + \mathbf{B}\mathbf{u}(t) - \mathbf{A}\hat{\mathbf{x}}(t) - \mathbf{B}\mathbf{u}(t) \\
&= \mathbf{A}\mathbf{x}(t) - \mathbf{A}\hat{\mathbf{x}}(t) \\
&= \mathbf{A}\tilde{\mathbf{x}}(t). \tag{7.172}
\end{aligned}$$

This shows that the error follows the same dynamics as the true states, which means that the error at any time t will be given by

$$\tilde{\mathbf{x}}(t) = e^{\mathbf{A}t}\tilde{\mathbf{x}}(0), \qquad (7.173)$$

where $\tilde{\mathbf{x}}(0)$ is the initial error. Now if the initial estimate is very close to the true state and all the eigenvalues of the system matrix $\mathbf{A}$ represent stable poles, then the error will keep decreasing and hence, the state estimate will converge to the true state vector. However, in most cases the system $\mathbf{A}$ is not stable (which is one of the reasons that the controller is to be designed). Therefore, means must be provided that will ensure that the plant poles in the estimation process allow the error to decay to zero. To accomplish this sort of state estimation, the model is designed to mimic the plant dynamics as

$$\dot{\hat{\mathbf{x}}}(t) = \mathbf{A}\hat{\mathbf{x}}(t) + \mathbf{B}\mathbf{u}(t), \qquad (7.174)$$

and is connected in parallel with the plant. The outputs of the model and the plant are compared and then the error is fed back through some estimation gain $\mathbf{L}$ in a way that the closed-loop estimation model matrix will have fast eigenvalues. The actual output measurement $\mathbf{y}(t)$ is due to the true state $\mathbf{x}(t)$ and the model output $\hat{\mathbf{y}}(t)$ is due to the estimated state $\hat{\mathbf{x}}(t)$. The difference between them gives the measurement error. In fact, the term *"measurement error"* as used here is misleading, as there are no errors in measurements insofar as an ideal noiseless system has been assumed. The correct term to be used here is the *measurement residual* or *innovations*.

$$\tilde{\mathbf{y}}(t) = \mathbf{y}(t) - \hat{\mathbf{y}}(t). \qquad (7.175)$$

Since

$$\mathbf{y}(t) = \mathbf{C}\mathbf{x}(t) + \mathbf{D}\mathbf{u}(t)$$
$$\hat{\mathbf{y}}(t) = \mathbf{C}\hat{\mathbf{x}}(t) + \mathbf{D}\mathbf{u}(t),$$

then

$$\tilde{\mathbf{y}}(t) = \mathbf{C}\mathbf{x}(t) - \mathbf{C}\hat{\mathbf{x}}(t)$$
$$= \mathbf{C}\tilde{\mathbf{x}}(t). \qquad (7.176)$$

The effect of feeding back this residual through a gain matrix $\mathbf{L}$ is to change the estimation model dynamics so that it can be expressed mathematically as

$$\dot{\hat{\mathbf{x}}}(t) = \mathbf{A}\hat{\mathbf{x}}(t) + \mathbf{B}\mathbf{u}(t) + \mathbf{L}\tilde{\mathbf{y}}(t)$$
$$= \mathbf{A}\hat{\mathbf{x}}(t) + \mathbf{B}\mathbf{u}(t) + \mathbf{L}\mathbf{C}\tilde{\mathbf{x}}(t) \qquad (7.177)$$

Now since $\tilde{\mathbf{x}}(t) = \mathbf{x}(t) - \hat{\mathbf{x}}(t)$, then

$$
\begin{aligned}
\dot{\hat{\mathbf{x}}}(t) &= \mathbf{A}\hat{\mathbf{x}}(t) + \mathbf{B}\mathbf{u}(t) + \mathbf{LC}\left[\mathbf{x}(t) - \hat{\mathbf{x}}(t)\right] \\
&= (\mathbf{A} - \mathbf{LC})\hat{\mathbf{x}}(t) + \mathbf{B}\mathbf{u}(t) + \mathbf{LCx}(t). \tag{7.178}
\end{aligned}
$$

Hence, on differentiating Equation 7.170 and using Equations 7.168 and 7.177, it follows that

$$
\begin{aligned}
\dot{\tilde{\mathbf{x}}}(t) &= \dot{\mathbf{x}}(t) - \dot{\hat{\mathbf{x}}}(t) \\
&= \mathbf{Ax}(t) + \mathbf{Bu}(t) - (\mathbf{A} - \mathbf{LC})\hat{\mathbf{x}}(t) - \mathbf{Bu}(t) - \mathbf{LCx}(t) \\
&= (\mathbf{A} - \mathbf{LC})\mathbf{x}(t) - (\mathbf{A} - \mathbf{LC})\mathbf{x}(t) \\
&= (\mathbf{A} - \mathbf{LC})\tilde{\mathbf{x}}(t). \tag{7.179}
\end{aligned}
$$

This is the modified state estimation error dynamics whose solution is

$$
\tilde{\mathbf{x}}(t) = e^{(\mathbf{A} - \mathbf{LC})t}\tilde{\mathbf{x}}(\mathbf{0}). \tag{7.180}
$$

As can be seen, the error dynamics depend only on the initial estimate $\tilde{\mathbf{x}}(\mathbf{0})$ and the three system model matrices $\mathbf{A}$, $\mathbf{L}$ and $\mathbf{C}$. It has no relationship with the control input $\mathbf{u}(t)$. For this error to converge to zero very fast, the matrix exponential $e^{(\mathbf{A} - \mathbf{LC})t}$ must decrement, which requires that the eigenvalues of $\mathbf{A} - \mathbf{LC}$ to be negative large (i.e., fast eigenvalues). This implies that the observation gain matrix should be made large enough. In fact, if the observation closed-loop poles are made large negative, the state estimate error will converge to zero irrespective of the initial condition. The selection of the estimator gain matrix $\mathbf{L}$, which results in fast eigenvalues of the closed-loop system $\mathbf{A} - \mathbf{LC}$, can be done in just the same way as in the determination of the controller gain matrix $\mathbf{K}$ through assignment of the eigenvalues of the estimation closed-loop model $\mathbf{A} - \mathbf{LC}$. Since the characteristic equation of this estimator model is

$$
\alpha(s) \triangleq \det\left[s\mathbf{I} - (\mathbf{A} - \mathbf{LC})\right] = 0. \tag{7.181}
$$

By using polynomial expansion, it is then possible to get values of the gain matrix $\mathbf{L}$.

Suppose that it is required to place the poles of the closed-loop observer model at

$$
s_i = \beta_1, \beta_2, \beta_3, \cdots \beta_n. \tag{7.182}
$$

Essentially, this means that the polynomial form of the characteristic Equation 7.181 becomes

$$
\alpha(s) = (s - \beta_1)(s - \beta_2)(s - \beta_3) \cdots (s - \beta_n). \tag{7.183}
$$

Now, since the model matrix $\mathbf{A}$ and the output matrix $\mathbf{C}$ are both known, by comparison of the coefficients of Equation 7.181 and those of Equation 7.183, it becomes possible to get the elements of $\mathbf{L}$. However, as it was shown in the controller design, Equation 7.181 might have many unknown elements than can be computed through this comparison. To be able to get all elements of $\mathbf{L}$, the estimator model $(\mathbf{A}, \mathbf{C})$ must be in observable canonical form, just the same as it was necessary for the control system $(\mathbf{A}, \mathbf{B})$ to be control canonical form.

Alternatively, one can use Ackermann's formula in almost the same way as it was used in determination of the controller gain matrix $\mathbf{K}$. Derivation of the observer Ackermann's formula will not be given here as it follows similar steps and reasoning as for the controller design. For the observer design, it is given as

$$\mathbf{L} = \alpha(\mathbf{A})\mathcal{O}^{-1}\mathbf{e}_1^T, \tag{7.184}$$

where $\alpha(\mathbf{A})$ is the Cayley-Hamilton matrix characteristic equation of the system $\mathbf{A}$, $\mathcal{O}$ is the observability matrix of the system $(\mathbf{A}, \mathbf{C})$ and $\mathbf{e}_1$ is the vector of the first column of an identity matrix.

7.9.2 Duality of Estimation and Control

The estimation and control problems are mathematically equivalent. The control problem is that of determining the gain matrix $\mathbf{K}$ so that the poles of the closed-loop system are fast eigenvalues of the matrix $\mathbf{A} - \mathbf{BK}$ and the estimation problem is that of determining the estimator gain that gives fast eigenvalues of the observation closed-loop system $\mathbf{A} - \mathbf{LC}$. Now, since the eigenvalues of $(\mathbf{A} - \mathbf{LC})$ are the same as those of the system $(\mathbf{A} - \mathbf{LC})^T$ where

$$(\mathbf{A} - \mathbf{LC})^T = \mathbf{A}^T - \mathbf{C}^T\mathbf{L}^T. \tag{7.185}$$

It follows that the method of determining the control gain matrix $\mathbf{K}$ in the system $\mathbf{A} - \mathbf{BK}$ is the same as that of determining the transpose of the estimation gain matrix $\mathbf{L}^T$ in the system $\mathbf{A}^T - \mathbf{C}^T\mathbf{L}^T$, where the matrices $\mathbf{A}$ and $\mathbf{B}$ in the control equations are replaced by $\mathbf{A}^T$ and $\mathbf{C}^T$. This equivalence is known as duality of estimation and control and is summarized in the following table

Controller	Estimator (Observer)
A	$\mathbf{A}^T$
B	$\mathbf{C}^T$
C	$\mathbf{B}^T$
D	D

This duality property allows us to use the same design tools for estimation as for control problems by using substitutions shown in the table above. For this reason, MATLAB commands "acker" and "place", which are used in the controller design, are also applicable to estimator design where the syntax becomes

$$L=\text{acker}(A', C', P_e)'$$ (7.186)

and

$$L=\text{place}(A', C', P_e)'.$$ (7.187)

Here, P_e is the vector that contains the desired estimator error poles.

7.9.3   Reduced-Order Estimator

The full-order estimator reconstructs the entire state vector using measurements of some of the state variables. This estimator gives some estimates that are redundant because some of the states can be available directly from the output measurements. The only advantage of this type of estimator is that, when the measurements are corrupted with noise, the full state estimator smoothens by filtering the measured states as well as reconstruction of the unmeasured states. However, on the other hand, it is associated with unnecessary computational load due to the fact that it computes even those states that can be available from measurements, for example, if the output matrix contains rows in which all entries are zero except one. The need to alleviate this load leads to the idea of reduced-order estimator, which separates the states that are available directly from the measurements in estimating the unknown states. This reduces the order of the estimator by the number of the measured outputs, and hence, the name "reduced-order estimator."

Suppose that for the linear system

$$\dot{\mathbf{x}}(t) = \mathbf{Ax}(t) + \mathbf{Bu}(t)$$
$$\mathbf{y}(t) = \mathbf{Cx}(t) + \mathbf{Du}(t),$$

the state vector is

$$\mathbf{x}(t) = [x_1(t) \quad x_2(t) \quad x_3(t) \quad x_4(t) \quad x_5(t)]^T,$$ (7.188)

and the measurement vector is

$$\mathbf{y}(t) = [\, y_1(t) \quad y_2(t) \quad y_3(t)]^T.$$ (7.189)

If the output and the feed-through matrices $\mathbf{C}$ and $\mathbf{D}$ are

$$\mathbf{C} = \begin{bmatrix} 1 & 0 & 0 & 0 & 0 \\ 0 & 1 & 0 & 0 & 0 \\ 0 & 1 & 1 & 1 & 1 \end{bmatrix}$$

$$D = \begin{bmatrix} 0 \\ 0 \\ 1 \end{bmatrix},$$

then the measurements will contain

$$\begin{bmatrix} y_1(t) \\ y_2(t) \\ y_3(t) \end{bmatrix} = \begin{bmatrix} 1\ 0\ 0\ 0\ 0 \\ 0\ 1\ 0\ 0\ 0 \\ 0\ 1\ 1\ 1\ 1 \end{bmatrix} \begin{bmatrix} x_1(t) \\ x_2(t) \\ x_3(t) \\ x_4(t) \\ x_5(t) \end{bmatrix} + \begin{bmatrix} 0 \\ 0 \\ 1 \end{bmatrix} u(t)$$

$$= \begin{bmatrix} x_1(t) \\ x_2(t) \\ x_2(t) + x_3(t) + x_4(t) + x_5(t) + u(t) \end{bmatrix}. \tag{7.190}$$

Clearly, from this measurement vector, the states $x_1(t)$ and $x_2(t)$ are directly available as measurements $y_1(t)$ and $y_2(t)$ respectively. However, from this same measurement vector, there is no sufficient information that can lead to determination of the states $x_3(t)$, $x_4(t)$, and $x_5(t)$. Therefore, an estimator that estimates only these unknown states will be a reduced estimator of order 3, i.e., the state vector of order 5 has been reduced by the order of the states that are directly available from the measurements.

Generally, if the measurement vector $\mathbf{y}$[1] is such that part of it contains measurements $\mathbf{y}_m$, which represent directly the states of the system and some measurements that do not, e.g., an under-determined linear combination of the states as in Equation 7.190, say $\mathbf{y}_R$, it is possible to partition it as

$$\mathbf{y} = \begin{bmatrix} \mathbf{y}_m \\ \cdots \\ \mathbf{y}_R \end{bmatrix}. \tag{7.191}$$

Corresponding to this partitioning, the output and the feed-through matrices can also be partitioned as

$$\mathbf{C} = \begin{bmatrix} \mathbf{C}_m \\ \cdots \\ \mathbf{C}_R \end{bmatrix} \qquad \mathbf{D} = \begin{bmatrix} \mathbf{0} \\ \cdots \\ \mathbf{D}_R \end{bmatrix} \tag{7.192}$$

so that the measurement equation becomes

$$\mathbf{y} = \begin{bmatrix} \mathbf{y}_m \\ \cdots \\ \mathbf{y}_R \end{bmatrix} = \begin{bmatrix} \mathbf{C}_m \\ \cdots \\ \mathbf{C}_R \end{bmatrix} \mathbf{x} + \begin{bmatrix} \mathbf{0} \\ \cdots \\ \mathbf{D}_R \end{bmatrix} \mathbf{u}. \tag{7.193}$$

---

[1] Direct dependance of the variables on time t will not be shown, for example, instead of writing $\mathbf{y}(t)$ it will be written as $\mathbf{y}$, {similarly for $\mathbf{x}$ instead of $\mathbf{x}(t)$, $\mathbf{z}$ instead of $\mathbf{z}(t)$ and $\mathbf{u}$ instead $\mathbf{u}(t)$} unless where the context requires otherwise.

Since the aim is to separate the states that are directly obtainable in these measurements from the rest of the state elements, a coordinate transformation $\mathbf{P}$ which transforms $\mathbf{x}$ to a new variable $\mathbf{z}$, is required so that the new state variable $\mathbf{z}$ can be partitioned to two components that correspond to the measurements $\mathbf{y}_m$, and the rest $\mathbf{z}_R$ is unknown, (to be estimated). The modal (diagonal) canonical form may be the best choice for this matter. Thus

$$\mathbf{x} = \mathbf{Pz}$$

where

$$\mathbf{z} = \begin{bmatrix} \mathbf{y}_m \\ \cdots \\ \mathbf{z}_R \end{bmatrix}.$$

The transformation matrix $\mathbf{T}$ can also be partitioned as

$$\mathbf{P} = \begin{bmatrix} \mathbf{P}_m & \vdots & \mathbf{P}_R \end{bmatrix},$$

so that the state vector $\mathbf{x}$ can be expressed as

$$\mathbf{x} = \begin{bmatrix} \mathbf{P}_m & \vdots & \mathbf{P}_R \end{bmatrix} \begin{bmatrix} \mathbf{y}_m \\ \cdots \\ \mathbf{z}_R \end{bmatrix}. \tag{7.194}$$

The transformation matrix $\mathbf{P}$ must be associated with a left inverse $\mathbf{T}$ where

$$\mathbf{T} = \begin{bmatrix} \mathbf{T}_m \\ \cdots \\ \mathbf{T}_R \end{bmatrix}$$

such that

$$\mathbf{TP} = \begin{bmatrix} \mathbf{I} & \mathbf{0} \\ \mathbf{0} & \mathbf{I} \end{bmatrix}. \tag{7.195}$$

Using this inverse on Equation 7.194 gives

$$\begin{bmatrix} \mathbf{T}_m \\ \cdots \\ \mathbf{T}_R \end{bmatrix} \mathbf{x} = \begin{bmatrix} \mathbf{y}_m \\ \cdots \\ \mathbf{z}_R \end{bmatrix}. \tag{7.196}$$

Therefore, the similarity transformation from $\mathbf{x}$ variable to the $\mathbf{z}$ variable gives

$$\dot{\mathbf{z}} = \mathbf{P}^{-1}\mathbf{APz} + \mathbf{P}^{-1}\mathbf{Bu}$$
$$\mathbf{y} = \mathbf{CPz} + \mathbf{Du},$$

which can be expressed as

$$
\begin{bmatrix} \dot{\mathbf{y}}_m \\ \cdots \\ \dot{\mathbf{z}}_R \end{bmatrix} = \begin{bmatrix} \mathbf{T}_m \\ \cdots \\ \mathbf{T}_R \end{bmatrix} \mathbf{A} \begin{bmatrix} \mathbf{P}_m \vdots \mathbf{P}_R \end{bmatrix} \begin{bmatrix} \mathbf{y}_m \\ \cdots \\ \mathbf{z}_R \end{bmatrix} + \begin{bmatrix} \mathbf{T}_m \\ \cdots \\ \mathbf{T}_R \end{bmatrix} \mathbf{Bu}
$$

$$
\begin{bmatrix} \mathbf{y}_m \\ \cdots \\ \mathbf{z}_R \end{bmatrix} = \begin{bmatrix} \mathbf{C}_m \\ \cdots \\ \mathbf{C}_R \end{bmatrix} \begin{bmatrix} \mathbf{P}_m \vdots \mathbf{P}_R \end{bmatrix} \begin{bmatrix} \mathbf{y}_m \\ \cdots \\ \mathbf{z}_R \end{bmatrix} + \begin{bmatrix} \mathbf{0} \\ \cdots \\ \mathbf{D}_R \end{bmatrix} \mathbf{u}.
$$

This structure can be simplified further and expressed in block matrix form as follows:

$$
\begin{bmatrix} \dot{\mathbf{y}}_m \\ \cdots \\ \dot{\mathbf{z}}_R \end{bmatrix} = \begin{bmatrix} \mathbf{A}_{mm} & \vdots & \mathbf{A}_{mR} \\ \cdots & \cdots & \cdots \\ \mathbf{A}_{Rm} & \vdots & \mathbf{A}_{RR} \end{bmatrix} \begin{bmatrix} \mathbf{y}_m \\ \cdots \\ \mathbf{z}_R \end{bmatrix} + \begin{bmatrix} \mathbf{B}_m \\ \cdots \\ \mathbf{B}_R \end{bmatrix} \mathbf{u} \tag{7.197}
$$

$$
\begin{bmatrix} \mathbf{y}_m \\ \cdots \\ \mathbf{y}_R \end{bmatrix} = \begin{bmatrix} \mathbf{C}_{mm} & \vdots & \mathbf{C}_{mR} \\ \cdots & \cdots & \cdots \\ \mathbf{C}_{Rm} & \vdots & \mathbf{C}_{RR} \end{bmatrix} \begin{bmatrix} \mathbf{y}_m \\ \cdots \\ \mathbf{z}_R \end{bmatrix} + \begin{bmatrix} \mathbf{0} \\ \cdots \\ \mathbf{D}_R \end{bmatrix} \mathbf{u}, \tag{7.198}
$$

where

$$
\begin{aligned}
\mathbf{A}_{mm} &= \mathbf{T}_m \mathbf{A} \mathbf{P}_m & \mathbf{A}_{mR} &= \mathbf{T}_m \mathbf{A} \mathbf{P}_R & \mathbf{A}_{Rm} &= \mathbf{T}_R \mathbf{A} \mathbf{P}_m \\
\mathbf{A}_{RR} &= \mathbf{T}_R \mathbf{A} \mathbf{P}_R & \mathbf{C}_{mm} &= \mathbf{C}_m \mathbf{P}_m & \mathbf{C}_{mR} &= \mathbf{C}_m \mathbf{P}_R \\
\mathbf{C}_{Rm} &= \mathbf{C}_R \mathbf{P}_m & \mathbf{C}_{RR} &= \mathbf{C}_R \mathbf{P}_R & \mathbf{B}_m &= \mathbf{T}_m \mathbf{B} \\
\mathbf{B}_R &= \mathbf{T}_R \mathbf{B}.
\end{aligned} \tag{7.199}
$$

Notice that from equations 7.193 and 7.196 it follows that

$$
\mathbf{C}_m = \mathbf{T}_m, \tag{7.200}
$$

and Equation 7.194 can be expanded as

$$
\mathbf{TP} = \begin{bmatrix} \mathbf{T}_m \\ \cdots \\ \mathbf{T}_R \end{bmatrix} \begin{bmatrix} \mathbf{P}_m \vdots \mathbf{P}_R \end{bmatrix}
$$

$$
= \begin{bmatrix} \mathbf{T}_m \mathbf{P}_m & \vdots & \mathbf{T}_m \mathbf{P}_R \\ \cdots & \cdots & \cdots \\ \mathbf{T}_R \mathbf{P}_m & \vdots & \mathbf{T}_R \mathbf{P}_R \end{bmatrix}.
$$

Therefore, according to Equation 7.195

$$\mathbf{T}_m\mathbf{P}_m = \mathbf{C}_m\mathbf{P}_m = \mathbf{C}_{mm} = \mathbf{I}$$
$$\mathbf{T}_m\mathbf{P}_R = \mathbf{C}_m\mathbf{P}_R = \mathbf{C}_{mR} = \mathbf{0}$$
$$\mathbf{T}_R\mathbf{P}_m = 0 \qquad\qquad (7.201)$$
$$\mathbf{T}_R\mathbf{P_R} = \mathbf{I}.$$

Since $\mathbf{y}_m$ is available as a measurement, it is only the unknown state vector $\mathbf{z}_R$ that needs to be determined. Expansion of Equation 7.197 gives the state equation for this unknown state vector as

$$\dot{\mathbf{z}}_R = \mathbf{A}_{Rm}\mathbf{y}_m + \mathbf{A}_{RR}\mathbf{z}_R + \mathbf{B}_R\mathbf{u}, \qquad (7.202)$$

and the measurements are

$$\mathbf{y}_m = \mathbf{C}_{mm}\mathbf{y}_m + \mathbf{C}_{mR}\mathbf{z}_R$$
$$= \mathbf{y}_m.$$

$$\mathbf{y}_R = \mathbf{C}_{Rm}\mathbf{y}_m + \mathbf{C}_{RR}\mathbf{z}_R + \mathbf{D}_R\mathbf{u},$$

which can be written as

$$\mathbf{y}_R - \mathbf{C}_{Rm}\mathbf{y}_m = \mathbf{C}_{RR}\mathbf{z}_R + \mathbf{D}_R\mathbf{u}, \qquad (7.203)$$

where $\mathbf{y}_m$ is known to be representing some of the untransformed states $\mathbf{x}_m$. Now, suppose that the unknown state vector $\mathbf{x}_R$, which has been transformed by $\mathbf{P}$ to $\mathbf{z}_R$, is estimated as $\hat{\mathbf{z}}_R$, this estimation will affect the measurement $\mathbf{y}_R$ only and the corresponding measurement residual $\tilde{\mathbf{y}}_r$ becomes

$$\tilde{\mathbf{y}}_r = (\mathbf{y}_R - \mathbf{C}_{Rm}\mathbf{y}_m) - \mathbf{C}_{RR}\hat{\mathbf{z}}_R. \qquad (7.204)$$

As done before, the estimation process requires that this residual be fed back to the estimation model through the estimator gain $\mathbf{L}_R$, which alters the model dynamics to

$$\dot{\hat{\mathbf{z}}}_R = \mathbf{A}_{Rm}\mathbf{y}_m + \mathbf{A}_{RR}\hat{\mathbf{z}}_R + \mathbf{B}_R\mathbf{u} + \mathbf{L}_R\tilde{\mathbf{y}}_r \qquad (7.205)$$

$$= \mathbf{A}_{Rm}\mathbf{y}_m + \mathbf{A}_{RR}\hat{\mathbf{z}}_R + \mathbf{B}_R\mathbf{u} + \mathbf{L}_R\left[(\mathbf{y}_R - \mathbf{C}_{Rm}\mathbf{y}_m) - \mathbf{C}_{RR}\hat{\mathbf{z}}_R\right]$$

$$= (\mathbf{A}_{RR} - \mathbf{L}_R\mathbf{C}_{RR})\hat{\mathbf{z}}_R + (\mathbf{A}_{Rm} - \mathbf{L}_R\mathbf{C}_{Rm})\mathbf{y}_m + \mathbf{B}_R\mathbf{u} + \mathbf{L}_R\mathbf{y}_R.$$

The transformed reduced state estimate $\hat{\mathbf{z}}_R$ is thus the solution of this equation. The estimator design is thus that of selecting the gain $\mathbf{L}_R$, which will ensure that the estimated vector $\hat{\mathbf{z}}_R$ gives a fair representation of the true states by minimizing the estimation error $\tilde{\mathbf{z}}_R$.

Defining the estimation error as

$$\tilde{\mathbf{z}}_R = \mathbf{z}_R - \hat{\mathbf{z}}_R. \tag{7.206}$$

It can be shown that by taking the time derivative of this error and using Equations 7.202 and 7.205, this error will have the dynamics described as

$$\dot{\tilde{\mathbf{z}}}_R = \mathbf{A}_{RR}\mathbf{z}_R - (\mathbf{A}_{RR} - \mathbf{L}_R\mathbf{C}_{RR})\,\hat{\mathbf{z}}_R + \mathbf{L}_R\mathbf{C}_{Rm}\mathbf{y}_m - \mathbf{L}_R\mathbf{y}_R, \tag{7.207}$$

which on replacing $\mathbf{y}_R$ and doing some simplifications and rearrangements, becomes

$$\begin{aligned} \dot{\tilde{\mathbf{z}}}_R &= (\mathbf{A}_{RR} - \mathbf{L}_R\mathbf{C}_{RR})\,\mathbf{z}_R - (\mathbf{A}_{RR} - \mathbf{L}_R\mathbf{C}_{RR})\,\hat{\mathbf{z}}_R - \mathbf{L}_R\mathbf{D}_R\mathbf{u} \\ &= (\mathbf{A}_{RR} - \mathbf{L}_R\mathbf{C}_{RR})\,\tilde{\mathbf{z}}_R - \mathbf{L}_R\mathbf{D}_R\mathbf{u}. \end{aligned} \tag{7.208}$$

Since $\mathbf{u}$ is a control forcing function, the homogeneous solution of this error dynamics equation is

$$\tilde{\mathbf{z}}_R(t) = e^{(\mathbf{A}_{RR} - \mathbf{L}_R\mathbf{C}_{RR})t}\tilde{\mathbf{z}}(0). \tag{7.209}$$

The estimate $\hat{\mathbf{z}}_R$ becomes more acceptable as a true estimate of the state $\mathbf{z}_R$ if the error $\tilde{\mathbf{z}}_R$ is minimized. For this error to decay to zero over a short time t, the closed-loop system $\mathbf{A}_{RR} - \mathbf{L}_R\mathbf{C}_{RR}$ must have eigenvalues that are large and negative, i.e., the roots of the characteristic equation

$$\alpha_r(s) \triangleq \det\left[s\mathbf{I} - (\mathbf{A}_{RR} - \mathbf{L}_R\mathbf{C}_{RR})\right] = 0 \tag{7.210}$$

should be large and negative. This calls for large gain $\mathbf{L}_R$. Reduced-order estimator thus becomes that of selecting the gain matrix $\mathbf{L}_R$ provide the reduced submatrices $\mathbf{A}_{RR}$ and $\mathbf{C}_{RR}$ are known according to Equation 7.210 where

$$\begin{aligned} \mathbf{A}_{RR} &= \mathbf{T}_R\mathbf{A}\mathbf{P}_R \\ \mathbf{C}_{RR} &= \mathbf{C}_R\mathbf{P}_R. \end{aligned}$$

The transformation matrices $\mathbf{P}_R, \mathbf{T}_R$ and the submatrix $\mathbf{C}_R$ are all predetermined as explained. Normally Equation 7.210 is written as

$$\alpha_r(s) \triangleq \det\left[s\mathbf{I} - (\mathbf{T}_R\mathbf{A} - \mathbf{L}_R\mathbf{C}_R)\,\mathbf{P}_R\right] = 0, \tag{7.211}$$

which essentially represent the same thing.

After this point, all that remains is the same as that of a full-order estimator, as explained previously. All methods that apply to the full-order estimator can also be applied to this reduced-order estimator. The estimator output in this case is $\hat{\mathbf{z}}_R$, however, it is required that the output should be $\hat{\mathbf{x}}_R$, therefore, the estimator output is then re-transformed back to the original state $\hat{\mathbf{x}}_R$ by

$$\hat{\mathbf{x}}_R = \mathbf{P}_R \hat{\mathbf{z}}_R. \qquad (7.212)$$

Notice that, if all measurements do not give the states directly, then $\mathbf{C}_m$ and hence $\mathbf{C}_{mm}$ will be dimensionless null matrices. Thus, $\mathbf{T}_m$ and $\mathbf{P}_m$ will also be dimensionless null matrices, which results in dimensionless matrices $\mathbf{A}_{mm}$, $\mathbf{A}_{Rm}$, $\mathbf{A}_{mR}$, $\mathbf{C}_{mm}$, $\mathbf{C}_{mR}$ and $\mathbf{C}_{Rm}$. This leaves matrix $\mathbf{A}_{RR}$ with the same dimension as that of $\mathbf{A}$ while matrix $\mathbf{C}_{RR}$ will have the dimension of $\mathbf{C}$. The reduced-order estimator then becomes the full-order estimator.

The main disadvantage with the estimator just described is that it requires that there must be some measurements $\mathbf{y}_R$ that do not provide sufficient information for determination of some of the elements in the state vector. If the whole of the measurements vector contains information that leads to direct determination of some of the state vector, but not the whole state, in which case $\mathbf{y}_R$ and hence $\mathbf{C}_R$ becomes zero, this method fails.

Consider, for example, for the state and measurement vector described by Equations 7.188 and 7.189, the output (measurement) matrices are

$$\mathbf{C} = \begin{bmatrix} 2 & 0 & 0 & 0 & 0 \\ 0 & 1 & 0 & 0 & 0 \\ 0 & 0 & 3 & 0 & 0 \end{bmatrix}$$

$$\mathbf{D} = [0],$$

and hence, the measurement vector is given by

$$\begin{bmatrix} y_1(t) \\ y_2(t) \\ y_3(t) \end{bmatrix} = \begin{bmatrix} 2 & 0 & 0 & 0 & 0 \\ 0 & 1 & 0 & 0 & 0 \\ 0 & 0 & 3 & 0 & 0 \end{bmatrix} \begin{bmatrix} x_1(t) \\ x_2(t) \\ x_3(t) \\ x_4(t) \\ x_5(t) \end{bmatrix}$$

$$= \begin{bmatrix} 2x_1(t) \\ x_2(t) \\ 3x_3(t) \end{bmatrix}. \qquad (7.213)$$

As can be seen, the whole of this measurement vector can be used to determine the states $x_1(t)$, $x_2(t)$ and $x_3(t)$ directly. There is no measurement component that is not directly linked (one-to-one link) to the state elements, as such $\mathbf{y}_R$ is zero, which then leads to $\mathbf{C}_R$ and $\mathbf{C}_{RR}$ being zero. This means the characteristic equation given in Equation 7.210 reduces to

$$\alpha_r(s) \triangleq \det [s\mathbf{I} - \mathbf{A}_{RR}] = 0, \qquad (7.214)$$

which does not help anything toward the estimator design. To resolve this problem, a different approach is used, although the fundamental principle of transforming the system so that the state vector can be shown as an augmented vector of the states which are directly available as measurements $\mathbf{y}_m$ and those which are not available directly from the measurements $\mathbf{z}_R$.

Since $\mathbf{C}_R$ is a null matrix of zero dimension, then both $\mathbf{T}_R$ and $\mathbf{P}_R$ and null matrices of zero dimension so that

$$\mathbf{T} = \mathbf{T}_m \qquad \text{and} \qquad \mathbf{P} = \mathbf{P}_m$$

Then

$$\begin{aligned} \mathbf{A}_{mm} &= \mathbf{T}_m \mathbf{A} \mathbf{P}_m \\ &= \mathbf{TAP} \end{aligned}$$

and

$$\begin{aligned} \mathbf{C}_{mm} &= \mathbf{C}_m \mathbf{P}_m \\ &= \mathbf{CP}, \end{aligned}$$

which can be partitioned as

$$\mathbf{C}_{mm} = \left[\, \mathbf{I}_m \;\vdots\; 0 \,\right], \tag{7.215}$$

where $\mathbf{I}_m$ is an identity matrix whose dimension is the same as that of the measurement vector. The transformed system matrix $\mathbf{A}_{mm}$ can be partitioned arbitrarily to conform with the partitioning of the transformed state vector $\mathbf{z}$ as

$$\mathbf{A} = \begin{bmatrix} \mathbf{A}_{11} & \vdots & \mathbf{A}_{12} \\ \cdots & \cdots & \cdots \\ \mathbf{A}_{21} & \vdots & \mathbf{A}_{22} \end{bmatrix} \tag{7.216}$$

and also

$$\mathbf{B} = \begin{bmatrix} \mathbf{B}_1 \\ \cdots \\ \mathbf{B}_2 \end{bmatrix}. \tag{7.217}$$

Therefore, the complete state description of the transformed system becomes

$$\begin{bmatrix} \dot{\mathbf{y}}_m \\ \cdots \\ \dot{\mathbf{z}}_R \end{bmatrix} = \begin{bmatrix} \mathbf{A}_{11} & \vdots & \mathbf{A}_{12} \\ \cdots & \cdots & \cdots \\ \mathbf{A}_{21} & \vdots & \mathbf{A}_{22} \end{bmatrix} \begin{bmatrix} \mathbf{y}_m \\ \cdots \\ \mathbf{z}_R \end{bmatrix} + \begin{bmatrix} \mathbf{B}_1 \\ \cdots \\ \mathbf{B}_2 \end{bmatrix} \mathbf{u} \tag{7.218}$$

$$\left[\mathbf{y}\right] = \left[\, \mathbf{I}_m \;\vdots\; 0 \,\right] \begin{bmatrix} \mathbf{y}_m \\ \cdots \\ \mathbf{z}_R \end{bmatrix}. \tag{7.219}$$

Expansion of these equations give

$$\dot{\mathbf{y}}_m = \mathbf{A}_{11}\mathbf{y}_m + \mathbf{A}_{12}\mathbf{z}_R + \mathbf{B}_1\mathbf{u} \qquad (7.220)$$

$$\dot{\mathbf{z}}_R = \mathbf{A}_{21}\mathbf{y}_m + \mathbf{A}_{22}\mathbf{z}_R + \mathbf{B}_2\mathbf{u} \qquad (7.221)$$

$$\mathbf{y} = \mathbf{y}_m. \qquad (7.222)$$

Now, if $\hat{\mathbf{z}}_R$ is a vector of the estimates of the unknown states, then the measurements residual $\mathbf{y}_r$ can be calculated directly from the measured state Equation 7.220 instead of the measurement Equation 7.222 as

$$\mathbf{y}_r = (\dot{\mathbf{y}}_m - \mathbf{A}_{11}\mathbf{y}_m - \mathbf{B}_1\mathbf{u}) - \mathbf{A}_{12}\hat{\mathbf{z}}_R. \qquad (7.223)$$

This is the basic difference between this approach and the previous approach. As usual, this residual is fed to the unknown state equation through the estimator gain $\mathbf{L}_R$ in the estimated dynamics of the following form

$$\dot{\hat{\mathbf{z}}}_R = \mathbf{A}_{21}\mathbf{y}_m + \mathbf{A}_{22}\hat{\mathbf{z}}_R + \mathbf{B}_2\mathbf{u} + \mathbf{L}_R\mathbf{y}_r. \qquad (7.224)$$

On replacing $\mathbf{y}_r$ from Equation 7.223, this equation can be expanded and rearranged as

$$\dot{\hat{\mathbf{z}}}_R = (\mathbf{A}_{22} - \mathbf{L}_R\mathbf{A}_{12})\,\hat{\mathbf{z}}_R + (\mathbf{A}_{21} - \mathbf{L}_R\mathbf{A}_{11})\,\mathbf{y}_m + (\mathbf{B}_2 - \mathbf{L}_R\mathbf{B}_1)\,\mathbf{u} + \mathbf{L}_R\dot{\mathbf{y}}_m. \qquad (7.225)$$

On replacing $\dot{\mathbf{y}}_m$ by Equation 7.220, and taking similar approaches as shown before for the other methods, the error dynamics can be expressed by using equations 7.221 and 7.225, which becomes

$$\dot{\tilde{\mathbf{z}}}_R = (\mathbf{A}_{22} - \mathbf{L}_R\mathbf{A}_{12})\,\tilde{\mathbf{z}}_R, \qquad (7.226)$$

which gives

$$\tilde{\mathbf{z}}_R(t) = e^{(\mathbf{A}_{22} - \mathbf{L}_R\mathbf{A}_{12})t}\tilde{\mathbf{z}}_R(0). \qquad (7.227)$$

According to the principle of estimation that requires this error to decay to almost zero, there is a need to select $\mathbf{L}$ large enough so that the roots of the equation

$$\alpha(s) \triangleq \det\left[s\mathbf{I} - (\mathbf{A}_{22} - \mathbf{L}_R\mathbf{A}_{12})\right] = 0 \qquad (7.228)$$

are large and negative. The steps that follow are the same as those discussed before. The estimator output will be $\hat{\mathbf{z}}_R$, and has to be re-transformed back to the original state of the system $\hat{\mathbf{x}}_R$ by the transformation

$$\hat{\mathbf{x}}_R = \mathbf{P}\hat{\mathbf{z}}_R. \qquad (7.229)$$

Thus, completing the estimation process.

7.9.4   Compensator Design: Control Law and Estimator

So far, the previous discussion on control and estimator design has been treating the two problems as independent of one another. No explicit mention of the effect of the dynamics of the estimator on that of the controller or vice versa, has been made. In both cases, it was assumed that there is no input reference signal $r(t)$ for the system to track, in which case the control system was a regulator. This section examines the combined system dynamics under the effect of both the controller and the estimator in the presence of the reference signal $r(t)$. Such a system is shown in Figure 7.8.

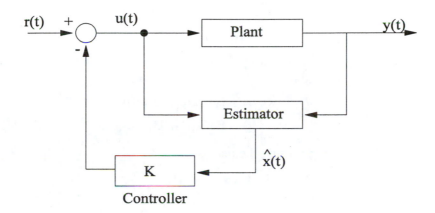

FIGURE 7.8
Combined Control Law and Estimator

Consider the plant given by

$$\dot{\mathbf{x}}(t) = \mathbf{A}\mathbf{x}(t) + \mathbf{B}\mathbf{u}(t)$$
$$\mathbf{y}(t) = \mathbf{C}\mathbf{x}(t) + \mathbf{D}\mathbf{u}(t),$$

and controlled by a full state feedback controller $\mathbf{K}$ in the presence of a reference signal $\mathbf{r}(t)$ such that the control signal becomes

$$\mathbf{u}(t) = -\mathbf{K}\mathbf{x}(t) + \mathbf{r}(t).$$

Since the actual state $\mathbf{x}(t)$ of the system for feedback is not available as mentioned in the previous sections, the estimated state $\hat{\mathbf{x}}(t)$ is the one that is fed back so that

$$\mathbf{u}(t) = -\mathbf{K}\hat{\mathbf{x}}(t) + \mathbf{r}(t), \tag{7.230}$$

where this estimated state is generated by an estimator that has a general form

$$\dot{\hat{\mathbf{x}}}(t) = \mathbf{A}_e\hat{\mathbf{x}}(t) + \mathbf{B}\mathbf{u}(t) + \mathbf{L}\mathbf{y}(t),$$

and $\mathbf{A}_e$ is the estimator closed-loop matrix. Recall that for a full-order estimator

$$\mathbf{A}_e = \mathbf{A} - \mathbf{LC} \qquad (7.231)$$

and for the reduced-order estimator, there are two forms used to represent $\mathbf{A}_e$, as discussed in the preceding section.

A popular design criterion for estimators is the reduction of the estimation error $\tilde{\mathbf{x}}(t)$ whose dynamics has been found to be

$$\begin{aligned}
\dot{\tilde{\mathbf{x}}}(t) &= \dot{\mathbf{x}}(t) - \dot{\hat{\mathbf{x}}}(t) \\
&= \mathbf{A}_e \left[\mathbf{x}(t) - \hat{\mathbf{x}}(t) \right] \\
&= \mathbf{A}_e \tilde{\mathbf{x}}(t). \qquad (7.232)
\end{aligned}$$

The closed-loop dynamics of the plant in the presence of both the controller and the estimator becomes

$$\begin{aligned}
\dot{\mathbf{x}}(t) &= \mathbf{A}\mathbf{x}(t) + \mathbf{B}\left[-\mathbf{K}\hat{\mathbf{x}}(t) + \mathbf{r}(t) \right] \\
&= \mathbf{A}\mathbf{x}(t) - \mathbf{BK}\hat{\mathbf{x}}(t) + \mathbf{B}\mathbf{r}(t), \qquad (7.233)
\end{aligned}$$

which, on eliminating $\hat{\mathbf{x}}$ using the relation

$$\tilde{\mathbf{x}}(t) = \mathbf{x}(t) - \hat{\mathbf{x}}(t)$$

gives

$$\dot{\mathbf{x}}(t) = (\mathbf{A} - \mathbf{BK})\mathbf{x}(t) + \mathbf{BK}\tilde{\mathbf{x}}(t) + \mathbf{B}\mathbf{r}(t). \qquad (7.234)$$

Similarly, the output equation becomes

$$\begin{aligned}
\mathbf{y}(t) &= \mathbf{C}\mathbf{x}(t) + \mathbf{D}\left[-\mathbf{K}\hat{\mathbf{x}}(t) + \mathbf{r}(t) \right] \\
&= \mathbf{C}\mathbf{x}(t) - \mathbf{DK}\hat{\mathbf{x}}(t) + \mathbf{D}\mathbf{r}(t) \\
&= (\mathbf{C} - \mathbf{DK})\mathbf{x}(t) + \mathbf{DK}\tilde{\mathbf{x}}(t) + \mathbf{D}\mathbf{r}(t). \qquad (7.235)
\end{aligned}$$

Combining Equations 7.232, 7.234 and 7.235 into matrix equations gives

$$\begin{bmatrix} \dot{\mathbf{x}}(t) \\ \dot{\tilde{\mathbf{x}}}(t) \end{bmatrix} = \begin{bmatrix} \mathbf{A} - \mathbf{BK} & \mathbf{BK} \\ 0 & \mathbf{A}_e \end{bmatrix} \begin{bmatrix} \mathbf{x}(t) \\ \tilde{\mathbf{x}}(t) \end{bmatrix} + \begin{bmatrix} \mathbf{B} \\ 0 \end{bmatrix} \mathbf{r}(t) \qquad (7.236)$$

$$\mathbf{y}(t) = \begin{bmatrix} \mathbf{C} - \mathbf{DK} & \mathbf{DK} \end{bmatrix} \begin{bmatrix} \mathbf{x}(t) \\ \tilde{\mathbf{x}}(t) \end{bmatrix} + [\mathbf{D}]\mathbf{r}(t). \qquad (7.237)$$

In most cases, the systems of interest are causal, in which case the feedfoward matrix $\mathbf{D}$ is zero, hence, the output equation can be written as

$$\mathbf{y}(t) = \begin{bmatrix} \mathbf{C} & 0 \end{bmatrix} \begin{bmatrix} \mathbf{x}(t) \\ \tilde{\mathbf{x}}(t) \end{bmatrix} \qquad (7.238)$$

The characteristic equation of this system is

$$\psi(s) = \begin{vmatrix} s\mathbf{I} - (\mathbf{A} - \mathbf{BK}) & -\mathbf{BK} \\ 0 & s\mathbf{I} - \mathbf{A}_e \end{vmatrix} = 0.$$

Since the augmented system is block triangular, the characteristic equation can be written as

$$\psi(s) = \det[s\mathbf{I} - (\mathbf{A} - \mathbf{BK})] \cdot \det[s\mathbf{I} - \mathbf{A}_e] = 0.$$

This indicates that the closed-loop poles of the whole system are just the poles of the plant that result from the selection of the feedback gain $\mathbf{K}$ and the desired estimator poles, as dictated by the estimator closed-loop system matrix $\mathbf{A}_e$, which depends on the choice of the estimator gain $\mathbf{L}$. Thus, the state feedback gain $\mathbf{K}$ and the estimator gain $\mathbf{L}$ can be selected separately for the desired closed-loop behavior. This means that the design of the feedback controller and that of the state estimator can be carried out separately. This is a special case of what is known as the separation principle, which holds when noise and other additive disturbances are included in the system.

The closed-loop transfer function $\mathbf{G}_{CL}(s)$ from the reference input signal $\mathbf{r}(t)$ to the output measurement $\mathbf{y}(t)$ can be determined using the same principles as discussed before.

Define

$$\mathbf{A}^* = \begin{bmatrix} \mathbf{A} - \mathbf{BK} & \vdots & \mathbf{BK} \\ \cdots & \vdots & \cdots \\ 0 & \vdots & \mathbf{A}_e \end{bmatrix} \qquad \mathbf{B}^* = \begin{bmatrix} \mathbf{B} \\ \cdots \\ 0 \end{bmatrix} \qquad (7.239)$$

$$\mathbf{C}^* = \begin{bmatrix} \mathbf{C} & \vdots & 0 \end{bmatrix} \qquad \mathbf{X}^* = \begin{bmatrix} \mathbf{x} \\ \cdots \\ \tilde{\mathbf{x}} \end{bmatrix}. \qquad (7.240)$$

The system and the output Equations 7.236 and 7.238 can then be written in compact form as

$$\dot{\mathbf{X}}^* = \mathbf{A}^*\mathbf{X}^* + \mathbf{B}^*\mathbf{r}$$
$$\mathbf{y} = \mathbf{C}^*\mathbf{X}^*$$

so that the closed-loop transfer function becomes

$$\mathbf{G}_{CL}(s) = \mathbf{C}^*(s\mathbf{I} - \mathbf{A}^*)^{-1}\mathbf{B}^*.$$

Reusing the original matrices given in 7.239 and 7.240 gives

$$\mathbf{G}_{CL}(s) = \begin{bmatrix} \mathbf{C} \vdots 0 \end{bmatrix} \begin{bmatrix} s\mathbf{I} - (\mathbf{A} - \mathbf{BK}) & \vdots & -\mathbf{BK} \\ \cdots & \vdots & \cdots \\ 0 & \vdots & s\mathbf{I} - \mathbf{A}_e \end{bmatrix}^{-1} \begin{bmatrix} \mathbf{B} \\ \cdots \\ 0 \end{bmatrix},$$

which, by the advantage of the triangularity of the block diagonal augmented matrix $\mathbf{A}^*$ reduces to

$$\mathbf{G}_{CL}(s) = \mathbf{C} \left[s\mathbf{I} - (\mathbf{A} - \mathbf{BK}) \right]^{-1} \mathbf{B}. \qquad (7.241)$$

This is exactly the same result as that obtained when full state feedback is applied.

In order for the estimator error to vanish quickly in the augmented system, it is required that the estimator poles be chosen to be faster than those of the closed-loop poles, at least twice as much.

7.10   Problems

Problem 7.1 *A linear system*

$$\dot{\mathbf{x}} = \mathbf{Ax} + \mathbf{Bu}$$
$$\mathbf{y} = \mathbf{Cx} + \mathbf{Du}$$

has $\mathbf{D} = \mathbf{0}$, *where the plant and input matrices* $(\mathbf{A}, \mathbf{B})$ *are given by*

$$(\mathbf{A}, \mathbf{B}) = \left(\begin{bmatrix} 3 & -2 & 0 \\ 0 & 0 & 1 \\ 4 & -3 & 0 \end{bmatrix}, \begin{bmatrix} 1 \\ 0 \\ 1 \end{bmatrix} \right)$$

(a) Discuss the controllability and stability of this system

(b) Determine the feedback matrix K that will place the poles of the system at $-1, -1 \pm j$.

Problem 7.2 *For the linear system*

$$\dot{\mathbf{x}} = \mathbf{Ax} + \mathbf{Bu}$$
$$\mathbf{y} = \mathbf{Cx} + \mathbf{Du},$$

$$\mathbf{A} = \begin{bmatrix} 0 & 0 & -1 \\ 1 & 0 & -3 \\ 0 & 1 & -3 \end{bmatrix}, \quad \mathbf{B} = \begin{bmatrix} 0 \\ 0 \\ 1 \end{bmatrix}, \quad \mathbf{C} = \begin{bmatrix} 0 & 1 & 0 \\ 0 & 0 & 1 \end{bmatrix}, \quad \mathbf{D} = [0].$$

(a) *Design a full-order state estimator for the system.*

(b) *Find an optimal control that minimizes the cost function*

$$= \int_0^\infty \left(\mathbf{X}^T \mathbf{Q} \mathbf{X} + \mathbf{u}^T \mathbf{R} \mathbf{u} \right) dt,$$

where

$$\mathbf{R} = \mathbf{I}, \quad \mathbf{Q} = \begin{bmatrix} 3 & 0 & 0 \\ 0 & 2 & 0 \\ 0 & 0 & 1 \end{bmatrix}.$$

Problem 7.3 *A linear system is described by*

$$\dot{\mathbf{x}} = \begin{bmatrix} -2 & 0 & 1 \\ 0 & -2 & 1 \\ 0 & -3 & -2 \end{bmatrix} \mathbf{x} + \begin{bmatrix} 1 \\ 2 \\ 2 \end{bmatrix} \mathbf{u}$$

$$\mathbf{y} = \begin{bmatrix} -1 & 1 & 0 \end{bmatrix} \mathbf{x}.$$

Using standard notation used in this chapter

(a) *Find the transfer function of this system and establish its controllability, observability and stability.*

(b) *Find a non-singular transformation T such that $T^{-1}AT$ is diagonal.*

(c) *Determine the state-transition matrix $\phi(s)$.*

Problem 7.4 *The state transfer function $G(s)$ is given by*

$$G(s) = \frac{s^2 + 7s + 10}{s^3 + 8s^2 + 19s + 122}.$$

(a) *Find the controllable and observable canonical forms of the system.*

(b) *Draw the signal flow graph for the canonical form in (a).*

Problem 7.5 *Optimization theory is a wide subject that was introduced in this chapter as a way toward optimal control theory. Use the given introduction to establish the minimum of*

$$f(\mathbf{x}) = x_1,$$

subject to the conditions

$$(x_1 - 1)^2 + x_2^2 = 1$$
$$(x_1 + 1)^2 + x_2^2 = 1,$$

where

$$\mathbf{x} = \begin{pmatrix} x_1 & x_2 \end{pmatrix}^T.$$

Problem 7.6 *Obtain a state-space representation for a system whose differential equation is given by*

$$\dddot{x} + 3\ddot{x} + 3\dot{x} + x = \dot{u} + u,$$

where the output is $y = x$.

(a) Use this result to determine the system transition matrix $\phi(t)$ and $\phi(s)$.

(b) Use Ackermann's formula to determine the controller K that places the roots of this system at $-1, -2 \pm 2j$.

Problem 7.7 *(a) Check the controllability and observability of the following two systems.*

$$\dot{x} = \begin{bmatrix} 1 & 4 & 3 \\ 0 & 2 & 16 \\ 0 & -25 & -20 \end{bmatrix} x + \begin{bmatrix} -1 \\ 0 \\ 0 \end{bmatrix} u, \qquad y = \begin{bmatrix} -1 & 3 & 0 \end{bmatrix} x$$

$$\dot{x} = \begin{bmatrix} 1 & 0 & 0 \\ 0 & 0 & 0 \\ -2 & -4 & -3 \end{bmatrix} x + \begin{bmatrix} 1 \\ 0 \\ -1 \end{bmatrix} u, \qquad y = \begin{bmatrix} 1 & 0 & 0 \end{bmatrix} x.$$

*(b) For each of the systems of part (a), find the controllable modes, uncontrollable modes, observable modes, and unobservable modes. (Hint: use the MATLAB commands **ctrbf** and **obsvf**.)*

(c) For each of the systems of part (a), assuming the states are available for feedback, determine whether a state feedback controller can be designed to stabilize the system.

(d) For each of the systems of part (a), assuming the states are not available for feedback, determine whether a controller together with an observer can be designed to stabilize the system.

Problem 7.8 *A certain system with state x is described by the state matrices*

$$A = \begin{bmatrix} -2 & 1 \\ -2 & 0 \end{bmatrix}, \quad B = \begin{bmatrix} 1 \\ 3 \end{bmatrix}$$

$$C = \begin{bmatrix} 1 & 0 \end{bmatrix}, \quad D = [0].$$

Find the transformation T so the if $x = Tz$, the state matrices describing the dynamics of z are in control canonical form. Compute the new matrices A_z, B_z, C_z, and D_z.

Problem 7.9 *Consider the control system shown below.*

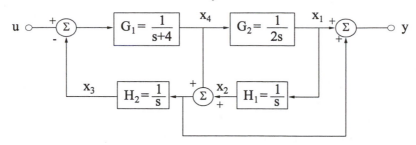

(a) *Find the transfer function from* $U(s)$ *to* $Y(s)$.

(b) *Write the state equations for the system using the state variables indicated.*

Problem 7.10 *Using the indicated state variables, obtain the state equations for the two systems shown below. Find the transfer function for each system using both block-diagram manipulation and matrix algebra.*

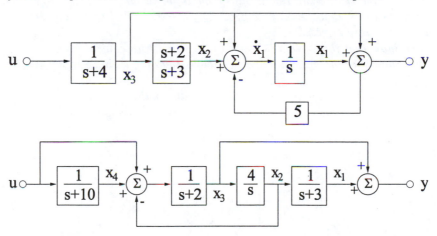

Problem 7.11 *For each of the transfer functions below, write the state equations in both control and observer canonical form. In each case, draw a block diagram and give the appropriate expressions for* **A, B,** *and* **C**

$$(a) \qquad \frac{s^2 + 1}{s^2(s^2 - 1)} \qquad (\textit{control of an inverted pendulum})$$

$$(b) \qquad \frac{3s + 4}{s^2 + 2s + 2}$$

Problem 7.12 *The linearized equations of motion for a satellite are*

$$\dot{\mathbf{x}} = \mathbf{Ax} + \mathbf{Bu}$$
$$\mathbf{y} = \mathbf{Cx} + \mathbf{Du}$$

where

$$\mathbf{A} = \begin{bmatrix} 0 & 1 & 0 & 0 \\ 3\omega^2 & 0 & 0 & 2\omega \\ 0 & 0 & 0 & 1 \\ 0 & -2\omega & 0 & 0 \end{bmatrix}, \quad \mathbf{B} = \begin{bmatrix} 0 & 0 \\ 1 & 0 \\ 0 & 0 \\ 0 & 1 \end{bmatrix}, \quad \mathbf{C} = \begin{bmatrix} 1 & 0 & 0 & 0 \\ 0 & 0 & 1 & 0 \end{bmatrix},$$

$$\mathbf{D} = [0]$$

$$\mathbf{u} = \begin{bmatrix} u_1 \\ u_2 \end{bmatrix}, \quad \mathbf{y} = \begin{bmatrix} y_1 \\ y_2 \end{bmatrix}.$$

The inputs u_1 and u_2 are the radial and tangential thrusts, the state variables x_1 and x_2 are the radial and angular deviations from the reference (circular) orbit, and the outputs y_1 and y_2 are the radial and angular measurements, respectively.

(a) Show that the system is controllable using both control inputs.

(b) Show that the system is controllable using only a single input. Which one is it?

(c) Show that the system is observable using both measurements.

(d) Show that the system is observable using only one measurement. Which one is it?

Problem 7.13 *An unstable robot system is described by the vector differential equation*

$$\frac{d}{dt} \begin{bmatrix} x_1 \\ x_2 \end{bmatrix} = \begin{bmatrix} 1 & 0 \\ -1 & 2 \end{bmatrix} \begin{bmatrix} x_1 \\ x_2 \end{bmatrix} + \begin{bmatrix} 1 \\ 1 \end{bmatrix} \mathbf{u}(t).$$

Both state variables are measurable, and so the control signal is set as

$$u(t) = -K(x_1 + x_2).$$

(a) Design gain K so that the performance index is minimized.

(b) Evaluate the minimum value of the performance index.

(c) Determine the sensitivity of the performance to a change in K. Assume that the initial conditions are

$$\mathbf{x}(0) = \begin{bmatrix} 1 \\ 1 \end{bmatrix}.$$

(d) Is the system stable without feedback signals due to $u(t)$?

Problem 7.14 *A feedback system has a plant transfer function*

$$G(s) = \frac{Y(s)}{R(s)} = \frac{K}{s(s+70)}$$

It is desired that the velocity error constant, K_v be 35 and the overshoot to a step be approximately 4% so that ξ is $1/\sqrt{2}$. The settling time (2% criterion) desired is 0.11 sec. Design an appropriate state variable feedback system.

Problem 7.15 *The following differential equations represent linear time-invariant systems. Write the dynamic equations (state equations and output equations) in vector form.*

(a)
$$\frac{d^2y(t)}{dt^2} + 4\frac{dy(t)}{dt} + y(t) = 5r(t)$$

(b)
$$2\frac{d^3y(t)}{dt^3} + 3\frac{d^2y(t)}{dt^2} + 5\frac{dy(t)}{dt} + 2y(t) = r(t)$$

(c)
$$\frac{d^3y(t)}{dt^3} + 5\frac{d^2y(t)}{dt^2} + 3\frac{dy(t)}{dt} + y(t) + \int_o^t y(\tau)d\tau = r(t)$$

(d)
$$\frac{d^4y(t)}{dt^4} + \frac{3}{2}\frac{d^3y(t)}{dt^3} + \frac{5}{2}\frac{dy(t)}{dt} + y(t) = 2r(t).$$

Problem 7.16 *A linear time-invariant system is described by the differential equation*

$$\frac{d^3y(t)}{dt^3} + 3\frac{d^2y(t)}{dt^2} + 3\frac{dy(t)}{dt} + y(t) = r(t).$$

(a) Let the state variables be defined as $x_1 = y$, $x_2 = dy/dt$, $x_3 = d^2y/dt^2$. Write the state equations of the system in vector-matrix form.
(b) Find the state-transition matrix $\phi(t)$ of $\mathbf{A}$.
(c) Let $y(0) = 1$, $dy(0)/dt = 0$, $d^2y(0)/dt^2 = 0$, and $r(t) = u_s(t)$. Find the state transition equation of the system.
(d) Find the characteristic equation and the eigenvalues of $\mathbf{A}$.

Chapter 8

Digital Control Systems

8.1 Introduction

With the breakthrough in computer technology (both hardware and software), there has been widespread use of computers as controllers for a broad range of dynamic systems. In fact, most control systems that are implemented today are based on computer control where digital computers (usually microprocessors) are used. However, a computer has three main features that distinguish its performance as a controller, from classical analog controllers. In the first place, the computer is a sequential machine that executes one instruction at a time, as such, the signals on which it works are discrete-time as opposed to continuous-time signals. On the other hand, the signals found from most physical systems are naturally continuous in time. Therefore, there arises a need to discretize the system signals so that they can be handled by the computer in digital control. Secondly, the computer works on finite digital signals (numbers) as opposed to analog signals, which are normally found in real-life systems. Again, this calls for signal digitization, which, together with the discretization, if not properly handled, will have some undesirable effects on the system being controlled. Furthermore, unlike analog electronics digital computers cannot integrate. Therefore, the differential equations describing system compensation must be approximated by reducing them to algebraic equations.

This chapter introduces the principles involved in digital control systems and discusses the issues that arise in the design and implementation of digital controllers. In particular, the motivation is to develop effective methods so that the digitization and discretization effects of continuous-time analog signals are eliminated or rather, minimized. The chapter proceeds by covering sampled-data systems, discrete-time systems, and the design of discrete time controllers. In particular, the discrete time PID controller is presented and appraised. Extensive examples, including MATLAB implementation, are used to illustrate the material.

8.2   Sampled Data Systems

8.2.1   General Structure

Under digital control, the controller is normally a digital computer, whereas the environment being controlled is usually characterized by continuous-time analog signals. This creates a system with a hybrid of signals, i.e., some points have analog signals while other points have digital signals. Such systems are generally known as sampled data systems, meaning that analog signals are sampled to represent discrete-time digital signals and vice versa. Some means of interfacing the digital computer to the analog environment must be provided. In most applications, the interfacing is provided through the *Digital to Analog* (D/A) and the *Analog to Digital* (A/D) converters.

The A/D converter takes analog measurements from analog environments and converts them to digital signals which can then be used by the computer. On the other hand, the D/A converter takes digital signals and converts them into equivalent analog signals, which can then be applied to the analog environment being controlled. Figure 8.1 shows the basic structure of the control system.

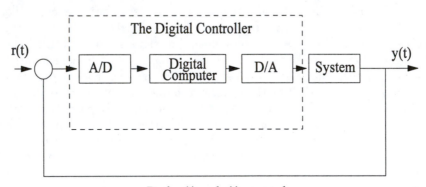

Both y(t) and r(t) are analog

FIGURE 8.1
Digital Control System: Analog Input and Output

8.2.2   Data Sampling

The process of data sampling can be viewed as an on-off switching process that takes on analog signals at some specific interval resulting in a train of

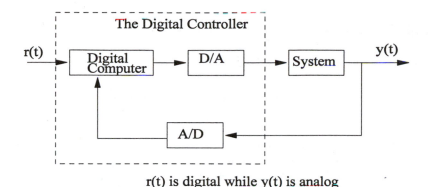

r(t) is digital while y(t) is analog

FIGURE 8.2
Control Systems Structure: Analog Output and Digital Input

impulses. At the end of each switching pulse, the sampled signal is held at that level until the next sampling interval. In digital control terminology, the interface that takes on analog signals and converts them to digital signals in this fashion is called the *sample and hold*. There are two elements in the sample-and-hold unit. The first unit is called the sampler, which does the switching, and the second unit is called the hold, which, as the name implies, holds the sampled signal at the same level until the next sampling instant. This section discusses the principles of the sample-and-hold process, which is divided into two parts, the *"sampler"* and the *"hold."*

8.2.2.1 Analysis of the Data Sampler

As noted before, the data sample can be viewed as an on and off switch that takes on analog signals, resulting in a train of impulses. A simple schematic diagram of the data sampler is shown in Figure 8.3.

FIGURE 8.3
Simple Data Sampler

The process sampling can be expressed as a product of the analog signal being sampled $r(t)$ and the train of impulses $\Sigma\delta(t - kT)$ where T is the sampling interval and k is a series of integers, 0,1,2,3,4..... so that the sampled signal $r^*(t)$ or the discrete signal can be expressed as

$$r^*(t) = \sum_{k=-\infty}^{\infty} r(t)\delta(t - kT), \tag{8.1}$$

where $\delta(t)$ is the Dirac delta function. This mathematical representation is technically known as impulse modulation and, in the analysis, it is represented as a sample-and-hold process. The time interval between the consecutive sampling instants is known as the sampling interval, and for periodic sampling, it is constant. The sampling frequency or simply the sample rate, expressed in samples per second (or Hertz), is the reciprocal of the sampling interval

$$f_s = \frac{1}{T} \quad (Hz),$$

where f_s is the sampling frequency. Note that if N_s is the number of discrete samples per unit time(say per second) then there will be N_s sampling intervals T in this unit time. Thus

$$N_s T = 1, \tag{8.2}$$

which shows that the frequency f_s equals the number of discrete-time samples N_s per unit time. Most often it becomes necessary to express the sampling frequency in radians per second. Mathematically,

$$\omega_s = 2\pi f_s = \frac{2\pi}{T},$$

where ω_s is the circular or sampling frequency in radians per second. Mathematical analysis of the sampled signal can be carried out using the Laplace transform analysis, Fourier analysis or $\mathcal{Z}$-transform analysis. The Fourier transform and the $\mathcal{Z}$-transform analysis will be described later, at this stage the Laplace transform (a more familiar technique) analysis is presented.

If the Laplace transform of the sampled signal is carried out on Equation 8.1 then

$$\mathcal{L}\left[r^*(t)\right] = R^*(s) = \int_{-\infty}^{\infty} r^*(\tau)e^{-s\tau}d\tau,$$

which gives

$$R^*(s) = \int_{-\infty}^{\infty} \sum_{k=-\infty}^{\infty} r(\tau)\delta(\tau - kT)e^{-s\tau}d\tau.$$

Interchanging the summation and the integration gives

$$R^*(s) = \sum_{k=-\infty}^{\infty} \int_{-\infty}^{\infty} r(\tau)e^{-s\tau}\delta(\tau - kT)d\tau. \tag{8.3}$$

If $\tau = kT$, this equation simplifies to

$$R^*(s) = \sum_{k=-\infty}^{\infty} e^{-skT} \int_{-\infty}^{\infty} r(\tau)\delta(\tau - kT)d\tau. \tag{8.4}$$

For all functions $f(t)$ which are continuous at ξ, the impulse function $\delta(t)$ has a shifting property given by

$$\int_{-\infty}^{\infty} f(t)\delta(t - \xi)dt = f(\xi).$$

This property, when used in Equation 8.4 gives

$$R^*(s) = \sum_{k=-\infty}^{\infty} e^{-skT} r(kT). \tag{8.5}$$

The sampled signal is normally defined for positive integers k. Hence, the range of integers from $-\infty$ to 0 (two-sided Laplace transform given in this equation) are rarely used. In most cases, it is the one-sided Laplace transform defined as

$$R^*(s) = \sum_{k=0}^{\infty} e^{-skT} r(kT) \tag{8.6}$$

that is used. This transform, Equation 8.6, gives the continuous-time model of the sampled data signal and also a mathematical model of the sampler.

8.2.2.2 Analysis of the Hold Operation.

The hold operation takes the impulses produced by the sampler to produce a piecewise constant signal of the sample-and-hold device. It is normally presented as a linear filter. Depending on the degree at which the sampled signal is held, several forms of hold units can be defined. However, the most common forms of hold are the *zero-order hold* (ZOH), and the *first-order hold* (FOH). The ZOH maintains the sampled signal at a constant level, while the FOH holds the signal linearly between the sampling instants. Since, in sampling processes, the next sampling value is never known, this type of hold (FOH) is never used in the sample-and-hold operation, though it might be used in the signal recovery (data extrapolation) process, which will be discussed in the next sections. The only feasible hold used in this case is the ZOH that is modeled as shown in Figure 8.4.

For any sampled signal input $r^*(t)$, the output of the ZOH is defined as

$$r_H(t) = r(kT) \qquad kT \le t \le (k+1)T.$$

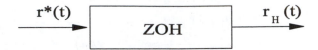

FIGURE 8.4
ZOH Model

Therefore, its transfer function can be defined as

$$ZOH(s) = \frac{\mathcal{L}[r_H(t)]}{\mathcal{L}[r^*(t)]}.$$ (8.7)

Considering single sampling instants, the ZOH receives only one impulse from the sampler, not the whole train of impulses, therefore,

$$r^*(t) = \delta(t).$$

If this impulse is of unit strength, then $r_H(t)$ will be a pulse of unit height and duration T (equal to sampling interval). Using the time delay function, the impulse response of the hold can be shown to be

$$r_H(t) = 1(t) - 1(t - T),$$

where $1(t)$ is the unit step function. Therefore, the transfer function in Equation 8.7 can be derived as follows:

$$ZOH(s) = \frac{\mathcal{L}[r_H(t)]}{\mathcal{L}[r^*(t)]}$$

$$= \frac{\mathcal{L}[1(t) - 1(t - T)]}{\mathcal{L}[\delta(t)]}$$

$$= \frac{\dfrac{1}{s} - \dfrac{e^{-sT}}{s}}{1},$$

so that

$$ZOH(s) = \frac{1 - e^{-sT}}{s}.$$ (8.8)

Thus, the whole A/D converter, which acts as a sample-and-hold unit, can be modeled as a combination of a sampler and a ZOH whose diagrammatic representation is shown in Figure 8.4.

   In real practice, there is no interest in the intermediate signal $r^*(t)$. Normally it is the output of the hold unit that is of interest, because even if one wanted $r^*(t)$, there is no way it could be obtained from the A/D

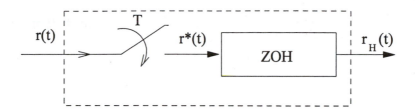

FIGURE 8.5
Sampler and ZOH

converter. For this reason, the transfer function of the whole A/D converter is regarded as just the transfer function of the ZOH unit. For notational convenience, the sampled signal $r(kT)$ where T is the sampling period will just be presented as $r(k)$ unless where the context makes it necessary to present it as $r(kT)$. Generally, signals of the form $r(kT \pm nT)$ for any $k, n = 1, 2, 3, 4...$ will be presented simply as $r(k \pm n)$.

Example 8.1 *A continuous-time signal*

$$r(t) = 2 \sin 4t + \cos 2t$$

is sampled at a sampling rate of 10 rad/sec using a ZOH. If the sampling starts at the time when $t = 0$, determine the sampling interval T, sample rate in samples per sec and the sampled value when $t = 4$ sec.

Solution 8.1 *Since,* $\omega_s = \dfrac{2\pi}{T}$, *the sampling interval can be calculated as*

$$T = \frac{2\pi}{\omega_s}$$

$$= \frac{2\pi}{10} = 0.628 \ sec.$$

Also the sampling rate in samples per sec is

$$N_s = f_s = \frac{1}{T} = 1.59 \quad samples/sec$$

at $t = 4$ sec, the complete samples covered will be

$$N = trunc\left(\frac{t}{T}\right)$$

$$= trunc\left(\frac{4}{0.628}\right)$$

$$= 6,$$

i.e., the whole number of the samples covered so far due to ZOH operation is $k = 6$. Since the sample value at any instant k is

$$r(k) = r(t)\delta(t - k)$$
$$= (2\sin 4t + \cos 2t)(\delta(t - k)).$$

The sample value at $k = 6$ becomes

$$r(6) = 2\sin(24) + \cos(12)$$
$$= -0.967.$$

8.2.3   Characteristics of Discrete Time Signals

8.2.3.1   Fourier Series Representation of Discrete Time Signals.

It is well known that any periodic signal has the property of repeating itself. For continuous-time signals, the periodic property is expressed as

$$r(t) = r(t + T), \tag{8.9}$$

where T is the period, in time units, of the signal when the signal has the same value as that in the previous period. For discrete-time signals the periodicity is measured in terms of the number of the signals. Normally the periodic property of discrete-time signals is given as

$$r(k) = r(k + N), \tag{8.10}$$

where N is the period given in number of samples that the discrete samples assume their previous values cyclically. By Fourier theory, any periodic signal can be represented as a sum of the harmonics of the fundamental frequency $\omega = \dfrac{2\pi}{N}$. (The reader should not confuse this period N and the sampling frequency N_s given in Equation 8.2). This is done for periodic discrete-time signals as

$$r(k) = \sum_{n=-\infty}^{\infty} a_n e^{2\pi j(\frac{n}{N})k}, \tag{8.11}$$

where $j = \sqrt{-1}$. To determine the coefficients a_n, let $n = m$ and multiply both sides of this expression by $e^{-j(\frac{2\pi n}{N})k}$ to get

$$r(k)e^{-j(\frac{2\pi n}{N})k} = \sum_{m=-\infty}^{\infty} a_m e^{2\pi j(\frac{m-n}{N})k}$$

Since $r(k) = r(k + N)$ and there are N distinct values of n in one period then summing both sides of this equation over one period $[0, N - 1]$ then

$$\sum_{k=0}^{N-1} r(k) e^{-j(\frac{2\pi n}{N})k} = \sum_{k=0}^{N-1} \sum_{m=-\infty}^{\infty} a_m e^{2\pi j(\frac{m-n}{N})k}$$

$$= \sum_{m=-\infty}^{\infty} a_m \sum_{k=0}^{N-1} e^{2\pi j(\frac{m-n}{N})k} \qquad (8.12)$$

If $(m - n)$ is an integer multiple of N, then

$$e^{2\pi j(\frac{m-n}{N})k} = e^{2\pi jc} \qquad \text{where} \quad c = 0, \pm 1, \pm 2, \cdots, \cdots$$

$$= \cos 2\pi c$$

$$= 1 \qquad \text{for all } c.$$

Therefore,

$$\sum_{k=0}^{N-1} e^{2\pi j(\frac{m-n}{N})k} = \sum_{k=0}^{N-1} 1 = N$$

On the other hand, if $\dfrac{m - n}{N}$ is not an integer multiple of N, i.e., $\dfrac{m - n}{N}$ $\neq c$ for some integer value c, then by letting

$$e^{2\pi j(\frac{m-n}{N})} = \xi$$

it follows that

$$\sum_{k=0}^{N-1} e^{2\pi j(\frac{m-n}{N})k} = \sum_{k=0}^{N-1} \xi^k$$

$$= \frac{1 - \xi^N}{1 - \xi}$$

$$= \frac{1 - e^{2\pi j(m-n)}}{1 - e^{2\pi j(\frac{m-n}{N})}}$$

$$= 0. \qquad (8.13)$$

These results can then be combined in a single expression

$$\sum_{k=0}^{N-1} e^{2\pi j(\frac{m-n}{N})k} = N\delta(m - n - cN)$$

so that finally

$$\sum_{k=0}^{N-1} r(k)e^{-j(\frac{2\pi n}{N})k} = \sum_{m=-\infty}^{\infty} Na_m \delta(m-n-cN).$$

The summation on the right is carried over N consecutive values of m for a fixed value of n, therefore, the value that c can take in the range of the summation is zero for which the only non-zero value in the sum is when

$$\delta(m-n) = 1$$

this requires that the non-zero value is when $m = n$ thus

$$\sum_{k=0}^{N-1} r(k)e^{-j(\frac{2\pi n}{N})k} = Na_n$$

or

$$a_n = \frac{1}{N}\sum_{k=0}^{N-1} r(k)e^{-j(\frac{2\pi n}{N})k} \qquad (8.14)$$

Since this summation is periodic with period N, it can be taken over N successive values of N. The same observation can be made for Equation 8.11. Based on these facts, for simplicity both these equations can be written as

$$r(k) = \sum_{n=0}^{N-1} a_n e^{2\pi j(\frac{n}{N})k}$$

$$a_n = \frac{1}{N}\sum_{k=0}^{N-1} r(k)e^{-j(\frac{2\pi n}{N})k}, \qquad (8.15)$$

which together form the discrete-time Fourier series pair expressing the discrete-time signal. Using the fact that

$$r(k) = r(k+N)$$

It can also be shown that

$$a_n = a_{n+N}$$

8.2.3.2   Spectrum of Sampled Signal

The spectrum of the sampled signal refers to the frequency components found in a sampled data signal. To study the spectrum of the sampled data signal, deep understanding of the Discrete Fourier transform (DFT) is required. Recall that the sampled signal is expressed as

$$r^*(t) = \sum_{k=-\infty}^{\infty} r(t)\delta(t - kT)$$

$$= r(t) \sum_{k=-\infty}^{\infty} \delta(t - kT) \tag{8.16}$$

If the sampling interval is constant, then by using the Fourier series the impulse train can be expressed as

$$\sum_{k=-\infty}^{\infty} \delta(t - kT) = \sum_{n=-\infty}^{\infty} C_n e^{2\pi j(\frac{n}{N})t} \tag{8.17}$$

where the Fourier coefficients C_n are given as

$$C_n = \frac{1}{T} \int_{\frac{-T}{2}}^{\frac{T}{2}} \sum_{k=-\infty}^{\infty} \delta(t - kT) e^{-jn(\frac{2\pi t}{T})} dt \tag{8.18}$$

Since the integral is taken within the range $\left[\dfrac{-T}{2}, \dfrac{T}{2}\right]$, then the impulses in the summation for which this integral holds occurs when $k = 0$, thus the expression for the Fourier coefficients reduce to

$$C_n = \frac{1}{T} \int_{\frac{-T}{2}}^{\frac{T}{2}} \delta(t) e^{-jn(\frac{2\pi t}{T})} dt, \tag{8.19}$$

and since

$$\int_{\frac{-T}{2}}^{\frac{T}{2}} \delta(t) e^{-jn(\frac{2\pi t}{T})} dt = 1$$

then the Fourier coefficients for the whole pulse train assume only one value given as

$$C_n = \frac{1}{T}. \tag{8.20}$$

Hence, the Fourier series of the sum of pulses becomes

$$\sum_{k=-\infty}^{\infty} \delta(t - kT) = \frac{1}{T} \sum_{n=-\infty}^{\infty} e^{2\pi j(\frac{n}{N})t}. \tag{8.21}$$

On introducing the sampling frequency $\omega_s = \dfrac{2\pi}{T}$ in radians per second, Equation 8.21 becomes

$$\sum_{k=-\infty}^{\infty} \delta(t - kT) = \frac{1}{T} \sum_{n=-\infty}^{\infty} e^{jn\omega_s t}. \tag{8.22}$$

Using this equation, the sampled signal in Equation 8.16 is expressed as

$$r^*(t) = \frac{r(t)}{T} \sum_{n=-\infty}^{\infty} e^{jn\omega_s t}. \tag{8.23}$$

The Laplace transform of Equation 8.23 is given by

$$R^*(s) = \mathcal{L}(r^*(t))$$

$$= \int_{-\infty}^{\infty} \frac{r(t)}{T} \sum_{n=-\infty}^{\infty} e^{jn\omega_s t} e^{-st} dt$$

$$= \frac{1}{T} \sum_{n=-\infty}^{\infty} \int_{-\infty}^{\infty} r(t) e^{-(s-jn\omega_s)t} dt.$$

Using the shift in frequency theorem for Laplace transforms it can be noted that

$$\int_{-\infty}^{\infty} r(t) e^{-(s-jn\omega_s)t} dt = R(s - jn\omega_s), \tag{8.24}$$

thus

$$R^*(s) = \frac{1}{T} \sum_{n=-\infty}^{\infty} R(s - jn\omega_s)$$

$$= \frac{1}{T} \sum_{n=0}^{\infty} R(s \pm jn\omega_s)$$

$$= \frac{1}{T} \sum_{n=0}^{\infty} R\left[j\left(\omega_o \pm n\omega_s\right)\right]. \tag{8.25}$$

This indicates that the sampled signal train has an infinite number of frequencies that are all integral multiples of the sampling frequency ω_s. This situation can be illustrated graphically.

8.2.3.3   Aliasing (Frequency Folding) and the Sampling Theorem

It has been shown that the sampled data signal contains frequencies that are multiples of the sampling frequency. If the fundamental signal $r(t)$

contains a frequency ω, then by defining

$$s = j\omega$$

the sampled signal $R^*(s)$ can be defined in the frequency domain as

$$R^*(\omega) = \frac{1}{T} \sum_{n=-\infty}^{\infty} R(j(\omega \pm n\omega_s)), \qquad (8.26)$$

which indicates that the sampled signal will have frequencies $\omega \pm n\omega_s$ for some integer n. However, generally the signal being sampled will not contain only one frequency; it will have several frequency components defining its natural frequency band (spectrum) as shown. Therefore, the sampled signal frequency spectrum becomes an infinite number of these individual spectra spaced by an interval ω_s corresponding to the sampling frequency. Now, if the sampling frequency is low, then the low- and the high-frequency components of each of these spectra will overlap with the low frequency components of the adjacent spectrum. This phenomenon of overlapping is known as aliasing or frequency folding. The lowest frequency component at which aliasing occurs is known as the aliasing frequency ω_a and is given by

$$\omega_a = |\omega_s \pm \omega|$$

Aliasing is a phenomenon that should be avoided when designing digital controllers because, when this occurs, it becomes impossible to recover all the fundamental frequency components. It should be recalled that although under digital control one deals with discrete signals, the ultimate goal is to recover the continuous-time (analog) signals that are natural using the D/A converter. This goal cannot be reached when aliasing occurs.

The phenomenon of aliasing has a clear meaning when explained in the time domain. Consider two sinusoidal signals at frequencies ω_1 and ω_2 where $\omega_1 \neq \omega_2$ (assume $\omega_1 = \dfrac{1}{2000}$, and $\omega_2 = \dfrac{1}{200}$ as shown in Figure 8.6). Clearly, one sees from Figure 8.6 that the sampled signal in both $r_1(t)$ and $r_2(t)$ are the same. In that case, recovery of both $r_1(t)$ and $r_2(t)$ is not uniquely possible. Generally, the two different frequency signals appear to be the same after sampling. It can be inferred from this observation that in order to preserve the signal information during the sampling operation, the sampling instants must be sufficiently close.

Several measures can be taken against the effects aliasing phenomena in digital control such as the use of anti-aliasing filters, and the application of Shannon's sampling theorem. These will be discussed in detail in the later sections. Presently, the sampling theorem that results will be discussed directly from the observation of the sampled signal spectrum. Before discussing the theorem, the term will be encountered mostly in the discussion of sampling is defined and its related theorem.

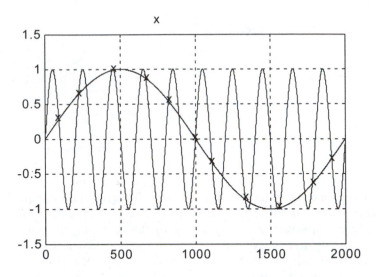

FIGURE 8.6
Two Sinusoidal Signals

Definition 8.1 *If, in the sampled data system, ω_s is the sampling frequency in radians per second, then half the value of this frequency is called the Nyquist frequency $\omega_N = \frac{\omega_s}{2}$. This is the maximum frequency in the fundamental signal that can be recovered completely without distortion after sampling.*

A corollary to this definition of the sampling theory is the following theorem:

Theorem 8.1 *A continuous-time signal $r(t)$ with frequency spectrum $(-\omega_0, \omega_0)$ can be represented uniquely by its values in equidistant points if the sampling frequency ω_s is higher than $2\omega_0$ as*

$$r(t) = \sum_{k=-\infty}^{\infty} r(k) \frac{sin\left((\omega_s(t - kT))/2\right)}{(\omega_s(t - kT)/2}. \tag{8.27}$$

Proof. This theorem has two parts: one that states the conditions under which sampling will not distort the signal and another that shows how to reconstruct the continuous-time signal. The proof for the second part will be given under the discussion on data extrapolators and impostors. Only the proof for the first part is given here. It is based on the observation that the samples $r(k)$ can be regarded as the coefficients of the Fourier series of the sampled signal and that the function being sampled is zero outside the frequency range $(-\omega_0, \omega_0)$

It is known that the Fourier transform of $r(t)$ is

$$R(\omega) = \int_{-\infty}^{\infty} r(t)e^{-j\omega t}dt$$

so that

$$r(t) = \int_{-\infty}^{\infty} R(\omega)e^{j\omega t}d\omega$$

where the frequency of the sampled signal is

$$\omega_d = \omega \pm n\omega_s$$
$$n = 0, 1, 2, \cdots$$

The following function can be defined

$$R_s(\omega_d) = \frac{1}{T}\sum_{n=-\infty}^{\infty} R(\omega + n\omega_s), \tag{8.28}$$

where $T = \dfrac{2\pi}{\omega_s}$ is the sampling interval. The Fourier expansion of this function is

$$R_s(\omega) = \sum_{n=-\infty}^{\infty} C_k e^{-j\omega kT}$$

with Fourier coefficients C_k given as

$$C_k = \int_0^{\omega_s} R_s(\omega)e^{j\omega kT}d\omega$$

which simplifies to

$$C_k = r(k).$$

The theorem requires that, for this result to exist, the function must be zero outside the frequency rage $(-\omega_0, \omega_0)$, therefore, it follows that

$$R(\omega_d) = \begin{cases} TR_s(\omega_d) & |\omega_d| \le \omega_0 \\ 0 & |\omega_d| > \omega_0 \end{cases}$$

Since $\omega_d = \omega_0 + n\omega_s$, for adjacent spectra, take $n = 1$ so that $\omega_d = \omega_0 + \omega_s$ which then is the condition for which data recovery for the sampled signal is possible. ■

8.2.3.4   Data Extrapolation and Impostors

The process of recovering the continuous-time signal from the sampled data is known as the data extrapolation process. If anti-aliasing measures are taken properly during the sampling process, it becomes possible to recover the information from the sampled data signal. The ideal element in the recovery would be a low pass filter that fills the gaps between the data samples by waves that have no frequencies above half the sampling frequency. However, such a filter is not causal in that it starts the extrapolation process at negative infinity time while the required signal occurs at zero time. For this and other reasons, a polynomial hold is normally used instead. There are two types of polynomial holds; the zero-order hold (ZOH) and the first-order hold (FOH). The ZOH is a model that maintains a constant signal between the samples. A/D converters discussed in the previous sections are good examples of ZOH and, as seen before, it is normally modeled as a ZOH.

The best alternative to ZOH is the FOH, which extrapolates data between sampling periods by a first-order polynomial. However, because of hardware complexity associated with FOH, the ZOH has remained popular in almost all control applications. Be it a ZOH or FOH extrapolator, the output goes with unwanted harmonics known as impostors. In practice, the impostors are equally well processed with the required signal and then filtered out in the final stages using a low pass filter that leaves only the fundamental signal.

8.2.3.5   Quantization Effects

Since the sampled data must be processed by a computer that handles finite numbers with finite accuracy, this signal must thus be presented to the computer in that way. However, in practice, the control variables, despite being inaccurate, they are not generally finite numbers as presented to the computer. This produces errors in the control computation that are generally known as errors due to quantization, or simply quantization effects. These quantization effects can sometimes be very detrimental to the performance of the controller if not properly addressed. In this section, some of the quantization effects, in particular the round-off errors, are discussed. Limit cycles that can also occur as quantization effects are introduced, but a detailed treatment will be given in the chapter that covers nonlinear control (Chapter 9).

8.3 Analysis of the Discrete Time Systems

8.3.1 Difference Equations

It has been seen that the A/D converter feeds the digital computer samples of discrete-time signals $r(k)$, which are sampled from the analog signal $r(t)$ of the physical system at discrete-times. The computer takes these sampled values and uses some predetermined rule to compute the output digital signals, which are fed to the physical system as analog signals through the D/A converter. The issue of concern to control engineers is the relationship between the digital computer discrete inputs and outputs. Normally, the computer generates the output signals based not only on its current input but also on some of the past input and output values. If the input signals to the computer up to the k-th sample are $r(0)$, $r(1)$, $r(3)$, $\cdots$, $\cdots$, $r(k)$ and the output signal prior to the sample times are $u(0)$, $u(1)$, $u(3)$, $\cdots$, $\cdots$, $u(k-1)$, then the computer output at the sample time $u(k)$ will be a function of both these sample values of the inputs and the outputs that can be expressed symbolically as

$$u(k) = f(r(0), r(1), r(3), \cdots , \cdots , r(k), u(0), u(1), u(3), \cdots , \cdots , u(k-1)).$$
$$(8.29)$$

The function f may be of any form, however, at present, it will be assumed to be linear depending only on a *finite* number of past values of the outputs and inputs. The word finite has been stressed to show that not <u>all</u> past values of inputs and outputs are actually used in the computation of the current output, though this looks like a more accurate approach. This is a consequence of the fact that computer memory is limited and cannot store all the possible past values. Imagine a process that continuously samples throughout a year at a sampling interval of 1 m sec.

In this way, suppose that only n-past values of the output signal $u(.)$ and m-past values of the input signal $r(.)$ are required to compute the output signal at some instant k and the corresponding relationship between the these values is linear. Equation 8.29 can thus be written as

$$u(k) = b_0 r(k) + b_1 r(k-1) + b_2 r(k-2) + \cdots\cdots + b_m r(k-m)$$
$$+ a_1 u(k-1) + a_2 u(k-2) + \cdots\cdots + a_n u(k-n). \qquad (8.30)$$

This equation is known as a linear recurrence equation, or, simply, the difference equation. In standard form, it is normally arranged to have all the like terms collected on one side and expressed in one term as a sum. This form is

$$\sum_{i=0}^{n} a_i u(k-i) = \sum_{j=0}^{m} b_j r(k-j), \qquad (8.31)$$

where a_0 is normalized to 1. The signal $u(k - i)$ is the system output (or response) and $r(k - j)$ is the forcing function. If the coefficients a_i and b_j are constants, the difference equation is known as the constant coefficient difference equation (CCDE) and has many similarities to the constant coefficient linear ordinary differential equation. It may be taken as the differential equivalent of a continuous-time system. The solution of this difference equation gives the value of the computer output at the sampling instant. In most cases, this computer output is the control signal required to provide the control action.

In discrete-time systems, the difference Equation 8.31 is one of the most fundamental system characteristics that must be known to carry out the analysis. There are several ways of presenting and analyzing this difference equation. Before discussing these alternate ways of presenting the difference Equation 8.31, a general approach to solving such an equation is presented.

8.3.1.1   Solution of the Difference Equation

As seen in the preceding sections, the control signal $u(k)$ computed by the digital computer is just the solution of the difference Equation 8.31. There are many other situations where the solution of this difference equation is important. In any case, some previous values prior to the sampling instant k must be known and, as noted before, these are the initial conditions that must be specified.

Similar to the solutions of linear ordinary differential equations, this solution has two components: the homogeneous solution $u_h(k)$, which depends only on the initial conditions, and the particular solution $u_p(k)$, which depends on the input forcing function. The general solution can be expressed by

$$u(k) = u_h(k) + u_p(k). \tag{8.32}$$

Like solutions of the ordinary differential equations, the homogeneous solution is also known as the free response of the system, while the particular solution is known as the forced response.

Consider a general expansion of the difference Equation 8.31, which can be given as

$$a_0 u(k) + a_1 u(k - 1) + a_2 u(k - 2) + \cdots + a_n u(k - n)$$
$$= b_0 r(k) + b_1 r(k - 1) + b_2 r(k - 2) + \cdots + b_m r(k - m)$$

The solution of this equation is

$$u(k) = \frac{1}{a_0} \left[\sum_{j=0}^{m} b_j r(k - j) - \sum_{i=1}^{n} a_i u(k - i) \right] \tag{8.33}$$

Now if for all $j = 0, 1, 2, \cdots m$ and $i = 1, 2, 3, \cdots n$ the values of $r(k-j)$ and $u(k-i)$ are known, and the coefficients a_i and b_j are also known, then the solution $u(k)$ can be determined easily by iterative substitution from these initial values. However, when the desired solution is at some instant far away from the initial time, this approach becomes very inefficient. In order to overcome this problem, more elegant methods that involve determination of both the homogeneous solution and the particular solution are normally employed.

Example 8.2 *Consider the difference equation in which the forcing function is exponential*

$$\frac{1}{4}u(k) - \frac{1}{2}u(k-1) + \frac{1}{4}u(k-2) = \left(\frac{1}{2}\right)^k$$

If $u(-1) = 1$ and $u(-2) = 0$ determine $u(k)$, $u(0)$, $u(2)$

Solution 8.2 *Using Equation 8.33 the solution can be written as*

$$u(k) = 4\left[\left(\frac{1}{2}\right)^k - \frac{1}{4}u(k-2) + \frac{1}{2}u(k-1)\right]$$

$$= 4\left(\frac{1}{2}\right)^k - u(k-2) + 2u(k-1).$$

Therefore,

$$u(0) = 4 - u(-2) + 2u(-1)$$
$$= 4 - 0 + 2$$
$$= 6$$

$$u(1) = 4\left(\frac{1}{2}\right)^1 - u(-1) + 2u(0)$$
$$= 2 - 1 + 2(6)$$
$$= 13$$

$$u(2) = 4\left(\frac{1}{2}\right)^2 - u(0) + 2u(1)$$
$$= 1 - 6 + 2(13)$$
$$= 21.$$

8.3.1.2   Homogeneous Solution of the Difference Equation

The homogeneous equation corresponding to the difference equation in 8.31 is given by

$$\sum_{i=0}^{n} a_i u(k-i) = 0.$$

The solution to this equation is given by the exponential function

$$u_h(k) = A\alpha^k$$

so that substitution into the difference equation yields

$$\sum_{i=0}^{n} a_i A\alpha^{k-i} = A\alpha^k \sum_{i=0}^{n} a_i \alpha^{-i} = 0.$$

Since the quantity $A\alpha^k$ is never zero, otherwise the solution becomes trivial, therefore, the homogeneous solution must satisfy the algebraic equation

$$\sum_{i=0}^{n} a_i \alpha^{-i} = 0. \qquad (8.34)$$

This equation is the characteristic equation of the difference equation and the values of α that satisfy it are called its characteristic roots or values. There are n-characteristic roots that may be either distinct or repeated. If the roots are distinct, the homogeneous solution is obtained as a linear combination of the terms of the type α_i^k such that

$$u_h(k) = A_1 \alpha_1^k + A_2 \alpha_2^k + \cdots + A_n \alpha_n^k.$$

If any of the roots are repeated, then n-independent solutions are generated by multiplying the corresponding characteristic solution by the appropriate power of k. For example, if α_1 has a multiplicity of p, while the other $n-p$ roots are distinct, it is assumed that the homogeneous solution is of the form

$$u_h(k) = A_1 \alpha_1^k + A_2 k \alpha_1^k + \cdots + A_p k^{p-1} \alpha_1^k + A_{p+1} \alpha_{p+1}^k$$
$$+ A_{p+2} \alpha_{p+2}^k + \cdots + A_n \alpha_n^k.$$

Simultaneous solution of this equation for the different initial conditions will give the values of A_i and hence the homogeneous solution.

Example 8.3 *Consider the difference equation*

$$20u(k) - 19u(k-1) + 5.5u(k-2) - 0.5u(k-3) = 0$$

with $u(-1) = 5$, $u(-2) = 11$, and $u(-3) = 13$. Determine the characteristic equation, characteristic roots, and the homogeneous solution and $u(7)$.

Solution 8.3 *Normalizing the equation (to make $a_0 = 0$) leads to*

$$u(k) - \frac{19}{20}u(k-1) + \frac{11}{40}u(k-2) - \frac{1}{40}u(k-3) = 0$$

Therefore, the characteristic equation is

$$1 - \frac{19}{20}\alpha^{-1} + \frac{11}{40}\alpha^{-2} - \frac{1}{40}\alpha^{-3} = 0$$

or

$$\alpha^3 - \frac{19}{20}\alpha^2 + \frac{11}{40}\alpha - \frac{1}{40} = 0,$$

which can be factored as

$$\left(\alpha - \frac{1}{2}\right)\left(\alpha - \frac{1}{4}\right)\left(\alpha - \frac{1}{5}\right) = 0.$$

so that the characteristic roots are $\alpha_1 = \frac{1}{2}$, $\alpha_2 = \frac{1}{4}$ and $\alpha_3 = \frac{1}{5}$. Since the roots are distinct, the homogeneous solution is of the form

$$u_h(k) = A_1\left(\frac{1}{2}\right)^k + A_2\left(\frac{1}{4}\right)^k + A_3\left(\frac{1}{5}\right)^k.$$

Substitution of the three initial conditions into the homogeneous solution leads to three equations that can be used to solve for A_1, A_2, and A_3.

$$2A_1 + 4A_2 + 5A_3 = 5$$
$$4A_1 + 16A_2 + 25A_3 = 11$$
$$8A_1 + 64A_2 + 125A_3 = 13.$$

From these equations $A_1 = \frac{21}{18}$, $A_2 = \frac{7}{4}$, and $A_3 = -\frac{39}{45}$. The homogeneous solution becomes

$$u_h(k) = 1.1667\left(\frac{1}{2}\right)^k + 1.75\left(\frac{1}{4}\right)^k - 0.8667\left(\frac{1}{5}\right)^k.$$

Using this expression, the solution at any instant k can be determined. For example,

$$u_h(7) = 1.1667\left(\frac{1}{2}\right)^7 + 1.75\left(\frac{1}{4}\right)^7 - 0.8667\left(\frac{1}{5}\right)^7$$
$$= 9.2106 \times 10^{-3}$$

As a check, note that when $k = -1$, -2, or -3, the initial conditions are obtained. For example for $k = -1$ the homogeneous solution is

$$u_h(-1) = 1.1667\left(\frac{1}{2}\right)^{-1} + 1.75\left(\frac{1}{4}\right)^{-1} - 0.8667\left(\frac{1}{5}\right)^{-1}$$
$$= 5.0$$

as expected.

In the following example a way of obtaining the homogeneous solution is demonstrated when the characteristic values are repeated.

Example 8.4 *Consider the following homogeneous difference equation*

$$12u(k) - 7u(k-1) + 3u(k-2) - u(k-3) = 0$$

with initial conditions $u(-1) = 0.5$, $u(-2) = 0.7$, *and* $u(-3) = 0.4$. *Determine the characteristic equation, characteristic roots and the homogeneous solution* $u(1)$ *and* $u(2)$.

Solution 8.4 *The normalized equation is*

$$u(k) - \frac{7}{12}u(k-1) + \frac{3}{12}u(k-2) - \frac{1}{12}u(k-3) = 0$$

so that the characteristic equation is

$$1 - \frac{7}{12}\alpha^{-1} + \frac{3}{12}\alpha^{-2} - \frac{1}{12}\alpha^{-3} = 0$$

with characteristic roots at $\alpha_1 = \dfrac{1}{2}$, $\alpha_2 = \dfrac{1}{2}$ *and* $\alpha_3 = \dfrac{1}{3}$. *Since the characteristic roots are not distinct, i.e.,* $\dfrac{1}{2}$ *is a repeated root, then the homogeneous solution becomes*

$$u_h(k) = A_1\left(\frac{1}{2}\right)^k + A_2 k\left(\frac{1}{2}\right)^k + A_3\left(\frac{1}{3}\right)^k,$$

which, on substitution of the initial conditions and solving the resulting system of linear equations, give $A_1 = 0.425$, $A_2 = -0.025$ *and* $A_3 = -0.133$. *Hence, subject to some rounding errors the homogeneous solution becomes*

$$u_h(k) = 0.425\left(\frac{1}{2}\right)^k - 0.025k\left(\frac{1}{2}\right)^k - 0.133\left(\frac{1}{3}\right)^k.$$

Thus,

$$u_h(1) = 0.425\left(\frac{1}{2}\right) - 0.025\left(\frac{1}{2}\right) - 0.133\left(\frac{1}{3}\right)$$

$$= 0.16$$

and

$$u_h(2) = 0.425\left(\frac{1}{2}\right)^2 - 0.5\left(\frac{1}{2}\right)^2 - 0.133\left(\frac{1}{3}\right)^2$$

$$= 0.09.$$

It can also be checked by using any of the given initial conditions (say $k = 2$) that

$$u_h(-2) = 0.425 \left(\frac{1}{2}\right)^{-2} - 0.025(-2) \left(\frac{1}{2}\right)^{-2} - 0.133 \left(\frac{1}{3}\right)^{-2}$$

$$= 0.70$$

as expected

8.3.1.3   The Particular Solution of the Difference Equation

Out of a number of available methods for determining the particular solution, only two will be presented here. These are the method of undetermined coefficients and the method of variation of parameters. Both require prior determination of the homogeneous solution $u_h(k)$. As for ordinary differential equations, the method of undetermined coefficients is restricted to equations with limited but fairly common forcing functions, while the method of variation of parameters is applicable to general forcing functions. On the other hand however, the method of undetermined coefficients is computationally efficient as compared with that of variation of parameters.

8.3.1.4   Method of Undetermined Coefficients

The method of undetermined coefficients is useful when the terms in the forcing function $r(k)$ have special forms as listed in the following table. Corresponding to each such term in $r(k)$ there is a trial solution containing a number of unknown constant coefficients that are then determined by substitution into the difference equation. The trial solutions used in each case are also shown in the table, where the constants A and B represent the unknown coefficients to be determined.

Terms in the Forcing Function	Trial Solution
αx^k	$A x^k$
$\sin(xk)$ or $\cos(xk)$	$A\cos(xk) + B$
$\displaystyle\sum_{j=0}^{m} \alpha_j k^j$	$\displaystyle\sum_{j=0}^{m} A_j k^j$
$\displaystyle\alpha^k \sum_{j=0}^{m} \beta_j k^j$	$\displaystyle\alpha^k \sum_{j=0}^{m} A_j k^j$
$\alpha^k \sin(xk)$ or $\alpha^k \cos(xk)$	$\alpha^k [A\cos(xk) + B\sin(xk)]$

The main requirement to be met by this method is that all terms that appear in the homogeneous solution should not be repeated in the trial solution. If the trial solution uses terms that are already in the homogeneous solution, they must be multiplied by a positive integer power of k large enough to push them out of the homogeneous solution range. In general, the method can be summarized in the following steps:

- Determine the homogeneous solution of the difference equation.

- Use the table to construct the trial solution, and if there exists common terms between the trial solution and the homogeneous solution, then these terms must be multiplied by some positive integer that is a power of k.

- Form the particular solution $u_p(k)$ as a linear combination of all trial solutions and determine the unknown coefficients by treating $u_p(k)$ as $u(k)$ and substituting into the original difference equation.

This method is illustrated in the following example:

Example 8.5 *Determine the particular solution of the following difference equation,*

$$u(k) - 3u(k-1) - 4u(k-2) = 2k^2 + 3 + 4^k,$$

with $u(-1) = 0.5$ and $u(-2) = 1.0$

Solution 8.5 *First, the homogeneous solution to the equation*

$$u(k) - 3u(k-1) - 4u(k-2) = 0$$

is determined. The solution is given by

$$u_h(k) = 4.8(4)^k + 0.7(-1)^k.$$

For the particular solution, the forcing function $2k^2 + 3 + 4^k$ suggests a trial solution of the form

$$u_p(k) = A_1 k^2 + A_2 k + A_3(4)^k + A_4$$

however, since the term 4^k also appears in the homogeneous solution, it is then multiplied by k getting a new form of the particular solution as

$$u_p(k) = A_1 k^2 + A_2 k + A_3 k(4)^k + A_4$$

Using this equation in the original difference equation gives

$$
\begin{aligned}
2k^2 + 3 + 4^k = {} & A_1 k^2 + A_2 k + A_3 k(4)^k + A_4 \\
& - 3\left(A_1(k-1)^2 + A_2(k-1) + A_3(k-1)(4)^{k-1} + A_4\right) \\
& - 4\left(A_1(k-2)^2 + A_2(k-2) + A_3(k-2)(4)^{k-2} + A_4\right),
\end{aligned}
$$

which, on expansion and equating the coefficients, gives

$$A_1 = -\frac{1}{3}, \qquad A_2 = -\frac{22}{18}, \qquad A_3 = \frac{4}{5}, \qquad A_4 = 1.6852,$$

so that finally the particular solution becomes

$$u_p(k) = -\frac{1}{3}k^2 - \frac{22}{18}k + \frac{4}{5}k(4)^k + 1.6852.$$

The general solution to this difference equation is

$$u(k) = 4.8(4)^k + 0.7(-1)^k - \frac{1}{3}k^2 - \frac{22}{18}k + \frac{4}{5}k(4)^k + 1.6852.$$

The Method of Variation of Parameters If the homogeneous solution of the difference equation has the general form,

$$u_h(k) = \sum_{j=0}^{n} C_j V_j(k)$$

where $V_j(k)$ includes the α root of the characteristic equation raised to the power k and any power of k for which α^k must be multiplied and C_j are combinatorial coefficients, then the complete solution of the difference equation may be assumed to have a general solution of the form

$$u(k) = \sum_{j=0}^{n} A_j(k) V_j(k)$$

where $A_j(k)$ are functions to be determined. There will be n such functions, and this requires n-initial conditions. But since the homogeneous solution must also be satisfied, this leaves only $(n-1)$ conditions, which can be arbitrarily imposed. The choice of these conditions is made so as to simplify the computations as much as possible. One of the simplest algorithms for computing $A_j(k)$ is

$$A_j(k) = A_j(0) + \sum_{m=1}^{k} \Delta A_j(m),$$

where $\Delta A_j(m)$ are components of matrix ΔA, which satisfies

$$V \Delta A = R,$$

where

$$V = \begin{bmatrix} V_1(k-1) & V_2(k-1) & \cdots & V_n(k-1) \\ V_1(k-2) & V_2(k-2) & \cdots & V_n(k-2) \\ \vdots & \vdots & \ddots & \vdots \\ V_1(k-n) & V_2(k-n) & \cdots & V_n(k-n) \end{bmatrix}$$

$$\Delta A = \begin{bmatrix} \Delta A_1(k) \\ \Delta A_2(k) \\ \vdots \\ \Delta A_n(k) \end{bmatrix},$$

and

$$R = \begin{bmatrix} \dfrac{r(k)}{a_0} \\ 0 \\ 0 \\ \vdots \\ 0 \end{bmatrix}.$$

Since

$$\Delta A = V^{-1}R$$

it can be seen that

$$\Delta A_j = V_{jn} \frac{r(k)}{a_0},$$

where V_{jn} is the element in row j and column n of the inverse of matrix V.

Because of the need to invert matrix V this method might seem to be very complex. However, in most cases, complete matrix inversion of V is not necessary since there is only one entry in matrix R, i.e., the first entry. This makes it possible to make use of the determinant repeatedly in the Gauss-Jordan complete procedure. This method is illustrated in the following example.

Example 8.6 *Solve for the particular solution of the following second-order difference equation*

$$y(k) - 5y(k-1) + 6y(k-2) = k^2.$$

Solution 8.6 *From the previous examples, it is known that the characteristic roots of this equation are 2 and 3. The general solution of this equation is*

$$y(k) = A_1 2^k + A_2 3^k.$$

Thus,

$$V_1(k) = 2^k \qquad V_2(k) = 3^k$$

so that

$$V = \begin{bmatrix} 2^{k-1} & 3^{k-1} \\ 2^{k-2} & 3^{k-2} \end{bmatrix}.$$

Using this result and forming the augmented Gaussian matrix leads to

$$\begin{bmatrix} 2^{k-1} & 3^{k-1} & \vdots & k^2 \\ 2^{k-2} & 3^{k-2} & \vdots & 0 \end{bmatrix},$$

which in row echelon form is given by

$$\begin{bmatrix} 2^{k-1} & 3^{k-1} & \vdots & k^2 \\ 0 & 3^{k-2}2^{k-1} - 3^{k-2}2^{k-1} & \vdots & -2^{k-2}k^2 \end{bmatrix}.$$

This leads to

$$\Delta A_2(k) = \frac{-2^{k-2}k^2}{3^{k-2}2^{k-1} - 3^{k-1}2^{k-2}}$$

$$= 9k^2 3^{-k}$$

and

$$\Delta A_1(k) = \left(k^2 - \frac{-2^{k-2}k^2 3^{k-1}}{3^{k-2}2^{k-1} - 3^{k-1}2^{k-2}} \right) \frac{1}{2^{k-1}}$$

$$= -4k^2 2^{-k}.$$

Proceeding with the computation of the $A_j(k)$ terms produces,

$$A_1(k) = A_1(0) + \sum_{m=1}^{k-1} (-4m^2 2^{-m})$$

$$= A_1(0) - 2 - 4 - \frac{9}{2} + \cdots\cdots + -4(k-1)^2 2^{-k+1}$$

$$= C_1 - 8 \left(\frac{1}{2} \right)^k [k^2 - 2k + 1]$$

where

$$C_1 = A_1(0) - 2 - 4 - \frac{9}{2} + \cdots\cdots$$

Similarly,

$$A_2(k) = C_2 + 27 \left(\frac{1}{3} \right)^k [k^2 - 2k + 1]$$

so that the complete solution becomes

$$u(k) = C_1 2^k + C_2 3^k + 19 [k^2 - 2k + 1]$$

8.3.2   The Difference Operator (Δ) and Shift Operator (q)

The name *"difference equation"* derives from the fact that it can be expressed in terms of the difference between consecutive instances. Thus, for the difference equation with a sequence of terms $\{u(k-i)\}_{i=0}^{n}$, the consecutive terms can be related by the difference

$$u(k-i) - u(k-i-1) = \Delta_b u(k-i),$$

which for $i = 0$ becomes

$$u(k) - u(k-1) = \Delta_b u(k). \tag{8.35}$$

The term $\Delta_b u(k)$ is known as the *backward difference* as it relates $u(k)$ to term which is back in time by one sampling interval. The two terms can also be related by

$$u(k-i) - u(k-i-1) = \Delta_f u(k-i-1),$$

which again for $i = 0$ reduces to

$$u(k) - u(k-1) = \Delta_f u(k-1). \tag{8.36}$$

In this case, the term $\Delta_f u(k-1)$ is known as the *forward difference* because it relates the term $u(k-1)$ with the term $u(k)$, which is ahead in time (forward) by one sampling interval. The symbol Δ represents the difference operation and is known as the difference operator. This, as will be shown, is closely related to the continuous-time differential operator $\frac{d}{dt}$.

The expressions in Equations 8.35 and 8.36 give the first differences between consecutive terms. To relate more than two terms, higher-order differences are used. Basically, the higher-order differences are differences of lower order (by one order). For example, the second-order backward difference is defined as

$$\Delta_b^2 u(k) = \Delta_b u(k) - \Delta_b u(k-1).$$

Since the definition of the first-order difference gives

$$\Delta_b u(k) = u(k) - u(k-1)$$

and

$$\Delta_b u(k-1) = u(k-1) - u(k-2).$$

It then follows that

$$\begin{aligned}
\Delta_b^2 u(k) &= [u(k) - u(k-1)] - [u(k-1) - u(k-2)] \\
&= u(k) - 2u(k-1) + u(k-2).
\end{aligned}$$

Using the same procedure, it can be shown that the third-order difference is given by

$$\Delta_b^3 u(k) = u(k) - 3u(k-1) + 3u(k-3) - u(k-3).$$

From examining the coefficients of these terms, it is clear that they are the binomial coefficients given by

$$\alpha_i = \binom{n}{i}$$

$$= \frac{n!}{(n-i)!i!}.$$

This will be made more clear after the introduction of the shift in the time operator later in this section. In general, it can be stated that the n-th-order backward difference equation is given by

$$\Delta_b^n u(k) = \Delta_b^{n-1} u(k) - \Delta_b^{n-1} u(k-1)$$

$$= \sum_{i=0}^{n} \frac{n! u(k-i)}{(n-i)!i!} (-1)^i. \tag{8.37}$$

Similarly, the forward difference can be shown to be

$$\Delta_f^n u(k-n) = \sum_{i=0}^{n} \frac{n! u(k-i)}{(n-i)!i!} (-1)^{n-i}. \tag{8.38}$$

In most of the digital control applications, the forward difference is more attractive than the backward difference. Hence, unless otherwise stated, it will be assumed that forward differences are used.

Counterpart to the difference operator is the shift in time operator q. The forward shift in time corresponding to forward differences can be expressed in terms of the q-operator as

$$qu(k) = u(k+1). \tag{8.39}$$

The corresponding backward shift in time is represented by the inverse time shift q^{-1} as

$$q^{-1}u(k) = u(k-1).$$

Notice that by this definition

$$q^{-1}u(k+1) = u(k+1-1)$$
$$= u(k).$$

Design and Analysis of Control Systems

Also from Equation 8.39, it follows that

$$q^{-1}u(k+1) = q^{-1}qu(k) = u(k)$$

so that

$$q^{-1}q = 1.$$

This proves that the backward shift in time is just the inverse of the forward shift in time. The general shift in time operation is expressed as

$$q^n u(k) = u(k+n),$$

or

$$q^{-n}u(k) = u(k-n).$$

Unlike the difference operator, which mainly indicates the relationship between the terms in a difference equation (except for numerical interpolation of the continuous-time signal that produced the difference equation, which will be covered at later stages of this chapter), the shift in time operation goes further to reduce the difference equation to a simple algebraic expression that can be handled by using normal algebraic rules. Consider, for example, the difference equation

$$\sum_{i=0}^{n} a_i u(k-i) = \sum_{j=0}^{m} b_j r(k-j) \qquad m \le n \qquad a_0 = 1. \tag{8.40}$$

Applying the shift in time operation to Equation 8.40 gives

$$\sum_{i=0}^{n} a_i q^{-i} u(k) = \sum_{j=0}^{m} b_j q^{-j} r(k)$$

so that

$$u(k) = \frac{\displaystyle\sum_{j=0}^{m} b_j q^{-j}}{\displaystyle\sum_{i=0}^{n} a_i q^{-i}} r(k).$$

By performing long division, the rational term can be reduced to

$$\frac{\displaystyle\sum_{j=0}^{m} b_j q^{-j}}{\displaystyle\sum_{i=0}^{n} a_i q^{-i}} = \sum_{\alpha=0}^{\infty} c_\alpha q^{-\alpha},$$

which gives

$$u(k) = \sum_{\alpha=0}^{\infty} c_{\alpha} q^{-\alpha} r(k)$$
$$= c_0 r(k) + c_1 q^{-1} r(k) + c_2 q^{-2} r(k) + \cdots$$
$$= c_0 r(k) + c_1 r(k-1) + c_2 r(k-2) + \cdots \qquad (8.41)$$

Hence, by knowing some of the terms of $r(k)$ (of course finite terms), it is obvious that one can approximate the solution of the difference equation using this expression.

8.3.2.1 The Δ-Operator and the q-Operator

There are many instances when an equation in the q-operator form is useful, while in other cases an equation in the Δ-operator is more useful. As such, one should be able to transform from one form to another by using the relationship between the two operators. To establish the relationship between the Δ_f-operator and the q-operator, recall the definitions

$$\Delta_f u(k-1) = u(k) - u(k-1)$$

and

$$q u(k-1) = u(k).$$

By combining these two definitions, it follows that

$$\Delta_f u(k-1) = q u(k-1) - u(k-1)$$
$$= [q-1] u(k-1)$$

so that

$$\Delta_f = q - 1.$$

Similarly, the backward difference operator Δ_b can be related to the shift in time operator by

$$\Delta_b = 1 - q^{-1}$$

For higher-order differences, these relationships develop respectively to

$$\Delta_f^n u(k-n) = (q-1)^n u(k-n)$$

and

$$\Delta_f^n u(k) = \left(1 - q^{-1}\right)^n u(k)$$

These relationships can explain why the coefficients of the terms in the higher-order difference operation are binomial coefficients.

8.3.3 Discrete Approximation of Continuous Processes

Most of the processes in real life are continuous in nature. The need to have these processes controlled using a digital computer gives rise to a need for their discretization or digitization. The sampling process previously discussed is indeed the sole process by which a continuous-time signal can be discretized. However, the sampling process as it was presented does not suffice to describe the discrete-time form (difference equations) of the continuous-time differential equations which describe the dynamics of most processes. This section extends the ideas presented under the sampling process section to generate difference equations from the continuous-time differential equations giving the discrete form of describing the dynamics of the system. There are three methods of discretization of continuous-time process, however the methods presented here are based only on the discrete-time approximation of the differential and integral equations.

8.3.3.1 Differential Equations and the δ-Transform

Euler's Forward Rule: Euler's methods provide the simplest ways of approximation of a continuous-time differential equation. There are basically three forms of Euler's approximation; namely the forward rectangular (difference) rule, backward rectangular (difference) rule, and the central (difference) rule. All these approximations are based on the fact that the differential

$$\left. \frac{dx(t)}{dt} \right|_{t=k} = \lim_{\delta t \to 0} \left. \frac{\delta x(t)}{\delta t} \right|_{t=k}$$

is just the gradient of the function $x(t)$ at $t = k$. The forward approximation takes the gradient of the function $x(t)$ at $t = k$ as

$$\left. \frac{dx(t)}{dt} \right|_{t=k} \approx \frac{x(k + \delta t) - x(k)}{\delta t}$$

therefore, if the sampling interval δt is specified as T, then the forward rectangular rule becomes

$$\left. \frac{dx(t)}{dt} \right|_{t=k} \approx \frac{x(k + T) - x(k)}{T} \tag{8.42}$$

Now, noting that by the forward difference operator

$$x(k + T) - x(k) = \Delta_f x(k)$$

then the approximate differential in Equation (8.42) can be written as

$$\left. \frac{dx(t)}{dt} \right|_{t=k} \approx \frac{\Delta_f x(k)}{T}$$

which gives the relationship between the differential operator $\dfrac{d}{dt}$ and the forward difference operator $\Delta f(t)$ as

$$\frac{df(t)}{dt} \cong \frac{\Delta f(t)}{T}$$

and since $\Delta f(t) = q - 1$, then

$$\frac{df(t)}{dt} \cong \frac{q-1}{T}(.)$$

which is known as the δ-transform of a given function $f(t)$, symbolized as δ, where

$$\delta = \frac{q-1}{T}$$

Thus, knowing the time derivative of the function $x(t)$, the approximate discrete-time approximation using Euler's forward rectangular rule becomes

$$\left.\frac{dx(t)}{dt}\right|_{t=k} \cong \frac{q-1}{T}x(k)$$
$$= \delta x(k)$$

Higher-order differential equations can also be defined where

$$\left.\frac{d^n x(t)}{dt^n}\right|_{t=k} \cong \left(\frac{q-1}{T}\right)^n x(k)$$
$$= \delta^n x(k).$$

Euler's Backward Rule: Alternative to the forward rectangular rule is the backward rectangular rule defined as

$$\left.\frac{dx(t)}{dt}\right|_{t=k} \cong \frac{x(k) - x(k-T)}{T}$$

Using the same approach as for the forward rectangular rule, this can be reduced by the use of the backward difference

$$x(k) - x(k-T) = \Delta_b x(k)$$

to give

$$\left.\frac{dx(t)}{dt}\right|_{t=k} \cong \frac{\Delta_b}{T}x(k)$$

or

$$\left.\frac{d(.)}{dt}\right|_{t=k} \cong \frac{\Delta_b}{T}(.)$$

Hence, the discrete-time approximation of the differential equation by backward rectangular rule becomes

$$\frac{dx(t)}{dt}\bigg|_{t=k} \cong \frac{1-q^{-1}}{T}x(k)$$

$$= \frac{q-1}{qT}x(k).$$

In terms of the δ-transform, this becomes

$$\frac{dx(t)}{dt}\bigg|_{t=k} \cong \frac{\delta}{q}x(k)$$

$$= \delta q^{-1}x(k)$$

Higher-order differential equations can be likewise be handled as

$$\frac{d^n x(t)}{dt^n}\bigg|_{t=k} \cong \left(\delta q^{-1}\right)^n x(k)$$

or simply

$$\frac{d^n x(t)}{dt^n}\bigg|_{t=k} \cong \left(\frac{1-q^{-1}}{T}\right)^n x(k). \qquad (8.43)$$

The Central (Difference) Rule: Finally, between the forward and backward rectangular rules is the central rule that spans half of the backward and half of the forward rules. It is defined as

$$\frac{dx(t)}{dt}\bigg|_{t=k} \cong \frac{x(k+0.5T) - x(k-0.5T)}{T} \qquad (8.44)$$

Due to the difficulty of sampling half intervals, it is sometimes expressed as

$$\frac{dx(t)}{dt}\bigg|_{t=k} \cong \frac{x(k+t) - x(k-T)}{2T}. \qquad (8.45)$$

Since the resulting range of gradient approximation is very wide, this method becomes very inaccurate, and for most applications, is seldom used.

8.3.3.2   Approximation of Integral Functions

Sometimes, the dynamics of the plant includes both the differential and the integral terms, for example the PID controller. The previous discussion focused on the discretization of the differential equations. In this section, the discretization of the integral functions is considered. The discrete-time approximation of the integral $\int_{k-T}^{k} r(t)dt$ comes from the fact that the integral is just the area enclosed by the $r(t)$ curve and the t-axis. Notice that the limits of an integral in a given sampling instant are normally chosen so that the upper limit is the current sampling instant (k). Thus, the discrete-time approximation is that of approximating this area. There could be many methods for this purpose, but the three most popular methods employed in control applications are: Euler's backward rectangular rule, Euler's forward rectangular rule and the trapezoidal rule. It will be seen that both the Euler's backward and forward rules are derived from the corresponding rules for approximation of differential equations.

Euler's Backward Rectangular Rule: Euler's backward rectangular rule for discrete approximation of integral functions between two sampling instants gives an approximate area of a rectangle whose base is the sampling interval and whose height is the value of the function at the final limit. Basically, this method assumes that the function is constant at its value at the end of the integration interval. Hence, the approximate integral becomes

$$\int_{k-T}^{k} r(t)dt = r(k)T$$

The backward rule gets its name from the fact that the value of the function $u(k)$ is used to determine the area backward from it.

Euler's Forward Rectangular Rule: The forward rule is just the opposite of the backward rule in that the discrete approximation of integral functions between two sampling instants gives an approximate area of a rectangle whose base is the sampling interval and whose height is the value of the function at the lower limit. It assumes that the function remains constant at its value at the lower end of the integration interval. By this rule, the approximate integral becomes

$$\int_{k-T}^{k} r(t)dt = r(k-T)T.$$

Simpson's Trapezoidal Rule: The main disadvantage of the Euler's methods described above is that either they overestimate or underestimate the integral. Simpson's trapezoidal approximation overcomes these disadvantages by taking the average of the limiting values of the integrand. As such, it treats the area between the two integration limits as a trapezium of width T with sides $r(k-T)$ and $r(k)$ that give the average height as $\frac{1}{2}[r(k-T)+r(k)]$. As such, the approximate integral becomes

$$\int_{k-T}^{k} r(t)dt = \frac{1}{2}[r(k-T)+r(k)]T$$

8.3.4   The $\mathcal{Z}$-Transforms

8.3.4.1   Definition of the $\mathcal{Z}$-Transform

The shift in time (q-operator) and the difference (δ-operator) are transformations that enable the system to be analyzed in time domain using algebraic rules. Most often, in control engineering there is much interest for the analysis to be carried out in both the time and the frequency domains. As such, similar to the Laplace transform for the case of continuous-time systems, which transforms the system from time domain to frequency domain, the $\mathcal{Z}$-transform is the transformation method that converts the discrete-time system from the time domain to the frequency domain. That is why the $\mathcal{Z}$-transform is regarded as the discrete-time equivalent of the Laplace transform. In this section, the basics of the $\mathcal{Z}$-transform and their properties are discussed. The application of the $\mathcal{Z}$-transforms in the analysis of the discrete-time systems will be apparent from the next sections.

Appendix A contains $\mathcal{Z}$-transforms and their properties. Given discrete-time signal values $\{r(k)\}_{k=-\infty}^{\infty}$ the $\mathcal{Z}$-transform is defined as follows:

$$R(z) \triangleq \mathcal{Z}\{r(k)\}$$

$$= \sum_{k=-\infty}^{\infty} r(k)z^{-k}, \tag{8.46}$$

where z is a complex variable with bounded magnitude such that for some values r_o and R_o

$$r_o < |z| < R_o$$

so that the series $\sum\limits_{k=-\infty}^{\infty} r(k)z^{-k}$ converges.

The definition is general and holds for all values of k, but since the sequence $\{r(k)\}$ is normally defined for positive integers k, the one sided $\mathcal{Z}$-transform is commonly used. This is defined as

$$R(z) = \sum_{k=0}^{\infty} r(k)z^{-k}, \tag{8.47}$$

where the z-variable is bounded such that for some value r_o

$$r_o < |z|.$$

8.3.4.2   The $\mathcal{Z}$-Transform and the Laplace Transform

To establish the relationship between the $\mathcal{Z}$-transform and the Laplace transform, consider a continuous-time signal $r(t)$ sampled at even intervals

T of time giving discrete-time signals $\{r(k)\}_{k=-\infty}^{\infty}$. It was shown that the Laplace transform of such discrete-time signals is

$$R^*(s) = \sum_{k=-\infty}^{\infty} r(k)e^{-skT}.$$

From the definition of the $\mathcal{Z}$-transform, the $\mathcal{Z}$-transform of this signal is given by

$$R(z) = \sum_{k=-\infty}^{\infty} r(k)z^{-k}.$$

Since both the Laplace transform and the $\mathcal{Z}$-transform map the same sampled signal from the time space to the complex frequency space, they are equal, i.e.,

$$F^*(s) = F(z)$$

or

$$\sum_{k=-\infty}^{\infty} r(k)e^{-skT} = \sum_{k=-\infty}^{\infty} r(k)z^{-k}. \tag{8.48}$$

From this equality, it follows that

$$e^{-skT} = z^{-k}$$

or simply

$$z = e^{sT}.$$

Using this relationship, it is possible to analyze the system in continuous-time domain using Laplace transforms and convert the results into the discrete-time domain and vice versa. The relationship is widely used in control design using the pole-zero matching technique, which will be discussed in later sections. The most obvious conclusion that can be drawn from this relationship is that if the system has a pole or zero at $s = -a$ in the s-plane, then in the z-plane this pole or zero will be at e^{-aT}. Similarly, a zero at b in the z-plane implies a zero at $\frac{1}{T}\ln b$ in the s-plane. The relationship between the two transforms described above can further be studied from the table of the Laplace and $\mathcal{Z}$-transforms for various continuous-time signals using a sampling interval of T.

8.3.4.3   Properties of the $\mathcal{Z}$-Transforms

The $\mathcal{Z}$-transforms along with their counterpart Laplace transforms for different discrete-time signal are shown in *Appendix* **A**. This table is useful in most digital control systems analysis and design applications. The reader is encouraged to become familiar with it. In this section, the fundamental properties of the $\mathcal{Z}$-transform are discussed. These help to simplify most of the transformation problems as dictated by the property itself.

Linearity: The $\mathcal{Z}$-transform is a linear operator. This means that if two discrete-time signals $f(k)$ and $g(k)$ are linear such that the principle of superposition holds, i.e., if for some scalars α and β

$$r(k) = \alpha f(k) + \beta g(k)$$

implies that

$$r(k+n) = \alpha f(k+n) + \beta g(k+n)$$

then the $\mathcal{Z}$-transform becomes

$$\mathcal{Z}\{r(k)\} = \mathcal{Z}\{\alpha f(k) + \beta g(k)\}$$
$$= \alpha \mathcal{Z}\{f(k)\} + \beta \mathcal{Z}\{g(k)\}$$

or simply

$$\mathcal{Z}\{\alpha f(k) + \beta g(k)\} = \alpha F(z) + \beta G(z) \qquad (8.49)$$

where $F(z) = \mathcal{Z}\{f(k)\}$ and $G(z) = \mathcal{Z}\{g(k)\}$

The Convolution Theorem: It has been shown before that in continuous-time systems for which the principle of superposition hold, the signal $u(t)$ is said to be a convolution of two signals $r(t)$ and $h(t)$ if

$$u(t) = \int_{-\infty}^{\infty} r(\tau)h(t-\tau)d\tau$$

and the time scale is valid for both positive and negative values. It was also shown that since in most applications time is counted only on the positive side to some time limit t, the convolution then becomes

$$u(t) = \int_{0}^{t} r(\tau)h(t-\tau)d\tau.$$

In either of these cases, the convolution is expressed symbolically as

$$u(t) = r(t) * h(t).$$

The discrete-time equivalent of this convolution is the convolution summation. If the discrete-time signals are available for any instant k in the range $(-\infty, \infty)$, then the discrete-time signal $u(k)$ is said to be convolution of two discrete-time signals $r(k)$ and $h(k)$ if

$$u(k) = \sum_{j=-\infty}^{\infty} r(j)h(k-j).$$

Since the sampled signal is normally only available at positive finite k instants of time, the convolution summation becomes

$$u(k) = \sum_{j=0}^{k} r(j)h(k-j).$$

This is also expressed as

$$u(k) = r(k) * h(k).$$

Now, if the $\mathcal{Z}$-transform is taken on both sides of this convolution summation the result is

$$U(z) = \mathcal{Z}\{u(k)\}$$

$$= \sum_{k=0}^{\infty} \left[\sum_{j=0}^{k} r(j)h(k-j) \right] z^{-k}. \tag{8.50}$$

Rearranging the terms in this expression (Equation 8.50) gives

$$U(z) = \sum_{j=0}^{\infty} r(j) \sum_{k=0}^{\infty} h(k-j)z^{-k}. \tag{8.51}$$

Letting $(k-j) = m$ and using this substitution in Equation 8.51 leads to

$$U(z) = \sum_{j=0}^{\infty} r(j) \sum_{m=-j}^{\infty} h(m)z^{-m-j}$$

$$= \sum_{j=0}^{\infty} r(j)z^{-j} \left[\sum_{m=-j}^{-1} h(m)z^{-m} + \sum_{m=0}^{\infty} h(m)z^{-m} \right]. \tag{8.52}$$

Considering the one sided $\mathcal{Z}$-transforms, where it is assumed that signals at negative instants are not defined, then

$$\sum_{m=-j}^{-1} h(m)z^{-m} = 0.$$

which reduces Equation 8.52 to

$$U(z) = \sum_{j=0}^{\infty} r(j)z^{-j} \sum_{m=0}^{\infty} h(m)z^{-m}. \qquad (8.53)$$

From the definition of the $\mathcal{Z}$-transform it follows that

$$\sum_{j=0}^{\infty} r(j)z^{-j} = \mathcal{Z}\{r(k)\} = R(z)$$

$$\sum_{m=0}^{\infty} h(m)z^{-m} = \mathcal{Z}\{h(k)\} = H(z). \qquad (8.54)$$

Thus, the $\mathcal{Z}$-transform of the convolution summation becomes

$$U(z) = R(z)H(z),$$

which means that the $\mathcal{Z}$-transform of a discrete-time signal that is a convolution of two discrete-time signal is just a product of the $\mathcal{Z}$-transform of the individual signals.

Differentiation with Respect to z : This property is very useful when dealing with reduced algebraic equations in z. Consider the definition of the $\mathcal{Z}$-transform

$$U(z) = \sum_{k=0}^{\infty} u(k)z^{-k}$$

Taking derivatives with respect to z leads to

$$\frac{dU(z)}{dz} = \sum_{k=0}^{\infty} (-k)u(k)z^{-k-1}$$

$$= -z^{-1}\sum_{k=0}^{\infty} ku(k)z^{-k}$$

or simply

$$-z\frac{dU(z)}{dz} = \sum_{k=0}^{\infty} ku(k)z^{-k}. \qquad (8.55)$$

As can be seen, the right-hand side of Equation 8.55 is just the $\mathcal{Z}$-transform of the function $ku(k)$, therefore, putting it together gives

$$\mathcal{Z}\{ku(k)\} = -z\frac{dU(z)}{dz}$$

This differentiation holds for both one-sided and two-sided $\mathcal{Z}$-transform and can be repeated successively to give

$$\mathcal{Z}\{k^n u(k)\} = \left[-z\frac{d}{dz}\right]^n U(z).$$

where

$$\left[-z\frac{d}{dz}\right]^n U(z) = -z\frac{d}{dz}\left(-z\frac{d}{dz}\left(-z\frac{d}{dz}\cdots\cdots\left(-z\frac{d}{dz}\right)\cdots\right)\right)U(z).$$

$$(8.56)$$

The differentiation is repeated n-times.

The Time Shifting Property: The time shifting property of the $\mathcal{Z}$-transform enables us to determine the $\mathcal{Z}$-transform of a discrete-time signal at any other instant of time, provided the transform at one instant is known. There are two types of time shifting: the backward shifting in time and the forward shifting in time. It will further be noted that both the backward and forward time shifting behave differently between the one-sided $\mathcal{Z}$-transform and the two-sided $\mathcal{Z}$-transform. First, this property is discussed in the more general two-sided $\mathcal{Z}$-transforms and then concludes with the special case one-sided $\mathcal{Z}$-transform. However, first a reminder on the concept of time shifting is given.

A sequence $\{r(k)\}_{k=p}^n$ for some number p is said to be time shifted if it can be expressed as another sequence $\{r(k)\}_{k=q}^{n-p+q}$ for another number q where $q \neq p$. If $q > p$, the sequence is said to be forward shifted in time while it becomes backward shifted in time if $q < p$. In other words, one can put it this way, a pulse $r(k)$ is forward shifted in time by q if it is defined as $r(k+q)$ and is said to backward shifted by the same amount if it becomes $r(k-q)$.

Consider a discrete-time signal sequence $\{r(k)\}_{k=-\infty}^\infty$ with a two-sided $\mathcal{Z}$-transform denoted by $R(z)$. From its definition

$$R(z) = \sum_{k=-\infty}^{\infty} r(k)z^{-k}.$$

Now if this sequence is time shifted by some finite interval n to become $\{r(k+n)\}_{k=-\infty}^\infty$ then its $\mathcal{Z}$-transform becomes

$$\mathcal{Z}\{r(k+n)\} = \sum_{k=-\infty}^{\infty} r(k+n)z^{-k}$$

For the right-hand side, let $k + n = p$ such that

$$\mathcal{Z}\{r(k+n)\} = \sum_{p=-\infty+n}^{\infty+n} r(p)z^{-p+n},$$

where $\infty + n = \infty$ and $-\infty + n = -\infty$. The above expression can be simplified as

$$\mathcal{Z}\{r(k+n)\} = z^n \sum_{p=-\infty}^{\infty} r(p)z^{-p}$$

but the summation in the right-hand side is just the $\mathcal{Z}$-transform $R(z)$ of the sequence $\{r(k)\}_{k=-\infty}^{\infty}$, therefore, it can be concluded that

$$\mathcal{Z}\{r(k+n)\} = z^n R(z), \tag{8.57}$$

which is the time shifting property of the two-sided $\mathcal{Z}$-transforms. Notice that when $n > 0$ it becomes the forward shift in time while $n < 0$ gives the backward shift in time.

When the one-sided $\mathcal{Z}$-transform is used, this property takes a different shape for the forward shifting in time. In fact it requires all intermediate conditions within the shifted interval to be known. Consider again the discrete-time signal sequence $\{r(k)\}_{k=0}^{\infty}$ with the one-sided $\mathcal{Z}$-transform $R(z)$ which is defined as

$$R(z) = \sum_{k=0}^{\infty} r(k)z^{-k}.$$

If this sequence is shifted in time by some finite interval n in time to become $\{r(k+n)\}_{k=0}^{\infty}$ then its $\mathcal{Z}$-transform becomes

$$\mathcal{Z}\{r(k+n)\} = \sum_{k=0}^{\infty} r(k+n)z^{-k},$$

which again by replacing $k + n$ by p in the right-hand term gives

$$\mathcal{Z}\{r(k+n)\} = \sum_{p=n}^{\infty} r(p)z^{-p+n}$$

$$= z^n \sum_{p=n}^{\infty} r(p)z^{-p}. \tag{8.58}$$

But

$$\sum_{p=n}^{\infty} r(p)z^{-k} = \sum_{p=0}^{\infty} r(p)z^{-p} - \sum_{p=0}^{n-1} r(p)z^{-p}$$

where

$$\sum_{p=0}^{\infty} r(p)z^{-p} = \mathcal{Z}\{r(p)\} = R(z).$$

Thus, these results can be combined together to give

$$\mathcal{Z}\{r(k+n)\} = z^n R(z) - z^n \sum_{p=0}^{n-1} r(p)z^{-p}. \qquad (8.59)$$

This is the expression for the forward shift in time of the one-sided $\mathcal{Z}$-transform. The terms in the summation are the intermediate values which must be known for the shifting property to work. Similarly, the backward shift in time can be done in the same way where

$$\mathcal{Z}\{r(k-n)\} = \sum_{k=0}^{\infty} r(k-n)z^{-k},$$

which, on replacing $(k-n)$ by p gives

$$\mathcal{Z}\{r(k-n)\} = \sum_{p=-n}^{\infty} r(p)z^{-p-n}$$

$$= z^{-n} \sum_{p=-n}^{\infty} r(p)z^{-p}$$

$$= z^{-n} \left[\sum_{p=-n}^{-1} r(p)z^{-p} + \sum_{p=0}^{\infty} r(p)z^{-p} \right]. \qquad (8.60)$$

Since

$$\sum_{p=0}^{\infty} r(p)z^{-p} = R(z)$$

then, the backward shift in time property becomes

$$\mathcal{Z}\{r(k-n)\} = z^{-n} \left[R(z) + \sum_{p=-n}^{-1} r(p)z^{-p} \right]. \qquad (8.61)$$

The summation term gives the sum of the terms in the negative time interval multiplied by the power of negative z corresponding to their position from the initial time.

The Initial Value Theorem: As noted before, it is the one sided $\mathcal{Z}$-transform that is used in many occasions. Given the one-sided $\mathcal{Z}$-transform

$R(z)$, it may become necessary to know the initial value $r(0)$ of the sequences that generated this transform $R(z)$. One possibility is to find the inverse of $R(z)$ (discussed in the next section) and then evaluate $f(0)$ directly. However, this method is not a very efficient one if the interest is just to find $r(0)$ only. The initial value theorem is a more elegant method that gives the initial value $r(0)$ of the sequence directly from its $\mathcal{Z}$-transform.

Recall that the one-sided $\mathcal{Z}$-transform $R(z)$ is defined as

$$R(z) = \sum_{k=0}^{\infty} r(k) z^{-k}$$

$$= r(0) + r(1)z^{-1} + r(2)z^{-2} + r(3)z^{-3} + \cdots, \qquad (8.62)$$

which converges uniformly for all $|z| > r_0$. Therefore, as $z \to \infty$, the terms multiplied by powers of z^{-1} disappear. This observation is what is stated in the initial value theorem, which is summarized as

$$\lim_{z \to \infty} R(z) = r(0).$$

Note that this theorem doesn't apply for the two-sided $\mathcal{Z}$-transforms.

The Final Value Theorem: The final value theorem may be viewed as the converse of the initial value theorem. It also applies only with the one-sided $\mathcal{Z}$-transform. This theorem enables us to determine the behavior of $r(k)$ as $k \to \infty$ from its $\mathcal{Z}$-transform. The final value theorem can be derived from the time shift theorem as follows:

Since for the one-sided $\mathcal{Z}$-transforms

$$\mathcal{Z}\{r(k-1)\} = z^{-1} R(z)$$

then

$$\mathcal{Z}\{r(k) - r(k-1)\} = [1 - z^{-1}]R(z). \qquad (8.63)$$

But by definition

$$\mathcal{Z}\{r(k) - r(k-1)\} = \sum_{k=0}^{\infty} [r(k) - r(k-1)]z^{-k}$$

$$= \lim_{N \to \infty} \sum_{k=0}^{N} [r(k) - r(k-1)]z^{-k}. \qquad (8.64)$$

So that from Equations 8.63 and 8.64.

$$[1 - z^{-1}]R(z) = \lim_{N \to \infty} \sum_{k=0}^{N} [r(k) - r(k-1)]z^{-k}$$

$$= \lim_{N \to \infty} \left\{ \left(r(0) + r(1)z^{-1} + r(2)z^{-2} + \cdots + r(N)z^{-N} \cdots \right) \right.$$

$$\left. - \left(r(-1) + r(0)z^{-1} + r(1)z^{-2} + \cdots + r(N-1)z^{-N} + \cdots \right) \right..$$

Further simplification can be made of this expression by multiplying by z to obtain

$$[z-1]R(z) = \lim_{N\to\infty} \{(r(0)z + r(1) + r(2)z^{-1} + \cdots + r(N)z^{-N+1}\cdots)$$
$$- (r(-1)z + r(0) + r(1)z^{-1} + \cdots + r(N-1)z^{-N+1} + \cdots)$$

$$(8.65)$$

Now if $z \to 1$, this will reduce to

$$\lim_{z\to 1}[z-1]R(z) = \lim_{N\to\infty} \{(r(0) + r(1) + r(2) + \cdots + r(N)\cdots)$$
$$- (r(-1) + r(0) + r(1) + \cdots + r(N-1) + \cdots)$$
$$= \lim_{N\to\infty} r(N) - r(-1).$$

$$(8.66)$$

Since for the one sided $\mathcal{Z}$-transforms $r(-1) = 0$, it follows that the final value theorem emerges as

$$\lim_{z\to 1}[z-1]R(z) = \lim_{N\to\infty} r(N).$$

$$(8.67)$$

Frequency Scaling (Multiplication by a^k): There are some situations when the samples in the sequence of a discrete-time signal are amplified by the power of the time instant. The $\mathcal{Z}$-transform of such signals are obtained through the frequency scaling property of the $\mathcal{Z}$-transform.

To determine the $\mathcal{Z}$-transform of the sequence $\{a^k r(k)\}_{k=0}^{\infty}$, which is $\{r(k)\}$ amplified by a factor a^k, one can start with the basic definition

$$\mathcal{Z}\{a^k r(k)\} = \sum_{k=0}^{\infty} a^k r(k) z^{-k}.$$

Rearrangement of the terms gives

$$\mathcal{Z}\{a^k r(k)\} = \sum_{k=0}^{\infty} r(k)(a^{-1}z)^{-k}.$$

This is essentially the $\mathcal{Z}$-transform, in which the complex variable z is scaled down by a factor a^{-1}, i.e.,

$$\mathcal{Z}\{a^k r(k)\} = R(a^{-1}z)$$

This property can also be used to show that the z-variable in the $\mathcal{Z}$-transform of the sequence $\{a^{-k}r(k)\}_{k=0}^{\infty}$ is scaled up by a factor a. This implies that

$$\mathcal{Z}\{a^{-k}r(k)\} = R(az).$$

The Inverse $\mathcal{Z}$-Transforms: For the $\mathcal{Z}$-transform to be useful in discrete-time systems analysis, there must be a way of obtaining the discrete-time signals that resulted in the $\mathcal{Z}$-transform. This is achieved by using the inverse of the $\mathcal{Z}$-transform.

$$r(k) = \mathcal{Z}^{-1}\{R(z)\}$$

The most direct method of finding the sequence $r(k)$ given its $\mathcal{Z}$-transform $R(z)$ is by using the inversion integral formula

$$r(k) = \frac{1}{2\pi j} \oint_\Gamma R(z)z^{k-1}dz \tag{8.68}$$

where $\oint_\Gamma$ is the integration in the complex z-plane along the closed curve Γ in the counterclockwise direction. The curve is taken to be the unit circle centered at the origin of the z-plane so that $R(z)$ is convergent.

The inversion integral formula is derived from the definition of the $\mathcal{Z}$-transform by using Cauchy integral theorem, which, in its simplest form, is given by

$$\oint_\Gamma z^{k-n-1}dz = \begin{cases} 2\pi j & \text{for } k = n \\ 0 & \text{for } k \neq n. \end{cases} \tag{8.69}$$

If both sides of the definition of the $\mathcal{Z}$-transform are multiplied by z^{k-1} and the closed integral taken over Γ then

$$\oint_\Gamma R(z)z^{n-1}dz = \oint_\Gamma \sum_{k=-\infty}^{\infty} r(k)z^{-k+n-1}dz$$

$$= \cdots + \oint_\Gamma r(-1)z^{-1+n-1}dz + \oint_\Gamma r(0)z^{0+n-1}dz$$

$$+ \cdots + \oint_\Gamma r(1)z^{1+n-1}dz + \cdots . \tag{8.70}$$

Since, in each of the terms on the right-hand side of this expression, the terms $r(i)$ are not functions of z, they can be taken out of the integral sign, and by using Cauchy's integral formula. Hence, all terms in this side will vanish to zero except when $k = n$ and this leaves the expression simplified to

$$\oint_\Gamma R(z)z^{k-1}dz = r(k).2\pi j, \tag{8.71}$$

from which

$$r(k) = \frac{1}{2\pi j} \oint_\Gamma R(z)z^{k-1}dz. \tag{8.72}$$

Now the inverse $\mathcal{Z}$-transform can be determined from this formula by using the residue theorem (Cauchy residual calculus), however for many cases in control engineering, this is not necessary. Thus, other methods that are more direct to applications in control engineering are discussed, where the one-sided $\mathcal{Z}$-transform $R(z)$ is used and assumed to be a rational function of z with the degree of the numerator being less than or equal to that of the denominator.

Inversion by Power Series Expansion (Direct Division) If $R(z)$ is a rational function in z of the form

$$R(z) = \frac{b_0 + b_1 z + b_2 z^2 + \cdots + b_m z^m}{a_0 + a_1 z + a_2 z^2 + \cdots + a_n z^n} \tag{8.73}$$

with $m \leq n$, $R(z)$ can be expressed as an infinite power series of z^{-1} by long division as

$$R(z) = c_0 z^{-0} + c_1 z^{-1} + c_2 z^{-2} + \cdots \cdots$$

or simply

$$R(z) = \sum_{k=0}^{\infty} c_k z^{-k}. \tag{8.74}$$

Obviously, by comparing this result and the definition of the $\mathcal{Z}$-transform, it follows that the value of $r(k)$ in the sequence that generated $R(z)$ will be given by the coefficients of z^{-k} in the infinite series. With this approach, it is possible to get the values of the discrete-time signals at any instant k. However, it is not easy to identify the general term from the first few sample values of the sequence.

Example 8.7 *Given*

$$R(z) = \frac{z}{z - 0.2} \qquad |z| > 0.2.$$

determine $r(0)$, $r(1)$, $r(2)$ *and* $r(3)$

Solution 8.7 *Perform long division*

$$
\begin{array}{r}
1 \;+0.2z^{-1} \quad 0.04z^{-2} \quad +0.008z^{-3} \quad \cdots\cdots\cdots \\
z - 0.2)\; z \\
z \quad -0.2 \\
0.2 \\
0.2 \quad -0.04z^{-1} \\
0.04z^{-1} \quad -0.008z^{-2} \\
0.008z^{-2} \quad -0.0016z^{-3}
\end{array}
$$

From this result, it becomes evident that

$$R(z) = 1 + 0.2z^{-1} - 0.04z^{-2} + 0.008z^{-3} + \cdots\cdots\cdots$$

so that

$$r(k) = (0.2)^n \, u(n).$$

From this equation it can be deduced that for unit $u(n)$

$$r(0) = 1$$
$$r(1) = 0.2$$
$$r(2) = 0.04$$
$$r(3) = 0.008.$$

Inversion by Partial Fraction Expansion: This is the most popular and direct method of finding the inverse of a $\mathcal{Z}$-transform, which gives not only the values of $r(k)$ but also the general sequence that produced the sequence. Like its counterpart Laplace transforms, it stems from the simple fact that every rational function can be expressed as a sum of partial fractions and that almost all fundamental $\mathcal{Z}$-transforms that appear as partial fractions of the given $R(z)$ are standard and their inverse $\mathcal{Z}$-transforms can be found from many standard mathematical handbooks and tables. These tables show the fundamental $\mathcal{Z}$-transforms and hence their inverses.

The idea behind this method is to obtain a partial fraction expansion of $R(z)$ over its poles just as it was shown in the case of Laplace transforms so that if

$$R(z) = \frac{N(z)}{D(z)},$$

with $\deg[N(z)] \le \deg[D(z)]$, under partial fraction expansion it becomes

$$R(z) = \hat{R}_1(z) + \hat{R}_2(z) + \hat{R}_3(z) + \cdots$$

where $\hat{R}_i(z)$ are the partial fractions. From the standard table, with use of the properties of the $\mathcal{Z}$-transform (Appendix A), the expression for $r(k)$ can be obtained as a sum of the inverse $\mathcal{Z}$-transforms of the elementary transforms of the partial fractions.

Methods of breaking $R(z)$ into its partial fractions are the same as those discussed under the Laplace transforms, as such they are not repeated here.

Example 8.8 *Use the method of partial fractions to calculate the inverse of*

$$Y(z) = \frac{4z}{z^2 - 1}$$

Determine $y(0)$, $y(1)$, $y(2)$ *and* $y(5)$.

Solution 8.8 *The given function can be presented in partial fractions as*

$$Y(z) = \frac{2}{z-1} + \frac{2}{z+1},$$

which, by using tables gives

$$y(k) = 2(1)^{k-1} + 2(-1)^{k-1},$$

or simply

$$y(k) = 2\left[1 - (-1)^k\right].$$

This means

$$y(0) = 0$$
$$y(1) = 4$$
$$y(2) = 0$$
$$y(5) = 4.$$

Using the z-Transform to Solve Difference Equations: At this point of interest the usefulness of the $\mathcal{Z}$-transforms in discrete-time analysis can be seen. Recall that the dynamics of the discrete-time system are described by using difference equations, and that the responses of such systems are obtained by the solution of the difference equation that describes its dynamics. Previously, methods for solving difference equations using iterative methods and other time-based methods were discussed. The main disadvantage with most of these approaches is the number of computations required to reach the solution. In this section, a method of solving such equations by using $\mathcal{Z}$-transforms is discussed. It is believed to be simpler computationally. Basically, like using the Laplace transforms on differential equations, the method involves two main steps; one that transforms the given difference equation into the complex z-plane, where the analysis is done using normal algebraic methods, and inverse transformation back to the time domain, which gives the required solution.

Consider the difference equation

$$\sum_{i=0}^{n} a_i u(k-i) = \sum_{j=0}^{m} b_j r(k-j) \qquad m \leq n \qquad a_0 = 1,$$

where it is desired to determine the response $u(k)$ at any instant k using $\mathcal{Z}$-transforms.

The first thing will be to get the $\mathcal{Z}$-transform of this equation and this is done by using the shift in time property of the $\mathcal{Z}$-transform.

$$\sum_{i=0}^{n} a_i z^{-i} U(z) = \sum_{j=0}^{m} b_j z^{-j} R(z) \qquad m \leq n.$$

Isolation of the $\mathcal{Z}$-transform of the required solution gives

$$U(z) = \frac{\displaystyle\sum_{j=0}^{m} b_j z^{-j} R(z)}{\displaystyle\sum_{i=0}^{n} a_i z^{-i}} \qquad m \leq n, \tag{8.75}$$

which is a rational function in z. With this $\mathcal{Z}$-transform of the required solution, it is possible to get $u(k)$ by taking its inverse $\mathcal{Z}$-transform using any of the methods described in the previous section.

Example 8.9 *Solve the difference equation*

$$u(k) - 3u(k-1) - 4u(k-2) = 2k^2 + 3 + 4^k$$

using initial conditions $u(-1) = 0.5$, $u(-2) = 1.0$

Solution 8.9 *Since the initial conditions are defined that are valid for $k <$ 0, for this problem, one can still use the one-sided $\mathcal{Z}$-transform as discussed previously. This will give*

$$U(z) - 3z^{-1}\left[U(z) + u(-1)z\right] - 4z^{-2}\left[U(z) + u(-1)z + u(-2)z^2\right]$$
$$= 2\frac{z(z+1)}{(z-1)^3} + 3\frac{z}{z-1} + \frac{z}{z-4},$$

which simplifies to

$$(1 - 3z^{-1} + 4z^{-2})U(z) = 2\frac{z(z+1)}{(z-1)^3} + 3\frac{z}{z-1} + \frac{z}{z-4} + 1.5 + 2z^{-1} + 4$$
$$= \frac{-111z^4 + 185z^3 - 125z^2 + 19z^5 - 8z + 16}{2(z-1)^3(z-4)z}.$$

Thus, the $\mathcal{Z}$-transform of the output $U(z)$ becomes

$$U(z) = \frac{(-111z^4 + 185z^3 - 125z^2 + 19z^5 - 8z + 16)z}{2(z-1)^3(z-4)(z^2 - 3z + 4)},$$

which can be broken into partial fractions as

$$U(z) = \frac{19}{2} + \frac{2}{(z-1)^3} + \frac{8}{(z-1)^2} + \frac{27}{2(z-1)} + \frac{8}{z-4} + \frac{18z}{z^2 - 3z + 4}.$$

Thus, the inverse $\mathcal{Z}$-transform, $u(k)$, can be obtained from the tables.

8.3.5 Dynamic System Response and Stability Analysis

In the previous section, different mathematical tools for analyzing discrete-time systems were discussed. In this section, methods of analysis of discrete-time (digital) control systems using these tools are discussed. The foremost of all to be considered in the analysis of any control system is the definition of the transfer function. In the following section, the description of the transfer function using both the $\mathcal{Z}$-transform and the time shift operator q is given. Once this is done, the system analysis method for the time response as well as stability will be given. Also the system description using state variables will be presented as well as the analysis using state variables.

8.3.5.1 The Pulse Transfer Function

The system transfer function is an expression that shows the relationship between the system input and its output. Just like the transfer function in the continuous-time systems where it is given by the ratio of the Laplace transform of the output to that of the input, the transfer function for discrete-time systems is the ratio of the $\mathcal{Z}$-transform of the output to that of the input. However, it is also possible to describe the transfer function in terms of the time delay operator q, as is shown shortly. Consider a discrete-time system described by a linear difference equation

$$\sum_{i=0}^{n} a_i u(k-i) = \sum_{j=0}^{m} b_j r(k-j).$$

Previously, it has been shown that the time shift operator q has the property

$$q^n f(k) = f(k+n)$$

or

$$q^{-n} f(k) = f(k-n).$$

This property can be applied on this difference equation to give

$$\sum_{i=0}^{n} a_i q^{-i} u(k) = \sum_{j=0}^{m} b_j q^{-j} r(k)$$

or simply

$$u(k) \sum_{i=0}^{n} a_i q^{-i} = r(k) \sum_{j=0}^{m} b_j q^{-j}. \tag{8.76}$$

The pulse transfer function $H(q)$ can be defined using the q variable as the ratio of the output pulse $u(k)$ to the input pulse $r(k)$, both taken at the same instant k, thus

$$H(q) = \frac{u(k)}{r(k)} = \frac{\sum_{j=0}^{m} b_j q^{-j}}{\sum_{i=0}^{n} a_i q^{-i}}, \tag{8.77}$$

which can be expanded as

$$H(q) = \frac{b_o + b_1 q^{-1} + b_2 q^{-2} + b_3 q^{-3} + \cdots + b_m q^{-m}}{a_o + a_1 q^{-1} + a_2 q^{-2} + a_3 q^{-3} + \cdots + a_n q^{-n}} \tag{8.78}$$

where a_o is normally normalized to 1 as explained before. Sometimes, instead of expressing the pulse transfer function as a rational function in ascending negative powers of q, it can be expressed in terms of ascending positive powers of q by dividing both the numerator and the denominator by the highest negative power q of the denominator. In that case, the above transfer function becomes

$$H(q) = \frac{q^{n-m}(b_o q^m + b_1 q^{m-1} + \cdots + b_{m-2} q^2 + b_{m-1} q + b_m)}{q^n + a_1 q^{n-1} + \cdots + a_{n-2} q^2 + a_{n-1} q + a_n}. \tag{8.79}$$

Example 8.10 *The input-output relationship of a certain discrete-time system is given by the difference equation*

$$u(k) - 4u(k-1) + 3u(k-2) = r(k) - 3r(k-1)$$

give an expression for its pulse transfer function using the q variable.

Solution 8.10 *The application of the q-operator on this difference equation gives*

$$u(k) - 4q^{-1}u(k) + 3q^{-2}u(k) = r(k) - 3q^{-1}r(k)$$

or

$$(1 - 4q^{-1} + 3q^{-2})u(k) = \left(1 - 3q^{-1}\right) r(k).$$

Hence, by its definition, the pulse transfer function becomes

$$H(q) = \frac{1 - 3q^{-1}}{1 - 4q^{-1} + 3q^{-2}},$$

which can also be presented as

$$H(q) = \frac{q^2 - 3q}{q^2 - 4q + 3}.$$

The transfer function can also be presented in terms of the complex variable z as a ratio of the $\mathcal{Z}$-transforms as

$$H(z) = \frac{U(z)}{R(z)}.$$

Again, the shift in time property of the $\mathcal{Z}$-transforms plays a big role in this case.

Consider again the system difference equation. The $\mathcal{Z}$-transform of this equation is

$$\sum_{i=0}^{n} a_i z^{-i} U(z) = \sum_{j=0}^{m} b_j z^{-j} R(z)$$

or

$$U(z) \sum_{i=0}^{n} a_i z^{-i} = R(z) \sum_{j=0}^{m} b_j z^{-j},$$

which gives the transfer function by its definition as

$$H(z) = \frac{U(z)}{R(z)} = \frac{\sum\limits_{j=0}^{m} b_j z^{-j}}{\sum\limits_{i=0}^{n} a_i z^{-i}}. \tag{8.80}$$

This can also be expanded as

$$H(z) = \frac{b_o + b_1 z^{-1} + b_2 z^{-2} + b_3 z^{-3} + \cdots + b_m z^{-m}}{a_o + a_1 z^{-1} + a_2 z^{-2} + a_3 z^{-3} + \cdots + a_n z^{-n}}.$$

Instead of expressing this transfer function as negative powers of z, it is also possible to express it in ascending positive powers of z by dividing both the numerator and dominator by the highest negative power of z, giving

$$H(z) = \frac{z^{n-m}(b_o z^m + b_1 z^{m-1} + \cdots + b_{m-2} z^2 + b_{m-1} z + b_m)}{z^n + a_1 z^{n-1} + \cdots + a_{n-2} z^2 + a_{n-1} z + a_n}. \tag{8.81}$$

As it has been seen, the transfer function is a rational function in either q or z. In any case, the denominator of this transfer function when equated to zero forms the system characteristic equation. This characteristic function plays the same role as was discussed for continuous-time systems, where it is used in the stability analysis of the respective systems. Stability analysis for discrete-time systems is given in the next sections. The roots of the characteristic equation are also known as the poles of the system, while the roots of the numerator of the transfer function are known as the zeros of the system. Both the zeros and poles play an important role in the determination of the system response.

Example 8.11 *Consider the following input-output relationship for an open-loop discrete-time system*

$$u(k) - 4u(k-1) + 3u(k-2) = r(k) - 3r(k-1).$$

Determine the poles of the closed-loop system with unity feedback.

Solution 8.11 *The $\mathcal{Z}$-transform of both sides of the given difference equation gives*

$$\left(1 - 4z^{-1} + 3z^{-2}\right) U(z) = \left(1 - 3z^{-1}\right) R(z)$$

from which the open-loop transfer function becomes

$$
\begin{aligned}
G(z) &= \frac{U(z)}{R(z)} \\
&= \frac{1 - 3z^{-1}}{1 - 4z^{-1} + 3z^{-2}} \\
&= \frac{z^2 - 3z}{z^2 - 4z + 3}.
\end{aligned}
$$

With unit feedback, the closed-loop transfer function is then

$$
\begin{aligned}
T(z) &= \frac{G(z)}{1 + G(z)} \\
&= \frac{z^2 - 3z}{2z^2 - 7z + 3},
\end{aligned}
$$

whose characteristic equation is

$$2z^2 - 7z + 3 = 0.$$

Hence, the closed-loop poles which are the solutions of this characteristic equation, can easily be computed and are $z = \frac{1}{2}$ and $z = 3$.

8.3.5.2   The System Response

So far, the previous discussion on the solution of the difference equation was aimed at determining the time response of the system whose dynamics is given by the difference equation, the solution of which is sought. The coverage of the system (time) response will thus be limited to systems whose transfer function is given rather than the difference equation. It will be noted that all this is about the same topic. This is because, as discussed in previous sections, the transfer function is just another way of presenting the system difference equation, whose solution, is the required time response. Nevertheless, the process of determining the time response from the difference equation is simpler because it involves fewer steps than the direct calculations than were done previously.

Consider a system whose pulse transfer function is described in compact form as

$$H(z) = \frac{\sum\limits_{j=0}^{m} b_j z^{-j}}{\sum\limits_{i=0}^{n} a_i z^{-i}},$$

where this also can be given as

$$H(z) = \frac{U(z)}{R(z)}.$$

The $\mathcal{Z}$-transform of the system response can be written in terms of the pulse transfer function $H(z)$ and the $\mathcal{Z}$-transform of the input function $R(z)$ as

$$U(z) = H(z)R(z).$$

The inverse $\mathcal{Z}$-transform of this equation will give the time response of the system. There are two possibilities for how to carry out this inversion; using the convolution theorem or by direct determination of the inverse transform.

If the inverse $\mathcal{Z}$-transform of both $H(z)$ and $R(z)$ are known, then the time response can be determined using the convolution theorem

$$u(k) = \sum_{j=0}^{k} h(j)r(k-j)$$

$$= \sum_{j=0}^{k} r(j)h(k-j), \tag{8.82}$$

where $r(.)$ and $h(.)$ are the inverse $\mathcal{Z}$-transforms of $R(z)$ and $H(z)$.

The main disadvantage of this approach is that it doesn't given a general form of the solution, and the values of $r(.)$ and $h(.)$ must be known for the whole interval from 0 to k. To determine the general form of time response, the inverse $\mathcal{Z}$-transform $H(z)R(z)$ must be determined using the method of partial fraction expansion as was discussed before.

If the product $H(z)R(z)$ could be written into its partial fractions as

$$H(z)R(z) = \hat{F}_1(z) + \hat{F}_2(z) + \hat{F}_3(z) + \cdots$$

then

$$U(z) = \hat{F}_1(z) + \hat{F}_2(z) + \hat{F}_3(z) + \cdots$$

Using standard tables, and some properties of the $\mathcal{Z}$-transforms, the inverse $\mathcal{Z}$-transform of $U(z)$ can be determined as the sum of the inverse $\mathcal{Z}$-transforms of the partial fractions.

$$u(k) = \hat{f}_1(k) + \hat{f}_2(k) + \hat{f}_3(k) + \cdots$$

Notice that, since the unit impulse (Kronecker delta) function,

$$\delta(k) = \begin{cases} 1 & k = 0 \\ 0 & k \neq 0 \end{cases},$$

has a unit $\mathcal{Z}$-transform

$$\Delta(z) = 1,$$

it will be seen that the transfer function $H(z)$ is just the $\mathcal{Z}$-transform of the response to a unit impulse.

$$\begin{aligned} U(z) &= H(z)\Delta(z) \\ &= H(z) \end{aligned}$$

Example 8.12 *A system is described by a discrete-time transfer function*

$$G(z) = \frac{0.5\,(z + 0.5)}{z^2 - 1.5z + 0.5}.$$

Determine the system response $y(k)$ to a series of unit step inputs $u(.)$

Solution 8.12 *The first step is to express the transfer function in ascending negative powers of z as*

$$G(z) = \frac{0.5\,\left(z^{-1} + 0.5z^{-2}\right)}{1 - 1.5z^{-1} + 0.5z^{-2}}.$$

However, the transfer function $G(z)$ is just the ratio

$$G(z) = \frac{Y(z)}{U(z)} = \frac{0.5\,\left(z^{-1} + 0.5z^{-2}\right)}{1 - 1.5z^{-1} + 0.5z^{-2}}.$$

Cross multiplication gives

$$\left(1 - 1.5z^{-1} + 0.5z^{-2}\right) Y(z) = 0.5\,\left(z^{-1} + 0.5z^{-2}\right) U(z).$$

Now, application of the time shift theorem and the definition of the $\mathcal{Z}$-transform reduces this expression to the difference equation

$$y(k) - 1.5y(k-1) + 0.5y(k-2) = 0.5u(k-1) + 0.25u(k-2)$$

so that the response at time instant k is

$$y(k) = 0.5u(k-1) + 0.25u(k-2) + 1.5y(k-1) - 0.5y(k-2).$$

Alternatively, one could use the inverse $\mathcal{Z}$-transform approach as

$$Y(z) = \frac{0.5(z+0.5)}{z^2 - 1.5z + 0.5}U(z)$$

$$= \left[\frac{1.5}{z-1.0} - \frac{1.0}{z-0.5} \right] U(z).$$

The inverse $\mathcal{Z}$-transform gives

$$y(k) = 1.5u(k).(1.0)^{k-1} - u(k)(0.5)^{k-1}$$

$$= \left[1.5 - 2(0.5)^k \right] u(k).$$

Notice that the two results are still the same, although the last result is more general, while the first result needs the knowledge of the initial conditions.

8.3.5.3 Stability Analysis

Stability From Time Response: The system stability is the center of most control system design. It is believed that the study of systems control emerged as a result of the need to combat system instability. System stability can be considered in two ways; internal stability when no input is applied, and external stability when some external input is applied. The internal stability dictates the system behavior when no input is applied, while the external stability dictates the behavior of the system under action of external input. Advanced analysis methods go beyond this categorization presented in the next chapter, however, all these methods finally converge to the same goal: system dynamic behavior. In this section, both the internal and external stability, starting with the internal stability, are discussed.

From the previous discussion on continuous-time systems it was pointed out that the stability of the system is established by its transient (free) response. The system is said to be stable if its transient response decays to zero as time grows to infinity. The discussion begins based on this fact.

External stability: System stability in relation to the input-output response is known as external stability. For such a stability, the common definition of appropriate response is that for every Bounded Input there should be a Bounded Output (BIBO). A test for BIBO stability can be given directly in terms of the unit pulse response and examining whether such an output response will be bounded within some limits.

A simple and direct test for the stability of a discrete-time system is that credited to Jury, known as Jury's test. This is an equivalent of the

Routh-Hurwitz test for continuous-time systems. It needs construction of an array composed of coefficients of the characteristic equation. Stability can only be ensured if the magnitude of the roots do not exceed 1. This is achieved if and only if the four necessary Jury's conditions are met.

8.3.6   Discretization in the Frequency Domain

Since most of the processes that are to be controlled have dynamics that are continuous-time, there is need to look at ways in that one can design the digital controllers to match such dynamics. Earlier, the methods of approximating the differential and integral functions to discrete-time difference equations in time domain were discussed. This section extends these ideas to come up with ways in which a continuous-time process in s-variable can be transformed into a discrete-time process in z-variable, i.e., in the frequency domain. This transformation enables us to design controllers using continuous principles and then transform them into their discrete-time equivalents. The most direct method of digitization of a continuous-time processes in the frequency domain is by direct application of the relationship between the Laplace transform and the $\mathcal{Z}$-transform

$$z = e^{sT},$$

which gives

$$s = \frac{1}{T} \ln (z). \tag{8.83}$$

However, in most cases, such transformation produces complex results because of the presence of logarithmic functions, so, alternative methods are sought. These alternative methods are based on discrete approximation of the integral functions that were presented earlier. Recall that a continuous-time integral signal that

$$u(k) = \int_0^k r(t)dt,$$

has a continuous-time transfer function from $r(t)$ to $u(t)$ given as

$$H(s) = \frac{U(s)}{R(s)} = \frac{1}{s}.$$

This signal can also be written as

$$u(k) = \int_0^{k-T} r(t)dt + \int_{k-T}^k r(t)dt$$

$$= u(k-T) + \int_{k-T}^k r(t)dt. \tag{8.84}$$

If the integral $\int_{k-T}^{k} r(t)dt$ can be expressed in discrete-time form, it becomes possible to determine the discrete-time transfer function from $r(k)$ to $u(k)$ which when compared with the continuous-time transfer function makes it possible to establish the relationship between s- and z-variables. Since the integral $\int_{k-T}^{k} r(t)dt$ can be approximated using one of the methods discussed, the balance of this section will use the results of this approximation to give the corresponding approximate relation between the s-variable and z-variable.

Euler's Backward Rectangular Rule: Euler's backward rectangular rule for discrete approximation of integral functions was given as

$$\int_{k-T}^{k} r(t)dt = r(k)T$$

so that

$$u(k) = u(k-T) + r(k)T.$$

Therefore, that its $\mathcal{Z}$-transform becomes

$$U(z) = z^{-1}U(z) + R(z)T$$

giving the corresponding transfer function as

$$H(z) = \frac{U(z)}{R(z)} = \frac{T}{1 - z^{-1}}$$
$$= \frac{Tz}{z - 1}. \tag{8.85}$$

Comparison of this transfer function with its continuous version, the equivalent continuous-time discretization of the s variable using the backward rectangular rule can thus be obtained as

$$s = \frac{z-1}{Tz}. \tag{8.86}$$

Euler's Forward Rectangular Rule: Under the forward rule, the approximate integral becomes

$$\int_{k-T}^{k} r(t)dt = r(k-T)T.$$

This approximation gives the solution as

$$u(k) = u(k-T) + r(k-T)T,$$

with the $\mathcal{Z}$-transform

$$U(z) = z^{-1}U(z) + z^{-1}R(z)T.$$

The transfer function from $r(k)$ to $u(k)$ is then

$$H(z) = \frac{U(z)}{R(z)} = \frac{z^{-1}T}{1 - z^{-1}}$$

$$= \frac{T}{z - 1}. \tag{8.87}$$

Again, comparison with the continuous-time transfer function gives the discrete approximation for s as

$$s = \frac{z - 1}{T}. \tag{8.88}$$

It is interesting to note the similarity between these approximations and those for the differential equations for the differential operator $\dfrac{d}{dt}$ and the shift-in-time operator q.

   The Trapezoidal Rule (Tustin's Approximation): The trapezoidal approximation gives an approximate integral as

$$\int_{k-T}^{k} r(t)dt = \frac{1}{2} \left[r(k - T) + r(k) \right] T.$$

The complete discrete-time solution becomes

$$u(k) = u(k - T) + \frac{1}{2} \left[r(k - T) + r(k) \right] T.$$

Again, the $\mathcal{Z}$-transform of this equation gives

$$U(z) = z^{-1}U(z) + \frac{1}{2}T \left[z^{-1}R(z) + R(z) \right].$$

Thus, the transfer function from $r(k)$ to $u(k)$ becomes

$$H(z) = \frac{U(z)}{R(z)} = \frac{1}{2}T\frac{z^{-1} + 1}{z^{-1} - 1}$$

$$= \frac{T(1 + z)}{2(1 - z)}. \tag{8.89}$$

Comparing this result with the continuous-time transfer function gives a discrete approximation for s as

$$s = \frac{2(1 - z)}{T(1 + z)}. \tag{8.90}$$

This approximation is commonly known as Tustin's approximation due to his work in this area, and the transformation from s to z using this approximation is known as the *bilinear transformation*.

Example 8.13 *A system is described by two blocks in series that have continuous-time transfer functions*

$$G_1(s) = 2\frac{\left(1 - e^{sT}\right)(s + 5)}{s^2}$$

and

$$G_2(s) = 25\frac{\left(1 - e^{sT}\right)}{s^2 + 26s + 25}.$$

Determine the closed-loop discrete-time transfer function for the system when unit feedback is applied. The sampling interval $T = 0.01s$. Wherever necessary, use Tustin's approximation.

Solution 8.13 *The combined transfer function of the system is*

$$G(s) = G_1(s)G_2(s)$$

$$= \left[2\frac{\left(1 - e^{sT}\right)(s + 5)}{s^2}\right]\left[25\frac{\left(1 - e^{sT}\right)}{s^2 + 26s + 25}\right].$$

Using the relationship between the $\mathcal{Z}$-transform and the Laplace transform, one can replace the exponential term by z, and by using Tustin's approximation

$$s = \frac{2(1 - z)}{T(1 + z)},$$

so that

$$s^2 = \left[\frac{2(1 - z)}{T(1 + z)}\right]^2$$

$$= \frac{4\left(1 - 2z + z^2\right)}{T^2\left(1 - 2z + z^2\right)}.$$

These replacements give

$$G(z) = \left\{50\frac{(1 - z)\left[\dfrac{2(1 - z)}{T(1 + z)} + 5\right]}{\left[\dfrac{4\left(1 - 2z + z^2\right)}{T^2\left(1 - 2z + z^2\right)}\right]}\right\}\left\{\frac{1 - z}{\left[\dfrac{4\left(1 - 2z + z^2\right)}{T^2\left(1 - 2z + z^2\right)}\right] + 26\left[\dfrac{2(1 - z)}{T(1 + z)}\right] + 25}\right\}$$

$$= \frac{25\,T^3\left((-2 + 5T)\,z^3 + (6 - 5T)\,z^2 + (-6 - 5T)\,z + 2 + 5T\right)}{2}\frac{}{\left(-52T + 4 + 25T^2\right)z + 52T + 4 + 25T^2},$$

which simplifies further by the use of the value of the sampling interval T=0.01 to become

$$G(z) = (-0.000\,25)\,\frac{39z^3 - 119z^2 + 121z - 41}{1393z + 1809}.$$

Now the closed-loop transfer function is given as

$$T(z) = \frac{G(z)}{1 + G(z)}$$

$$= \frac{(-0.000\,25)\left(39z^3 - 119z^2 + 121z - 41\right)}{1393z + 1809 + (-0.000\,25)\left(39z^3 - 119z^2 + 121z - 41\right)}$$

$$= \frac{39z^3 - 119z^2 + 121z - 41}{39z^3 - 119z^2 - 5.5719 \times 10^6 z - 7.236 \times 10^6}.$$

8.3.7 The State-Space Analysis

As was shown in the previous sections, mathematical models that describe the dynamics of sampled data systems in time domain are almost always finite-order difference equations whose solutions exist, and in most cases, these solutions are unique. Therefore, it is possible to predict the behavior of such systems for any time instant $t > t_o$ expressed at sampling instants k if an appropriate set of initial conditions at $t = t_o$ is specified. These initial conditions contain all the past behavior of the system necessary to calculate the future behavior. This is also true for continuous-time systems, which are described by differential equations. It was further shown in Chapter 7 that the higher-order differential equations can be broken down into a set of simple first-order equations which are then presented in matrix form, using state variables. In this section, the methods of breaking higher-order difference equations are introduced. These equations represent a discrete-time system into first-order difference equations and hence combine them into state variable presentation. Generally, the discrete-time state-space model can be obtained in two ways; by reduction of the higher-order system difference equation (which may be obtained from the corresponding continuous-time differential equation) or by conversion of the continuous-time state-space model. Both methods are discussed starting with the reduction of the higher-order difference equation.

8.3.7.1 Discrete-Time State-Space Model

Consider the difference equation

$$y(k) + 2\,y(k - 1) + y(k - 2) = u(k)$$

in which the initial conditions $y(-1)$ and $y(-2)$ are specified and the output of interest is known to be $y(k)$. In order to break this difference equation into a set of first-order difference equations, the system states must be specified first. Since the output is defined to be $y(k)$, then the system states at any instant k will be the past values of $y(k)$, i.e., $y(k-1)$ and $y(k-2)$. Thus, the system state vector becomes

$$\mathbf{x}(k) = \begin{bmatrix} x_1(k) \\ x_2(k) \end{bmatrix} = \begin{bmatrix} y(k-2) \\ y(k-1) \end{bmatrix}.$$

The first-order difference equations corresponding to each of these states are then

$$x_1(k+1) = y(k-1) = x_2(k)$$
$$x_2(k+1) = y(k) = u(k) - y(k-2) - 2\,y(k-1)$$
$$= u(k) - x_1(k) - 2x_2(k).$$

Putting these equations together in matrix form yields

$$\begin{bmatrix} x_1(k+1) \\ x_2(k+1) \end{bmatrix} = \begin{bmatrix} 0 & 1 \\ -1 & -2 \end{bmatrix} \begin{bmatrix} x_1(k) \\ x_2(k) \end{bmatrix} + \begin{bmatrix} 0 \\ 1 \end{bmatrix} u(k). \tag{8.91}$$

The output $y(k)$ can also be presented in terms of the system states as

$$y(k) = u(k) - y(k-2) - 2\,y(k-1),$$

which in matrix form becomes

$$y(k) = \begin{bmatrix} -1 & -2 \end{bmatrix} \begin{bmatrix} x_1(k) \\ x_2(k) \end{bmatrix} + \begin{bmatrix} 0 \\ 1 \end{bmatrix} u(k). \tag{8.92}$$

Equations 8.91 and 8.92 together form the vector-matrix difference equation that describes the evolution of the state and an output equation instead of the second-order difference equation. Algebraically, these equations are written in compact form as

$$\mathbf{x}(k+1) = \mathbf{F}\mathbf{x}(k) + \mathbf{G}\mathbf{u}(k) \tag{8.93}$$
$$\mathbf{y}(k) = \mathbf{H}\mathbf{x}(k) + \mathbf{J}\mathbf{u}(k), \tag{8.94}$$

where matrices $\mathbf{F}$, $\mathbf{G}$, $\mathbf{H}$, and $\mathbf{J}$ are the plant (also known as the state-transition matrix) input, output and feedfoward matrices similar to matrices $\mathbf{A}$, $\mathbf{B}$, $\mathbf{C}$, and $\mathbf{D}$ for continuous systems.

This discussion can be generalized for any system described by a general difference equation of the form

$$\sum_{i=0}^{n} a_i y(k-i) = \sum_{j=0}^{m} b_j u(k-j),$$

which can be expressed as

$$y(k) = \sum_{j=0}^{m} b_j u(k-j) - \sum_{i=1}^{n} a_i y(k-i) \qquad (8.95)$$

where, in this case, the output of interest is $y(k)$. By identifying the system states as

$$x_\alpha(k) = y(k-\alpha) \qquad \alpha = 1, 2, \cdots\cdots, k-1,$$

the following is true for the state equation

$$x_1(k+1) = x_2(k)$$
$$x_2(k+1) = x_3(k)$$
$$\cdots\cdots = \cdots\cdots$$
$$x_{n-1}(k+1) = x_n(k)$$
$$x_n(k+1) = -\sum_{i=1}^{n} a_i x_i(k-i) + \sum_{j=0}^{m} b_j u(k-j).$$

In matrix form this becomes

$$\begin{bmatrix} x_1(k+1) \\ x_2(k+1) \\ \cdots\cdots \\ x_n(k+1) \end{bmatrix} = \begin{bmatrix} 0 & 1 & \vdots & 0 \\ 0 & 0 & \vdots & 0 \\ \cdots & \cdots & & \cdots \\ -a_1 & -a_2 & \vdots & -a_n \end{bmatrix} \begin{bmatrix} x_1(k) \\ x_2(k) \\ \cdots \\ x_n(k) \end{bmatrix} + \begin{bmatrix} 0 & 0 & \vdots & 0 \\ 0 & 0 & \vdots & 0 \\ \cdots & \cdots & & \cdots \\ 1 & 1 & \vdots & 1 \end{bmatrix} \begin{bmatrix} u_1(k) \\ u_2(k) \\ \cdots \\ u_m(k) \end{bmatrix},$$

where $u_i(k) = u(k-i)$. Also, the output equation

$$y(k) = -\sum_{i=1}^{n} a_i x_i(k) + \sum_{j=0}^{m} b_j u(k-j),$$

becomes

$$y(k) = \begin{bmatrix} -a_1 & -a_2 & \vdots & -a_n \end{bmatrix} \begin{bmatrix} x_1(k) \\ x_2(k) \\ \cdots \\ x_n(k) \end{bmatrix} + \begin{bmatrix} 1 & 1 & \vdots & 1 \end{bmatrix} \begin{bmatrix} u_1(k) \\ u_2(k) \\ \cdots \\ u_m(k) \end{bmatrix}. \qquad (8.96)$$

Again in compact form it can be written as shown earlier. If the transfer function $G(z)$ is given the transformation to the state-variable form can be carried out indirectly as shown in the following example.

Example 8.14 *Derive the state description of the system with the following transfer function*

$$G(z) = \frac{0.5\,(z+1)}{(z-1)^2}.$$

Solution 8.14 *Express the transfer function as a rational function in which both the numerator and the denominator are in ascending powers of z. This gives*

$$G(z) = \frac{0.5(z+1)}{z^2 - 2z + 1}$$
$$= \frac{0.5(1 + z^{-1})}{1 - 2z^{-1} + z^{-2}}.$$

By defining an intermediate variable $X(z)$ such that

$$G(z) = \frac{Y(z)}{U(z)} = \frac{Y(z)}{X(z)}\frac{X(z)}{U(z)}$$

and letting

$$\frac{X(z)}{U(z)} = \frac{0.5}{1 - 2z^{-1} + z^{-2}}$$

gives

$$\left(1 - 2z^{-1} + z^{-2}\right) X(z) = 0.5U(z).$$

Hence,

$$x(k) - 2x(k-1) + x(k-2) = 0.5u(k)$$

Defining the states as

$$x_1(k) = x(k)$$
$$x_2(k) = x_1(k-1),$$

leads to

$$x_2(k-1) = x(k-2)$$
$$= -x_1(k) + 2x_2(k) + 0.5u(k).$$

In matrix form this becomes

$$\begin{bmatrix} x_1(k-1) \\ x_2(k-1) \end{bmatrix} = \begin{bmatrix} 0 & 1 \\ -1 & 2 \end{bmatrix} \begin{bmatrix} x_1(k) \\ x_2(k) \end{bmatrix} + \begin{bmatrix} 0 \\ 0.5 \end{bmatrix} u(k)$$

The remaining part of the transfer function,

$$\frac{Y(z)}{X(z)} = 1 + z^{-1}$$

gives

$$Y(z) = \left(1 + z^{-1}\right) X(z),$$

or

$$
\begin{aligned}
y(k) &= x(k) + x(k-1) \\
     &= x_1(k) + x_2(k).
\end{aligned}
$$

In matrix form it becomes

$$y(k) = \begin{bmatrix} 1 & 1 \end{bmatrix} \begin{bmatrix} x_1(k) \\ x_2(k) \end{bmatrix}$$

completing the state-space description.

8.3.7.2   Discrete-Time State Model from Continuous-Time Model.

Alternatively, if the continuous-time state-space model

$$
\begin{aligned}
\dot{\mathbf{x}}(t) &= \mathbf{A}\mathbf{x}(t) + \mathbf{B}\mathbf{u}(t) \\
\mathbf{y}(t) &= \mathbf{C}\mathbf{x}(t) + \mathbf{D}\mathbf{u}(t)
\end{aligned}
$$

is known, the discrete-time state-space model can still be established directly from this continuous-time model. Recall that the solution of the linear model is

$$\mathbf{x}(t) = e^{\mathbf{A}(t-t_o)}\mathbf{x}(t_o) + \int_{t_o}^{t} e^{\mathbf{A}(t-\tau)}\mathbf{B}\mathbf{u}(\tau)d\tau.$$

This solution can be used to transfer the continuous-time system model to discrete-time model as follows:

Let $t_o = k$ and $t = k + T$ where T is the sampling interval. Applying the continuous-time solution over this sample interval gives

$$\mathbf{x}(k+T) = e^{\mathbf{A}T}\mathbf{x}(k) + \int_{k}^{k+T} e^{\mathbf{A}(k+T-\tau)}\mathbf{B}\mathbf{u}(\tau)d\tau.$$

When using Zero Order Hold (ZOH) with no delays in the system, it follows that

$$\mathbf{u}(\tau) = \mathbf{u}(k) \quad \text{for } k \leq \tau \leq (k+T).$$

By letting

$$k + T - \tau = \eta$$

it follows that for $\tau = k$ the value of η will be T, and for $\tau = k + T$, the value of η becomes 0. Differentiation with respect to τ gives

$$d\tau = -d\eta.$$

Substitution of these values results in

$$\mathbf{x}(k + T) = e^{\mathbf{A}T}\mathbf{x}(k) - \int_T^0 e^{\mathbf{A}\eta}\mathbf{Bu}(k)d\eta,$$

which can be rewritten as

$$\mathbf{x}(k + T) = e^{\mathbf{A}T}\mathbf{x}(k) + \int_0^T e^{\mathbf{A}\eta}d\eta\mathbf{Bu}(k). \tag{8.97}$$

Now if

$$\mathbf{F} = e^{\mathbf{A}T} \tag{8.98}$$

$$\mathbf{G} = \int_0^T e^{\mathbf{A}\eta}d\eta\mathbf{B} \tag{8.99}$$

the above equation can then be expressed as

$$\mathbf{x}(k + T) = \mathbf{Fx}(k) + \mathbf{Gu}(k), \tag{8.100}$$

which is the state-space equation, the same as Equation 8.93. At any time $t = k$ the output equation can be derived to be almost the same as that of the continuous-time system, which will be

$$\mathbf{y}(k) = \mathbf{Hx}(k) + \mathbf{Ju}(k) \tag{8.101}$$

These two equations, 8.100 and 8.101, are the ones that describe the discrete-time state-space model of the system.

Most often, it is required to establish the state-transition matrix $\mathbf{F} = e^{\mathbf{A}T}$. This is a matrix exponential whose evaluation was shown in Chapter 7 as

$$\mathbf{F} = e^{\mathbf{A}T}$$

$$= \mathbf{I} + \mathbf{A}T + \frac{\mathbf{A}^2 T^2}{2!} + \frac{\mathbf{A}^3 T^3}{3!} + \cdots + \frac{\mathbf{A}^n T^n}{n!} + \cdots\cdots$$

$$= \sum_{n=0}^{\infty} \frac{\mathbf{A}^n T^n}{n!}. \tag{8.102}$$

The output matrix $\mathbf{G}$ can be obtained by evaluating term by term of the integral in Equation 8.99 after expanding the matrix exponential $e^{\mathbf{A}\eta}$. Since

$$e^{\mathbf{A}\eta} = \mathbf{I} + \mathbf{A}\eta + \frac{\mathbf{A}^2\eta^2}{2!} + \frac{\mathbf{A}^3\eta^3}{3!} \cdots + \frac{\mathbf{A}^n\eta^n}{n!} + \cdots \cdots \qquad (8.103)$$

then

$$
\begin{aligned}
\mathbf{G} &= \int_0^T \left(\mathbf{I} + \mathbf{A}\eta + \frac{\mathbf{A}^2\eta^2}{2!} + \frac{\mathbf{A}^3\eta^3}{3!} \cdots + \frac{\mathbf{A}^n\eta^n}{n!} + \cdots \cdots \right) d\eta \mathbf{B} \\
&= \left(\mathbf{I}T + \frac{\mathbf{A}T}{2} + \frac{\mathbf{A}^2 T^3}{3!} + \frac{\mathbf{A}^3 T^4}{4!} \cdots + \frac{\mathbf{A}^n T^{n+1}}{(n+1)!} + \cdots \cdots \right) \mathbf{B} \\
&= \sum_{n=0}^{\infty} \frac{\mathbf{A}^n T^{n+1}}{(n+1)!} \mathbf{B}. \qquad\qquad (8.104)
\end{aligned}
$$

If the sampling interval is small enough, only very few terms are needed in these infinite summations. Normally two to four terms are sufficient.

Example 8.15 *The continuous-time state-space description of a system is given by*

$$\dot{\mathbf{x}}(t) = \mathbf{A}\mathbf{x}(t) + \mathbf{B}\mathbf{u}(t)$$
$$\mathbf{y}(t) = \mathbf{C}\mathbf{x}(t) + \mathbf{D}\mathbf{u}(t),$$

where

$$\mathbf{A} = \begin{bmatrix} 2 & 4 \\ 1 & 5 \end{bmatrix} \qquad \mathbf{B} = \begin{bmatrix} 1.0 \\ 2.0 \end{bmatrix}$$
$$\mathbf{C} = \begin{bmatrix} 1 & 1 \end{bmatrix} \qquad \mathbf{D} = 0$$

Give the corresponding discrete-time state-space description of this system using a sampling interval of $0.001sec$

Solution 8.15 *The solution is obtained by direct application of the equations, where*

$$\mathbf{F} = \sum_{n=0}^{\infty} \frac{\mathbf{A}^n (0.001)^n}{n!}.$$

With the given sampling interval, only two terms may be sufficient, so that

$$\mathbf{F} = \sum_{n=0}^{2} \frac{\mathbf{A}^n (0.001)^n}{n!}$$

$$= \mathbf{I} + \mathbf{A}(0.001) + \frac{\mathbf{A}^2 (0.001)^2}{2}$$

$$= \begin{bmatrix} 1 & 0 \\ 0 & 1 \end{bmatrix} + \begin{bmatrix} 2 & 4 \\ 1 & 5 \end{bmatrix} (0.001) + \begin{bmatrix} 2 & 4 \\ 1 & 5 \end{bmatrix}^2 \frac{(0.001)^2}{2}$$

$$= \begin{bmatrix} 1.002 & 4.002 \times 10^{-3} \\ 1.0005 \times 10^{-3} & 1.005 \end{bmatrix}.$$

For matrix $\mathbf{G}$ *even only one term is enough*

$$\mathbf{G} = \sum_{n=0}^{\infty} \frac{\mathbf{A}^n T^{n+1}}{(n+1)!} \mathbf{B}$$

$$= \sum_{n=0}^{2} \frac{\mathbf{A}^n T^{n+1}}{(n+1)!} \mathbf{B}$$

$$= \left[\mathbf{I} + \frac{\mathbf{A} (0.001)^2}{2} + \frac{\mathbf{A}^2 (0.001)^3}{3!} \right] \mathbf{B}$$

$$= \begin{bmatrix} 1.0 \\ 2.0 \end{bmatrix}$$

The matrices $\mathbf{H}$ *and* $\mathbf{J}$ *remain the same as* $\mathbf{C}$ *and* $\mathbf{D}$ *of the continuous-time system*

$$\mathbf{H} = \begin{bmatrix} 1 & 1 \end{bmatrix} \qquad \mathbf{J} = [0]$$

and the discrete-time state-space description becomes

$$\mathbf{x}(k+1) = \mathbf{F}\mathbf{x}(k) + \mathbf{G}\mathbf{u}(k) \qquad (8.105)$$
$$\mathbf{y}(k) = \mathbf{H}\mathbf{x}(k) + \mathbf{J}\mathbf{x}(k). \qquad (8.106)$$

As can be seen in this example, the higher the sampling rate, the closer the system is to the continuous-time transfer function. The reader can try for the case when the sampling interval is 0.1sec.

8.3.7.3 Controllability, Observability, Canonical Forms, and the Transfer Matrix

When the system is presented in discrete-time form using state variables as shown above, the analysis of controllability, observability, similarity transformation and canonical forms are done in exactly the same way as it was shown for continuous-time systems in Chapter 7. More importantly, it can be shown without going into details that by taking the $\mathcal{Z}$-transform of both Equations 8.105 and 8.106 give

$$z\mathbf{X}(z) = \mathbf{F}\mathbf{X}(z) + \mathbf{G}\mathbf{U}(z) \qquad (8.107)$$

$$\mathbf{Y}(z) = \mathbf{H}\mathbf{X}(z) + \mathbf{J}\mathbf{U}(z). \qquad (8.108)$$

From Equation 8.107 can be seen that

$$\mathbf{X}(z) = (z\mathbf{I} - \mathbf{F})^{-1}\mathbf{G}\mathbf{U}(z),$$

which, when used in, Equation 8.108 leads to

$$\mathbf{Y}(z) = \mathbf{H}(z\mathbf{I} - \mathbf{F})^{-1}\mathbf{G}\mathbf{U}(z) + \mathbf{J}\mathbf{U}(z)$$

$$= \left(\mathbf{H}(z\mathbf{I} - \mathbf{F})^{-1}\mathbf{G} + \mathbf{J}\right)\mathbf{U}(z).$$

From this equation, the transfer matrix is obtained as

$$\mathbf{G}(z) = \mathbf{H}(z\mathbf{I} - \mathbf{F})^{-1}\mathbf{G} + \mathbf{J}, \qquad (8.109)$$

which is similar in form to the one obtained for continuous-time systems. For SISO systems where G is a column vector and H is a row vector, this transfer matrix becomes a scalar function, known as the transfer function.

8.3.8   The Root Locus in the $\mathcal{Z}$-Plane

For a long time now, the root locus has been a very powerful method for analysis and design of both continuous-time and digital control systems. It gives a graphical representation of the variation of the roots of the closed-loop characteristic equation as a function of gain parameter variation. It was shown in earlier chapters for continuous-time systems that the root locus enables the design of the controller that meets a number of time domain as well as frequency-domain specifications, which include the damping ratio, speed of response, settling time and natural frequency. The same fact applies for discrete-time systems with some very minor modifications.

Consider a closed-loop transfer function of a discrete-time system

$$T(z) = \frac{Y(z)}{U(z)} = \frac{KG(z)}{1 + KH(z)G(z)}, \qquad (8.110)$$

whose characteristic equation is

$$1 + KH(z)G(z) = 0. \qquad (8.111)$$

As the gain K varies, the roots of this characteristic equation will also vary. The plot of this variation in the z-plane is known as the discrete-time root locus. Since $H(z)G(z)$ is normally a rational function in z, for

$K = 0$, the roots of the characteristic equation will be the roots of the denominator of $H(z)G(z)$, and for $K = \infty$, the roots of the characteristic equation correspond to the zeros of $H(z)G(z)$. In short, the root locus will start from the poles of $H(z)G(z)$ and end at the zeros of $H(z)G(z)$ in just the same way as for the continuous-time systems. In particular, if $H(z) = 1$, i.e., unity feedback, then the roots of the characteristic equation starts from the poles of $G(z)$ and end at the zeros of $G(z)$. Generally, the techniques for plotting the discrete-time are basically the same as those of continuous-time systems, and they will not be repeated here.

Though the plotting techniques are the same for continuous-time systems as for the discrete-time systems root loci, the interpretation of the results differ. While for continuous-time systems the stability zone is enclosed in the left-hand half plane, for discrete-time systems, the stability zone is enclosed in the unit circle about the origin of the z-plane. This is due to the transformation of the s-pane into the z-plane by

$$z = e^{sT}$$

where for $\text{Re}\,(s) \leq 0$, the magnitude of z is bound as $|z| \leq 1$.

Since the lines of constant damping ratio for continuous-time systems are determined by the angle θ where

$$\xi = \cos\theta$$

such that for any value r,

$$s = r\left(-\cos\theta + j\sin\theta\right).$$

In the z-plane, these lines are mapped to be

$$z = e^{r(-\cos\theta + j\sin\theta)T}$$

which is a curve starting at $z = 1$ for $r = 0$ to any value between 0 and -1 depending on the value of the damping ratio. For $\xi = 0$, the constant ξ curve is a semicircle from 1 to -1. Because of the symmetry of $\cos\theta$ and $\cos\left(-\theta\right)$ about real axis, the lines of constant ξ are also symmetrical about the $\text{Re}(z)$. Similarly, lines of constant damped frequency ω_d, which in the continuous-time system are circular about the origin of the s-plane, are given by

$$\omega_d = \omega_n\sqrt{1 - \xi^2}.$$

Depending on the natural frequency of the system ω_n, in the discrete-time domain are mapped according to

$$z = e^{\omega_n \xi \pm j\omega_n\sqrt{1 - \xi^2}}. \tag{8.112}$$

Figure 8.7 is a continuous-time s-plane showing lines of constant ξ as well as circles of constant ω_d in the stable zone and Figure 8.8 shows how this plane is mapped into the z-plane.

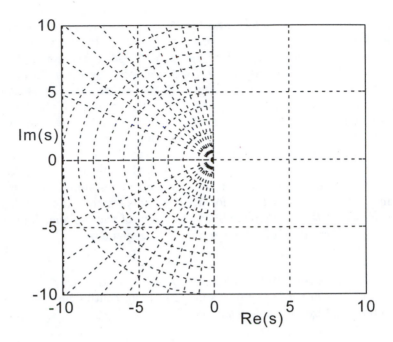

FIGURE 8.7
Root Locus in the Z-Plane

8.3.9   The Frequency-Response Analysis

8.3.9.1   Exact Frequency Response

Just like continuous-time systems, time response methods are sometimes inadequate in handling higher-order systems, while frequency-response methods are versatile in this aspect. Using the advantages of the correlation between the frequency response and the time response of systems, it becomes necessary to study the frequency response of dynamic systems so that controllers can be designed from this domain.

The frequency response analysis of discrete-time systems can be viewed as an extension of the frequency response of continuous-time systems, which gives the behavior of the system when a sinusoidal input $r(t, \omega_o)$ where

$$r(t, \omega_o) = A \sin (\omega_o t),$$

is applied for varying the frequency, ω_o, in the range $[0, \infty]$. If the open-loop transfer function of the system is $G(s)$, the common procedure used for continuous-time system in studying the frequency response is letting

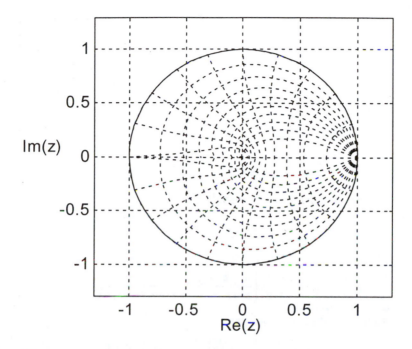

FIGURE 8.8
Root Locus in the Z-Plane

$s = j\omega_o$ so that $G(s) = G(j\omega_o)$ can be expressed in terms of magnitude and phase angle φ as

$$G(s) = G(j\omega_o) = |G(j\omega_o)| \, e^{j\varphi(\omega_o)}$$

as discussed in Chapter 6. The frequency response of such a system is determined by observing how the magnitude and phase vary with the frequency ω_o. Methods of examining this variation include the Bode plots, Nyquist plots, polar plots, and the Nichols charts as described in detail in Chapter 6.

For digital systems, the frequency response is obtained by using similar formulation, which includes discretization of continuous-time signals using appropriate means, as discussed before. The most direct discretization method is using the analytical relationship between the z-variable and s-variable

$$z = e^{sT}.$$

Since for the continuous-time systems the frequency response is obtained by letting $s = j\omega_o$, then the corresponding formulation for discrete-time system $G(z)$ is to let

$$z = e^{j\omega_o T}$$

so that

$$G(z) = G(e^{j\omega_o T}) = \left| G(e^{j\omega_o T}) \right| e^{j\phi(\omega_o T)}. \qquad (8.113)$$

This can be shown analytically by considering a sinusoidal discrete-time signal $r(k, \omega_o)$ applied to a system $G(z)$, which gives the system response as

$$U(z) = G(z)R(z),$$

where the $\mathcal{Z}$-transform for a sinusoidal signal,

$$r(k, \omega_o) = A \cos(\omega_o T k),$$

is

$$R(z) = \frac{A}{2} \left[\frac{z}{z - e^{j\omega_o T}} + \frac{z}{z - e^{-j\omega_o T}} \right].$$

The system response then becomes

$$U(z) = \frac{A}{2} \left[\frac{zG(z)}{z - e^{j\omega_o T}} + \frac{zG(z)}{z - e^{-j\omega_o T}} \right]. \qquad (8.114)$$

Now, the steady state response $u_{ss}(k)$ corresponds to $U(z)$ when z has stable poles. In this case, $G(z)$ takes on the values $G(z = e^{j\omega_o T})$ and $G(z = e^{-j\omega_o T})$. This fact can also be verified by expansion of the equation into partial fractions and applying the final value theorem. Thus,

$$U_{ss}(z) = \frac{A}{2} \left[\frac{zG(e^{j\omega_o T})}{z - e^{j\omega_o T}} + \frac{zG(e^{-j\omega_o T})}{z - e^{-j\omega_o T}} \right].$$

Since

$$G(e^{j\omega_o T}) = \left| G(e^{j\omega_o T}) \right| e^{j\phi(\omega_o T)}$$

$$G(e^{-j\omega_o T}) = \left| G(e^{-j\omega_o T}) \right| e^{-j\phi(\omega_o T)}$$

and

$$\left| G(e^{j\omega_o T}) \right| = \left| G(e^{-j\omega_o T}) \right|$$

then

$$U_{ss}(z) = \frac{A \left| G(e^{j\omega_o T}) \right|}{2} \left[\frac{ze^{j\phi(\omega_o T)}}{z - e^{j\omega_o T}} + \frac{ze^{-j\phi(\omega_o T)}}{z - e^{-j\omega_o T}} \right]$$

whose inverse $\mathcal{Z}$-transform gives

$$u_{ss}(k) = \frac{A\left|G(e^{j\omega_o T})\right|}{2} \left[e^{j\omega_o Tk} e^{j\phi(\omega_o T)} + e^{-j\omega_o Tk} e^{-j\phi(\omega_o T)}\right].$$

This can be simplified to

$$u_{ss}(k) = A\left|G(e^{j\omega_o T})\right| \cos\left(\omega_o Tk + \phi(\omega_o T)\right), \qquad (8.115)$$

indicating that the steady state response to a discrete-time sinusoidal signal can be expressed in terms of the magnitude and phase angle of $G(z = e^{j\omega_o T})$.

However, because of aliasing inherent with discrete-time systems, the frequency may not be varied indefinitely as it is done for continuous-time systems. Instead, the frequency variation is taken in the range

$$\omega_o \in \left[0, \tfrac{1}{2}\omega_s\right].$$

Notice that, as ω_o is varied from 0 to $\tfrac{1}{2}\omega_s$, the value of z, which is given by

$$z = e^{j\omega_o T}$$
$$= \cos\left(\omega_o T\right) + j\sin\left(\omega_o T\right)$$
$$= \cos\left(\frac{2\pi\omega_o}{\omega_s}\right) + j\sin\left(\frac{2\pi\omega_o}{\omega_s}\right), \qquad (8.116)$$

and will vary along a circle of unit radius. Beyond this frequency range, the value of z keeps repeating itself about this unit circle.

With knowledge of the phase angle and magnitude of $G(z)$ in the frequency range defined above, Bode plots as well as Nyquist plots can be made and will look similar to those for continuous-time systems. Computer Aided Systems Analysis and Design (CASD) packages are available that can accomplish this task. In MATLAB, the routines "dbode," "dnyquist," and "dnichols" are available for the same. Details of using these commands can, as usual, be found in MATLAB Control Toolbox manual, here a brief description of each of these commands is given as follows:

The syntax for using the "dbode" is

$$\text{dbode(N,D,Ts)}$$

where N and D are vectors containing the polynomial coefficients of $N(z)$ and $D(z)$ in ascending powers of z when the system is given in the form

$$G(z) = \frac{N(z)}{D(z)},$$

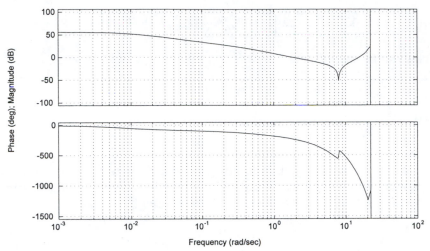

FIGURE 8.9
Discrete Time Bode Plot

and T_s is the sampling interval. However, if the system is given in discrete-time state-space form

$$\mathbf{x}(k+1) = \mathbf{Fx}(k) + \mathbf{Gu}(k)$$

$$\mathbf{y}(k) = \mathbf{Hx}(k) + \mathbf{Ju}(k),$$

then the command dbode is used as

$$\text{dbode}(\mathsf{F}, \mathsf{G}, \mathsf{H}, \mathsf{J}, \mathsf{Ts}, \mathsf{Iu}).$$

This function plots the Bode plot from the single input Iu to all outputs of the discrete-time state-space system. In both cases of the *dbode* command above, the frequency is chosen automatically. When the interest is to get the Bode plot in specific frequency range, it becomes necessary to include a vector W in the *dbode* command that describes the frequencies in radians per second at which the Bode response is to be evaluated. The syntax in this case takes the form

$$\text{dbode}(\mathsf{N},\mathsf{D},\mathsf{Ts},\mathsf{W})$$

$$\text{dbode}(\mathsf{F}, \mathsf{G}, \mathsf{H}, \mathsf{J}, \mathsf{Ts}, \mathsf{Iu}, \mathsf{W}).$$

A typical discrete-time system Bode plot is shown in Figure 8.9.

The MATLAB "dnichols" command produces a Nichols plot for a discrete-time linear system described in either polynomial transfer function $G(z)$ or in state-space form $(\mathbf{F}, \mathbf{G}, \mathbf{H}, \mathbf{J})$. The general usage of the command is similar to the "dbode" command described above. Its syntax is

$$\text{dnichols}(N,D,Ts)$$
$$\text{dnichols}(F, G, H, J, Ts, Iu)$$

or

$$\text{dnichols}(N,D,Ts,W)$$
$$\text{dnichols}(F, G, H, J, Ts, Iu, W),$$

where all variables have same meaning as described for the "dbode" command. Figure 8.10 shows a typical discrete-time Nichols chart.

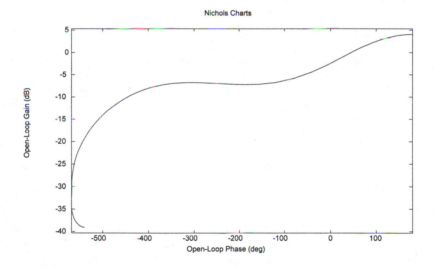

FIGURE 8.10
A Typical Digital Nichols Chart

The discrete-time Nyquist plot is generated by MATLAB by command "dnyquist." Similar to the "dnichols" command, this command also has a general form and usage as is for the "dbode" command, with all parameters taking on the usual values. Thus, the syntax for the "dnyquist" command becomes

$$\text{dnyquist}(N,D,Ts)$$
$$\text{dnyquist}(F, G, H, J, Ts, Iu)$$

or

$$\text{dnyquist(N,D,Ts,W)}$$
$$\text{dnyquist(F, G, H, J, Ts, Iu, W)}.$$

A typical Nyquist plot is shown in Figure 8.11.

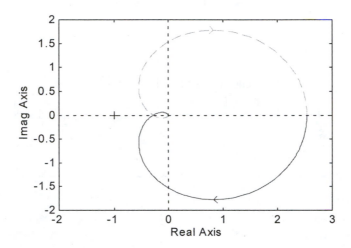

FIGURE 8.11
A Typical Nyquist Plot

   Recall that the measures of performance in frequency response domain
are the gain margin, phase margin and the bandwidth. A study of the
discrete-time Nyquist plot for varying sampling rates is given in Figure
8.12. As can be seen in this figure, the sampling rate affects the frequency
response of the system by decreasing both the gain margin and the phase
margin. For high sampling rates, the frequency response of the system
approximates the continuous-time plot, however, as the sampling rate decreases, both the gain margin and the phase margin decrease leading to the
eventuality of instability. Thus, the frequency response also dictates the
required sampling rate for the system to remain stable.

**8.3.9.2   Approximate Frequency Response: the ω-Plane and Bi-
             linear Transformation.**

   The exact frequency response discussed in the preceding section involves
functions of the form $e^{j\omega_o T}$, which makes the analysis too complex to be
done manually. Thus, use of computers becomes indispensable. However,
there are many cases where the design engineer must do the analysis manually, in which case a simple alternative method to the one presented before

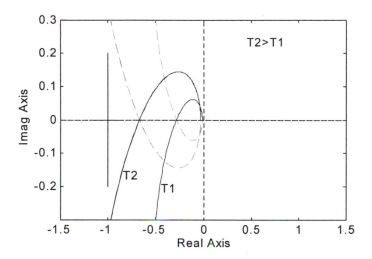

FIGURE 8.12
Nyquist Plot for Varying Sampling Rates

needs to be available. The simpler approach to frequency response, which does not need evaluation of the complex $e^{j\omega_o T}$, is an approximation that transforms the $\mathcal{Z}$-plane to another plane called the ω-plane which is similar (conformally equivalent) to the complex s-plane. The transformation employed is called the bilinear transformation, defined as

$$z = \frac{1+\omega}{1-\omega}, \tag{8.117}$$

which was partly derived. The ω-plane is another complex plane which defines the ω-variable as

$$w = \sigma_w + jv_w. \tag{8.118}$$

In this w-plane, the true frequency response ω_o in the z-plane is presented using an approximate frequency v_w, whose value can be determined directly from the relationship between the z-variable and the s-variable. Equation 8.117 can be rearranged to give

$$w = \frac{z-1}{z+1} \tag{8.119}$$

where

$$z = e^{sT}$$
$$= e^{(\sigma+j\omega_o)T}$$

It then follows that

$$w = \frac{e^{(\sigma+j\omega_o)T} - 1}{e^{(\sigma+j\omega_o)T} + 1}. \qquad (8.120)$$

Now, for frequency response analysis, consider only the complex (frequency axis) where $\sigma = 0$. This gives the expression for the frequency response in the ω-plane as

$$[w]_{w=jv_w} = \left[\frac{e^{j\omega_o T} - 1}{e^{j\omega_o T} + 1}\right]_{z=e^{j\omega_o T}}$$

$$= \left[\frac{\cos\omega_o T + j\sin\omega_o T - 1}{\cos\omega_o T + j\sin\omega_o T - 1}\right].$$

Simplification of this gives

$$v_w = \frac{j\sin\omega_o T}{1 + \cos\omega_o T}$$

or simply

$$v_w = \frac{\sin\omega_o T}{1 + \cos\omega_o T}.$$

By expressing

$$\sin\omega_o T = 2\sin\left(\tfrac{1}{2}\omega_o T\right)\cos\left(\tfrac{1}{2}\omega_o T\right)$$
$$\cos\omega_o T = \cos^2\left(\tfrac{1}{2}\omega_o T\right) - \sin^2\left(\tfrac{1}{2}\omega_o T\right),$$

the approximated frequency in the w-plane becomes

$$v_w = \tan\left(\tfrac{1}{2}\omega_o T\right).$$

This indicates that the frequency v_w in the w-plane is distorted. To counter this distortion, the result is normally multiplied by $\dfrac{2}{T}$ so that

$$v'_w = \tfrac{2}{T}\tan\left(\tfrac{1}{2}\omega_o T\right).$$

Notice that this corresponds to doing the analysis in another w'-plane, which results from Tustin's approximation

$$w' = \frac{2}{T}\left[\frac{z-1}{z+1}\right].$$

Under this approximation, for small T, the frequency in the w'-plane is almost equal to that in the z-plane, however, as T is made larger, several deviations are noticed. For most practical applications, the sampling frequency is taken to be in the range

$$0 \leq \omega_o \leq \frac{\omega_s}{4}$$

This approximation can further be improved by employing pre-warping techniques, which will not be discussed here.

After this approximation is done, the analysis can then be carried out using continuous-time frequency analysis methods including all parameters of interest such as the gain margin, phase margin, bandwidth, and sensitivity.

8.4 Design of Discrete Time Controllers

8.4.1 Controller Feasibility: the Concept of Causality

Some controller configurations might be theoretically attractive but practically infeasible. They become theoretically attractive if from a theoretical point of view they present very good system response as per specifications. On the other hand, they may be practically infeasible if their difference equation quantities, which are ahead of time, are included. Consider a controller of the form

$$D(z) = \frac{U(z)}{E(z)} = \frac{a_o + a_1 z + a_2 z^2 + \cdots + a_n z^n}{1 + b_1 z + b_2 z^2 + \cdots + b_m z^m} \qquad n > m.$$

This controller can be written in negative powers of z as

$$\frac{U(z)}{E(z)} = \frac{a_o z^{-m} + a_1 z^{-m+1} + a_2 z^{-m+2} + \cdots + a_n z^{-m+n}}{z^{-m} + b_1 z^{-m+1} + b_2 z^{-m+2} + \cdots + b_m},$$

which gives the following difference equation

$$u(k-m) + b_1 u(k-m+1) + b_2 u(k-m+2) + \cdots + b_m u(k)$$
$$= a_o e(k-m) + a_1 e(k-m+1) + \cdots + a_n e(k-m+n).$$

Thus, the control signal at instant k will be given by

$$u(k) = b_m^{-1} \{ a_o e(k-m) + a_1 e(k-m+1) +$$
$$\cdots + a_n e(k-m+n) - u(k-m) - \cdots - b_1 u(k-2) - b_2 u(k-1) \}.$$
$$(8.121)$$

Now, since $n > m$ the error term $e(k-m+n)$ will be ahead of time k by $(n-m)$, i.e., at the time k, the controller will require error signals that are at time $(k+n-m)$, which are practically not available. This renders the controller infeasible. Such a controller is known technically as a *non-causal* controller. In order to illustrate this concept further, let us assume that in the controller given above, $m = 1$ and $n = 2$ so that the control difference equation becomes

$$u(k) = b_m^{-1} \{ a_o e(k-1) + a_1 e(k) + a_n e(k+1) - u(k-1) \}, \qquad (8.122)$$

which needs the error value at time $k+1$ in order to general the control signal required at time k.

In control applications, non-causal controllers should be avoided, even if they offer very attractive analytical results.

8.4.2   Pole-Zero Matching Control Technique

The pole-zero matching design technique is just an application of the continuous-time design methods in s-domain to the discrete-time systems in the z-domain. This technique allows the controller to be designed in the s-plane and then transferred in the z-plane. In some applications, this technique is known as the emulation technique. Basically, when the controller is designed in the continuous-time domain, it is then digitized by using the relationship

$$z = e^{sT}.$$

The technique requires the continuous-time poles and zeros of the controller to be preserved by the discrete-time controller, and hence the name *"pole-zero matching."* The technique assumes that for every continuous-time controller there is an equal number of poles as the zeros. Therefore, if the degree of the numerator is less than that of the denominator, then some of the zeros are infinity. Thus, for the continuous-time controller

$$D_c(s) = K_c \frac{\prod_{i=1}^{n}(s - z_i)}{\prod_{j=1}^{m}(s - p_j)} \qquad n < m, \qquad (8.123)$$

where $\{z_i\}_{i=1}^{n}$ and $\{p_j\}_{j=1}^{m}$ are the zeros and poles respectively, there are $(m - n)$ zeros at infinity. Now, the discrete-time controller which emulates this continuous-time controller must map all poles according to

$$\mathrm{P}_j = e^{p_j T}$$

and zeros according to

$$\mathrm{Z}_i = e^{z_i T},$$

whereas those zeros and poles (if any) that are at infinity are mapped to digital zeros at $z = -1$ and the zeros at $s = 0$ are mapped into corresponding digital zeros at $z = 1$. Finally, the DC gains for both the continuous-time controller and the discrete-time controller must be matched according to

$$D_c(s)|_{s=0} = D_D(z)|_{z=1}$$

Thus, the digitized version of the continuous controller in Equation 8.123 is given by

$$D_D(z) = K_D \, (z+1)^{n-m} \left[\frac{\prod\limits_{i=1}^{n} (z - e^{z_i T})}{\prod\limits_{j=1}^{m} (z - e^{p_j T})} \right]. \qquad (8.124)$$

While this method is the fastest and simplest of all because of the fact that most systems found in real life are continuous-time in nature, so that analysis and design in the continuous-time domain seem more appealing. The presence of the ZOH element in the actual digital system, which is not taken into account when designing the continuous-time controller and hence the resulting digitized controller, causes functional problems with controllers designed this way. For most applications, this approach is only suitable if the sampling rate is very high to the level that the whole system can be considered as a continuous-time system. The use of this method is illustrated in the following example.

Example 8.16 *Employ the pole-zero matching technique to derive a discrete-time controller equivalent to the following continuous-time controller*

$$D_c(s) = \frac{4s + 3}{3s^2 + 5s + 2},$$

for a general sampling interval T and at a sampling rate of $0.005s$.

Solution 8.16 *Write the controller in the pole-zero form as*

$$D_c(s) = \frac{4(s + 0.75)}{3(s + 1)(s + \frac{2}{3})},$$

which can be seen to have zeros at $s = \infty$, and $s = -0.75$, while the poles are at $s = -1$, and $s = -\frac{2}{3}$. The continuous-time DC gain $K_c = \frac{2}{3}$.

Therefore, the equivalent discrete-time controller will have corresponding zeros at $z = -1$, and $z = e^{-0.75T}$ while the zeros will be at $z = e^{-T}$ and at $z = e^{-\frac{2}{3}T}$. Therefore, the resulting discrete-time controller will be

$$D_D(z) = \frac{K_D(z + 1)(z - e^{-0.75T})}{(z - e^{-T})\left(z - e^{-\frac{2}{3}T}\right)}.$$

To determine the steady state gain, set

$$D_c(s)|_{s=0} = D_D(z)|_{z=1}$$

which gives

$$\left.\frac{4(s+0.75)}{3(s+1)\left(s+\frac{2}{3}\right)}\right|_{s=0} = \left.\frac{K_D(z+1)(z-e^{-0.75T})}{(z-e^{-T})\left(z-e^{-\frac{2}{3}T}\right)}\right|_{z=1}$$

$$1.5 = 2K_D \frac{1-e^{-.75T}}{(1-e^{-T})\left(1-e^{-\frac{2}{3}T}\right)},$$

so that

$$K_D = 0.75\left(1-e^{-T}\right)\left(\frac{1-e^{-\frac{2}{3}T}}{1-e^{-.75T}}\right).$$

This gives the equivalent discrete-time controller for a general sampling interval T as

$$D_D(z) = 0.75\left(1-e^{-T}\right)\left(\frac{1-e^{-\frac{2}{3}T}}{1-e^{-.75T}}\right)\frac{(z+1)(z-e^{-0.75T})}{(z-e^{-T})\left(z-e^{-\frac{2}{3}T}\right)}$$

and at a sampling rate of 0.005*s it becomes*

$$D_D(z) = 3.3257 \times 10^{-3}\frac{(z+1)(z-0.99626)}{(z-0.99501)(z-0.99667)}.$$

8.4.3   The Pole Placement Methods in the z-Domain

Pole placement methods are useful when the required positions of poles in the controlled system are known. For a given pole location, the controller is designed and included in the system so that the poles of the controlled system coincide with those of the required system dynamics. To achieve proper pole placement, three methods are discussed as follows:

8.4.3.1   The Root-Locus Method

Just like the root locus method for continuous-time systems, this method gives the values of the same control parameters that place the poles of the transfer function within a unit circle with damping and natural frequency, as required in the design specification. In general, this technique is not much different from the one that has been seen for continuous tine systems in that it looks for the values of the controller parameters, which together brings about the required system dynamics from the root locus of the system.

8.4.3.2 Polynomial Method

The polynomial method compares the characteristic polynomial of the required system dynamics to that for the controlled system by equating the coefficients of the two polynomials. The controller parameters are selected so that the coefficients of the two polynomials will be equal.

8.4.3.3 State-Space Method

From the discrete time state-space presentation of a system pole placement is then carried out using the methods similar to those discussed in Chapter 7 with very minor differences, depending on the difference in the domain of analysis.

8.4.4 The Frequency-Response Design Methods

8.4.4.1 Bilinear transformation methods

This is a discrete-time transform that is introduced in the frequency methods so that many of the design features that are found in continuous-time systems are preserved. The basic idea is to introduce a new variable w with a defined bilinear mapping that allows the compensator to be designed in the w-plane and the final result converted back to the z-plane. This transformation technique is discussed in detail, pointing out its advantages in the frequency-response design.

8.4.4.2 Compensator design

When the frequency-response design specifications such as the Gain margin and Phase margin are known, one can design the phase lead, phase lag, and the lead lag compensators just as is done in the continuous time systems. A discussion on the discrete-time compensator design is thus given with examples.

8.4.5 The Discrete Time PID Controller

The general performance of the discrete-time PID controller is the same as the continuous-time PID controller. All characteristics of the PID controller discussed on the continuous-time systems apply well to discrete-time systems subject to some adjustments of the discrete-time version. Ways of implementing the discrete-time PID controller are discussed.

It was shown that the continuous-time three term PID control has the general form

$$D(s) = K_P + \frac{K_I}{s} + K_D s, \tag{8.125}$$

where the three parameters or gains K_P, K_I, and K_D are chosen to give the desired system dynamics. These gains are normally known respectively as the proportional, integral, and derivative gain. As a matter of convenience, this controller is normally presented as

$$D(s) = K_P \left(1 + \frac{1}{T_I s} + T_D s\right), \tag{8.126}$$

where the three parameters K_P, T_I, and T_D give a complete description of the controller. While K_P is the *proportional gain*, T_I and T_D are known as the *integral* (or *reset*) *time* and *derivative time* respectively.

The function of each of these terms in controlling the system has been well explained in Chapter 4, where it was shown that the proportional gain improves the system responsiveness, but at a cost of introducing steady state error and some degree of instability (oscillations). The derivative term introduces some damping in the system dynamics to curb the oscillations created by the proportional gain while the integral term reduces the steady state error produced by the proportional gain. While it is possible to employ only the proportional controller, the derivative and the integral controllers cannot stand alone in controlling the system. The methods of tuning these gains have also been discussed, in particular the Ziegler-Nichols method.

In this section, ideas presented for the continuous-time system will be extended to cover the discrete-time systems. Recall that the transfer function for the continuous-time system is defined as

$$D(s) = \frac{U(s)}{E(s)}, \tag{8.127}$$

where the $U(s)$ and $E(s)$ are the Laplace transforms of the control signal and the error signal respectively.

8.4.5.1 Proportional Controller (P)

The continuous-time proportional controller is just the amplification of the control error signal. Proportional control implementation involves multiplication of the error signal with appropriate amplification factor as the design may require. In general, it has been shown that the transfer function for the continuous-time proportional controller is

$$D(s) = K_P,$$

which gives the control signal as

$$U(s) = D(s)E(s)$$
$$= K_P E(s).$$

The corresponding time domain representation is simply

$$u(t) = K_P e(t).$$

Therefore, at any sampling instant k, the control signal in the discrete-time environment becomes

$$u(k) = K_P e(k). \tag{8.128}$$

In order to derive the discrete-time equivalent of the proportional controller $D(z)$, first take the $\mathcal{Z}$-transform of Equation 8.128, which gives

$$U(z) = K_P E(z),$$

from which the proportional controller is derived as

$$D_P$$
$$D(z) = \frac{U(z)}{E(z)}$$
$$= K_P.$$

This shows that the discrete-time proportional controller has the same gain as that of the continuous-time system.

8.4.5.2 Integral Controller (I)

Implementation of the digital integral controller involves numerical integration of the error signal or numerical differentiation of the control signal. The common numerical differentiation and integration methods, which include the trapezoidal rule (also known as the Tustin's rule or the bilinear transformation) and Euler's forward as well as the backward rectangular rules, have been described.

If the continuous time derivative controller were to stand alone, its transfer function would have been

$$D(s) = \frac{K_P}{T_I s},$$

so that

$$U(s) = \frac{K_P}{T_I s} E(s). \tag{8.129}$$

This can also be written as

$$U(s)s = \frac{K_P}{T_I}E(s). \qquad (8.130)$$

In time domain, Equations 8.129 and 8.130 can be written respectively as

$$u(t) = \frac{K_P}{T_I}\int_0^t e(t)dt$$

and

$$\frac{du(t)}{dt} = \frac{K_P}{T_I}e(t). \qquad (8.131)$$

Employment of different numerical methods on these two equations leads to different versions of the discrete-time integral controller, however, the most popular method employed is the one that requires that the control signal be a function of all past values of the control and error signal as well as the current error signal. Derivation of such a controller stems from the application of Euler's backward approximation of the differential equation, which gives

$$\frac{u(k) - u(k-1)}{T} = \frac{K_P}{T_I}e(k),$$

such that

$$u(k) = u(k-1) + \frac{K_P T}{T_I}e(k), \qquad (8.132)$$

where T is the sampling interval. Notice that, since all past values of $e(.)$ up to $(k-1)$, as well as the past values of $u(.)$ up to $(k-2)$, are embedded in $u(k-1)$, then this controller will be employing information from all past values of $u(.)$ up to $(k-1)$ as well as $e(.)$ up to k. In implementation, this poses a problem known as the integral wind-up effect, or simply the reset wind-up in which the total error at time k grows beyond some allowable limits.

The $\mathcal{Z}$-transform then gives

$$U(z) = z^{-1}U(z) + \frac{K_P}{T_I}E(z)$$

whose rearrangement gives

$$U(z)\left[1 - z^{-1}\right] = \frac{K_P}{T_I}E(z)$$

so that

$$D(z) = \frac{U(z)}{E(z)}$$

$$= \frac{K_P}{T_I \left[1 - z^{-1}\right]}$$

$$= \frac{K_P z}{T_I (z - 1)}. \tag{8.133}$$

8.4.5.3   Derivative Controller (D)

If the continuous-time derivative controller were to stand alone, its transfer function would have been

$$D(s) = \frac{U(s)}{E(s)}$$

$$= K_P T_D s.$$

The inverse Laplace transfer gives this control signal in the time domain as

$$u(t) = K_P T_D \frac{de(t)}{dt}. \tag{8.134}$$

Again, the intention here is to derive the control signal as a function of all past error signal up to time k. However, in this case, it doesn't need all past values, only the current and the previous error signals are sufficient as shown below. By employing Euler's backward rule of numerical differentiation, the error signal at time k gives

$$\left. \frac{de(t)}{dt} \right|_{t=k} = \frac{e(k) - e(k - 1)}{T},$$

where T is the sampling interval. Therefore, by using this approximation at any sampling instant k, it follows that

$$u(k) = K_P T_D \frac{e(k) - e(k - 1)}{T}$$

$$= \frac{K_P T_D}{T} \left[e(k) - e(k - 1)\right].$$

Now the $\mathcal{Z}$-transform of this expression gives

$$U(z) = \frac{K_P T_D}{T} \left[E(z) - z^{-1} E(z)\right]$$

$$= \frac{K_P T_D}{T} E(z) \left[1 - z^{-1}\right]$$

so that

$$D(z) = \frac{U(z)}{E(z)}$$

$$= \frac{K_P T_D}{T} \left[1 - z^{-1}\right]$$

$$= \frac{K_P T_D (z - 1)}{T z}.$$

Notice that, although by using Euler's forward rule, one may be tempted to do the same derivation, such a controller will have practical limits, as it will require error signals that are ahead of time.

8.4.5.4 The Complete PID Controller

Using the results in the previous sections, a complete PID controller can now be constructed as follows:

$$D(z) = K_P \left(1 + \frac{Tz}{T_I(z-1)} + \frac{T_D(z-1)}{Tz}\right)$$

$$= K_P \frac{T_I(z-1)Tz + T^2 z^2 + T_D T_I (z-1)^2}{T_I(z-1)Tz}$$

$$= K_P \frac{(T_I T + T^2 + T_D T_I) z^2 + (-T_I T - 2T_D T_I) z + T_D T_I}{T_I T z^2 - T_I T z}$$

or

$$D(z) = K_P \frac{(T_I T + T^2 + T_D T_I) + (-T_I T - 2T_D T_I) z^{-1} + T_D T_I z^{-2}}{T_I T (1 - z^{-1})}$$

$$(8.135)$$

The control difference equation can then be evaluated from

$$D(z) = \frac{U(z)}{E(z)},$$

which gives

$$\left(1 - z^{-1}\right) U(z)$$

$$= \frac{K_P}{T_I T} \left[(T_I T + T^2 + T_D T_I) + (-T_I T - 2T_D T_I) z^{-1} + T_D T_I z^{-2}\right] E(z).$$

Hence, the inverse $\mathcal{Z}$-transform gives

$$u(k) - u(k-1)$$

$$= \frac{K_P}{T_I T} \left[(T_I T + T^2 + T_D T_I) e(k) + (-T_I T - 2T_D T_I) e(k-1) + T_D T_I e(k-2)\right].$$

Hence, the control signal becomes

$$u(k) = u(k-1) +$$

$$\frac{K_P}{T_I T} \left[(T_I T + T^2 + T_D T_I) \, e(k) + (-T_I T - 2T_D T_I) \, e(k-1) + T_D T_I e(k-2) \right]$$

$$= u(k-1) + Ae(k) + Be(k-1) + Ce(k-2), \tag{8.136}$$

where

$$A = \frac{K_P}{T_I T} \left(T_I T + T^2 + T_D T_I \right)$$

$$B = \frac{K_P}{T_I T} \left(-T_I T - 2T_D T_I \right)$$

$$C = \frac{K_P T_D}{T}.$$

8.4.6 Implementation of Digital Control Systems

In the implementation of digital control systems two distinct issues have to be addressed, hardware issues and software issues. This section gives a brief overview of such issues in real application.

The hardware to be used in the construction of digital control systems needs to be fast enough to go with the real system dynamics. The qualities of the hardware for control are discussed under this section. There must be a cost balance between the hardware and the controlled system in general. Memory requirements and the single board computers (microcontrollers) are discussed.

Transformation of the controller difference equations into software is explained. Software for control must be fast to match with system dynamics, hence, items of interest in implementing software for control are discussed. The real time computing techniques should be addressed and particular emphasis placed on multitasking policies. Techniques for handling integral and reset wind-up effects are essential.

Normally after designing the software and the hardware, the two will have to be integrated. Before the system is put into operation, it must be tested and debugged where necessary. In any software and the hardware must be compatible. All such issues should be discussed.

8.5   Problems

Problem 8.1 *Assess the controllability, observability, and stability of the following digital system.*

$$\mathbf{x}(k+1) = \begin{bmatrix} -1 & 0.8 \\ 0.5 & -1.6 \end{bmatrix} \mathbf{x}(k) + \begin{bmatrix} 1 \\ 2 \end{bmatrix} \mathbf{u}(k)$$

$$\mathbf{y}(k) = \begin{bmatrix} 1 & 0 \end{bmatrix} \mathbf{x}(k) + [0]\, \mathbf{u}(k).$$

Problem 8.2 *Represent the SISO system with the following transfer function using state-space*

$$\frac{Y(z)}{U(z)} = \frac{z + 0.3}{z^2 - 0.6z - 0.16}.$$

Problem 8.3 *The following open-loop continuous-time transfer function represents a second-order system that is to be controlled using a digital computer with ZOH.*

$$G(s) = \frac{1}{(s+1)(s+10)}.$$

Choose a suitable sampling interval T and design a digital PID controller that ensures that the settling time $t_s \leq 1sec$, damping ratio $\xi \geq 0.5$ and there is zero steady state error to a step input.

Problem 8.4 *A unit feedback digital control system at a sampling interval of $0.05sec$. If the plant transfer function is given by*

$$G(z) = \frac{z + 0.8}{(z-1)(z-0.8)}.$$

Design the controller $D(s)$ using root locus methods so that the closed-loop system satisfies the following time domain specification:

$$settling\ time \ \leq \ 0.4sec$$
$$damping\ ratio \ \geq \ 0.7$$
$$steady\ state\ error = zero\ (step\ input).$$

Problem 8.5 *Find the time function $y(kT)$ corresponding to the following* *closed form z-transforms Y(z):*

$$\text{(a)} \qquad Y(z) = \frac{4z}{z^2 - 1}$$

$$\text{(b)} \qquad Y(z) = \frac{2z}{z^2 - 0.5z - 0.5}$$

$$\text{(c)} \qquad Y(z) = \frac{0.522z^2 + 0.361z - 0.203}{z^3 - 2.347z^2 + 1.797z - 0.449}$$

For each of these, use the following methods:
(i) Inverse transform method
(ii) Partial fraction expansion and use of z-transform table
(iii) Power series expansion

Problem 8.6 *Suppose that the following continuous-time lead compensator*

$$D(s) = \frac{10(s + 1)}{s + 10}$$

is to be used in controlling the plant

$$G(s) = \frac{20}{s(s + 1)(s + 10)}.$$

If this system is to be implemented using a digital computer at a sampling interval $0.1sec$, *determine the transfer function* $D(z)$ *of the controller using:*
(a) Pole-zero matching method
(b) Tustin's conversion method
(c) Bilinear transformation method

Problem 8.7 *(a) Compute the poles and zeros of the discrete-time systems that result from discretizing the continuous-time system with transfer function*

$$G(s) = \frac{10(s^2 + 0.2s + 2)}{(s^2 + 0.5s + 1)(s + 10)},$$

for sampling intervals ranging from $T_s = 1sec$ *to* $T_s = 0.01sec$.
(b) Create a root locus plot showing the evolution of the poles and zeros computed in part (a) as a function of T_s.
(c) Repeat parts (a) and (b) for

$$G(s) = \frac{(s + 0.1 + 2i)(s + 0.1 - 2i)}{(s + 0.1 + i)(s + 0.1 - i)(s + 0.1 + 0.5i)(s + 0.1 - 0.5i)}.$$

(d) Repeat parts (a), (b), and (c) for different discretization methods.

Problem 8.8 *(a) Transform*

$$G(z) = \frac{(z+0.3)(z-0.3)}{(z-0.1)(z-0.5+0.5i)(z-0.5-0.5i)}$$

into a continuous-time equivalent system by means of ZOH method of sampling intervals ranging from $T_s = 0.01sec$ to $T_s = 1sec$.

(b) Determine the poles and zeros of the resulting continuous-time systems.

(c) Plot the root locus for each of the continuous-time systems.

Problem 8.9 *The z-transform of a discrete-time filter $h(k)$ at a $1Hz$ sample rate is*

$$H(z) = \frac{1 + (1/2)z^{-1}}{[1 - (1/2)z^{-1}][1 + (1/2)z^{-1}]}.$$

(a) Let $u(k)$ and $y(k)$ be the discrete input and output of this filter. Find a difference equation relating $u(k)$ and $y(k)$.

(b) Find the natural frequency and the damping coefficient of the filter's poles.

(c) Is the filter stable?

Problem 8.10 *Use the $\mathcal{Z}$-transform to solve the difference equation*

$$y(k) - 3y(k-1) + 2y(k-2) = 2u(k-1) - 2u(k-2)$$

where

$$u(k) = k, \quad k \geq 0$$
$$= 0, \quad k < 0$$
$$y(k) = 0, \quad k > 0.$$

Problem 8.11 *A unity feedback system has an open-loop transfer function given by*

$$G(s) = \frac{250}{s[(s/10) + 1]}.$$

The following lag compensator added in series with the plant yields a phase margin of $50°$,

$$D_c(s) = \frac{(s/1.25) + 1}{50s + 1}.$$

(a) Using the matched pole-zero approximation, determine an equivalent digital realization of this compensator.

(b) *The following transfer function is a lead network designed to add about $60°$ of phase at $\omega_1 = 3\ rad/sec$,*

$$H(s) = \frac{s+1}{0.1s+1}.$$

Assume a sampling period of $T = 0.25$ sec, and compute and plot in the z-plane the pole and zero locations of the digital implementations of $H(s)$ obtained using (1) Tustin's method and (2) pole-zero mapping. For each case, compute the amount of phase lead provided by the network at $z_1 = e^{j\omega_1 T}$.

(c) *Using log-scale for the frequency range $\omega = 0.1$ to $\omega = 100\ rad/sec$, plot the magnitude Bode plots for each of the equivalent digital systems found in part (a), and compare with $H(s)$. (hint: Magnitude Bode plots are given by $|H(z)| = |H(e^{j\omega T})|$.*

Problem 8.12 (a) *The following transfer function is a lag network designed to introduce a gain attenuation of 10 ($-20dB$) at $\omega = 3\ rad/sec$,*

$$H(s) = \frac{10s+1}{100s+1}.$$

Assume a sampling period of $T = 0.25$ sec, and compute and plot in the z-plane the pole and zero locations of the digital implementations of $H(s)$ obtained using (1) Tustin's method and (2) pole-zero mapping. For each case, compute the amount of gain attenuation provided by the network at $z_1 = e^{j\omega T}$.

(b) *For each of the equivalent digital systems in part (a), plot the Bode magnitude curves over the frequency range $\omega = 0.01$ to $10 rad/sec$.*

Problem 8.13 *Write a computer program to compute Φ and Γ from A, B, and the sample period T. Use the program to compute Φ and Γ when*
(a)

$$\mathbf{A} = \begin{bmatrix} -1 & 0 \\ 0 & -2 \end{bmatrix}, \quad \mathbf{B} = \begin{bmatrix} 1 \\ 1 \end{bmatrix}, \quad T = 0.2 sec.$$

(b)

$$\mathbf{A} = \begin{bmatrix} -3 & -2 \\ 1 & 0 \end{bmatrix}, \quad \mathbf{B} = \begin{bmatrix} 1 \\ 0 \end{bmatrix}, \quad T = 0.2 sec.$$

Problem 8.14 *Consider the following discrete-time system in state-space form:*

$$\begin{bmatrix} x_1(k+1) \\ x_2(k+1) \end{bmatrix} = \begin{bmatrix} 0 & -1 \\ 0 & -1 \end{bmatrix} \begin{bmatrix} x_1(k) \\ x_2(k) \end{bmatrix} + \begin{bmatrix} 0 \\ 10 \end{bmatrix} u(k).$$

Use state feedback to relocate all of the system's poles to 0.5.

Problem 8.15 *The characteristic equation of a sampled system is*

$$z^2 + (K - 1.5)z + 0.5 = 0.$$

Find the range of K so that the system is stable. (Answer: $0 < K < 3$)

Problem 8.16 *A unit ramp $r(t) = t$, $t > 0$, is used as an input to a process where*

$$G(s) = \frac{1}{(s + 1)},$$

as shown in the diagram below.

Sampling System

Determine the output $y(kT)$ for the first four sampling instants.

Problem 8.17 *A closed-loop system has a hold circuit, as shown in Problem 8.16. Determine $G(z)$ when $T = 1\,\text{sec}$ and*

$$G_p(s) = \frac{2}{s + 2}.$$

Problem 8.18 *Determine which of the following digital transfer functions are physically realizable.*

$$(a) \quad G_c(z) = \frac{10(1 + 0.2z^{-1} + 0.5z^{-2})}{z^{-1} + z^{-2} + 1.5z^{-3}}$$

$$(b) \quad G_c(z) = \frac{1.5z^{-1} - z^{-2}}{1 + z^{-1} + 2z^{-2}}$$

$$(c) \quad G_c(z) = \frac{z + 1.5}{z^3 + z^2 + z + 1}$$

$$(d) \quad G_c(z) = \frac{z^{-1} + 2z^{-2} + 0.5z^{-3}}{z^{-1} + z^{-2}}$$

$$(e) \quad G_c(z) = 0.1z + 1 + z^{-1}$$

$$(f) \quad G_c(z) = z^{-1} + z^{-2}$$

Problem 8.19 *Consider the digital control system*

$$\mathbf{x}[(k + 1)T] = \mathbf{A}\mathbf{x}(kT) + \mathbf{B}u(kT)$$

where

$$\mathbf{A} = \begin{bmatrix} 0 & -1 \\ -1 & -1 \end{bmatrix}, \qquad \mathbf{B} = \begin{bmatrix} 0 \\ 1 \end{bmatrix}.$$

The state feedback control is described by $u(kT) = -\mathbf{K}\mathbf{x}(kT)$, *where*

$$\mathbf{K} = \begin{bmatrix} k_1 & k_2 \end{bmatrix}.$$

Find the values of k_1 and k_2 so that the roots of the characteristic equation of the closed-loop system are at 0.5 and 0.7.

Chapter 9

Advanced Control Systems

9.1 Introduction

In this chapter advanced topics and issues involved in the design and analysis of control systems are addressed. In particular, the subjects of discrete-time estimation (both state-space and information space), optimal stochastic control, and nonlinear control systems are presented. Adaptive control systems and robust control are briefly introduced.

A multisensor system may employ a range of different sensors, with different characteristics, to obtain information about an environment. The diverse and sometimes conflicting information obtained from multiple sensors gives rise to the problem of how the information can be combined in a consistent and coherent manner. This is the *data fusion* problem. Multisensor fusion is the process by which information from a multitude of sensors is combined to yield a coherent description of the system under observation. All data fusion problems involve an estimation process. An estimator is a decision rule that takes as an argument a sequence of observations and computes a value for the parameter or state of interest. General recursive estimation is presented and, in particular, the *Kalman filter* is discussed. A Bayesian approach to probabilistic information fusion is outlined and the notion and measures of information are defined. This leads to the derivation of the algebraic equivalent of the Kalman filter, the (linear) *Information filter*. State estimation for systems with nonlinearities is considered and the *extended* Kalman filter treated. Linear information space is then extended to *nonlinear* information space by deriving the *extended* Information filter. This filter forms the basis of decentralized estimation and control methods for nonlinear systems. The estimation techniques are then extended to LQG stochastic control problems including systems involving nonlinearities, that is, the nonlinear stochastic control systems.

In most of the work in the previous eight chapters it has been assumed that the dynamics of systems to be controlled can be described completely by a set of linear differential equations and that the principle of superposition holds. Such systems are known as linear dynamic systems. However, in most applications, these assumptions are not valid, and the systems are termed nonlinear dynamic systems. The nonlinearity of dynamic systems can be inherent or deliberately added to improve the control action. This chapter addresses the whole concept of nonlinear systems, their analysis and control design.

9.2   State-Space Estimation

In this section, the principles and concepts of estimation are introduced. An estimator is a decision rule that takes as an argument a sequence of observations and computes a value for the parameter or state of interest. The Kalman filter is a recursive linear estimator that successively calculates a minimum variance estimate for a state that evolves over time, on the basis of periodic observations that are linearly related to this state. The Kalman filter estimator minimizes the mean squared estimation error and is optimal with respect to a variety of important criteria under specific assumptions about process and observation noise. The development of linear estimators can be extended to the problem of estimation for nonlinear systems. The Kalman filter has found extensive applications in such fields as aerospace navigation, robotics and process control.

9.2.1   System Description

A very specific notation is adopted to describe systems throughout this chapter [3]. The state of nature is described by an n-dimensional vector $\mathbf{x}=[x_1, x_2, ..., x_n]^T$. Measurements or observations are made of the state of $\mathbf{x}$. These are described by an m-dimensional observation vector $\mathbf{z}$.

A linear discrete-time system is described as follows:

$$\mathbf{x}(k) = \mathbf{F}(k)\mathbf{x}(k-1) + \mathbf{B}(k)\mathbf{u}(k-1) + \mathbf{w}(k-1), \qquad (9.1)$$

where $\mathbf{x}(k)$ is the state of interest at time k, $\mathbf{F}(k)$ is the state-transition matrix from time $(k-1)$ to k, while $\mathbf{u}(k)$ and $\mathbf{B}(k)$ are the input control vector and matrix, respectively. The vector, $\mathbf{w}(k) \sim N(\mathbf{0}, \mathbf{Q}(k))$, is the associated process noise modeled as an uncorrelated, zero mean, white sequence with process noise covariance,

$$\mathrm{E}[\mathbf{w}(i)\mathbf{w}^T(j)] = \delta_{ij}\mathbf{Q}(i).$$

The system is observed according to the linear discrete equation

$$\mathbf{z}(k) = \mathbf{H}(k)\mathbf{x}(k) + \mathbf{v}(k), \tag{9.2}$$

where $\mathbf{z}(k)$ is the vector of observations made at time k. $\mathbf{H}(k)$ is the observation matrix or model and $\mathbf{v}(k) \sim N(\mathbf{0}, \mathbf{R}(k))$ is the associated observation noise modeled as an uncorrelated white sequence with measurement noise covariance,

$$E[\mathbf{v}(i)\mathbf{v}^T(j)] = \delta_{ij}\mathbf{R}(i).$$

It is assumed that the process and observation noises are uncorrelated, i.e.,

$$E[\mathbf{v}(i)\mathbf{w}^T(j)] = \mathbf{0}.$$

The notation due to Barshalom [3] is used to denote the vector of estimates of the states $\mathbf{x}(j)$ at time i given information up to and including time j by

$$\hat{\mathbf{x}}(i \mid j) = E\left[\mathbf{x}(i) \mid \mathbf{z}(1), \cdots \mathbf{z}(j)\right].$$

This is the conditional mean, the minimum mean square error estimate. This estimate has a corresponding variance given by

$$\mathbf{P}(i \mid j) = E\left[(\mathbf{x}(i) - \hat{\mathbf{x}}(i \mid j))(\mathbf{x}(i) - \hat{\mathbf{x}}(i \mid j))^T \mid \mathbf{z}(1), \cdots \mathbf{z}(j)\right].$$

9.2.2 Kalman Filter Algorithm

A great deal has been written about the Kalman filter and estimation theory in general [3], [4], [13]. An outline of the Kalman filter algorithm is presented here without derivation. Figure 9.1 summarizes its main functional stages. For a system described by Equation 9.1 and being observed according to Equation 9.2, the Kalman filter provides a recursive estimate $\hat{\mathbf{x}}(k \mid k)$ for the state $\mathbf{x}(k)$ at time k, given all information up to time k in terms of the predicted state $\hat{\mathbf{x}}(k \mid k-1)$ and the new observation $\mathbf{z}(k)$. The *one-step-ahead* prediction, $\hat{\mathbf{x}}(k \mid k-1)$, is the estimate of the state at a time k given only information up to time $(k-1)$. The Kalman filter algorithm can be summarized in two stages:

Prediction

$$\hat{\mathbf{x}}(k \mid k-1) = \mathbf{F}(k)\hat{\mathbf{x}}(k-1 \mid k-1) + \mathbf{B}(k)\mathbf{u}(k) \tag{9.3}$$

$$\mathbf{P}(k \mid k-1) = \mathbf{F}(k)\mathbf{P}(k-1 \mid k-1)\mathbf{F}^T(k) + \mathbf{Q}(k). \tag{9.4}$$

Estimation

$$\hat{\mathbf{x}}(k \mid k) = [\mathbf{1} - \mathbf{W}(k)\mathbf{H}(k)]\,\hat{\mathbf{x}}(k \mid k-1) + \mathbf{W}(k)\mathbf{z}(k) \tag{9.5}$$

$$\mathbf{P}(k \mid k) = \mathbf{P}(k \mid k-1) - \mathbf{W}(k)\mathbf{S}(k)\mathbf{W}^T(k), \tag{9.6}$$

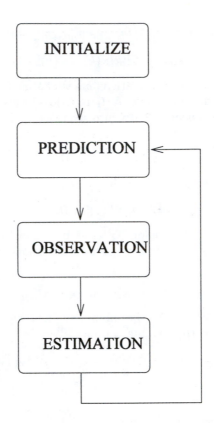

FIGURE 9.1
The Kalman Filter Algorithm

where $\mathbf{W}(k)$ and $\mathbf{S}(k)$ known as the gain and innovation covariance matrices, respectively, are given by

$$\mathbf{W}(k) = \mathbf{P}(k \mid k-1)\mathbf{H}^T(k)\mathbf{S}^{-1}(k), \qquad (9.7)$$

$$\mathbf{S}(k) = \mathbf{H}(k)\mathbf{P}(k \mid k-1)\mathbf{H}^T(k) + \mathbf{R}(k). \qquad (9.8)$$

The matrix $\mathbf{1}$ represents the identity matrix. From Equation 9.5, the Kalman filter state estimate can be interpreted as a linear weighted sum of the state prediction and observation. The weights in this averaging process are $\{\mathbf{1} - \mathbf{W}(k)\mathbf{H}(k)\}$ associated with the prediction and $\mathbf{W}(k)$ associated with the observation. The values of the weights depend on the balance of confidence in prediction and observation as specified by the process and observation noise covariances.

9.3   The Information Filter

The *Information filter* is essentially a Kalman filter expressed in terms of measures of *information* about the parameters (*states*) of interest rather than direct state estimates and their associated covariances [15]. This filter has also been called the *inverse covariance* form of the Kalman filter [4], [13]. In this section, the contextual meaning of information is explained and the Information filter is derived.

9.3.1   Information Space

Bayesian Theory

The probabilistic information contained in $\mathbf{z}$ about $\mathbf{x}$ is described by the probability distribution function, $p(\mathbf{z}|\mathbf{x})$, known as the *likelihood function*. Such information is considered *objective* because it is based on observations. The likelihood function contains all the relevant information from the observation $\mathbf{z}$ required in order to make inferences about the true state $\mathbf{x}$. This leads to the formulation of the *likelihood principle,* which states that all that is known about the unknown state is what is obtained through experimentation. Thus, the likelihood function contains all the information needed to construct an estimate for $\mathbf{x}$. However, the likelihood function does not give the complete picture, if, before measurement, information about the state $\mathbf{x}$ is made available exogenously. Such *a priori* information about the state is encapsulated in the prior distribution function $p(\mathbf{x})$ and is regarded as *subjective* because it is not based on any observed data. How such prior information and the likelihood information interact to provide a *posteriori* (combined prior and observed) information, is solved by *Bayes theorem,* which gives the posterior conditional distribution of $\mathbf{x}$ given $\mathbf{z}$,

$$p(\mathbf{x}, \mathbf{z}) = p(\mathbf{x}|\mathbf{z})p(\mathbf{z})$$
$$= p(\mathbf{z}|\mathbf{x})p(\mathbf{x})$$
$$\Leftrightarrow p(\mathbf{x}|\mathbf{z}) = \frac{p(\mathbf{z}|\mathbf{x})p(\mathbf{x})}{p(\mathbf{z})}. \tag{9.9}$$

where $p(\mathbf{z})$ is the marginal distribution.

To reduce uncertainty several measurements can be taken over time before constructing the posterior. The set of all observations up to time k is defined as

$$\mathbf{Z}^k \triangleq \{\mathbf{z}(1), \mathbf{z}(2), ..., \mathbf{z}(k)\}. \tag{9.10}$$

The corresponding likelihood function is given by

$$\Lambda_k(\mathbf{x}) \triangleq p(\mathbf{Z}^k|\mathbf{x}). \tag{9.11}$$

This is a measure of how *"likely"* a parameter value $\mathbf{x}$ is, given that all the observations in $\mathbf{Z}^k$ are made. Thus, the likelihood function serves as a measure of *evidence from data*. The posterior distribution of $\mathbf{x}$, given the set of observations $\mathbf{Z}^k$, is now computed as

$$p(\mathbf{x}|\mathbf{Z}^k) = \frac{p(\mathbf{Z}^k|\mathbf{x})p(\mathbf{x})}{p(\mathbf{Z}^k)}. \qquad (9.12)$$

It can also be computed recursively after each observation $\mathbf{z}(k)$ as follows:

$$p(\mathbf{x}|\mathbf{Z}^k) = \frac{p(\mathbf{z}(k)|\mathbf{x})p(\mathbf{x}|\mathbf{Z}^{k-1})}{p(\mathbf{z}(k)|\mathbf{Z}^{k-1})}. \qquad (9.13)$$

In this recursive form there is no need to store all the observations. Only the current observation $\mathbf{z}(k)$ at step k is considered. This recursive definition has reduced memory requirements and hence it is the most commonly implemented form of Bayes theorem.

Measures of Information

The term information·is employed in the *Fisher* sense, that is, a measure of the amount of information about a *random* state $\mathbf{x}$ present in the set of observations $\mathbf{Z}^k$, up to time k. The *score function*, $\mathbf{s}_k(\mathbf{x})$, is defined as the gradient of the log-likelihood function,

$$\mathbf{s}_k(\mathbf{x}) \triangleq \boldsymbol{\nabla}_x ln\, p(\mathbf{Z}^k, \mathbf{x}) = \frac{\boldsymbol{\nabla}_x p(\mathbf{Z}^k, \mathbf{x})}{p(\mathbf{Z}^k, \mathbf{x})}. \qquad (9.14)$$

By considering $\mathbf{s}_k(\mathbf{x})$ as a random variable, its mean is obtained from

$$\mathrm{E}\left[\mathbf{s}_k(\mathbf{x})\right] = \int \frac{\boldsymbol{\nabla}_x p(\mathbf{Z}^k, \mathbf{x})}{p(\mathbf{Z}^k, \mathbf{x})} p(\mathbf{Z}^k, \mathbf{x}) dz$$

$$= \boldsymbol{\nabla}_x \int p(\mathbf{Z}^k, \mathbf{x}) d\mathbf{z} = 0.$$

The *Fisher information matrix* $\mathcal{J}(k)$ is then defined as the covariance of the score function,

$$\mathcal{J}(k) \triangleq \mathrm{E}\left[\{\boldsymbol{\nabla}_x ln\, p(\mathbf{Z}^k, \mathbf{x})\}\{\boldsymbol{\nabla}_x ln\, p(\mathbf{Z}^k, \mathbf{x})\}^T\right]. \qquad (9.15)$$

Expressing this result as the negative expectation of the Hessian of the log-likelihood gives

$$\mathcal{J}(k) = -\mathrm{E}\left[\boldsymbol{\nabla}_x \boldsymbol{\nabla}_x^T ln\, p(\mathbf{Z}^k, \mathbf{x})\right]. \qquad (9.16)$$

For a *non-random* state $\mathbf{x}$ the expression of the Fisher information matrix becomes

$$\mathcal{J}(k) = -\mathrm{E}\left[\boldsymbol{\nabla}_x \boldsymbol{\nabla}_x^T ln\, p(\mathbf{Z}^k|\mathbf{x})\right]. \qquad (9.17)$$

The notion of Fisher information is useful in estimation and control. It is consistent with information in the sense of the *Cramer-Rao lower bound* (**CRLB**) [4]. According to the CRLB, the mean squared error corresponding to the estimator of a parameter cannot be smaller than a certain quantity related to the likelihood function. Thus, the CRLB bounds the mean squared error vector of any unbiased estimator $\hat{\mathbf{x}}(k \mid k)$ for a state vector $\mathbf{x}(k)$ modeled as random.

$$\mathrm{E}[\{\mathbf{x}(k) - \hat{\mathbf{x}}(k \mid k)\}\{\mathbf{x}(k) - \hat{\mathbf{x}}(k \mid k)\}^T \mid \mathbf{Z}^k] \geq \mathcal{J}^{-1}(k). \qquad (9.18)$$

In this way, the covariance matrix of an unbiased estimator is bounded from below. It follows from Equation 9.18 that the CRLB is the inverse of the Fisher information matrix, $\mathcal{J}(k)$. This is a very important relationship. A necessary condition for an estimator to be *consistent* in the mean square sense is that there must be an increasing amount of information (in the sense of Fisher) about the parameter in the measurements, i.e., the Fisher information has to tend to infinity as $k \to \infty$. The CRLB then converges to zero as $k \to \infty$ and thus the variance can also converge to zero. Furthermore, if an estimator's variance is equal to the CRLB, then such an estimator is called *efficient*.

Consider the expression for the Fisher information matrix in Equations 9.15 or 9.16. In the particular case where the likelihood function, $\Lambda_k(\mathbf{x})$, is *Gaussian*, it can be shown that the Fisher information matrix, $\mathcal{J}(k)$, is equal to the inverse of the covariance matrix $\mathbf{P}(k \mid k)$, that is, the CRLB is the covariance matrix. This is done by considering the probability distribution function of a *Gaussian* random vector $\mathbf{x}(k)$ whose mean and associated covariance matrix are $\hat{\mathbf{x}}(k \mid k)$ and $\mathbf{P}(k \mid k)$, respectively. In particular,

$$
\begin{aligned}
p(\mathbf{x}(k) \mid \mathbf{Z}^k) &= \mathcal{N}\left(\mathbf{x}(k), \hat{\mathbf{x}}(k \mid k), \mathbf{P}(k \mid k)\right) \\
&\triangleq \frac{1}{\mathbf{A}} \exp\left\{ -\frac{[\mathbf{x}(k) - \hat{\mathbf{x}}(k \mid k)]^T \, \mathbf{P}^{-1}(k \mid k) \, [\mathbf{x}(k) - \hat{\mathbf{x}}(k \mid k)]}{2} \right\},
\end{aligned}
$$

where $\mathbf{A} = \sqrt{det(2\pi \mathbf{P}(k \mid k))}$.

Substituting this distribution into Equation 9.16 leads to

$$
\begin{aligned}
\mathcal{J}(k) &= -\mathrm{E}\left[\boldsymbol{\nabla}_x \boldsymbol{\nabla}_x^T ln\ p(\mathbf{x}(k)|\mathbf{Z}^k)\right] \\
&= \mathrm{E}\left[\boldsymbol{\nabla}_x \boldsymbol{\nabla}_x^T \left\{ \frac{\left[\mathbf{x}(k)-\hat{\mathbf{x}}(k\mid k)\right]^T \mathbf{P}^{-1}(k\mid k)\left[\mathbf{x}(k)-\hat{\mathbf{x}}(k\mid k)\right]}{2} + ln\ \mathbf{A}\right\}\right] \\
&= \mathrm{E}\left[\boldsymbol{\nabla}_x \boldsymbol{\nabla}_x^T \left(\frac{\left[\mathbf{x}(k)-\hat{\mathbf{x}}(k\mid k)\right]^T \mathbf{P}^{-1}(k\mid k)\left[\mathbf{x}(k)-\hat{\mathbf{x}}(k\mid k)\right]}{2} \right)\right] \\
&= \mathrm{E}\left[\mathbf{P}^{-1}(k\mid k)\left\{\left[\mathbf{x}(k)-\hat{\mathbf{x}}(k\mid k)\right]\left[\mathbf{x}(k)-\hat{\mathbf{x}}(k\mid k)\right]^T\right\}\mathbf{P}^{-1}(k\mid k)\right] \\
&= \mathbf{P}^{-1}(k\mid k)\mathbf{P}(k\mid k)\mathbf{P}^{-1}(k\mid k) \\
&= \mathbf{P}^{-1}(k\mid k) \hspace{4cm} (9.19) \\
&= (CRLB)^{-1}. \hspace{4cm} (9.20)
\end{aligned}
$$

Thus, assuming Gaussian noise and minimum mean squared error estimation, the Fisher information matrix is equal to the inverse of the covariance matrix.

This *information matrix* is central to the filtering techniques employed in this chapter. Although the filter constructed from this information space is algebraically equivalent to the Kalman filter, it has been shown to have advantages over the Kalman filter in multisensor data fusion applications. These include reduced computation, algorithmic simplicity, and easy initialization. In particular, these attributes make the Information filter easier to decouple, decentralize, and distribute. These are important filter characteristics in multisensor data fusion systems.

9.3.2 Information Filter Derivation

The two key information-analytic variables are the *information matrix* and *information state vector*. The information matrix has already been derived above as the inverse of the covariance matrix,

$$
\mathbf{Y}(i\mid j) \triangleq \mathbf{P}^{-1}(i\mid j). \hspace{2cm} (9.21)
$$

The information state vector is a product of the inverse of the covariance matrix (information matrix) and the state estimate,

$$
\begin{aligned}
\hat{\mathbf{y}}(i\mid j) &\triangleq \mathbf{P}^{-1}(i\mid j)\hat{\mathbf{x}}(i\mid j) \\
&= \mathbf{Y}(i\mid j)\hat{\mathbf{x}}(i\mid j) \hspace{2cm} (9.22)
\end{aligned}
$$

The variables, $\mathbf{Y}(i\mid j)$ and $\hat{\mathbf{y}}(i\mid j)$, form the basis of the information space ideas that are central to the material presented in this chapter.

The Information filter is derived from the Kalman filter algorithm by post-multiplying the term $\{\mathbf{1} - \mathbf{W}(k)\mathbf{H}(k)\}$ from Equation 9.5, by the term $\{\mathbf{P}(k \mid k-1)\mathbf{P}^{-1}(k \mid k-1)\}$ (i.e., post-multiplication by the identity matrix $\mathbf{1}$),

$$
\begin{aligned}
\mathbf{1} - \mathbf{W}(k)\mathbf{H}(k) &= \left[\mathbf{P}(k \mid k-1) - \mathbf{W}(k)\mathbf{H}(k)\mathbf{P}(k \mid k-1)\right]\mathbf{P}^{-1}(k \mid k-1) \\
&= \left[\mathbf{P}(k \mid k-1) - \mathbf{W}(k)\mathbf{S}(k)\mathbf{S}^{-1}(k)\mathbf{H}(k)\mathbf{P}(k \mid k-1)\right] \times \\
&\quad \mathbf{P}^{-1}(k \mid k-1) \\
&= \left[\mathbf{P}(k \mid k-1) - \mathbf{W}(k)\mathbf{S}(k)\mathbf{W}^T(k)\right]\mathbf{P}^{-1}(k \mid k-1) \\
&= \mathbf{P}(k \mid k)\mathbf{P}^{-1}(k \mid k-1).
\end{aligned}
\tag{9.23}
$$

Substituting the expression of the innovation covariance $\mathbf{S}(k)$, given in Equation 9.8, into the expression of the filter gain matrix $\mathbf{W}(k)$, from Equation 9.7 gives

$$
\begin{aligned}
\mathbf{W}(k) &= \mathbf{P}(k \mid k-1)\mathbf{H}^T(k)[\mathbf{H}(k)\mathbf{P}(k \mid k-1)\mathbf{H}^T(k) + \mathbf{R}(k)]^{-1} \\
\Leftrightarrow \mathbf{W}(k)[\mathbf{H}(k)\mathbf{P}(k \mid k-1)\mathbf{H}^T(k) + \mathbf{R}(k)] &= \mathbf{P}(k \mid k-1)\mathbf{H}^T(k) \\
\Leftrightarrow \mathbf{W}(k)\mathbf{R}(k) &= [\mathbf{1} - \mathbf{W}(k)\mathbf{H}(k)]\mathbf{P}(k \mid k-1)\mathbf{H}^T(k)
\end{aligned}
$$

$$
\Leftrightarrow \mathbf{W}(k) = [\mathbf{1} - \mathbf{W}(k)\mathbf{H}(k)]\mathbf{P}(k \mid k-1)\mathbf{H}^T(k)\mathbf{R}^{-1}(k).
\tag{9.24}
$$

Substituting Equation 9.23 into Equation 9.24 gives

$$
\mathbf{W}(k) = \mathbf{P}(k \mid k)\mathbf{H}^T(k)\mathbf{R}^{-1}(k).
\tag{9.25}
$$

Substituting Equations 9.23 and 9.25 into Equation 9.5 and pre-multiplying through by $\mathbf{P}^{-1}(k \mid k)$ gives the update equation for the information state vector as

$$
\mathbf{P}^{-1}(k \mid k)\hat{\mathbf{x}}(k \mid k) = \mathbf{P}^{-1}(k \mid k-1)\hat{\mathbf{x}}(k \mid k-1) + \mathbf{H}^T(k)\mathbf{R}^{-1}(k)\mathbf{z}(k),
$$

or

$$
\hat{\mathbf{y}}(k \mid k) = \hat{\mathbf{y}}(k \mid k-1) + \mathbf{H}^T(k)\mathbf{R}^{-1}(k)\mathbf{z}(k).
\tag{9.26}
$$

A similar expression can be found for the information matrix associated with this estimate. From Equations 9.6, 9.7 and 9.23 it follows that

$$
\begin{aligned}
\mathbf{P}(k \mid k) &= [\mathbf{1} - \mathbf{W}(k)\mathbf{H}(k)]\mathbf{P}(k \mid k-1)[\mathbf{1} - \mathbf{W}(k)\mathbf{H}(k)]^T \\
&\quad + \mathbf{W}(k)\mathbf{R}(k)\mathbf{W}^T(k).
\end{aligned}
\tag{9.27}
$$

Substituting in Equations 9.23 and 9.25 gives

$$
\begin{aligned}
\mathbf{P}(k \mid k) &= \left[\mathbf{P}(k \mid k)\mathbf{P}^{-1}(k \mid k-1)\right]\mathbf{P}(k \mid k-1)\left[\mathbf{P}(k \mid k)\mathbf{P}^{-1}(k \mid k-1)\right]^T \\
&\quad + \left[\mathbf{P}(k \mid k)\mathbf{H}^T(k)\mathbf{R}^{-1}(k)\right]\mathbf{R}(k)\left[\mathbf{P}(k \mid k)\mathbf{H}^T(k)\mathbf{R}^{-1}(k)\right]^T.
\end{aligned}
\tag{9.28}
$$

Pre- and post-multiplying by $\mathbf{P}^{-1}(k \mid k)$ then simplifying, gives the information matrix update equation as

$$\mathbf{P}^{-1}(k \mid k) = \mathbf{P}^{-1}(k \mid k-1) + \mathbf{H}^T(k)\mathbf{R}^{-1}(k)\mathbf{H}(k) \qquad (9.29)$$

or

$$\mathbf{Y}(k \mid k) = \mathbf{Y}(k \mid k-1) + \mathbf{H}^T(k)\mathbf{R}^{-1}(k)\mathbf{H}(k). \qquad (9.30)$$

The information state contribution $\mathbf{i}(k)$ from an observation $\mathbf{z}(k)$, and its associated information matrix $\mathbf{I}(k)$ are defined, respectively, as follows:

$$\mathbf{i}(k) \triangleq \mathbf{H}^T(k)\mathbf{R}^{-1}(k)\mathbf{z}(k), \qquad (9.31)$$

$$\mathbf{I}(k) \triangleq \mathbf{H}^T(k)\mathbf{R}^{-1}(k)\mathbf{H}(k). \qquad (9.32)$$

The information propagation coefficient $\mathbf{L}(k \mid k-1)$, which is independent of the observations made, is given by the expression

$$\mathbf{L}(k \mid k-1) = \mathbf{Y}(k \mid k-1)\mathbf{F}(k)\mathbf{Y}^{-1}(k-1 \mid k-1). \qquad (9.33)$$

With these information quantities well defined, the linear Kalman filter can now be written in terms of the information state vector and the information matrix.

Prediction

$$\hat{\mathbf{y}}(k \mid k-1) = \mathbf{L}(k \mid k-1)\hat{\mathbf{y}}(k-1 \mid k-1) \qquad (9.34)$$

$$\mathbf{Y}(k \mid k-1) = \left[\mathbf{F}(k)\mathbf{Y}^{-1}(k-1 \mid k-1)\mathbf{F}^T(k) + \mathbf{Q}(k)\right]^{-1}. \qquad (9.35)$$

Estimation

$$\hat{\mathbf{y}}(k \mid k) = \hat{\mathbf{y}}(k \mid k-1) + \mathbf{i}(k) \qquad (9.36)$$

$$\mathbf{Y}(k \mid k) = \mathbf{Y}(k \mid k-1) + \mathbf{I}(k). \qquad (9.37)$$

This is the information form of the Kalman filter [15]. Despite its potential applications, it is not widely used and it is thinly covered in literature. Barshalom [4] and Maybeck [13] briefly discuss the idea of information estimation, but do not explicitly derive the algorithm in terms of information as done above, nor do they use it as a principal filtering method.

9.3.3   Filter Characteristics

By comparing the implementation requirements and performance of the Kalman and Information filters, a number of attractive features of the latter are identified:

- The information estimation Equations 9.36 and 9.37 are computationally simpler than the state estimation Equations 9.5 and 9.6. This can be exploited in partitioning these equations for decentralized multisensor estimation.

- Although the information prediction Equations 9.34 and 9.35 are more complex than Equations 9.3 and 9.4, prediction depends on a propagation coefficient that is independent of the observations. It is thus again easy to decouple and decentralize.

- There are no gain or innovation covariance matrices and the maximum dimension of a matrix to be inverted is the state dimension. In multisensor systems the state dimension is generally smaller than the observation dimension, hence it is preferable to employ the Information filter and invert smaller information matrices than use the Kalman filter and invert larger innovation covariance matrices.

- Initializing the Information filter is much easier than for the Kalman filter. This is because information estimates (matrix and state) are easily initialized to zero information. However, in order to implement the Information filter, a start-up procedure is required where the information matrix is set with small non-zero diagonal elements to make it invertible.

These characteristics are useful in the development of decentralized data fusion and control systems. Consequently, this chapter employs information space estimation as the principal filtering technique.

9.3.4   An Example of Linear Estimation

To compare the Kalman and the Information filter and illustrate the issues discussed above, the following example of a linear estimation problem is considered. Consider two targets moving with two different but constant velocities, v_1 and v_2. The state vector describing their true positions and velocities can be represented as follows:

$$\mathbf{x}(k) = \begin{bmatrix} x_1(k) \\ x_2(k) \\ \dot{x}_1(k) \\ \dot{x}_2(k) \end{bmatrix} = \begin{bmatrix} v_1 k \\ v_2 k \\ v_1 \\ v_2 \end{bmatrix}. \tag{9.38}$$

The objective is to estimate the entire state vector $\mathbf{x}(k)$ in Equation 9.38 after obtaining observations of the two target positions, $x_1(k)$ and $x_2(k)$.

The discrete-time state equation with sampling interval ΔT is given by

$$\mathbf{x}(k) = \mathbf{F}(k)\mathbf{x}(k-1) + \mathbf{w}(k-1), \qquad (9.39)$$

where $\mathbf{F}(k)$ is the state-transition matrix. This matrix is obtained by the series method (discussed in Chapter 7) as follows:

$$\mathbf{F}(k) = e^{\mathbf{A}\Delta T} \approx \mathbf{1} + \Delta T\mathbf{A}$$

$$= \begin{bmatrix} 1 & 0 & \Delta T & 0 \\ 0 & 1 & 0 & \Delta T \\ 0 & 0 & 1 & 0 \\ 0 & 0 & 0 & 1 \end{bmatrix},$$

where $\mathbf{1}$ is an identity matrix and A is given by

$$\mathbf{A} = \begin{bmatrix} 0 & 0 & 1 & 0 \\ 0 & 0 & 0 & 1 \\ 0 & 0 & 0 & 0 \\ 0 & 0 & 0 & 0 \end{bmatrix}.$$

Since only linear measurements of the two target positions are taken, the observation matrix is given by

$$\mathbf{H}(k) = \begin{bmatrix} 1 & 0 & 0 & 0 \\ 0 & 1 & 0 & 0 \end{bmatrix}.$$

In order to complete the construction of models, the measurement error covariance matrix $\mathbf{R}(k)$ and the process noise $\mathbf{Q}(k)$ are then obtained as follows:

$$\mathbf{R}(k) = \begin{bmatrix} \sigma^2_{meas\_noise} & 0 \\ 0 & \sigma^2_{meas\_noise} \end{bmatrix},$$

$$\mathbf{Q}(k) = \begin{bmatrix} \sigma^2_{pos\_noise} & 0 & 0 & 0 \\ 0 & \sigma^2_{pos\_noise} & 0 & 0 \\ 0 & 0 & \sigma^2_{vel\_noise} & 0 \\ 0 & 0 & 0 & \sigma^2_{vel\_noise} \end{bmatrix}.$$

The terms σ_{pos_noise} and σ_{vel_noise} represent the system modeling errors in target position and velocity, respectively, while σ_{meas_noise} represents

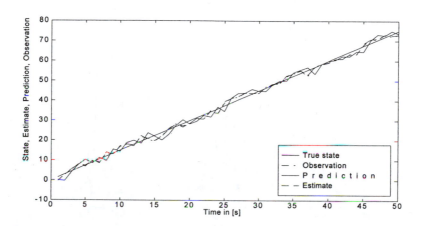

FIGURE 9.2
Performance of the Kalman and Information Filters

the error in measuring a target position. The corresponding process and measurement noise vectors are defined and generated as follows:

$$\mathbf{w}(k) = \begin{bmatrix} |\text{rand num}| \leq 2\sigma_{pos\_noise} \\ |\text{rand num}| \leq 2\sigma_{pos\_noise} \\ |\text{rand num}| \leq 2\sigma_{vel\_noise} \\ |\text{rand num}| \leq 2\sigma_{vel\_noise} \end{bmatrix},$$

$$\mathbf{v}(k) = \begin{bmatrix} |\text{rand num}| \leq 2\sigma_{meas\_noise} \\ |\text{rand num}| \leq 2\sigma_{meas\_noise} \end{bmatrix}.$$

These system modeling matrices and vectors are then used in the algorithms of the Kalman and Information filters to carry out estimation. In both cases the simulations are implemented using the same models with process and observation noises generated by the same random generators. The results are discussed in the next section.

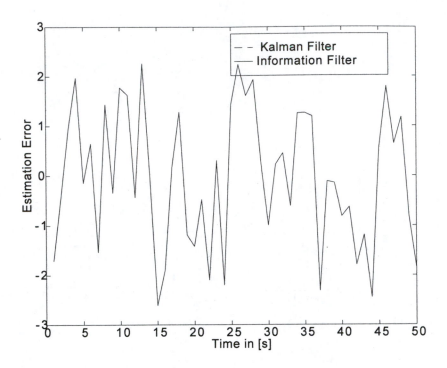

FIGURE 9.3
Estimation Error for the Kalman and Information Filters

9.3.5   Comparison of the Kalman and Information Filters

The Kalman and Information filters are compared by simulating the constant velocity system described above. In order to study and compare the performance of the filters, estimation of the same state is considered for the two filters; the position of the first target, $x_1(k)$. Figure 9.2 compares the target's true position, predicted position, estimated position, and observed position for both the Kalman and Information filters.

The curves depicting the same variables are identical and indistinguishable for the two filters. They lie on top of each other. This illustrates the algebraic equivalence of the two filters. From Figure 9.2 it can be observed, for both filters, that the state estimate is always well placed between the observation and state prediction. This means that there is balanced confidence in observations and predictions. Since, as the time k goes to infinity, the process noise variance $\mathbf{Q}(k)$ governs the confidence in predictions and the observation noise variance $\mathbf{R}(k)$ governs the confidence in observations, the results are an indication that the noise variances were well chosen.

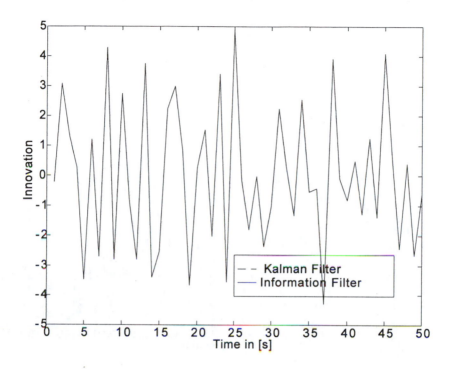

FIGURE 9.4
Innovations for the Kalman and Information Filters

Figure 9.3 shows the state estimation errors for both the Kalman and Information filters, while the innovations are similarly shown in Figure 9.4. The estimation errors and the innovations are identical for the two filters, again demonstrating the algebraic equivalence of the two filters. By inspection and computing the sequence mean, the innovations are shown to be zero mean with variance $\mathbf{S}(k)$. Practically, it means the noise level in the filter is of the same order as the true system noise. There is no visible correlation of the innovations sequences. This implies that there are no significant higher-order unmodeled dynamics nor excessive observation noise-to-process ratio. The innovations also satisfy the 95% confidence rule. This implies that the filters are consistent and well-matched.

Since the curves in Figures 9.2, 9.3, and 9.4 look indistinguishable for the two filters, it is prudent to plot parameter differences between the filters to confirm the algebraic equivalence. Figure 9.5 shows the difference between the state estimates for the filters. The difference is very small (*lies within* 10^{-13}%) and hence, attributable to numerical and computational

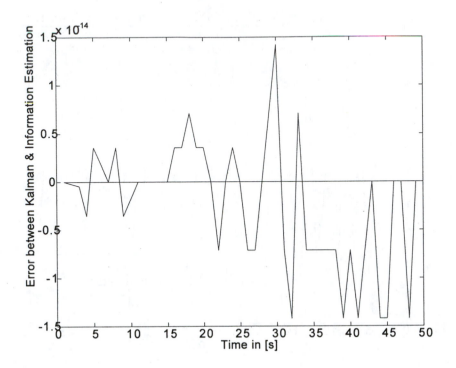

FIGURE 9.5
The Difference between Kalman and Information Filters' State Estimates

errors such as truncation and rounding off errors. Thus, the Kalman and Information filters are demonstrably equivalent. This confirms the algebraic equivalence, which is mathematically proven and established in the derivation of the Information filter from the Kalman filter.

9.4   The Extended Kalman Filter (EKF)

In almost all real data fusion problems, the state or environment of interest does not evolve linearly. Consequently, simple linear models will not be adequate to describe the system. Furthermore, the sensor observations may not depend linearly on the states that describe the environment. A popular approach to solve nonlinear estimation problems has been to use the extended Kalman filter (**EKF**) [3], [15]. This is a *linear* estimator for a nonlinear system obtained by *linearization* of the nonlinear state and observations equations. For any nonlinear system, the EKF is the best linear unbiased estimator with respect to minimum mean squared error criteria.

The EKF is conceptually simple and its derivation follows from arguments of linearization and the Kalman filter algorithm. The difficulty arises in implementation. It can be made to work well, but may perform badly or even become unstable with diverging estimates. This is most often due to lack of careful modeling of sensors and environment. Failure to understand the limitations of the algorithm exacerbates the problem.

9.4.1 Nonlinear State-Space

The system of interest is described by a nonlinear discrete-time state transition equation in the form

$$\mathbf{x}(k) = \mathbf{f}\left(\mathbf{x}(k-1), \mathbf{u}(k-1), (k-1)\right) + \mathbf{w}(k), \qquad (9.40)$$

where $\mathbf{x}(k-1)$ is the state vector and $\mathbf{u}(k-1)$ is a known input vector, both at time $(k-1)$. The vectors $\mathbf{x}(k)$ and $\mathbf{w}(k)$ represent the state vector and some additive process noise vector, respectively, both at time-step k. The nonlinear function $\mathbf{f}(\cdot, \cdot, k)$ is the nonlinear state transition function mapping the previous state and current control input to the current state. It is assumed that observations of the state of this system are made according to a nonlinear observation equation of the form

$$\mathbf{z}(k) = \mathbf{h}\left(\mathbf{x}(k), k\right) + \mathbf{v}(k), \qquad (9.41)$$

where $\mathbf{z}(k)$ is the observation made at time k, $\mathbf{x}(k)$ is the state at time k, $\mathbf{v}(k)$ is some additive observation noise, and $\mathbf{h}(\cdot, k)$ is a nonlinear observation model mapping current state to observations.

It is assumed that both noise vectors $\mathbf{v}(k)$ and $\mathbf{w}(k)$ are linearly additive Gaussian, temporally uncorrelated with zero mean, which means

$$\mathrm{E}[\mathbf{w}(k)] = \mathrm{E}[\mathbf{v}(k)] = \mathbf{0}, \qquad \forall k,$$

with the corresponding covariances being given by

$$\mathrm{E}[\mathbf{w}(i)\mathbf{w}^T(j)] = \delta_{ij}\mathbf{Q}(i), \quad \mathrm{E}[\mathbf{v}(i)\mathbf{v}^T(j)] = \delta_{ij}\mathbf{R}(i).$$

It is assumed that process and observation noises are uncorrelated, i.e.,

$$\mathrm{E}[\mathbf{w}(i)\mathbf{v}^T(j)] = \mathbf{0}, \qquad \forall i, j.$$

9.4.2   EKF Derivation

The derivation of the EKF follows from that of the linear Kalman filter, by linearizing state and observation models using Taylor's series expansion [4], [13].

State Prediction

It is assumed that there exists an estimate at time $(k-1)$ that is approximately equal to the conditional mean,

$$\hat{\mathbf{x}}(k-1 \mid k-1) \approx \mathrm{E}[\mathbf{x}(k-1) \mid Z^{k-1}]. \tag{9.42}$$

The objective is to find a prediction $\hat{\mathbf{x}}(k \mid k-1)$ for the state at the next time k based only on the information available up to time $(k-1)$. Expanding Equation 9.40 as a Taylor series about the estimate $\hat{\mathbf{x}}(k-1 \mid k-1)$, the following expression is obtained.

$$
\begin{aligned}
\mathbf{x}(k) = {}& \mathbf{f}\left(\hat{\mathbf{x}}(k-1 \mid k-1), \mathbf{u}(k-1), (k-1)\right) \\
& + \nabla \mathbf{f}_x(k)\left[\mathbf{x}(k-1) - \hat{\mathbf{x}}(k-1 \mid k-1)\right] \\
& + O\left(\left[\mathbf{x}(k-1) - \hat{\mathbf{x}}(k-1 \mid k-1)\right]^2\right) + \mathbf{w}(k)
\end{aligned} \tag{9.43}
$$

where $\nabla \mathbf{f}_x(k)$ is the Jacobian of $\mathbf{f}$ evaluated at $\mathbf{x}(k-1) = \hat{\mathbf{x}}(k-1 \mid k-1)$.

Truncating Equation 9.43 at first-order, and taking expectations conditioned on the first $(k-1)$ observations, gives an equation for the state prediction.

$$
\begin{aligned}
\hat{\mathbf{x}}(k \mid k-1) = {}& \mathrm{E}\left[\mathbf{x}(k) \mid Z^{k-1}\right] \\
\approx {}& \mathrm{E}\left[\mathbf{f}\left(\hat{\mathbf{x}}(k-1 \mid k-1) + \mathbf{A} + \mathbf{w}(k) \mid Z^{k-1}, \mathbf{u}(k-1), (k-1)\right)\right] \\
& (\textit{where} \quad \mathbf{A} = \nabla \mathbf{f}_x(k)\left[\mathbf{x}(k-1) - \hat{\mathbf{x}}(k-1 \mid k-1)\right]) \\
= {}& \mathbf{f}\left(\hat{\mathbf{x}}(k-1 \mid k-1), \mathbf{u}(k-1), (k-1)\right).
\end{aligned} \tag{9.44}
$$

This follows from the assumption that the estimate $\hat{\mathbf{x}}(k-1 \mid k-1)$ is approximately equal to the conditional mean (Equation 9.42) and that the process noise $\mathbf{w}(k)$ has zero mean. The state estimate error at a time i, given all observations up to time j, is defined as

$$\tilde{\mathbf{x}}(i \mid j) \overset{\triangle}{=} \mathbf{x}(i) - \hat{\mathbf{x}}(i \mid j), \tag{9.45}$$

and the state covariance is defined as the outer product of this error with itself conditioned on the observations made

$$\mathbf{P}(i \mid j) \overset{\triangle}{=} \mathrm{E}\left[\tilde{\mathbf{x}}(i \mid j)\tilde{\mathbf{x}}^T(i \mid j) \mid Z^j\right]. \tag{9.46}$$

In particular, the prediction error $\tilde{\mathbf{x}}(k \mid k-1)$ can be found by subtracting the true state $\mathbf{x}(k)$ given in Equation 9.43 from the prediction given in

Equation 9.44

$$\tilde{\mathbf{x}}(k \mid k-1)$$
$$= \mathbf{x}(k) - \hat{\mathbf{x}}(k \mid k-1)$$
$$= \mathbf{f}\left(\hat{\mathbf{x}}(k-1 \mid k-1), \mathbf{u}(k), k\right) + \nabla \mathbf{f}_x(k) \left[\mathbf{x}(k-1) - \hat{\mathbf{x}}(k-1 \mid k-1)\right]$$
$$\quad + O\left(\left[\mathbf{x}(k-1) - \hat{\mathbf{x}}(k-1 \mid k-1)\right]^2\right) + \mathbf{w}(k)$$
$$\quad - \mathbf{f}\left(\hat{\mathbf{x}}(k-1 \mid k-1), \mathbf{u}(k), k\right)$$
$$\approx \nabla \mathbf{f}_x(k) \left[\mathbf{x}(k-1) - \hat{\mathbf{x}}(k-1 \mid k-1)\right] + \mathbf{w}(k)$$
$$= \nabla \mathbf{f}_x(k)\tilde{\mathbf{x}}(k-1 \mid k-1) + \mathbf{w}(k). \tag{9.47}$$

The prediction is unbiased when the previous estimate is unbiased and the condition that the noise sequences are zero mean and white hold.

Taking expectations conditioned on the observations made up to time $(k-1)$ of the outer product of the prediction error gives an expression for the prediction covariance in terms of the covariance of the previous estimate.

$$\mathbf{P}(k \mid k-1) \quad \triangleq \mathrm{E}\left[\tilde{\mathbf{x}}(k \mid k-1)\tilde{\mathbf{x}}^T(k \mid k-1) \mid Z^{k-1}\right]$$
$$\approx \mathrm{E}\left[\left(\nabla \mathbf{f}_x(k)\tilde{\mathbf{x}}(k-1 \mid k-1) + \mathbf{w}(k)\right) \mathbf{A} \mid Z^{k-1}\right]$$
$$\left(where \quad \mathbf{A} = \left(\nabla \mathbf{f}_x(k)\tilde{\mathbf{x}}(k-1 \mid k-1) + \mathbf{w}(k)\right)^T\right)$$
$$= \nabla \mathbf{f}_x(k)\mathrm{E}\left[\tilde{\mathbf{x}}(k-1 \mid k-1)\tilde{\mathbf{x}}^T(k-1 \mid k-1) \mid Z^{k-1}\right]\nabla \mathbf{f}_x^T(k)$$
$$\quad + \mathrm{E}\left[\mathbf{w}(k)\mathbf{w}^T(k)\right]$$
$$= \nabla \mathbf{f}_x(k)\mathbf{P}(k-1 \mid k-1)\nabla \mathbf{f}_x^T(k) + \mathbf{Q}(k). \tag{9.48}$$

The last two lines follow from the fact that the estimate and true state at time $(k-1)$ are statistically dependent only on the noise terms $\mathbf{v}(j)$ and $\mathbf{w}(j)$, $j \leq (k-1)$. Hence, by assumption, they are uncorrelated with the *current* process noise $\mathbf{w}(k)$.

Observation Prediction and Innovation

The next objective is to obtain a predicted observation and its corresponding innovation to be used in updating the predicted state. This is achieved by expanding Equation 9.41, describing the observations made, as a Taylor series about the state prediction $\hat{\mathbf{x}}(k \mid k-1)$.

$$\mathbf{z}(k) = \mathbf{h}\left(\mathbf{x}(k)\right) + \mathbf{v}(k)$$
$$= \mathbf{h}\left(\hat{\mathbf{x}}(k \mid k-1)\right) + \nabla \mathbf{h}_x(k)\left[\hat{\mathbf{x}}(k \mid k-1) - \mathbf{x}(k)\right] +$$
$$\quad O\left(\left[\hat{\mathbf{x}}(k \mid k-1) - \mathbf{x}(k)\right]^2\right) + \mathbf{v}(k) \tag{9.49}$$

where $\nabla \mathbf{h}_x(k)$ is the Jacobian of $\mathbf{h}$ evaluated at $\mathbf{x}(k) = \hat{\mathbf{x}}(k \mid k-1)$. Again, ignoring second- and higher-order terms and taking expectations conditioned on the first $(k-1)$ observations gives an equation for the predicted observation.

$$\hat{\mathbf{z}}(k \mid k - 1)$$

$$\triangleq \mathrm{E}\left[\mathbf{z}(k) \mid Z^{k-1}\right]$$

$$\approx \mathrm{E}\left[\mathbf{h}\left(\hat{\mathbf{x}}(k \mid k - 1)\right) + \nabla\mathbf{h}_x(k)\left[\hat{\mathbf{x}}(k \mid k - 1) - \mathbf{x}(k)\right] + \mathbf{v}(k) \mid Z^{k-1}\right]$$

$$= \mathbf{h}\left(\hat{\mathbf{x}}(k \mid k - 1)\right), \tag{9.50}$$

where the last two lines follow from the fact that the state prediction error and the observation noise both have zero mean. After taking an observation $\mathbf{z}(k)$, the innovation can be found by subtracting the predicted observation as

$$\boldsymbol{\nu}(k) = \mathbf{z}(k) - \mathbf{h}\left(\hat{\mathbf{x}}(k \mid k - 1)\right). \tag{9.51}$$

The innovation covariance can now be found from the mean squared error in the predicted observation. The error in the predicted observation can be approximated by subtracting this prediction from the Taylor's series expansion of the observation in Equation 9.49 as

$$\tilde{\mathbf{z}}(k \mid k - 1) \triangleq \mathbf{z}(k) - \hat{\mathbf{z}}(k \mid k - 1)$$

$$= \mathbf{h}\left(\hat{\mathbf{x}}(k \mid k - 1)\right) + \nabla\mathbf{h}_x(k)\left[\hat{\mathbf{x}}(k \mid k - 1) - \mathbf{x}(k)\right]$$

$$+ O\left(\left[\hat{\mathbf{x}}(k \mid k - 1) - \mathbf{x}(k)\right]^2\right) + \mathbf{v}(k)$$

$$- \mathbf{h}\left(\hat{\mathbf{x}}(k \mid k - 1)\right)$$

$$\approx \nabla\mathbf{h}_x(k)\left[\hat{\mathbf{x}}(k \mid k - 1) - \mathbf{x}(k)\right] + \mathbf{v}(k). \tag{9.52}$$

A clear distinction is made between the "estimated" observation error $\tilde{\mathbf{z}}(k \mid k - 1)$ and the measured observation error, the innovation, $\boldsymbol{\nu}(k)$. Squaring the expression for the estimated observation error and taking expectation conditions on the first $(k - 1)$ measurements gives an equation for the innovation covariance.

$$\mathbf{S}(k) = \mathrm{E}\left[\tilde{\mathbf{z}}(k \mid k - 1)\tilde{\mathbf{z}}^T(k \mid k - 1)\right]$$

$$= \mathrm{E}\left[\mathbf{A}\mathbf{A}^T\right]$$

$$(where \quad \mathbf{A} = \left(\nabla\mathbf{h}_x(k)\left[\hat{\mathbf{x}}(k \mid k - 1) - \mathbf{x}(k)\right] + \mathbf{v}(k)\right))$$

$$= \nabla\mathbf{h}_x(k)\mathbf{P}(k \mid k - 1)\nabla\mathbf{h}_x^T(k) + \mathbf{R}(k). \tag{9.53}$$

This follows from the fact that the state prediction is dependent only on the noise terms $\mathbf{v}(j)$ and $\mathbf{w}(j)$, $j \leq (k - 1)$. Consequently, by assumption, it is statistically uncorrelated with the current observation noise $\mathbf{v}(k)$.

Update Equations

Equipped with the prediction and innovation equations, a *linearized* estimator can be proposed. It gives a state estimate vector $\hat{\mathbf{x}}(k \mid k)$ for the state vector $\mathbf{x}(k)$, which is described by the nonlinear state transition of Equation 9.40, and which is being observed according to the nonlinear observation Equation 9.41. It is assumed that a prediction $\hat{\mathbf{x}}(k \mid k-1)$ for the state at time k has been made on the basis of the first $(k-1)$ observations Z^{k-1} according to Equation 9.44. The current observation is $\mathbf{z}(k)$. The estimator is assumed to be in the form of a linear *unbiased* average of the prediction and innovation so that,

$$\hat{\mathbf{x}}(k \mid k) = \hat{\mathbf{x}}(k \mid k-1) + \mathbf{W}(k)\left[\mathbf{z}(k) - \mathbf{h}(\hat{\mathbf{x}}(k \mid k-1))\right]. \qquad (9.54)$$

The gain matrix $\mathbf{W}(k)$ is chosen so as to minimize conditional mean squared estimation error. This error is equal to the trace of the estimate covariance $\mathbf{P}(k \mid k)$, which itself is simply the expected value of the state error $\tilde{\mathbf{x}}(k \mid k)$ squared.

From Equation 9.54 and the approximate observation error given in Equation 9.52, the state error becomes

$$
\begin{aligned}
&\tilde{\mathbf{x}}(k \mid k) \\
&= \hat{\mathbf{x}}(k \mid k) - \mathbf{x}(k) \\
&= [\hat{\mathbf{x}}(k \mid k-1) - \mathbf{x}(k)] + \mathbf{W}(k)\left[\mathbf{h}(\mathbf{x}(k)) - \mathbf{h}(\hat{\mathbf{x}}(k \mid k-1))\right] + \mathbf{W}(k)\mathbf{v}(k) \\
&\approx [\hat{\mathbf{x}}(k \mid k-1) - \mathbf{x}(k)] - \mathbf{W}(k)\nabla\mathbf{h}_x(k)\left[\hat{\mathbf{x}}(k \mid k) - \mathbf{x}(k)\right] + \mathbf{W}(k)\mathbf{v}(k) \\
&= [\mathbf{1} - \mathbf{W}(k)\nabla\mathbf{h}_x(k)]\,\tilde{\mathbf{x}}(k \mid k-1) + \mathbf{W}(k)\mathbf{v}(k). \qquad (9.55)
\end{aligned}
$$

The estimate is unbiased when the prediction is unbiased and the condition that the noise sequences are zero mean and white hold.

Taking the expectation condition on the observations made up to time k of the outer product of the state error gives an expression for the state covariance in terms of the prediction covariance.

$$
\begin{aligned}
\mathbf{P}(k \mid k) &\triangleq \mathrm{E}\left[\tilde{\mathbf{x}}(k \mid k)\tilde{\mathbf{x}}^T(k \mid k) \mid Z^k\right] \\
&\approx [\mathbf{1} - \mathbf{W}(k)\nabla\mathbf{h}_x(k)]\,\mathrm{E}\left[\tilde{\mathbf{x}}(k \mid k-1)\tilde{\mathbf{x}}^T(k \mid k-1) \mid Z^{k-1}\right] \times \\
&\quad [\mathbf{1} - \mathbf{W}(k)\nabla\mathbf{h}_x(k)]^T + \mathbf{W}(k)\mathrm{E}\left[\mathbf{v}(k)\mathbf{v}^T(k)\right]\mathbf{W}^T(k) \\
&\approx [\mathbf{1} - \mathbf{W}(k)\nabla\mathbf{h}_x(k)]\,\mathbf{P}(k \mid k-1)[\mathbf{I} - \mathbf{W}(k)\nabla\mathbf{h}_x(k)]^T + \\
&\quad \mathbf{W}(k)\mathbf{R}(k)\mathbf{W}^T(k). \qquad (9.56)
\end{aligned}
$$

The gain matrix $\mathbf{W}(k)$ is now chosen to minimize the mean squared estimation error $L(k)$, which is defined as

$$L(k) = \mathrm{E}[\tilde{\mathbf{x}}^T(k \mid k)\tilde{\mathbf{x}}(k \mid k)] = \mathrm{trace}[\mathbf{P}(k \mid k)]. \qquad (9.57)$$

Minimization of this error calls for

$$\frac{\partial L}{\partial \mathbf{W}(k)} = -2(\mathbf{1} - \mathbf{W}(k)\nabla \mathbf{h}_x(k))\mathbf{P}(k \mid k-1)\nabla \mathbf{h}_x^T(k) + 2\mathbf{W}(k)\mathbf{R}(k) = \mathbf{0},$$
(9.58)

which, on simplification and rearrangement, provides an expression for the gain matrix as

$$\mathbf{W}(k) = \mathbf{P}(k \mid k-1)\nabla \mathbf{h}_x^T(k)\left[\nabla \mathbf{h}_x(k)\mathbf{P}(k \mid k-1)\nabla \mathbf{h}_x^T(k) + \mathbf{R}(k)\right]^{-1}$$
$$= \mathbf{P}(k \mid k-1)\nabla \mathbf{h}_x^T(k)\mathbf{S}^{-1}(k).$$
(9.59)

With this gain matrix, Equation 9.54 becomes the best (*minimum mean squared error*) linear unbiased estimator for the state $\mathbf{x}(k)$ under the stated conditions. This completes the derivation of the extended Kalman filter.

9.4.3   Summary of the EKF Algorithm

Prediction

$$\hat{\mathbf{x}}(k \mid k-1) = \mathbf{f}\left(\hat{\mathbf{x}}(k-1 \mid k-1), \mathbf{u}(k-1), (k-1)\right) \tag{9.60}$$

$$\mathbf{P}(k \mid k-1) = \boldsymbol{\nabla}\mathbf{f}_x(k)\mathbf{P}(k-1 \mid k-1)\boldsymbol{\nabla}\mathbf{f}_x{}^T(k) + \mathbf{Q}(k-1). \tag{9.61}$$

Estimation

$$\hat{\mathbf{x}}(k \mid k) = \hat{\mathbf{x}}(k \mid k-1) + \mathbf{W}(k)\left[\mathbf{z}(k) - \mathbf{h}(\hat{\mathbf{x}}(k \mid k-1))\right] \tag{9.62}$$

$$\mathbf{P}(k \mid k) = \mathbf{P}(k \mid k-1) - \mathbf{W}(k)\mathbf{S}(k)\mathbf{W}^T(k). \tag{9.63}$$

The gain and innovation covariance matrices are given by

$$\mathbf{W}(k) = \mathbf{P}(k \mid k-1)\boldsymbol{\nabla}\mathbf{h}_x{}^T(k)\mathbf{S}^{-1}(k) \tag{9.64}$$

$$\mathbf{S}(k) = \boldsymbol{\nabla}\mathbf{h}_x(k)\mathbf{P}(k \mid k-1)\boldsymbol{\nabla}\mathbf{h}_x{}^T(k) + \mathbf{R}(k). \tag{9.65}$$

The Jacobians $\boldsymbol{\nabla}\mathbf{f}_x(k)$ and $\boldsymbol{\nabla}\mathbf{h}_x(k)$ are typically not constant, being functions of both the state and time-step. It is clearly evident that the EKF is very similar to the Kalman filter algorithm, with the substitutions $\mathbf{F} \to \boldsymbol{\nabla}\mathbf{f}_x(k)$ and $\mathbf{H} \to \boldsymbol{\nabla}\mathbf{h}_x(k)$ being made in the equations for the variance and gain propagation.

It is prudent to note a number of problematic issues specific to the EKF. Unlike the linear filter, the covariances and gain matrix must be computed online as estimates and predictions are made available, and will not, in general, tend to constant values. This significantly increases the amount of computation that must be performed online by the algorithm. Also, if the nominal (predicted) trajectory is too far away from the true trajectory, then the true covariance will be much larger than the estimated covariance and the filter will become poorly matched. This might lead to severe filter instabilities. Last, the EKF employs a linearized model that must be computed from an approximate knowledge of the state. Unlike the linear algorithm, this means that the filter must be accurately initialized at the start of operation to ensure that the linearized models obtained are valid. All these issues must be taken into account to achieve acceptable performance for the EKF.

9.5 The Extended Information Filter (EIF)

9.5.1 Nonlinear Information Space

In this section, the linear Information filter is extended to a linearized estimation algorithm for nonlinear systems. The general approach is to apply the principles of the EKF and those of the linear Information filter to construct a new estimation method for nonlinear systems. This generates a filter that predicts and estimates information about *nonlinear* state parameters given *nonlinear* observations and *nonlinear* system dynamics. All the computation and tracking is in information space. The new filter will be termed the extended Information filter (**EIF**) [15]. In addition to providing a solution to the nonlinear estimation problem, the EIF also has all the advantages of the Information filter and resolves some of the problems associated with the EKF.

In particular, information space allows easy initialization of filter matrices and vectors. Given the importance of accurate initialization when using linearized models, this is an extremely desirable characteristic. As discussed before, a major drawback of the EKF is excessive computational burden. Carrying out the prediction and estimation processes in terms of information will significantly reduce this load by simplifying the prediction and estimation equations. These equations are then easily partitioned and decentralized. It is proposed that estimation for nonlinear systems, in particular multisensor systems, is best carried out using information variables rather than state variables.

9.5.2   EIF Derivation

The derivation of the extended Information filter uses principles from both the derivations of the Information filter and the EKF. The EIF cannot be extrapolated from these two filters in an obvious manner. This is because, in the nonlinear case, the function operator **h** *cannot* be separated from $\mathbf{x}(k)$ in the nonlinear observation equation

$$\mathbf{z}(k) = \mathbf{h}\left(\mathbf{x}(k), k\right) + \mathbf{v}(k), \tag{9.66}$$

and yet the derivation of the Information filter *depends* on this separation, which is possible in the linear observation equation.

The derivation of the EIF proceeds by considering equations from the derivation of the EKF algorithm. Post-multiplying $\{\mathbf{1} - \mathbf{W}(k)\boldsymbol{\nabla}\mathbf{h}_x(k)\}$ from Equation 9.55 by the term $\{\mathbf{P}(k \mid k-1)\mathbf{P}^{-1}(k \mid k-1)\}$, i.e., post-multiplication by the identity matrix $\mathbf{1}$ leads to

$$
\begin{aligned}
\mathbf{1} - \mathbf{W}(k)\boldsymbol{\nabla}\mathbf{h}_x(k) &= [\mathbf{P}(k \mid k-1) - \mathbf{W}(k)\boldsymbol{\nabla}\mathbf{h}_x(k)\mathbf{P}(k \mid k-1)] \times \\
&\quad \mathbf{P}^{-1}(k \mid k-1) \\
&= [\mathbf{P}(k \mid k-1) - \mathbf{W}(k)\{\mathbf{S}(k)\mathbf{S}^{-1}(k)\}\boldsymbol{\nabla}\mathbf{h}_x(k) \times \\
&\quad \mathbf{P}(k \mid k-1)]\mathbf{P}^{-1}(k \mid k-1) \\
&= [\mathbf{P}(k \mid k-1) - \mathbf{W}(k)\mathbf{S}(k)\mathbf{W}^{T}(k)]\mathbf{P}^{-1}(k \mid k-1) \\
\Leftrightarrow \mathbf{1} - \mathbf{W}(k)\boldsymbol{\nabla}\mathbf{h}_x(k) &= \mathbf{P}(k \mid k)\mathbf{P}^{-1}(k \mid k-1).
\end{aligned}
\tag{9.67}
$$

Substituting the expression of the EKF innovation covariance matrix from Equation 9.65 in the EKF gain matrix given in Equation 9.64 produces

$$
\begin{aligned}
\mathbf{W}(k) &= \mathbf{P}(k \mid k-1)\boldsymbol{\nabla}\mathbf{h}_x{}^{T}(k)[\boldsymbol{\nabla}\mathbf{h}_x(k)\mathbf{P}(k \mid k-1)\boldsymbol{\nabla}\mathbf{h}_x{}^{T}(k) + \mathbf{R}(k)]^{-1} \\
&\Leftrightarrow \mathbf{W}(k)[\boldsymbol{\nabla}\mathbf{h}_x(k)\mathbf{P}(k \mid k-1)\boldsymbol{\nabla}\mathbf{h}_x{}^{T}(k) + \mathbf{R}(k)] = \mathbf{P}(k \mid k-1)\boldsymbol{\nabla}\mathbf{h}_x{}^{T}(k) \\
&\Leftrightarrow \mathbf{W}(k)\mathbf{R}(k) = [\mathbf{1} - \mathbf{W}(k)\boldsymbol{\nabla}\mathbf{h}_x(k)]\mathbf{P}(k \mid k-1)\boldsymbol{\nabla}\mathbf{h}_x{}^{T}(k).
\end{aligned}
$$

Now, substituting Equation 9.67 into this expression gives

$$\mathbf{W}(k) = \mathbf{P}(k \mid k)\boldsymbol{\nabla}\mathbf{h}_x{}^{T}(k)\mathbf{R}^{-1}(k). \tag{9.68}$$

In order to use Equations 9.67 and 9.68 to derive the EIF, the EKF state estimation Equation 9.62 must be expressed in the same form as that for the conventional Kalman filter. This is done by adding and subtracting the term $\mathbf{W}(k)\boldsymbol{\nabla}\mathbf{h}_x(k)\hat{\mathbf{x}}(k \mid k-1)$ to the left-hand side of Equation 9.62

(addition of zero):

$$\hat{\mathbf{x}}(k \mid k)$$
$$= \hat{\mathbf{x}}(k \mid k-1) + \mathbf{W}(k)\left[\mathbf{z}(k) - \mathbf{h}(\hat{\mathbf{x}}(k \mid k-1))\right]$$
$$= \hat{\mathbf{x}}(k \mid k-1) - \mathbf{W}(k)\boldsymbol{\nabla}\mathbf{h}_x(k)\hat{\mathbf{x}}(k \mid k-1) + \mathbf{W}(k)\left[\mathbf{z}(k) - \mathbf{h}(\hat{\mathbf{x}}(k \mid k-1))\right]$$
$$\quad + \mathbf{W}(k)\boldsymbol{\nabla}\mathbf{h}_x(k)\hat{\mathbf{x}}(k \mid k-1)$$
$$= [\mathbf{1} - \mathbf{W}(k)\boldsymbol{\nabla}\mathbf{h}_x(k)]\hat{\mathbf{x}}(k \mid k-1) + \mathbf{W}(k)\left[\boldsymbol{\nu}(k) + \boldsymbol{\nabla}\mathbf{h}_x(k)\hat{\mathbf{x}}(k \mid k-1)\right]$$
$$= [\mathbf{1} - \mathbf{W}(k)\boldsymbol{\nabla}\mathbf{h}_x(k)]\hat{\mathbf{x}}(k \mid k-1) + \mathbf{W}(k)\mathbf{z}'(k), \tag{9.69}$$

where $\mathbf{z}'(k)$ is the *"equivalent" linearized* observation vector,

$$\mathbf{z}'(k) = \boldsymbol{\nu}(k) + \boldsymbol{\nabla}\mathbf{h}_x(k)\hat{\mathbf{x}}(k \mid k-1), \tag{9.70}$$

and the innovation vector is given by

$$\boldsymbol{\nu}(k) = \mathbf{z}(k) - \mathbf{h}(\hat{\mathbf{x}}(k \mid k-1)).$$

Equation 9.69 is now in a form similar to that of a linear Kalman filter.

The derivation of the EIF then proceeds by substituting Equations 9.67 and 9.68 into Equation 9.69.

$$\hat{\mathbf{x}}(k \mid k)$$
$$= [\mathbf{1} - \mathbf{W}(k)\boldsymbol{\nabla}\mathbf{h}_x(k)]\hat{\mathbf{x}}(k \mid k-1) + \mathbf{W}(k)\mathbf{z}'(k)$$
$$= [\mathbf{P}(k \mid k)\mathbf{P}^{-1}(k \mid k-1)]\hat{\mathbf{x}}(k \mid k-1) + [\mathbf{P}(k \mid k)\boldsymbol{\nabla}\mathbf{h}_x{}^T(k)\mathbf{R}^{-1}(k)]\mathbf{z}'(k).$$

Pre-multiplying both sides by $\mathbf{P}^{-1}(k \mid k)$ gives

$$\mathbf{P}^{-1}(k \mid k)\hat{\mathbf{x}}(k \mid k) = [\mathbf{P}^{-1}(k \mid k)\mathbf{P}(k \mid k)\mathbf{P}^{-1}(k \mid k-1)]\hat{\mathbf{x}}(k \mid k-1) +$$
$$[\mathbf{P}^{-1}(k \mid k)\mathbf{P}(k \mid k)\boldsymbol{\nabla}\mathbf{h}_x{}^T(k)\mathbf{R}^{-1}(k)]\mathbf{z}'(k)$$
$$= \mathbf{P}^{-1}(k \mid k-1)\hat{\mathbf{x}}(k \mid k-1) + \boldsymbol{\nabla}\mathbf{h}_x{}^T(k)\mathbf{R}^{-1}(k)\mathbf{z}'(k)$$
$$\Leftrightarrow \hat{\mathbf{y}}(k \mid k) = \hat{\mathbf{y}}(k \mid k-1) + \mathbf{i}(k). \tag{9.71}$$

This is the extended information estimation equation where the *information contribution* from nonlinear observation $\mathbf{z}(k)$ is given by

$$\mathbf{i}(k) = \boldsymbol{\nabla}\mathbf{h}_x{}^T(k)\mathbf{R}^{-1}(k)\mathbf{z}'(k),$$

where $\mathbf{z}'(k)$ is the *"equivalent"* linearized observation given by Equation 9.70. The vector $\mathbf{z}'(k)$ gives an expression of the system observations if the nonlinear system is replaced by an equivalent linearized system. It depends on the innovation, the state prediction, and the Jacobian evaluated at this prediction.

To compute the information matrix update, Equations 9.67 and 9.68 are substituted into the EKF variance update equation,

$$\mathbf{P}(k \mid k)$$
$$= [\mathbf{1} - \mathbf{W}(k)\boldsymbol{\nabla}\mathbf{h}_x(k)]\mathbf{P}(k \mid k-1)[\mathbf{1} - \mathbf{W}(k)\boldsymbol{\nabla}\mathbf{h}_x(k)]^T + \mathbf{W}(k)\mathbf{R}(k)\mathbf{W}^T(k)$$
$$= [\mathbf{P}(k \mid k)\mathbf{P}^{-1}(k \mid k-1)]\mathbf{P}(k \mid k-1)[\mathbf{P}(k \mid k)\mathbf{P}^{-1}(k \mid k-1)]^T +$$
$$[\mathbf{P}(k \mid k)\boldsymbol{\nabla}\mathbf{h}_x{}^T(k)\mathbf{R}^{-1}(k)]\mathbf{R}(k)[\mathbf{P}(k \mid k)\boldsymbol{\nabla}\mathbf{h}_x{}^T(k)\mathbf{R}^{-1}(k)]^T. \qquad (9.72)$$

Pre- and post-multiplying both sides by $\mathbf{P}^{-1}(k \mid k)$ gives

$$\Leftrightarrow \mathbf{P}^{-1}(k \mid k) = \mathbf{P}^{-1}(k \mid k-1) + \boldsymbol{\nabla}\mathbf{h}_x{}^T(k)\mathbf{R}^{-1}(k)\boldsymbol{\nabla}\mathbf{h}_x(k) \qquad (9.73)$$
$$\Leftrightarrow \mathbf{Y}(k \mid k) = \mathbf{Y}(k \mid k-1) + \mathbf{I}(k). \qquad (9.74)$$

This is the linearized information matrix update equation where the associated matrix contribution is given by

$$\mathbf{I}(k) = \boldsymbol{\nabla}\mathbf{h}_x{}^T(k)\mathbf{R}^{-1}(k)\boldsymbol{\nabla}\mathbf{h}_x(k). \qquad (9.75)$$

To obtain the corresponding prediction equations, consider the EKF state and variance prediction equations. Pre-multiplying the state prediction Equation 9.60 by $\mathbf{P}^{-1}(k \mid k-1)$ and inverting the variance prediction Equation 9.61 gives the information vector prediction as

$$\hat{\mathbf{x}}(k \mid k-1) = \mathbf{f}\left(\hat{\mathbf{x}}(k-1 \mid k-1), \mathbf{u}(k-1), (k-1)\right)$$
$$\Leftrightarrow \mathbf{P}^{-1}(k \mid k-1)\hat{\mathbf{x}}(k \mid k-1) = \mathbf{P}^{-1}(k \mid k-1) \times$$
$$\mathbf{f}\left(\hat{\mathbf{x}}(k-1 \mid k-1), \mathbf{u}(k-1), (k-1)\right)$$
$$\Leftrightarrow \hat{\mathbf{y}}(k \mid k-1) = \mathbf{Y}(k \mid k-1)\ \mathbf{f}\ \left(\hat{\mathbf{x}}(k-1 \mid k-1), \mathbf{u}(k-1), (k-1)\right).$$

The linearized information matrix prediction is obtained as follows:

$$\mathbf{P}(k \mid k-1) = \boldsymbol{\nabla}\mathbf{f}_x(k)\mathbf{P}(k-1 \mid k-1)\boldsymbol{\nabla}\mathbf{f}_x{}^T(k) + \mathbf{Q}(k-1)$$
$$\Leftrightarrow \mathbf{Y}(k \mid k-1) = \left[\boldsymbol{\nabla}\mathbf{f}_x(k)\mathbf{Y}^{-1}(k-1 \mid k-1)\boldsymbol{\nabla}\mathbf{f}_x{}^T(k) + \mathbf{Q}(k)\right]^{-1}.$$

This completes the derivation of the EIF; the entire algorithm can be summarized as follows:

9.5.3   Summary of the EIF Algorithm

Prediction

$$\hat{\mathbf{y}}(k \mid k-1) = \mathbf{Y}(k \mid k-1)\mathbf{f}\left(k, \hat{\mathbf{x}}(k-1 \mid k-1), \mathbf{u}(k-1), (k-1)\right) \qquad (9.76)$$
$$\mathbf{Y}(k \mid k-1) = \left[\boldsymbol{\nabla}\mathbf{f}_x(k)\mathbf{Y}^{-1}(k-1 \mid k-1)\boldsymbol{\nabla}\mathbf{f}_x{}^T(k) + \mathbf{Q}(k)\right]^{-1}. \qquad (9.77)$$

Estimation

$$\hat{\mathbf{y}}(k \mid k) = \hat{\mathbf{y}}(k \mid k-1) + \mathbf{i}(k) \tag{9.78}$$

$$\mathbf{Y}(k \mid k) = \mathbf{Y}(k \mid k-1) + \mathbf{I}(k). \tag{9.79}$$

The information state contribution and its associated information matrix are given, respectively, as

$$\mathbf{I}(k) = \boldsymbol{\nabla}\mathbf{h}_x{}^T(k)\mathbf{R}^{-1}(k)\boldsymbol{\nabla}\mathbf{h}_x(k) \tag{9.80}$$

$$\mathbf{i}(k) = \boldsymbol{\nabla}\mathbf{h}_x{}^T(k)\mathbf{R}^{-1}(k)\left[\boldsymbol{\nu}(k) + \boldsymbol{\nabla}\mathbf{h}_x(k)\hat{\mathbf{x}}(k \mid k-1)\right], \tag{9.81}$$

where $\boldsymbol{\nu}(k)$ is the innovation given by

$$\boldsymbol{\nu}(k) = \mathbf{z}(k) - \mathbf{h}\left(\hat{\mathbf{x}}(k \mid k-1)\right). \tag{9.82}$$

9.5.4 Filter Characteristics

This filter has several attractive practical features, in particular:

- The filter solves, in information space, the linear estimation problem for systems with both nonlinear dynamics and observations. In addition to having all the attributes of the Information filter, it is a more practical and general filter.

- The information estimation Equations 9.78 and 9.79 are computationally simpler than the EKF estimation equations. This makes the partitioning of these equations for decentralized systems easy.

- Although the information prediction Equations 9.76 and 9.77 are of the same apparent complexity as the EKF ones, they are easier to distribute and fuse because of the *orthonormality* properties of information space parameters.

- Since the EIF is expressed in terms of information matrices and vectors, it is more easily initialized than the EKF. Accurate initialization is important where linearized models are employed.

Some of the drawbacks inherent in the EKF still affect the EIF. These include the nontrivial nature of Jacobian matrix derivation (and computation) and linearization instability.

9.6 Examples of Estimation in Nonlinear Systems

In order to compare the extended Kalman filter and the extended Information filter and illustrate the characteristics discussed above, three estimation problems in nonlinear systems are considered. These examples are

chosen so that all possible combinations of nonlinearities in observations
and nonlinearities in the system evolution are exhausted.

9.6.1   Nonlinear State Evolution and Linear Observations

Consider a two-dimensional radar tracking system that tracks a missile
traveling vertically in an xy plane with known vertical velocity v and acceleration a, such that $\dot{x}(k) = 0$, $\dot{y}(k) = v$ and $\ddot{y}(k) = a$. The missile is fired
vertically, in the positive y-axis direction, from some point on the x-axis.
The radar is located at the origin of the xy plane such that it measures the
polar coordinates of the missile, that is, the radial position $r(k)$, and the
angular displacement from the horizontal $\theta(k)$ where

$$r(k) = \sqrt{x^2(k) + y^2(k)} \quad and \quad \theta(k) = arctan\left[\frac{y(k)}{x(k)}\right].$$

Using the polar measurements, the objective is to estimate the entire
missile state vector $\mathbf{x}(k)$ given by

$$\mathbf{x}(k) = \begin{bmatrix} x_1(k) \\ x_2(k) \\ x_3(k) \\ x_4(k) \end{bmatrix} = \begin{bmatrix} r(k) \\ \dot{r}(k) \\ \theta(k) \\ \dot{\theta}(k) \end{bmatrix}.$$

The equations of motion with respect to the polar coordinates, $r(k)$ and
$\theta(k)$, are obtained by taking first and second derivatives. The result can be
expressed in a nonlinear state vector form as follows:

$$\dot{\mathbf{x}}(k) = \begin{bmatrix} \dot{r}(k) \\ \ddot{r}(k) \\ \dot{\theta}(k) \\ \ddot{\theta}(k) \end{bmatrix} = \begin{bmatrix} v \sin \theta(k) \\ a \sin \theta(k) + \dfrac{v^2}{r(k)} \cos^2 \theta(k) \\ \dfrac{v}{r(k)} \cos \theta(k) \\ \dfrac{ar(k) - v^2 \sin \theta(k)}{r^2(k)} \cos \theta(k) - \dfrac{v^2}{r^2(k)} \sin \theta(k) \cos \theta(k) \end{bmatrix}.$$

Using the definition of the derivative of a generic state vector element $x_i(k)$
leads to

$$\dot{x}_i(k-1) = \frac{x_i(k) - x_i(k-1)}{\Delta T},$$
$$\Leftrightarrow x_i(k) = x_i(k-1) + \Delta T \dot{x}_i(k-1). \tag{9.83}$$

The discrete-time nonlinear system model is then obtained by employing Equation 9.83 for all elements of the state vector $\mathbf{x}(k)$, while using the state element derivatives from the vector $\dot{\mathbf{x}}(k)$ (assuming $\Delta T = 1$).

$$\begin{aligned}
&\mathbf{x}(k) \\
&= \mathbf{f}\left(\mathbf{x}(k-1), (k-1)\right) + \mathbf{w}(k)
\end{aligned}$$

$$= \begin{bmatrix} x_1(k-1) + v\sin x_3(k-1) \\[2mm] x_2(k-1) + a\sin x_3(k-1) + \dfrac{v^2}{x_1(k-1)}\cos^2 x_3(k-1) \\[2mm] x_3(k-1) + \dfrac{v}{x_1(k-1)}\cos x_3(k-1) \\[2mm] x_4(k-1) + \dfrac{\left[ax_1(k-1)-v^2\sin x_3(k-1)\right]\cos x_3(k-1) - v^2\sin x_3(k-1)\cos x_3(k-1)}{x_1^2(k-1)} \end{bmatrix}$$

$$+\, \mathbf{w}(k).$$

The Jacobian of this nonlinear model is given by

$$\nabla \mathbf{f}(x) = \begin{bmatrix} a_{11} & a_{12} & a_{13} & a_{14} \\ a_{21} & a_{22} & a_{23} & a_{24} \\ a_{31} & a_{32} & a_{33} & a_{34} \\ a_{41} & a_{42} & a_{43} & a_{44} \end{bmatrix},$$

where

$$a_{11} = 1.0$$
$$a_{12} = 0.0$$
$$a_{13} = v\cos x_3(k-1)$$
$$a_{14} = 0.0$$
$$a_{21} = -\frac{v^2\cos^2 x_3(k-1)}{x_1^2(k-1)}$$
$$a_{22} = 1.0$$
$$a_{23} = a\cos x_3(k-1) - \frac{v^2}{x_1(k-1)}\sin 2x_3(k-1)$$
$$a_{24} = 0.0$$
$$a_{31} = -\frac{v\cos x_3(k-1)}{x_1^2(k-1)}$$

$a_{32} = 0.0$

$a_{33} = 1 - \dfrac{v \sin x_3(k-1)}{x_1(k-1)}$

$a_{34} = 0.0$

$a_{41} = \dfrac{\left[2v^2 x_1(k-1)\sin 2x_3(k-1)\right] - \left[ax_1{}^2(k-1)\cos x_3(k-1)\right]}{x_1^4(k-1)}$

$a_{42} = 0.0$

$a_{43} = -\left[\dfrac{ax_1(k-1)\sin x_3(k-1) + 2v^2 \cos 2x_3(k-1)}{x_1^2(k-1)}\right]$

$a_{44} = 1.0.$

Since the radial position $r(k)$ and the angular displacement $\theta(k)$ are linearly measured by the radar, the observation matrix for the tracking system is given by

$$\mathbf{H}(k) = \begin{bmatrix} 1 & 0 & 0 & 0 \\ 0 & 0 & 1 & 0 \end{bmatrix}.$$

These system modeling matrices and vectors are then used in the algorithms of the EKF and EIF to carry out estimation of the missile state vector. The process and observation noises are generated by random number generators, as has already been explained. The results are discussed later.

9.6.2   Linear State Evolution with Nonlinear Observations

A system might involve linear system dynamics and nonlinear measurement equations. An example of such a system is a radar station that makes measurements of the radial position $r(k)$ and the angular displacement $\theta(k)$ of an aircraft, from which it is desired to obtain the estimated values of the horizontal and vertical positions and velocities of the aircraft. The motion of the aircraft is such that the horizontal velocity v_x and the vertical velocity v_y are constant, which means the aircraft is executing linear motion. This is a four-dimensional problem involving two positions and two velocities.

As in the previous radar and missile example, the polar coordinates and the Cartesian coordinates are related by the equations

$$r(k) = \sqrt{x^2(k) + y^2(k)} \quad and \quad \theta(k) = arctan\left[\frac{y(k)}{x(k)}\right].$$

The state vector of interest consists of the two positions and the two

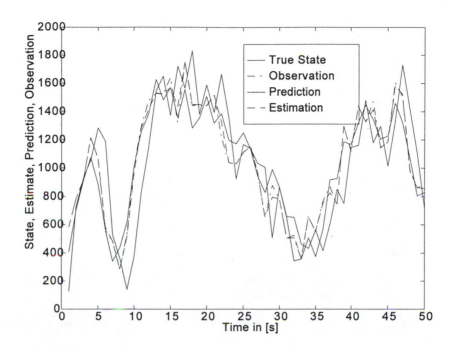

FIGURE 9.6
EKF & EIF: Nonlinear State Evolution with Linear Observations

velocities such that

$$\mathbf{x}(k) = \begin{bmatrix} x_1(k) \\ x_2(k) \\ x_3(k) \\ x_4(k) \end{bmatrix} = \begin{bmatrix} x(k) \\ y(k) \\ \dot{x}(k) \\ \dot{y}(k) \end{bmatrix} = \begin{bmatrix} v_x k \\ v_y k \\ v_x \\ v_y \end{bmatrix}.$$

The system models are established as in previous examples. In particular,

$$\mathbf{F}(k) = \begin{bmatrix} 1 & 0 & \Delta T & 0 \\ 0 & 1 & 0 & \Delta T \\ 0 & 0 & 1 & 0 \\ 0 & 0 & 0 & 1 \end{bmatrix}$$

$$\mathbf{z}(k) = \begin{bmatrix} r(k) \\ \theta(k) \end{bmatrix} + \mathbf{v}(k)$$

$$= \mathbf{h}\left(\mathbf{x}(k), k\right) + \mathbf{v}(k)$$

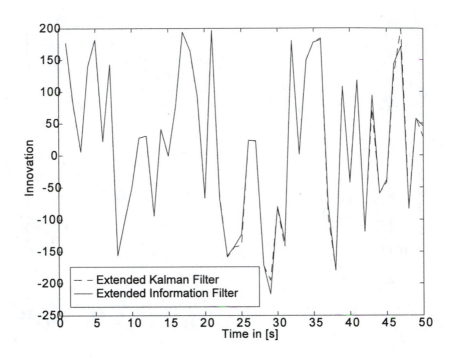

FIGURE 9.7
Innovations for EKF and EIF: Nonlinear State Evolution with Linear Observations

$$
= \begin{bmatrix} \sqrt{x_1{}^2(k) + x_2{}^2(k)} \arctan \left[\dfrac{x_2(k)}{x_1(k)} \right] \\[3mm] \arctan \left[\dfrac{x_2(k)}{x_1(k)} \right] \end{bmatrix} + \mathbf{v}(k)
$$

$$
\nabla \mathbf{h}_x(k) = \begin{bmatrix} \dfrac{x_1(k)}{\sqrt{x_1^2(k) + x_2^2(k)}} & \dfrac{x_2(k)}{\sqrt{x_1^2(k) + x_2^2(k)}} & 0 & 0 \\[4mm] \dfrac{-x_2(k)}{x_1^2(k) + x_2^2(k)} & \dfrac{x_1(k)}{\sqrt{x_1^2(k) + x_2^2(k)}} & 0 & 0 \end{bmatrix}.
$$

These system models are then used in the algorithms of the EKF and EIF to estimate the state vector of the aircraft's horizontal and vertical positions and velocities. The results are also presented and discussed later.

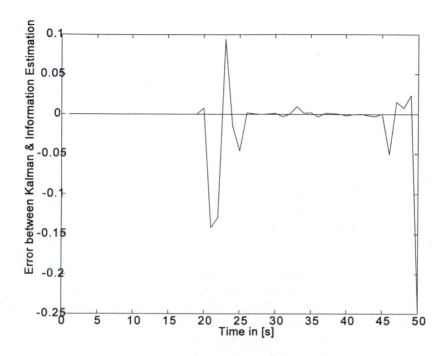

FIGURE 9.8
Difference Between EKF and EIF State Estimates: Nonlinear State Evolution with Linear Observations

9.6.3 Both Nonlinear State Evolution and Observations

A highly nonlinear system has nonlinearities in both system state evolution and observations. An example of such a system is a wheeled mobile robot (WMR) vehicle moving in a plane. The state vector of the vehicle at any time instant k is determined by its location and orientation such that

$$\mathbf{x}(k) = \begin{bmatrix} x(k) \\ y(k) \\ \phi(k) \end{bmatrix},$$

where $x(k)$ and $y(k)$ denote the WMR positions along the x and y axes of the plane, respectively, and $\phi(k)$ is the WMR orientation. Control is extended over the WMR vehicle motion through a demanded velocity $v(k)$ and direction of travel $\psi(k)$,

$$\mathbf{u}(k) = \begin{bmatrix} v(k) \\ \psi(k) \end{bmatrix}.$$

The motion of the vehicle can now be described in terms of the simple nonlinear state transition equation,

$$
\begin{bmatrix} x(k) \\ y(k) \\ \phi(k) \end{bmatrix} = \begin{vmatrix} x(k-1) + \Delta T v(k) \cos\left[\phi(k-1) + \psi(k)\right] \\ y(k-1) + \Delta T v(k) \sin\left[\phi(k-1) + \psi(k)\right] \\ \phi(k-1) + \Delta T \dfrac{v(k)}{B} \sin \psi(k) \end{vmatrix} + \begin{bmatrix} w_x(k) \\ w_y(k) \\ w_\phi(k) \end{bmatrix},
$$

where B is the wheel base line, ΔT is the time in travel between time steps, and $\mathbf{w}(k) = [w_x(k) \quad w_y(k) \quad w_\phi(k)]^T$ is the random vector describing the noise in the process due to both modeling errors and uncertainty in control. It is assumed that the vehicle is equipped with a sensor that can measure the range and bearing to a moving beacon with motion described by two parameters, $B_i = [x_i \quad y_i]^T$, such that x_i varies linearly with time, i.e., $x_i = 0.5k$. Assuming that the beacon is moving in circular motion of radius 10 units about the vehicle, then y_i is given by the expression $y_i = \sqrt{100 - x_i^2}$. The observation equations for the beacon are given by the nonlinear measurement model,

$$
\begin{bmatrix} z_r{}^i(k) \\ z_\theta{}^i(k) \end{bmatrix} = \begin{bmatrix} \sqrt{[x_i - x(k)]^2 + [y_i - y(k)]^2} \\ \arctan\left[\dfrac{y_i - y(k)}{x_i - x(k)}\right] - \phi(k) \end{bmatrix} + \begin{bmatrix} v_r(k) \\ v_\theta(k) \end{bmatrix},
$$

where the random vector $\mathbf{v}(k) = [v_r(k) \quad v_\theta(k)]^T$ describes the noise in the observation process. The system models are defined and established as before. In particular,

$$
\nabla \mathbf{f_x}(k) = \begin{bmatrix} 1 & 0 & -\Delta T v(k) \sin\left[\hat{\phi}(k-1 \mid k-1) + \psi(k)\right] \\ 0 & 1 & \Delta T v(k) \cos\left[\hat{\phi}(k-1 \mid k-1) + \psi(k)\right] \\ 0 & 0 & 1 \end{bmatrix}
$$

$$
\nabla \mathbf{h_x}(k) = \begin{bmatrix} \dfrac{\hat{x}(k|k-1) - x_i}{d} & \dfrac{\hat{y}(k|k-1) - y_i}{d} & 0 \\ -\dfrac{\hat{y}(k|k-1) - y_i}{d^2} & \dfrac{\hat{x}(k|k-1) - x_i}{d^2} & -1 \end{bmatrix},
$$

where $d = \sqrt{[x_i - \hat{x}(k|k-1)]^2 + [y_i - \hat{y}(k|k-1)]^2}$.

These system models are then used in the algorithms of the EKF and EIF to estimate the WMR vehicle's state vector $\mathbf{x}(k)$, that is, estimate the location and orientation of the vehicle. The results are also presented and discussed in the next section.

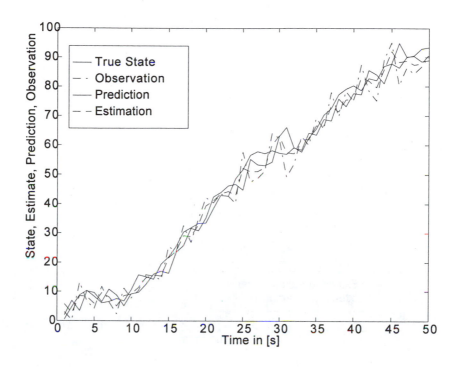

FIGURE 9.9
Linear State Evolution with Nonlinear Observations

9.6.4   Comparison of the EKF and EIF

The EKF and EIF are compared in the same way as the linear filters. However, the three nonlinear estimation examples, outlined in the sections above, are implemented for each filter. These examples were chosen to allow exhaustive investigation of nonlinearities in both system evolution and observations. In order to study and compare the performance of the filters, for each example, estimation of the same state is considered.

The general filter performance for these examples is shown in Figures 9.6 and 9.9. As was the case with the linear filters, the state observations, predictions, and estimates are identical for the EKF and EIF. The curves depicting the same variables are indistinguishable because they lie on top of each other. The same equivalence is observed for estimation errors and innovations. There are still slight differences between the two filters, attributable to numerical errors. As the number of nonlinearities increases, the errors tend to increase. However, even in the worst case, the errors are still bounded and inconsequential.

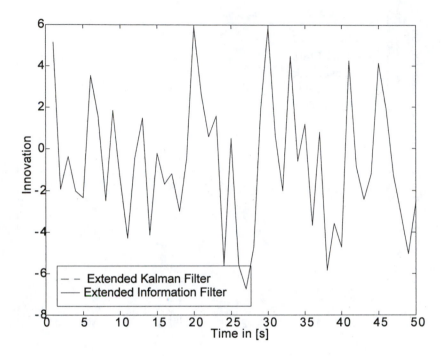

FIGURE 9.10
Linear State Evolution with Nonlinear Observations

The nature and trend of the difference between the EKF and EIF state estimates is shown in Figures 9.8 and 9.13. The errors are worse than those for linear systems because of the need to compute the Jacobians, $\nabla \mathbf{f}_x(k)$ and $\nabla \mathbf{h}_x(k)$. The Jacobians are not constants; they are functions of both time-step and state. As a result the covariances and system models must be computed online. This increases the number of computations performed and hence the numerical errors between the two filters are greater. The greater the complexity of the nonlinearities, the greater the number and complexity of the Jacobians, which leads to more computational costs. This tends to produce more numerical and rounding-off errors. In spite of these errors, the equivalence of the EKF and EIF is amply demonstrated in the three examples. This confirms the algebraic equivalence, which is mathematically proven and established in the derivation of the EIF from the EKF and the Information filter.

In terms of filter performance, both the EKF and EIF filters show unbiasedness, consistency, efficiency, and good matching. In all three examples, the state estimate is always well placed between the observation and state

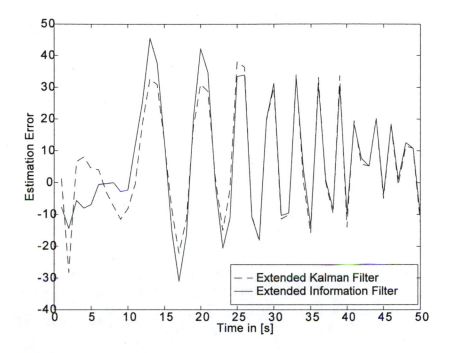

FIGURE 9.11
**Estimation Errors for EKF and EIF: Nonlinear State Evolution
with Nonlinear Observations**

prediction. This means there is balanced confidence in observations and predictions. By inspection and computing the sequence mean, the innovations (Figures 9.7, 9.10 and 9.12) are shown to be zero mean with variance $\mathbf{S}(k)$. There is no visible correlation of the innovations sequences and they satisfy the 95% confidence rule. However, in general, the performance of the EKF and EIF is not as good as that of the linear filters. This is because of the nontrivial nature of the Jacobian matrix computation and the general instability inherent in linearized filters.

9.6.5 Decentralized Estimation

These estimation techniques that have been developed and discussed form the basis of decentralized estimation. The notation and system description have been introduced and explained. Estimation theory and its use were discussed, in particular, the Kalman filter algorithm was outlined. The Information filter was then derived as an algebraic equivalent to the tra-

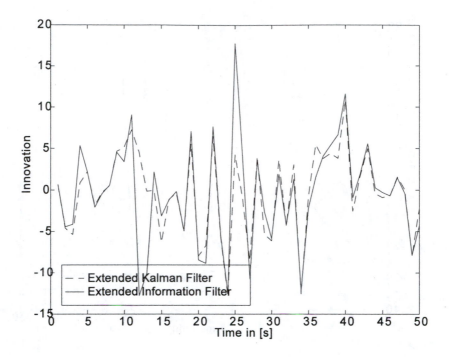

FIGURE 9.12
Innovations for EKF and EIF: Nonlinear State Evolution with Nonlinear Observations

ditional Kalman filter. Its attributes were outlined and discussed. The extended Kalman filter was then presented as a state-space solution to the estimation problem for a system characterized by both nonlinear system evolution and nonlinear measurements. The original and novel contribution of this chapter is the extended Information filter, EIF. This algorithm provides an estimation technique in extended information space for nonlinear systems. It was derived from first principles, explained, and appraised. It has all the attributes of the linear Information filter and fewer of the problems associated with the EKF. The simulated examples of estimation in linear and nonlinear systems validated the Information filter and EIF algorithms with respect to those of the Kalman filter and EKF. For the EIF and EKF, examples involving nonlinearities in both system evolution and observations were considered. The key benefit of information estimation theory is that it makes fully decentralized estimation for multisensor systems attainable [15].

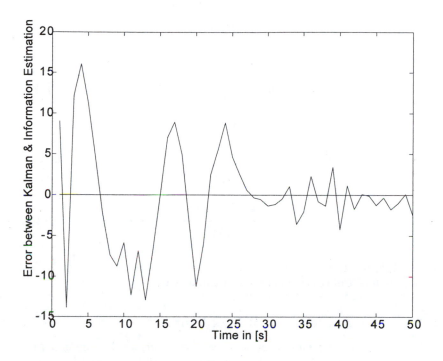

FIGURE 9.13
Difference Between EKF and EIF State Estimates: Nonlinear State Evolution with Nonlinear Observations

9.7 Optimal Stochastic Control

This section extends the estimation algorithms to the problem of sensor-based control by introducing stochastic control ideas. In particular, the LQG control problem and its solution are outlined. For systems involving nonlinearities, the nonlinear stochastic control problem is discussed. This section describes the *optimal stochastic control* problem and its solution. The practical design of stochastic controllers for problems described by the **LQG** assumptions, *Linear* system model, *Quadratic* cost criterion for optimality, and *Gaussian* white noise inputs are briefly discussed. Problems involving nonlinear models are then considered.

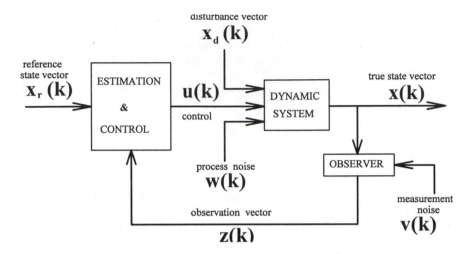

FIGURE 9.14
Stochastic Control System Configuration

9.7.1 Stochastic Control Problem

Most control problems of interest can be described by the general system configuration in Figure 9.14. There is some *dynamic system* of interest whose behavior is to be affected by an applied *control* input vector $\mathbf{u}(k)$ in such a way that the *controlled* state vector $\mathbf{x}(k)$ exhibits desirable characteristics. These characteristics are prescribed, in part, as the controlled state vector $\mathbf{x}(k)$ matching a *reference* state vector $\mathbf{x_r}(k)$ as closely and quickly as possible. The simplest control problem is in the LQG form and hence it is important to understand the meaning of the LQG assumptions.

- **Linear System Model**: Linearity is assumed where a linear system obeys the principle of superposition and its response is the convolution of the input with the system impulse response.

- **Quadratic Cost Function**: A quadratic cost criterion for optimality is assumed such that the control is optimal in the sense of minimizing the expected value of a quadratic performance index associated with the control problem.

- **Gaussian Noise Model**: White Gaussian noise process corruption is assumed.

Problem Statement

Let the system of interest be described by the n-dimensional stochastic discrete-time difference equation

$$\mathbf{x}(k) = \mathbf{F}(k)\mathbf{x}(k-1) + \mathbf{B}(k)\mathbf{u}(k-1) + \mathbf{D}(k)\mathbf{w}(k), \qquad (9.84)$$

where $\mathbf{u}(k)$ is the r-dimensional control input to be applied and $\mathbf{w}(k)$ is the zero mean white Gaussian discrete-time noise. The objective is to determine the control vector $\mathbf{u}(k)$ that minimizes the quadratic cost function

$$\mathbf{J}(N) = \mathrm{E}\left[\sum_{i=1}^{N}\left\{\mathbf{e}_r^T(k)\mathbf{X}(k)\mathbf{e}_r(k) + [\mathbf{u}(k) - \mathbf{u}_r(k)]^T\mathbf{U}(k)[\mathbf{u}(k) - \mathbf{u}_r(k)]\right\}\right],$$

where $\mathbf{e}_r(k) = [\mathbf{x}(k) - \mathbf{x}_r(k)]$ and $\mathbf{X}(k)$ is an n-by-n real and positive semi-definite cost weighting matrix, reflecting the relative importance of maintaining individual state component deviations at small values. $\mathbf{U}(k)$ is an r-by-r real, symmetric, and positive definite cost weighting matrix reflecting the relative importance of maintaining individual control component deviations at small values [14].

There are several reasons for the use of a quadratic cost function of states and control:

- Quadratics are a good description of many control objectives, such as minimizing mean squared error or energy.

- Inherently, such a function enhances the adequacy of the linear perturbation model.

- This combination of modeling assumptions yields a *tractable* problem whose solution is in the form of a readily synthesized, efficiently implemented, feedback control law.

9.7.2 Optimal Stochastic Solution

In this subsection, the solution to the LQG control problem outlined above is presented. Deterministic methods cannot be used to solve for an optimal control vector $\mathbf{u}(k)$ from the function $\mathbf{J}(N)$ because of the stochastic nature of the problem [14]. The dynamic driving noise term $\mathbf{w}(k)$ prevents perfect, ahead-of-time knowledge of where the system will be at time $(k+1)$. There is no single optimal history of states and controls, but an entire family of trajectories. Two closely related techniques are employed in determining an optimal stochastic control solution [14].

- **Optimality principle**: An optimal policy has the property that for any initial states and decision *(control law)*, all remaining decisions must constitute an optimal policy with regard to the state that results from the first decision.

- **Stochastic dynamic programming**: This is a technique of stepping backward in time to obtain optimal control. It is dependent on the *Markov* nature of the discrete-time process.

Two further structural properties are essential for the solution to be realized. These are *separation* and *certainty equivalence principles*. A control problem is said to be separable if its optimal control depends only on an estimate $\hat{\mathbf{x}}(k \mid k)$ of the state $\mathbf{x}(k)$ and not at all on the accuracy of the estimate. It is also said to be certainty equivalent if, being separable, the control is exactly the same as it would be in a related deterministic problem. The two principles imply that the problem of seeking a linear control law for a linear dynamical system with Gaussian measurement noise subject to a quadratic performance index can be cast in terms of two separate problems:

- Optimal deterministic control

- Optimal stochastic estimation

These two problems can be solved separately to yield an optimal solution to the combined problem. The optimal stochastic estimation problem has been solved in previous sections. The basis of these algorithms is the Kalman filter and its algebraic equivalent, the Information filter. Although only the information space algorithms are extended to stochastic control algorithms in this section, the state-space estimation algorithms can be similarly extended.

The cost minimizing control function is given by

$$\mathbf{u}(k) = -\mathbf{G}(k)[\hat{\mathbf{x}}(k \mid k) - \mathbf{x}_r(k)], \tag{9.85}$$

where $\mathbf{G}(k)$ is the associated optimal deterministic control gain. Its value is generated from the solution to the *Backward Riccati recursion* [14],

$$\mathbf{G}(k) = [\mathbf{U}(k) + \mathbf{B}^T(k)\mathbf{K}(k)\mathbf{B}(k)]^{-1}[\mathbf{B}^T(k)\mathbf{K}(k)\mathbf{F}(k)], \tag{9.86}$$

where $\mathbf{K}(k)$ is the n-by-n symmetric matrix satisfying the *Backward Riccati difference* equation [14],

$$\begin{aligned}
\mathbf{K}(k) &= \mathbf{X}(k) + \mathbf{F}^T(k)\mathbf{K}(k+1)\mathbf{F}(k) - \left[\mathbf{F}^T(k)\mathbf{K}(k+1)\mathbf{B}(k)\mathbf{G}(k)\right] \\
&= \mathbf{X}(k) + \left[\mathbf{F}^T(k)\mathbf{K}(k+1)\right]\left[\mathbf{F}(k) - \mathbf{B}(k)\mathbf{G}(k)\right]. \tag{9.87}
\end{aligned}$$

This equation is solved backward from the terminal condition, $\mathbf{K}(N+1) = \mathbf{X}_f(k)$. The *untracked* state estimate $\hat{\mathbf{x}}(k \mid k)$ is reconstructed from the *tracked* information estimate and the (information matrix),

$$\hat{\mathbf{x}}(k \mid k) = \mathbf{Y}^{-1}(k \mid k)\hat{\mathbf{y}}(k \mid k). \tag{9.88}$$

Solution Statement
The optimal stochastic control for a problem described by linear system models driven by white Gaussian noise, subject to a quadratic cost criterion,

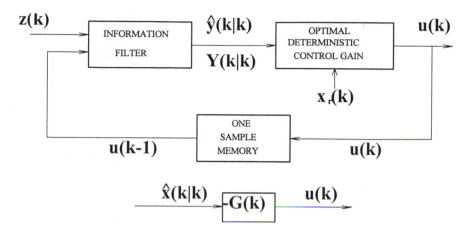

FIGURE 9.15
Optimal Stochastic Control

consists of an optimal linear Information filter cascaded with the optimal feedback gain matrix of the corresponding deterministic optimal control problem. This means the optimal stochastic control function is equivalent to the associated optimal deterministic control function with the true state replaced by the conditional mean of the state given the measurements. Illustration of this stochastic control solution is shown in Figure 9.15.

The importance of this result is the synthesis capability it yields. Under the LQG assumptions, the design of the optimal stochastic controller can be completely separated into the design of the appropriate Information filter and the design of an optimal deterministic controller associated with the original problem. The feedback control gain matrix is *independent* of all uncertainty, so a controller can be designed assuming that $\mathbf{x}(k)$ is known perfectly all the time. Similarly, the filter is *independent* of the matrices that define the controller performance measures. The estimation algorithm can thus be developed ignoring the fact that a control problem is under consideration.

Algorithm Summary

Estimation is carried out according to the Information filter Equations 9.36 and 9.37. The information estimate $\hat{\mathbf{y}}(k \mid k)$ is used to generate the state estimate and then the control signal.

$$\hat{\mathbf{x}}(k \mid k) = \mathbf{Y}^{-1}(k \mid k)\hat{\mathbf{y}}(k \mid k) \tag{9.89}$$

$$\mathbf{u}(k) = -\mathbf{G}(k)\left[\hat{\mathbf{x}}(k \mid k) - \hat{\mathbf{x}}_r(k \mid k)\right]. \tag{9.90}$$

The control law is generated as follows:

$$\mathbf{G}(k) = \left[\mathbf{U}(k) + \mathbf{B}^T(k)\mathbf{K}(k)\mathbf{B}(k)\right]^{-1} \left[\mathbf{B}^T(k)\mathbf{K}(k)\mathbf{F}(k)\right] \qquad (9.91)$$

$$\mathbf{K}(k) = \mathbf{X}(k) + \left[\mathbf{F}^T(k)\mathbf{K}(k+1)\right]\left[\mathbf{F}(k) - \mathbf{B}(k)\mathbf{G}(k)\right]. \qquad (9.92)$$

This is the optimal stochastic LQG control solution for single-sensor and single actuator-system. Before extending it to multisensor and multiactuator systems, the case of stochastic control problems with nonlinearities is considered.

9.7.3   Nonlinear Stochastic Control

The *separation* and *certainty* equivalence principles do not hold for nonlinear systems. Several methods have been employed in literature to attempt to solve this problem [14]. These include *linear perturbation control* (LQG direct synthesis), *closed-loop controller* ("dual control" approximation) and *stochastic adaptive control.*

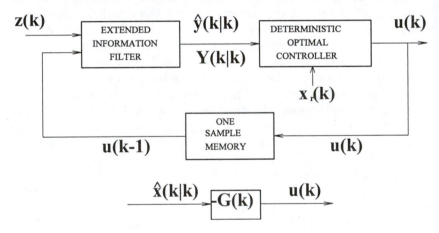

FIGURE 9.16
Nonlinear Stochastic Control

In this chapter, *assumed certainty equivalence* design is used. This is a synthesis technique that *separates* the stochastic controller into the cascade of an estimator and a deterministic optimal control function even when the *optimal* stochastic controller does *not* have the certainty equivalence property. It must be emphasized that, by definition, certainty equivalence assumes that the separation principle holds. Thus, the first objective is to solve the associated deterministic optimal control, ignoring the uncertainties and assuming perfect access to the entire state. Deterministic dynamic programming is used to generate the control law as a feedback law. The second objective is to solve the *nonlinear estimation* problem. This has already been done by deriving the EKF and EIF. In order to utilize the advantages of information space, the EIF is used.

Finally, the assumed certainty equivalence control law is computed by substituting the *linearized* information estimate from the EIF in the deterministic control law. This is the assumed certainty equivalence nonlinear stochastic control algorithm, illustrated in Figure 9.16. One important special case of this design methodology is the cascading of an EIF equivalent of a *constant gain* EKF to a *constant gain* linear quadratic state feedback controller. The constant gain EKF has the basic structure of an EKF, except that the constant gain is precomputed based on linearization about the *nominal* trajectory. This filter is robust against divergence. However, there is no fundamental reason to limit attention to constant gain designs other than computational burden of the resulting algorithms.

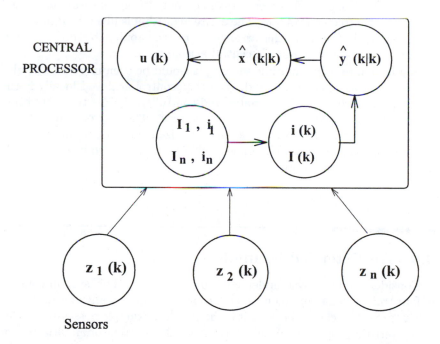

FIGURE 9.17
Centralized Control

Equipped with both *single* sensor *LQG* and nonlinear stochastic control algorithms, the next step is to extend them to multisensor and multiactuator control systems.

9.7.4   Centralized Control

Just as in the multisensor data fusion case, there are three broad categories of multiactuator control architectures: centralized, hierarchical and decentralized. A centralized control system consists of multiple sensors forming a *decentralized observer*. The control realization remains centrally placed, whereby information from the sensors is globally fused to generate a control law. Figure 9.17 shows such a system. Only observations are locally taken and sent to a center where estimation and control occurs centrally. The information prediction equations are the same as those of a single-sensor system, Equations 9.34 and 9.35.

Control Generation

Global information estimates are centrally computed from global information predictions and observations generated by the different sensors. The state estimate is reconstructed from the tracked central information vector and matrix. The control vector is then computed from the state error and globally generated control law. The entire algorithm is illustrated in Figure 9.17. The main feature of this arrangement is the ability to employ several sensors while retaining a single central actuator.

In addition to using multiple sensors, it would be even more beneficial if multiple actuators could be used, such that control achieved locally is the same as that achieved with a centralized controller. This is the motivation behind decentralized multisensor-based control. The approach adopted is to initially derive a *fully connected* decentralized control system, and then proceed to eliminate the full connection constraint to produce a scalable decentralized control system.

9.8   An Extended Example

Consider the mass system depicted in Figure 9.18. This system consists of four trolley masses interconnected by springs. Input in any one affects the other three. It is chosen because it is the simplest example possessing all the characteristic properties of an interconnected, multisensor, multiactuator dynamic system. It is used here to show how nodal transformations can be derived by using the system model $\mathbf{F}(k)$ to identify those states that are locally relevant.

First, the case is developed where the local states are unscaled, locally relevant, individual global states. The case of local states proportionally dependent on individual global states is then considered. Finally, the case of linear combinations of global states as local states is developed. The

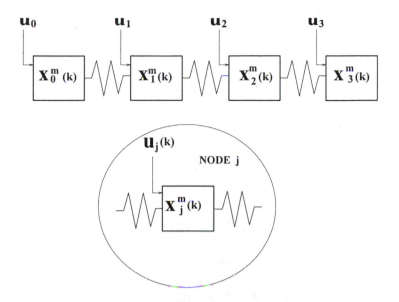

FIGURE 9.18
Coupled Mass System

results are then generalized to cover any coupled system.

9.8.1 Unscaled Individual States

The position and velocity of a general mass j are denoted by

$$\mathbf{x}_j^m(k) = \begin{bmatrix} x_j \\ \dot{x}_j \end{bmatrix}. \tag{9.93}$$

The global (central) state vector $\mathbf{x}(k)$, which consists of all the states of the system, can be defined as follows:

$$\begin{bmatrix} \mathbf{x}_0^m(k) \\ \mathbf{x}_1^m(k) \\ \mathbf{x}_2^m(k) \\ \mathbf{x}_3^m(k) \end{bmatrix} = \begin{bmatrix} x_0 \\ \dot{x}_0 \\ x_1 \\ \dot{x}_1 \\ x_2 \\ \dot{x}_2 \\ x_3 \\ \dot{x}_3 \end{bmatrix}. \tag{9.94}$$

Rearranging the states so that the positions and velocities are grouped together gives the global state vector

$$\mathbf{x}(k) = \begin{bmatrix} x_0 \\ x_1 \\ x_2 \\ x_3 \\ x_4 \\ x_5 \\ x_6 \\ x_7 \end{bmatrix}, \qquad (9.95)$$

where the last four states are the velocities, i.e.,

$$\begin{bmatrix} x_4 \\ x_5 \\ x_6 \\ x_7 \end{bmatrix} = \begin{bmatrix} \dot{x}_0 \\ \dot{x}_1 \\ \dot{x}_2 \\ \dot{x}_3 \end{bmatrix}. \qquad (9.96)$$

9.8.2   Continuous Time Models

Each of the four masses is conceptually decoupled and the forces operating on each is analyzed. As a result, the following equations are derived:

$$\ddot{x}_1 + (b/m_1)\dot{x}_1 + (k/m_1)(x_1 - x_2) = (u_1/m_1)$$
$$\ddot{x}_2 + (b/m_2)\dot{x}_2 + (k/m_2)(2x_2 - x_1 - x_3) = (u_2/m_2)$$
$$\ddot{x}_3 + (b/m_3)\dot{x}_3 + (k/m_3)(2x_3 - x_2 - x_4) = (u_3/m_3)$$
$$\ddot{x}_4 + (b/m_4)\dot{x}_4 + (k/m_4)(x_4 - x_3) = (u_4/m_4).$$

Rearranging the free body equations,

$$\dot{x}_1 = x_5$$
$$\dot{x}_2 = x_6$$
$$\dot{x}_3 = x_7$$
$$\dot{x}_4 = x_8$$
$$\dot{x}_5 = -(k/m_1)x_1 + (k/m_1)x_2 - (b/m_1)x_5$$
$$\dot{x}_6 = (k/m_2)x_1 - 2(k/m_2)x_2 + (k/m_2)x_3 - (b/m_2)x_6$$
$$\dot{x}_7 = (k/m_3)x_2 - 2(k/m_3)x_3 + (k/m_3)x_4 - (b/m_3)x_7$$
$$\dot{x}_8 = (k/m_4)x_3 - (k/m_4 x_4 - (b/m_4)x_8.$$

The continuous-time system models in the equation

$$\dot{\mathbf{x}}(t) = \mathbf{A}\mathbf{x}(t) + \mathbf{B}\mathbf{u}(t) + \mathbf{w}(t) \qquad (9.97)$$

are then given by

$$
A = \begin{bmatrix}
0 & 0 & 0 & 0 & 1 & 0 & 0 & 0 \\
0 & 0 & 0 & 0 & 0 & 1 & 0 & 0 \\
0 & 0 & 0 & 0 & 0 & 0 & 1 & 0 \\
0 & 0 & 0 & 0 & 0 & 0 & 0 & 1 \\
-k/m_1 & k/m_1 & 0 & 0 & -b/m_1 & 0 & 0 & 0 \\
k/m_2 & -2k/m_2 & k/m_2 & 0 & 0 & -b/m_2 & 0 & 0 \\
0 & k/m_3 & -2k/m_3 & k/m_3 & 0 & 0 & -b/m_3 & 0 \\
0 & 0 & k/m_4 & -k/m_4 & 0 & 0 & 0 & -b/m_4
\end{bmatrix}
$$

$$
B = \begin{bmatrix}
0 & 0 & 0 & 0 \\
0 & 0 & 0 & 0 \\
0 & 0 & 0 & 0 \\
0 & 0 & 0 & 0 \\
(1/m_1) & 0 & 0 & 0 \\
0 & (1/m_2) & 0 & 0 \\
0 & 0 & (1/m_3) & 0 \\
0 & 0 & 0 & (1/m_4)
\end{bmatrix}.
$$

Using the following numerical data,

$$
k_1 = k_2 = k_3 = k_4 = 50N/m
$$
$$
m_1 = m_2 = m_3 = m_4 = 1.0Kg
$$
$$
b_1 = b_2 = b_3 = b_4 = 0.1N/(m/s)
$$
$$
u_1 = u_2 = u_3 = u_4 = 10N,
$$

gives the following models:

$$
A = \begin{bmatrix}
0 & 0 & 0 & 0 & 1 & 0 & 0 & 0 \\
0 & 0 & 0 & 0 & 0 & 1 & 0 & 0 \\
0 & 0 & 0 & 0 & 0 & 0 & 1 & 0 \\
0 & 0 & 0 & 0 & 0 & 0 & 0 & 1 \\
-50 & 50 & 0 & 0 & -0.1 & 0 & 0 & 0 \\
50 & -100 & 50 & 0 & 0 & -0.1 & 0 & 0 \\
0 & 50 & -100 & 50 & 0 & 0 & -0.1 & 0 \\
0 & 0 & 50 & -50 & 0 & 0 & 0 & -0.1
\end{bmatrix}
$$

$$\mathbf{B} = \begin{bmatrix} 0\ 0\ 0\ 0 \\ 0\ 0\ 0\ 0 \\ 0\ 0\ 0\ 0 \\ 0\ 0\ 0\ 0 \\ 1\ 0\ 0\ 0 \\ 0\ 1\ 0\ 0 \\ 0\ 0\ 1\ 0 \\ 0\ 0\ 0\ 1 \end{bmatrix}.$$

9.8.3   Discrete Time Models

The state-transition matrix $\mathbf{F}(k)$ and the input control matrix $\mathbf{B}(k)$ are derived from the continuous-time model matrices $\mathbf{A}$ and $\mathbf{B}$. The state-transition matrix $\mathbf{F}(k)$ is computed by the series method (discussed in Chapter 7), where for linear time-invariant systems,

$$\mathbf{F}(k) = e^{\mathbf{A}(\triangle T)} = \mathbf{I} + \sum_{i=1}^{n} \{(\triangle T)^i \mathbf{A}^i\}/i!$$

$$\mathbf{B}(k) = \int_{k}^{k+1} B e^{\mathbf{A}\{ (k+1) - \triangle T\}}.$$

A discrete-time approximation can be applied if $\triangle T$ is sufficiently small compared with the *time constants* of the system.

$$\mathbf{F}(k) = \mathbf{I} + (\triangle T)\mathbf{A}$$
$$\mathbf{B}(k) = (\triangle T)\mathbf{B}.$$

For the mass system, both the approximation and the general method give the same results. This is because $\triangle T$, which was taken as 1.0 sec, is sufficiently small compared with the time constants of the system. The following system and observation models are obtained:

$$\mathbf{F}(k) = \begin{bmatrix} 1 & 0 & 0 & 0 & 1 & 0 & 0 & 0 \\ 0 & 1 & 0 & 0 & 0 & 1 & 0 & 0 \\ 0 & 0 & 1 & 0 & 0 & 0 & 1 & 0 \\ 0 & 0 & 0 & 1 & 0 & 0 & 0 & 1 \\ -50 & 50 & 0 & 0 & 0.9 & 0 & 0 & 0 \\ 50 & -100 & 50 & 0 & 0 & 0.9 & 0 & 0 \\ 0 & 50 & -100 & 50 & 0 & 0 & 0.9 & 0 \\ 0 & 0 & 50 & -50 & 0 & 0 & 0 & 0.9 \end{bmatrix}$$

$$\mathbf{B}(k) = \begin{bmatrix} 0\,0\,0\,0 \\ 0\,0\,0\,0 \\ 0\,0\,0\,0 \\ 0\,0\,0\,0 \\ 1\,0\,0\,0 \\ 0\,1\,0\,0 \\ 0\,0\,1\,0 \\ 0\,0\,0\,1 \end{bmatrix}$$

$$\mathbf{H}(k) = \begin{bmatrix} 1\,0\,0\,0\,0\,0\,0\,0 \\ 0\,1\,0\,0\,0\,0\,0\,0 \\ 0\,0\,1\,0\,0\,0\,0\,0 \\ 0\,0\,0\,1\,0\,0\,0\,0 \end{bmatrix}$$

$$\mathbf{R}(k) = \begin{bmatrix} 2 & 0 & 0 & 0 \\ 0 & 1.5 & 0 & 0 \\ 0 & 0 & 4 & 0 \\ 0 & 0 & 0 & 2.5 \end{bmatrix}.$$

9.9 Nonlinear Control Systems

In most of the material covered in the previous eight chapters, it has been assumed that the dynamics of systems to be controlled can be described completely by a set of linear differential equations and that the principle of superposition holds. Such systems are known as linear dynamic systems. However, in most applications, these assumptions are not valid, and the systems are termed nonlinear dynamic systems. The nonlinearity of dynamic systems can be inherent or deliberately added to improve the control action. The material in the next sections addresses the whole concept of nonlinear systems, their analysis, and control design.

9.9.1 Nonlinear Dynamic Systems

The main characteristic of nonlinear dynamic systems is their failure to follow the principle of superposition. That is, the combined output of two or more inputs is not necessarily equal to the resultant of the outputs due to the individual inputs. There are two reasons for this behavior to manifest. The foremost reason is that the dynamics of such a system is described by a set of nonlinear differential equations, and the other reason is the energy losses, response delays and size limitations in the system itself. Nonlinear

dynamic behavior that follows nonlinear differential equations can be easy
to handle if it can be modeled completely. However, the nonlinearities due
to energy losses, delays, and size limitations are normally difficult to model
precisely and hence, the control is quite difficult in some situations. In general, any nonlinear differential equation which describes nonlinear dynamics
can be decomposed into a linear differential equation and a hard nonlinear
element due to either energy loss, size limitation or response delays. In this
section, different hard nonlinearities are discussed. Mathematical models
for such nonlinearities will be given.

   Like linear systems, the analysis of nonlinear systems is mainly concerned
with the study of the dynamics of such systems, thereby identifying stability zones and other parameters of interest for control purposes depending
on the problem. In this section, three methods for the analysis of nonlinear
systems will be discussed. These are phase plane methods, describing function analysis methods and Lyapunov stability method. Nonlinear problems
usually arise because the structure or the fixed elements of the system are
inherently nonlinear. Another source of nonlinearities is nonlinear compensation introduced into the system for purposes of improving system
behavior. Figure 9.19 shows some examples of nonlinear elements.

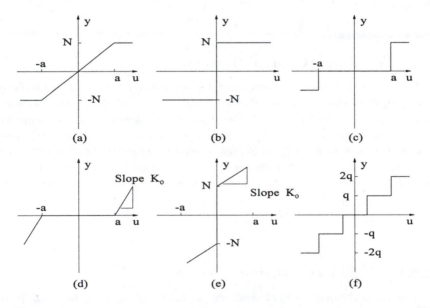

FIGURE 9.19
Examples of Nonlinear Elements

Figure 9.20 shows a nonlinear element in a control system.

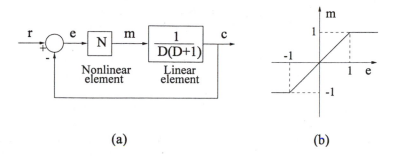

(a)   (b)

FIGURE 9.20
Nonlinear Element in a Control System

9.9.2 Describing Function Analysis

When studying the frequency response of linear dynamic systems in Chapter 6, the input was treated as a sinusoid. In that case the output was also a sinusoid with the same frequency as the input signal, but differing in amplitude and phase. For nonlinear systems subjected to sinusoid input, the output will be periodic but not sinusoid. Fourier expansion of such output gives an infinite number of sinusoids with frequencies that are harmonics of the input frequency. Using the advantage of the complex representation of the signals the describing function is defined as the ratio of the fundamental component of the nonlinear output and input signals.

The describing-functions technique is an extension of frequency-response methods to nonlinear systems. Classically, it is applicable only to nonlinear elements whose outputs in response to sinusoidal inputs with period T (i.e., of the form $A \sin 2\pi t/T$) are also periodic with period T. The output can then be written as a Fourier series:

$$\sum_{n=1}^{\infty} B_n \sin(n_\omega t + \phi).$$

The describing function is the ratio of the complex Fourier coefficient $B_1 e^{j\phi_1}$, which is essentially a frequency response function of an approximation of the nonlinear element. In general, B_1 and ϕ_1 are functions of both input frequency $\omega = 2\pi/T$ and the input amplitude A. Therefore, the following can be written: $B_1 = B_1(A, \omega)$ and $\phi_1 = \phi_1(A, \omega)$.

To apply the method, the describing functions must first replace the nonlinearities in the block diagram. Then the frequency domain techniques can be used to analyze the system, with some modification to account for the dependence of B_1 and ϕ_1 on A.

9.9.3   Phase Plane Methods

For any second-order system, there are two states x_1 and x_2. A plane formed by these states is called the phase plane. Given some initial conditions of the system dynamics, one can plot the variation of the two states in this phase plane. This plot is known as the phase motion trajectory of the system for the given initial condition. Different initial conditions result in different motion trajectories forming system phase portraits. Phase plane analysis is a graphical method that uses the information from the phase portrait of the system to analyze the dynamics of the system in question. It enables one to visualize the dynamics of the system without having to solve nonlinear differential equations analytically. One of the major advantages is that it can handle all types of nonlinearities, i.e., smooth, strong, and hard nonlinearities. On the other hand, however, it has the limitation that it cannot handle higher-order systems because of the computational complexity as well as the complexity associated with graphical presentation of higher-order systems. The only way phase plane methods can be applied to higher-order systems is by approximation of these systems to second-order equivalents. This section discuses the fundamentals of phase plane analysis, presenting the theoretical basics of the method that eventually lays a strong groundwork for understanding the stability of systems. The methods for plotting and interpretation of the system phase plots are presented.

9.9.3.1   Theory Behind Nonlinear Phase Plots

Although the phase plane methods are basically designed for second-orders systems only, the material in this section has been presented to assume any system (even a higher-order system) that can at most be presented using only two states, as for second-order systems. This makes it possible to apply the method even to higher-order ones without the need to approximate them, providing the controllable states of interest are only two. The study of nonlinear phase plane basics starts by the discussion of time dependency structure of nonlinear systems.

The time dependence of nonlinear systems is classified as being either autonomous or non-autonomous, depending on the appearance of the time t in its system differential equation. For an autonomous system, the time t does not appear explicitly in its differential equation. It is only the time differential dt that appears and such a system is normally described by a nonlinear function, say f, as

$$\ddot{x} = f(x, \dot{x}). \tag{9.98}$$

Non-autonomous systems have direct dependence on the time and as such, the time t appears directly in their differential equation as

$$\ddot{x} = f(t, x, \dot{x}). \tag{9.99}$$

Phase plane analysis of non-autonomous systems is complex because of the need to add an extra axis for the time element. For generality, however, non-autonomous systems can also be handled as autonomous systems by expressing them as

$$\ddot{x}(t) = f\left(x(t), \dot{x}(t)\right). \tag{9.100}$$

In the following discussion, the assumption is made that the system is autonomous, with the understanding that non-autonomous systems can be handled this way. Consider the system described by Equation 9.98. With choice of the state variables as

$$x_1 = x$$
$$x_2 = \dot{x}.$$

This system can be represented in state-variable form as

$$\dot{x}_1 = x_2$$
$$\dot{x}_2 = f(x_1, x_2), \tag{9.101}$$

or in general form as

$$\dot{x}_1 = f_1(x_1, x_2)$$
$$\dot{x}_2 = f_2(x_1, x_2). \tag{9.102}$$

The plane describing the relationship between the two states with coordinate axes x_1 and x_2 is known as the phase plane. It can be regarded as a special case of the state-space with two states only. A point in the phase plane represents the state of the system. Now, as the system state changes, a point in the phase plane moves, thereby generating a curve that is known as the state (or phase) motion trajectory, or simply the trajectory. The time rate at which the state changes its position is called the state velocity or trajectory velocity. A family of state trajectories corresponding to various initial conditions in one plane is called the phase portrait of the system.

Phase plane analysis is based on the observation that the slope of the trajectory through any point is given by the expression

$$\frac{dx_2}{dx_1} = \frac{dx_2}{dt}\frac{dt}{dx_1}$$

$$= \frac{\dot{x}_2}{\dot{x}_1} \qquad\qquad (9.103)$$

$$= \frac{f_2(x_1, x_2)}{f_1(x_1, x_2)} = \frac{f(x_1, x_2)}{x_2}$$

with the following properties

$$\frac{dx_1}{dt} > 0 \qquad \text{for } x_2 > 0$$

$$\frac{dx_1}{dt} < 0 \qquad \text{for } x_2 < 0. \qquad (9.104)$$

Thus, motion trajectories are directed to the right in the upper half plane and to the left in the lower half plane. Typical phase plane trajectories are shown in Figure 9.21.

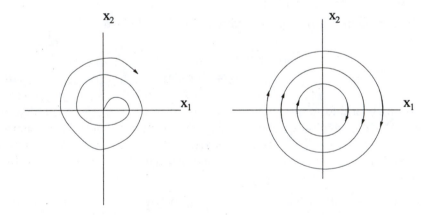

FIGURE 9.21
Typical Phase Plane Trajectories

For the autonomous system in 9.98, one can easily observe that if x_1 represents the system position, then x_2 represents the system velocity so that the system acceleration can thus be given as a function of both the position and velocity, i.e.

$$\ddot{x} = \dot{x}_2 = f(x_1, x_2). \qquad (9.105)$$

Based on these observations then, it can be inferred that the gradient of the motion trajectory in the phase plane gives the ratio of the system acceleration to velocity. As such, horizontal trajectories for which the gradient is zero either represents infinity velocity or zero acceleration. However, because of the physical difficulty of realizing an infinity velocity, it is generally accepted that the horizontal trajectories represent zero acceleration, i.e., constant velocity. In most systems, zero acceleration occurs at either the vertical turning points or at the origin when all quantities are zero. Similarly, vertical trajectories that might represent either infinity acceleration or zero velocity are taken to represent zero velocity, which also occurs either at the origin or at the horizontal turning point. It can be seen that the origin has dual characteristics as far as the system motion is concerned. This makes the origin a special point in phase plane analysis such that a detailed discussion of the system behavior at the origin is given in later sections.

It should be noted with great care that the notion of *system position and velocity* discussed here is different from that of the *state position and velocity*, which will later be used to describe the system equilibrium and hence stability. While the system velocity and position can have a physical quantity associated with them, the state position and velocity are graphical quantities that do not necessarily need to have physical quantities associated with them.

9.9.3.2 Techniques for Plotting Phase Trajectories

If the variables x_1 and x_2 in Equations 9.101 and 9.102 can be separated, the normal integration methods can give explicitly the motion trajectories. However, for most nonlinear systems, this is not a realistic possibility, hence graphical or semi-graphical construction (geometric) methods are normally used. The common graphical methods used are the Reinhardt construction method, the method of isoclines and the method of isocentrics. In this section, each of these methods is discussed.

9.9.3.2.1 The Reinhardt Construction Method

The Reinhardt construction method applies when a system can be described as

$$f(x, \dot{x}) = -\lambda x - g(\dot{x}) \tag{9.106}$$

so that the trajectory gradient becomes

$$\frac{dx_2}{dx_1} = \frac{-x_1 - \left(\frac{1}{\lambda}\right) g(x_2)}{x_2}. \tag{9.107}$$

The first step under this method is to identify and plot out all points of zero trajectory gradient by setting

$$-x_1 - \left(\frac{1}{\lambda}\right) g(x_2) = 0$$

or simply

$$x_1 = -\left(\frac{1}{\lambda}\right) g(x_2). \tag{9.108}$$

These points, as it has been stated before, are the vertical turning points of the trajectories (i.e., maxima and minima). Once this is done, the next step is to identify the directions of all trajectories in the phase plane, i.e., the direction field. Generally this is a geometrical process whose description is as shown in Figure 9.22.

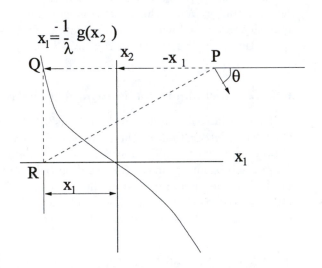

FIGURE 9.22
Description of Geometrical Process

Consider point $P(x_1, x_2)$ on the phase plane. The gradient of the trajectory that passes through this point is given by Equation 9.107. This is the negative tangent of the angle θ in the fourth quadrant (see Figure 9.22). According to Reinhardt (which can be easily verified from the above figure), this angle θ equals angle QRP if a line is drawn from R on the x_1- axis to point P, making a right angle with the locus of zero gradient.

Therefore, the following relations are valid

$$\tan(\theta) = \tan \angle QRP$$
$$= \frac{PQ}{QR}$$
$$= \frac{-x_1 - \left(\frac{1}{\lambda}\right) g(x_2)}{x_2}. \tag{9.109}$$

Thus, by a pair of compasses, it is possible to construct all direction fields using short segments for all points $P(x_1, x_2)$ as small arcs of a circle whose centre R is given as

$$R = x_1 = -\left(\frac{1}{\lambda}\right) g(x_2)$$

and radius RP is

$$RP = \sqrt{\left[x_1 + \left(\frac{1}{\lambda}\right) g(x_2)\right]^2 + x_2^2}. \tag{9.110}$$

By this approach for all points of interest in the phase plane, the whole plane will be covered by short line segments all pointing in the directions of the trajectories. Depending on their density, by joining these segments the motion trajectory can be traced as shown in Figure 9.22.

Method of Isoclines: In the phase plane, isoclines are lines of constant trajectory gradient. The method of isoclines, as the name implies, traces the phase portrait of the system by identifying the points of the same gradient. This method can be seen to be more general than that of Reinhardt, but it involves more computations than the Reinhardt construction method. It applies to any general system of the form

$$\ddot{x} = f(x, \dot{x}).$$

The method of isoclines sets the gradient constant by assigning it a definite number, say α, and then solving the gradient equation for the curve.

$$\frac{dx_2}{dx_1} = \frac{f(x_1, x_2)}{x_2}$$
$$= \phi(x_1, x_2) = \alpha \tag{9.111}$$

This will give all points at which the trajectory gradient is α.

Practically, this is done by determining the gradients at every point $P(x_1, x_2)$ in the phase plane, and then drawing a line segment that points in the direction corresponding to the gradient at the point. Finally, the whole phase plane will be covered by line segments that represent direction fields, and the trajectory can be traced out by joining these segments.

Method of Isocentrics: By definition, isocentric circles are circles with a common centre. Closely related to this definition, the method of isocentrics is inspired by the Reinhardt construction, where it was observed that whenever the trajectory crosses any point $P(x_1, x_2)$ in the phase plane, it is tangential to a clockwise arc of centre $R(c, 0)$ where

$$c = x_1 + \left(\frac{1}{\lambda}\right) g(x_2).$$

The isocentric method generates the direction fields in the phase plane by using this equation for a constant value of c. As can be seen, this method is just a variant of Reinhardt's construction method, but it is simpler than both the Reinhardt method and the method of isoclines. When using this method, the radius of construction is set so that

$$R = \sqrt{[x_1 - c]^2 + x_2^2},$$

where the centre $(c, 0)$ is determined in advance.

9.9.3.3   Equilibria and Singular Points in the Phase Plane

Before delving further into the subject of this section, the difference between the system velocity and the state velocity is clarified first. In the previous section, it was mentioned that, for a second-order system,

$$\ddot{x} = f(x, \dot{x}).$$

The quantities x and $\dot{x}$ are referred to as the system position and velocity respectively, while $\ddot{x}$ is the system acceleration. The system position could be temperature and the corresponding system velocity becomes the rate of change of this temperature. Now, to introduce the idea of the state velocity, recall that it has been defined before that the state (or trajectory) velocity is the rate at which the system changes its states. No physical meaning can be attached to the state velocity, rather than that it is a vector sum of the rates of change of the system states. Since, in most cases, the states have different physical meanings, and hence different units, the state or simply trajectory velocity cannot have proper units. It is just a parameter introduced for analytical purposes because it is more useful in describing the system stability than the system parameters.

The state velocity $\underline{v}$ is given as the vector sum of the system states

$$\underline{v} = \underline{\dot{x}}_1 + \underline{\dot{x}}_2$$

thus the magnitude of this velocity, normally known as the state speed, is

$$v = \sqrt{\dot{x}_1^2 + \dot{x}_2^2}.$$

When the magnitude of this state speed is zero, which means the system states are not changing, the equilibrium point is said to be reached. In that situation, both the system velocity and the system acceleration are zero, so that the states remain stationary at one point in the phase plane. By virtue of this definition, one can see that the equilibrium point must occur at least along the vertical axis, i.e.,

$$x_2 = \dot{x}_1 = 0.$$

And since the system acceleration can be inferred from the trajectory gradient, which has been defined before as the ratio of the system acceleration and system velocity, i.e.,

$$\frac{dx_2}{dx_1} = \frac{f(x_1, x_2)}{x_2}$$

The system acceleration can then be written as

$$\dot{x}_2 = f(x_1, x_2) = x_2 \frac{dx_2}{dx_1}, \tag{9.112}$$

which indicates that for the system acceleration $\dot{x}_2$ to be zero, then either the system velocity x_2 is zero or the trajectory gradient is zero. For most analytical applications, the former condition is normally assumed, because even if the latter is the prevailing condition, it still can be transformed to the former condition by coordinate transformation procedures. From the observations given above, the system equilibrium point is defined to be at the origin.

Another useful parameter in the phase plane plots is related to the continuity of the trajectory. All points in the trajectory for which the gradient is well defined according to Equation 9.103 are non-singular. Singular points are the ones for which the gradient is indeterminate. In most cases, these refer to discontinuities in the phase trajectory, but the other singular occurs when both x_2 and $f(x_1, x_2)$ are zero, in this situation, the trajectory gradient

$$\frac{dx_2}{dx_1} = \frac{f(x_1, x_2)}{x_2} = \frac{0}{0}. \tag{9.113}$$

is indeterminate (neither zero, infinity nor any number). It can easily be observed that this point also corresponds to the origin of the phase plane, which, as has been seen before, is the equilibrium point of the system. Therefore, it can be concluded that the origin is both a singular point and an equilibrium point of the system. For some systems, there might be more than one equilibria and singular points, but it is only the origin that is both the equilibrium and the singular point of the system. In dynamic systems, stability depends highly on both the equilibrium and singular points, and for that reason, the origin makes a good operating point for most systems because it has both the singularity and equilibria characteristics. In the next section, the relation between the system equilibrium and stability are discussed.

9.9.3.4   Stability of Equilibria in Phase plane

To be able to establish the stability of a nonlinear system using phase plane analysis, it is worth studying linear systems first and then extending the results to nonlinear systems. This approach is necessitated by the ease with which linear systems can be understood.

Phase Plane Analysis of Linear Systems (Overview): Consider an autonomous second-order linear system with the general form

$$\ddot{x}(t) + a\dot{x}(t) + cx(t) = 0, \tag{9.114}$$

together with some initial conditions $\dot{x}(0) = \dot{x}_o$ and $x(0) = x_o$, where $\dot{x}_o$ and x_o are constants. By choosing states $\dot{x}(t)$ and $x(t)$, this system can be modeled in state-variable form as

$$\dot{\mathbf{x}}(t) = \mathbf{A}\mathbf{x}(t)$$
$$\mathbf{x}(0) = \mathbf{x}_o.$$

The solution for such a system was discussed in Chapter 7 and is given by

$$\mathbf{x}(t) = e^{\mathbf{A}t}\mathbf{x}_o,$$

which, in component form is

$$x_1(t) = e^{\lambda_1 t} x_o$$
$$x_2(t) = e^{\lambda_2 t} \dot{x}_o.$$

where λ_1 and λ_2 are eigenvalues of the system matrix. In the phase plane, the gradient of the trajectory can be determined using Equation 9.103, which gives

$$\frac{dx_2}{dx_1} = \frac{\dot{x}_o \lambda_2}{x_o \lambda_1} e^{(\lambda_2 - \lambda_1)t}. \tag{9.115}$$

Thus, this system has only one singular and equilibrium point, which is the origin when both initial states are zero.

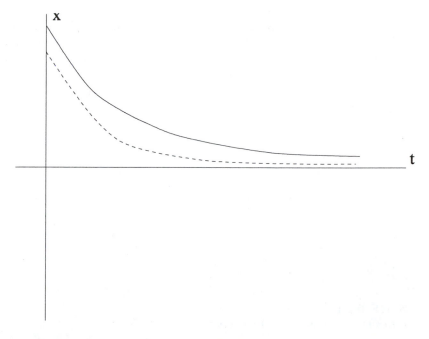

FIGURE 9.23
Both States Increasing or Decreasing

In the vicinity of the equilibrium point, the states (solutions of state-variable equation) can take any of the following forms, depending on the eigenvalues λ_1 and λ_2 of the system matrix A:

- Both states are exponentially increasing or decreasing corresponding to the fact that both eigenvalues of A are real with the same sign, (positive or negative). This situation is depicted in Figure 9.23.

- One state is exponentially increasing while the other is exponentially decreasing. This corresponds to the situation when the eigenvalues are real with opposite sign.

- Both oscillate with decaying or growing amplitude. This situation corresponds to the eigenvalues of A being complex conjugate with a non-zero real part. Growing amplitude indicates that the real part of the eigenvalue is positive while a decaying amplitude occurs when the real part is negative. Figure 9.25 depicts the variation of the states with time for this case.

- Both states oscillating with a constant amplitude, but maybe with different phase angles, as shown in Figure 9.26 This case happens when the eigenvalues are complex conjugate with zero real-part.

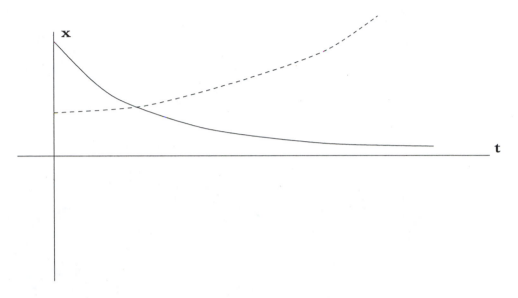

FIGURE 9.24
Exponential Increase or Decrease of State

As has been shown in previous chapters, the system is said to be stable when the real parts of the eigenvalues of A are all negative corresponding to one of the possibilities in the first and the third of these four cases. The fourth case corresponds to the marginal (neutral) stability of the system.

Nodes, Foci, Saddle, and Centre Points: When both eigenvalues of *A* are real and have the same sign, the equilibrium point is called a node. The node has such characteristics that all trajectories either start or end there, depending on the sign of the eigenvalues of A. If both eigenvalues are positive, the trajectories start from the node and end at infinity, indicating a growing magnitude of the state. Such a node is said to be an unstable node. On the other hand, if the eigenvalues are negative, the trajectories starting anywhere within the vicinity of the node will end at that node, indicating decay in magnitude. Such a node is said to be a stable node. Because of the direction of the trajectories relative to the node, the stable node is also known as an attractive node, while an unstable node is called a repulsive node. Figure 9.27 shows the typical phase portraits of a stable and an unstable node.

Saddle Points: A saddle point corresponds to the case when the matrix *A* has eigenvalues of different signs. A saddle point is characterized by the antagonizing nature of the states. To get a clear understanding of the saddle points, and also for other cases of phase planes, it is imperative to define

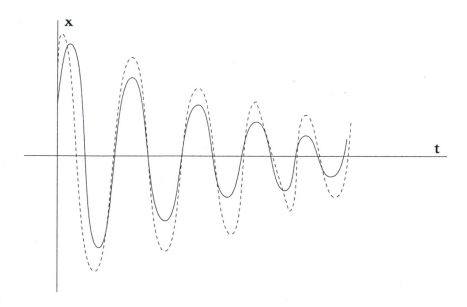

FIGURE 9.25
Decaying Amplitude

the phase angle ϕ as

$$\phi = \tan^{-1}\left(\frac{x_2}{x_1}\right), \tag{9.116}$$

i.e., the tangent of the angle substended by the two states of the system at a particular point and the origin.

Foci: When eigenvalues of A are complex conjugate, the corresponding trajectory has a spiral shape with the spiral center at the equilibrium point, indicating the changing signs as well as the magnitudes of the states. This corresponds to the sinusoidal nature of the states. There are two possibilities for the shape of this spiral, depending on the sign of the real part of the eigenvalue. When the real part is positive, then the magnitude of the states will be increasing exponentially, hence, the spiral will be such that the trajectories start from the equilibrium and run outward to infinity. This situation represents an unstable focus. The second case is when the real part is negative, in which case, the amplitude of the states will be decaying and hence the trajectories run from infinity to the equilibrium point. Similarly, to stable and unstable nodes, the foci can also be described as repulsive or attractive.

Centre Points: For a system with complex conjugate eigenvalues, with zero real parts, the states are purely sinusoidal with constant amplitudes and hence, the trajectories are cyclic. In a general form, they will be elliptic

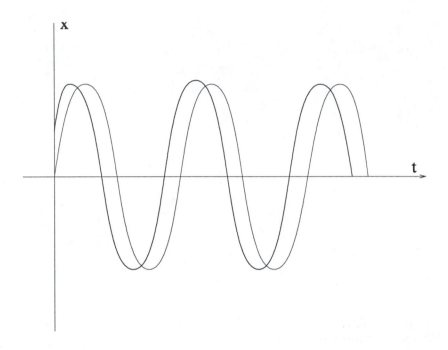

FIGURE 9.26
Constant Amplitude

with major and minor axes corresponding to initial conditions, and they
become circular when both sides have the same initial conditions. In this
case, the equilibrium point is the centre of the elliptic (circular) trajectories,
and hence the name, center point.

With these characteristics, it is possible to describe the stability behavior
of the linear system completely where a stable equilibrium corresponds to
either a stable focus or a stable node. A center point is an equilibrium point
for a uniformly stable system such as an undamped spring-mass system.

9.9.3.5 Nonlinear Systems

Having seen the phase plane portraits of linear systems, attention should
be turned to nonlinear systems, which are the focus of this chapter. Nonlinear phase plane analysis can be carried out by considering only a small
range from the equilibrium point where the nonlinear system can be approximated as a linear one. For this reason, it is sometimes known as local
analysis. This is basically the equivalent of nonlinear linearization, which
will be discussed later. Far from the equilibrium points, the nonlinear system may display other features that are not common to linear systems.
The main features of nonlinear phase planes that are not found in linear

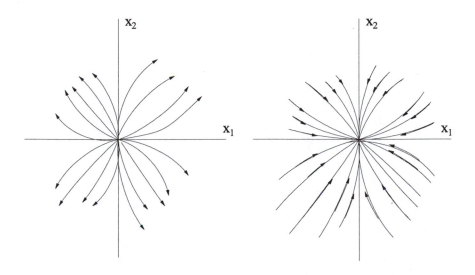

FIGURE 9.27
A Stable Node and an Unstable Node

systems are limit cycles and the presence of multiple equilibrium points.

9.9.3.6 Systems with Multiple Equilibrium Points

Consider a system described by

$$\ddot{x} = f(x) \tag{9.117}$$

where f is a nonlinear function. Since

$$\ddot{x}dx = \dot{x}d\dot{x}$$

the nonlinear Equation 9.117 can thus be written as

$$\dot{x}d\dot{x} = f(x)dx$$

so that, on integration, it yields

$$\frac{1}{2}\dot{x}^2 = \int_0^x f(x)dx + \text{constant} \tag{9.118}$$

or

$$\frac{1}{2}\dot{x}^2 - \int_0^x f(x)dx = \text{constant}. \tag{9.119}$$

By denoting $\frac{1}{2}\dot{x}^2$ as the kinetic energy per unit mass and $\int_0^x f(x)dx$ as the potential energy, it can be easily seen that this system is a conservative

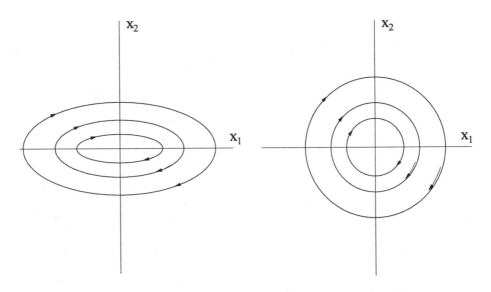

FIGURE 9.28
Nonlinear Phase Plane

system in which the total energy is constant, depending on the initial conditions. Different initial conditions together result in a phase plane portrait of the shape shown in Figure 9.29 which is characterized by the presence of multiple equilibrium points. The different equilibrium points that can be seen in the phase portrait of Figure 9.29 include the center points and the saddle points. There are many cases where the phase portrait will exhibit multiple equilibrium points, however, this typical case presented here is just enough to serve as an illustration of the subject.

9.9.3.7 Systems with Limit Cycles.

The limit cycle is the most interesting and unique feature for most nonlinear systems. It was first observed by B. Van der Pol (1926) during his study of vacuum tube oscillators, which lead to a nonlinear equation known as van der Pol's equation

$$\ddot{x} - \mu(1 - x^2)\dot{x} + x = 0. \tag{9.120}$$

The phase portrait for such a nonlinear system, shown in Figure 9.29, has one feature that all trajectories, irrespective of their initial conditions, converge to one closed cycle known as the limit cycle. For different values of μ, the shape of this phase portrait and hence, the limit cycle change, but maintain the system position x_1 to some value close to 2. When $|x_1| < 1$, the dynamics of such a system will keep x_1 increasing (corresponding to

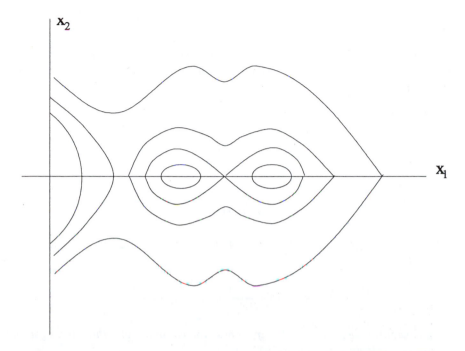

FIGURE 9.29
Multiple Equilibrium Points

positive real eigenvalues) until $|x_1| = 1$. For $|x_1| > 1$ the amplitude will decrease with time and reach a point where there is no further decrease, but also no increase. Hence, it remains trapped in this cycle, which is the limit cycle. On the other hand, if $x \gg 1$ (say far beyond 2), the amplitude will decrease up to some point where it cannot decrease any more, but cannot increase, i.e., the limit cycle. Therefore, it can be seen that trajectories inside the circle and those outside the circle all tend to this limit cycle. Trajectories which started on this limit cycle remain there circling periodically about the origin forever. It should be noted that van der Pol's system is just one of the many nonlinear systems that exhibit limit cycle phenomenon. For one to determine whether the system has limit cycles, there have been many proposed theorems, some of which are discussed next.

9.9.3.8   Existence of Limit Cycles

The existence or non-existence of limit cycles is normally summarized in a number of theorems as stated before. There are many theorems for the same purpose, however, this section will present only three of them. For further information, it is recommended to consult specialized texts on nonlinear control theory.

9.9.3.9   Bendixson's Theorem

Bendixson's theorem presents a sufficient condition to guarantee that a given nonlinear system does not have limit cycles. The theorem is stated as follows:

Theorem 9.1 *If a nonlinear second-order autonomous system*

$$\ddot{x} = f(x, \dot{x})$$

or in state-variable form

$$\dot{\mathbf{x}} = \mathbf{f}(\mathbf{x})$$

where $\mathbf{x}=[x_1 \ x_2]^T$ *and* $\mathbf{f}=[f_1 \ f_2]^T$ *is defined over a region* Ω *such that the quantity*

$$\nabla \mathbf{f}(\mathbf{x}) = \frac{\partial f_1(x_1, x_2)}{\partial x_1} + \frac{\partial f_1(x_1, x_2)}{\partial x_2} \tag{9.121}$$

is not identically zero over Ω *and does not change sign, then the system phase portrait does not have limit cycles.*

The proof of this theorem requires some knowledge of the calculus of vector fields, in particular line and surface integrals, and is done as follows: First, it is noted that, since the gradient of the trajectory is

$$\frac{dx_2}{dx_1} = \frac{f_2(x_1, x_2)}{f_1(x_1, x_2)},$$

then the equation

$$f_2(x_1, x_2)dx_1 - f_1(x_1, x_2)dx_2 = 0, \tag{9.122}$$

is satisfied for all trajectories of the system including the limit cycle. Suppose J is a limit cycle in this phase plane, i.e., is a closed trajectory, then for each point $P(x_1, x_2)$ on J the vector field $\mathbf{f}(\mathbf{x})$ is tangent to J. therefore, the closed-loop line integration

$$\int_J f_2(x_1, x_2)dx_1 - f_1(x_1, x_2)dx_2 = 0, \tag{9.123}$$

over J or simply

$$\int_J \mathbf{f}(\mathbf{x}).\mathbf{n}(\mathbf{x})dl = 0, \tag{9.124}$$

where $\mathbf{n}(\mathbf{x})$ is an outward normal to J at (x_1, x_2). By the divergence theorem, this can be expressed as

$$\int_J \mathbf{f}(\mathbf{x}).\mathbf{n}(\mathbf{x})dl = \int\int_S \nabla\mathbf{f}(\mathbf{x})\mathbf{dx} = 0 \qquad (9.125)$$

where S is the area enclosed by J. In order for this to hold, either $\nabla\mathbf{f}(\mathbf{x})$ must be identically zero over S or else $\nabla\mathbf{f}(\mathbf{x})$ must change sign. Now if S is a subset of Ω, then the hypothesis of the theorem rules out both possibilities.

9.9.3.10 The Poincare-Bendixson Theorem

This theorem can be used to prove the existence of the limit cycles, provided a domain of the phase plane that satisfies certain conditions can be found. Basically, this is concerned with asymptotic behavior of trajectories of second-order systems. It states as follows:

Theorem 9.2 *If a trajectory of a second-order autonomous system remains in a finite region Ω, then one of the following is true: 1)The trajectory goes to the equilibrium point 2). The trajectory tends to an asymptotically stable limit cycle. 3)The trajectory is itself a limit cycle.*

The proof of this theorem is omitted here as is the case with many other standard textbooks because of its mathematical involvement with algebraic topology, which is beyond the scope of this book. However, roughly speaking, this theorem states that if one can find a closed region Ω in the phase plane such that Ω does not contain any equilibria and such that all points of the trajectory are contained in Ω, then either the trajectory itself is a limit cycle or the trajectory goes to an equilibrium point or an asymptotically stable point.

9.9.3.11 The Poincare Theorem

Poincare theorem presents a simple relationship between the existence of limit cycles and the number of equilibrium points it encloses. It states that:

Theorem 9.3 *If N is the number of nodes, centres, and foci enclosed by a limit cycle, and S is the number of saddle points, then if the limit cycle really exists in the second-order autonomous system then*

$$N = S + 1. \qquad (9.126)$$

Like the previous one, this theorem is also mathematically involving and is not proved here. It is just one of the many theorems that postulate the presence of limit cycles, which together are known as index theorems. A number of index theorems can be found in literature on the subject.

9.9.4   Lyapunov's Theory

Lyapunov's theory addresses mainly the question of stability of any system, be it linear or nonlinear. It gets more application in nonlinear systems because it is the only tool available for the purpose where the other methods discussed in the previous chapters fail to work. It is well known that for any stable system, there is always a point where the motion of such a system will converge after some disturbance, whereas there is none for an unstable one, thereby resulting in motion divergence even under very slight disturbances. In dealing with stability problems, Lyapunov's stability theory appears in two versions, the Lyapunov linearization method (first method) and the Lyapunov direct method (second method). Before delving into Lyapunov's stability theories, the basics behind the theory are introduced.

9.9.4.1   The Basics of Lyapunov's Theories

The question of stability can be better described in terms of the motion trajectory and equilibrium points in the phase plane as discussed before. If any point in the phase plane is taken, one can identify the state $\mathbf{x}(t)$ as being stable or unstable depending on certain conditions, as described in the following discussion. First, consider linear systems. The reader is assumed to be familiar with the mathematical notation used here because of the mathematical nature of the topic itself, as they simplify the presentation. These symbols can be revisited in the list of symbols at the beginning of this book.

Recall that a linear system

$$\dot{\mathbf{x}}(t) = \mathbf{A}(t)\mathbf{x}(t)$$
$$\mathbf{x}(t) \in \mathbb{R}^n \tag{9.127}$$

has a solution

$$\mathbf{x}(t) = \Phi(t_0, t)\mathbf{x}(t_o) \tag{9.128}$$

where $\Phi(t_0, t)$ is the state-transition matrix of the system. This solution $\mathbf{x}(t)$ is an equilibrium if

$$\mathbf{x}(t) = \Phi(\tau, t)\mathbf{x}(t_o) = \mathbf{x}(t_o) \tag{9.129}$$
$$\forall \tau \geq t_0.$$

This proposes two possibilities that either $\Phi(\tau, t)$ is an identity matrix or $\mathbf{x}(t_o)$ is zero. If $\Phi(\tau, t)$ is an identity matrix, then the entire subspace $\mathbb{R}^n$ consists of equilibria for the system which is a trivial possibility. The only reasonable possibility is for the $\mathbf{x}(t)$ to be zero. Further on, it will be recalled that the equilibrium point $\mathbf{x}(t)$ of a dynamic system in 9.127 is such that $\dot{\mathbf{x}}(t) = 0$, i.e.,

$$\mathbf{A}(t)\mathbf{x}(t) = 0 \qquad (9.130)$$
$$t \in \mathbb{R}^+.$$

In other words, the equilibrium point of the system is the null-space of the system matrix $\mathbf{A}(t)$. This suggests that for an unforced process that starts in a state of rest, which is its equilibrium point, will remain in that state of rest indefinitely unless disturbed by some external forces. This requires that small perturbations in the initial state result in arbitrarily small perturbations in the corresponding solution trajectories. That is, the norm of the solution $||x||$ does not exceed a pre-specified number ε. Theoretically, this assumes that it is possible to determine *a priori* bound $\delta(t_0, \varepsilon)$ on the norm of the initial condition $||x(t_0)||$ in such a way that any solution trajectory of the system starting at t_0 from an initial state inside a sphere of radius $\delta(t_0, \varepsilon)$ will always remain inside this sphere at position $r \leq \varepsilon$.

From this observation, three conditions pertaining to the stability of an equilibrium can be defined, all based on the fact that a stable system has the property that, if it starts from some state $\mathbf{x}(t + \delta t) \ncong \mathbf{x}^*(t)$, then in the long term it remains in that state or tends to the equilibrium point $\mathbf{x}^*(t)$. These are summarized as follows:

- A system $\dot{\mathbf{x}}(t) = \mathbf{A}(t)\mathbf{x}(t)$ with equilibrium point $\mathbf{x}^*(t)$ is said to be stable if for every $\varepsilon > 0$, there exists for each $t_0 \in \mathbb{R}_+$ a term $\delta > 0$, depending on t_0 such that the error between the initial state $\mathbf{x}(t_0)$ and the equilibrium point $\mathbf{x}^*(t)$ is bounded by δ, i.e.,

$$||\mathbf{x}(t_0) - \mathbf{x}^*(t)|| \leq \delta. \qquad (9.131)$$

It then follows that

$$||\Phi(t, t_0)\mathbf{x}(t_0) - \mathbf{x}^*(t)|| \leq \varepsilon \qquad (9.132)$$
$$\forall t > t_0. \qquad (9.133)$$

- If the error bound in the stable system described above is independent of time it becomes a uniformly stable system.

- The system will finally be termed as being unstable if it is not stable in both senses described above.

These conditions constitute what is known as the Lyapunov stability. Another notion found frequently in the description of the Lyapunov stability of systems is that of asymptotic stability. Briefly, the asymptotic stability refers to long-term stability as time $t \to \infty$. An asymptotically stable system is a stable system in the Lyapunov sense, but, when perturbed from the equilibrium position, in the long term, it will converge to its original position. Generally, it requires that, in addition to Lyapunov stability, there exist a number $R > 0$ depending on the initial condition at t_0 such that if

$$\|\mathbf{x}(t_0) - \mathbf{x}^*(t)\| \leq R, \tag{9.134}$$

then

$$\lim_{t \to \infty} \Phi(t, t_0) x(t_0) = x^*(t). \tag{9.135}$$

It becomes uniformly asymptotically stable if it is uniformly stable, asymptotically stable and converges to $x^*(t)$ independent of the initial condition t_0, and the initial error is also independent of time, such that for every $\varepsilon > 0$, there exists $T > 0$ and $R > 0$ such that

$$\|\Phi(t_0 + T, t_0)\mathbf{x}(t_0) - \mathbf{x}^*(t)\| \leq \varepsilon.$$

Now consider a nonlinear system described by a differential equation

$$\dot{\mathbf{x}}(t) = \mathbf{f}(t, \mathbf{x}(t)) \tag{9.136}$$

$$t > 0; \qquad \mathbf{x}(t) \in \mathbb{R}^n; \qquad f : \mathbb{R}_+ \times \mathbb{R}^n \to \mathbb{R}^n$$

satisfying a global Lipschitz condition so that it has a unique solution for each initial condition. If the solution to such an equation is $\mathbf{s}(t, \mathbf{x}(t_0))$ corresponding to some initial condition $\mathbf{x}(t_0)$, i.e.

$$\frac{d\mathbf{s}(t, \mathbf{x}(t_0))}{dt} = \mathbf{f}(t, \mathbf{s}(t, \mathbf{x}(t_0)) \tag{9.137}$$

$$\forall t \geq t_0 \qquad \mathbf{s}(t, \mathbf{x}(t_0)) = \mathbf{x}(t_0).$$

Then the equilibrium condition is

$$\dot{\mathbf{x}}(t) = \mathbf{0} \tag{9.138}$$

$$\mathbf{f}(t, \mathbf{x}(t_0)) = \mathbf{x}(t_0) \tag{9.139}$$

$$\forall t \geq t_0 \geq 0; \qquad \forall \mathbf{x}(t_0) \in \mathbb{R}^n.$$

This translates to meaning that if the system starts in equilibrium, it stays there. In most cases, the equilibrium point is assumed to be the origin ($\mathbf{x}(t_0) = \mathbf{0}$) as has been discussed in the phase plane analysis, and if a different value ($\mathbf{x}(t_0) \neq \mathbf{0}$) is given, coordinate transformation is done so that this given $\mathbf{x}(t_0)$ becomes the origin. Thus, the equilibrium is at

$$\mathbf{f}(t, 0) = 0 \tag{9.140}$$
$$\mathbf{s}(t, \mathbf{x}(t_0)) = 0 \tag{9.141}$$
$$\forall t \geq 0 \qquad \forall t_0 \geq 0.$$

Now, for nonlinear systems, the Lyapunov theory is concerned with the function $\mathbf{s}(t, \mathbf{x}(t_0))$ when $\mathbf{x}(t_0) \neq 0$ but close to 0. Before a thorough discussion on the Lyapunov theory is given, it is important to define a few stability terms that define the Lyapunov stability. They are essentially the same as those for the linear system and can be listed as:

- The equilibrium $x(t) = 0$ is stable if for each $\varepsilon > 0$, at each $t \in \mathbb{R}_+$ there exists a number $\delta = \delta(\varepsilon, t_o)$, depending on the initial conditions at t_0 such that

$$\|\mathbf{s}(t, x(t_0))\| < \varepsilon \tag{9.142}$$
$$\text{for} \quad \|\mathbf{x}(t_0)\| < \delta(\varepsilon, t_0) \qquad \forall t \geq t_0.$$

- If for each $\varepsilon > 0$ there exists a number $\delta = \delta(\varepsilon)$ independent of the initial conditions at t_0 such that

$$\|\mathbf{s}(t, x(t_0))\| < \varepsilon \tag{9.143}$$
$$\text{for} \quad \|\mathbf{x}(t_0)\| < \delta(\varepsilon) \qquad \forall t \geq t_0,$$

then the equilibrium is said to be uniformly stable.

- The equilibrium is unstable if it is not stable, i.e., has neither stability in the first sense nor second sense.

Similarly, the asymptotic stability of nonlinear systems can be defined as the long-term measure of the Lyapunov stability as time $t \to \infty$. Basically, this requires that the states that started close to a stable equilibrium will converge to the equilibrium in the long term. Since under the Lipschitz continuity of $\mathbf{f}$ the solution of the nonlinear dynamics is a continuous function of the initial condition, given any $t_0 \geq 0$, and any finite T, the map $s(., x(t_0))$ that takes the initial condition into the solution trajectory in $C^n[t_0, T]$ is continuous, then in the long term as $t \to \infty$, $C^n[t_0, \infty]$ becomes a linear space of continuous n-vector valued function on $[t_0, \infty]$. A Banach space $BC^n[t_0, \infty]$ can be defined as a subset of $C^n[t_0, T]$ consisting of bounded continuous functions such that

$$\|x(.)\|_s = \begin{array}{c} max \\ t \in [t_0, \infty] \end{array} \|x(t)\| \tag{9.144}$$

Therefore, the asymptotic stability can then be defined as follows:

- For each $t_0 \geq 0$, there is a number $d(t_0)$ that defines the boundary of stability such that $s(., x(t_0)) \in BC^n[t_0, \infty]$ whenever $x(t_0) \in B_{d(t_0)}$ where B_d is the radius of the bounding sphere defined as

$$B_d = \{\mathbf{x} \in \mathbb{R}^n : \|\mathbf{x}\| < d\}. \qquad (9.145)$$

- The map $s(., \mathbf{x}(t_0))$ that maps initial condition $\mathbf{x}(t_0) \in B_{d(t_0)}$ into the corresponding solution trajectory in $BC^n[t_0, \infty]$ is continuous at $x_0 = 0$ for each $t_0 \geq 0$.

It should be noted that normally the stability of an equilibrium point depends on some criterion number defined as a number $\delta = \delta(t, \varepsilon)$ that is greater than zero. This number may depend on both some prespecified criterion ε and time t. However, if it happens that δ is independent of time, then the equilibrium point is said to exhibit uniform stability. Now, since autonomous systems are independent of time, the distinction of stability and uniform stability is not there.

Another measure employed hand in hand with the exponential stability in defining the long term Lyapunov stability is the exponential stability. The concept of exponential stability is an expression of the measure of the rate at which a perturbed equilibrium goes back to its original state $x(t_0) = 0$. If the perturbed state vector of the system converges back to its equilibrium faster than an exponential function, the equilibrium is said to have an exponential stability. This type of stability is stronger in a sense than the asymptotic stability, because it not only gives the convergence property of the equilibrium that is expressed by the asymptotic stability, but also goes further to express how fast the convergence will occur. Mathematically, the exponential stability can be described as follows:

- If there exists some strictly positive numbers α and λ such that

$$\|s(t_0 + t, x(t_0))\| \leq \alpha \|x(t_0)\| e^{-\lambda t} \qquad (9.146)$$
$$\forall t, t_0 \geq 0$$
$$\forall x(t_0) \in B_r$$

where B_r is the range or the radius of the Lyapunov stability, then the equilibrium $x(t_0)$ is said to be exponentially stable. This gives a stronger and explicit boundary on the state of the system at any time $t \geq t_0$.

Finally, consider the concept of local and global stability. So far, the previous discussion on stability has been focusing on the behavior of the system at some state close to the equilibrium position. In other words, this has been a study of the local behavior of the system in the neighborhood of the equilibrium position, hence it constitutes what is known as the local stability of the system. When attention is paid to the behavior of the system when the initial state is at some distance away from the equilibrium position, the concept of global stability comes in. If asymptotic (or exponential) stability holds for any initial states, irrespective of their position from the equilibrium position, then the equilibrium is said to be globally asymptotically (or exponentially) stable.

It is emphasized that these refined concepts of stability are more applicable to nonlinear systems than linear systems because while it is possible for a nonlinear to be locally asymptotically stable but not globally asymptotically stable, the same is not the case with linear systems. Linear time-invariant systems that are asymptotically stable are automatically exponential and globally stable. A distinction among these forms of stability for linear systems is nonexistent.

9.9.4.2 Lyapunov's Linearization Method

This method looks for the local stability of a nonlinear system under justifiable assumption that any nonlinear system can be linearized over a small range of motions. It is an analysis tool that enables us to study the dynamics of a nonlinear system using a model of a linear system. Although there are various methods of linearizing a nonlinear system, Lyapunov's linearization method serves as a fundamental justification of using linear control techniques on nonlinear systems.

To describe the concept of linearization, consider the autonomous system

$$\dot{\mathbf{x}}(t) = \mathbf{f}(\mathbf{x}(t)) \tag{9.147}$$

with an equilibrium $\mathbf{x}(t_0) = 0$ so that $\mathbf{f}(0) = 0$ and $\mathbf{f} : \mathbb{R}^n \to \mathbb{R}^m$ is continuously differentiable. Let $\mathbf{A}$ be the Jacobian matrix of $\mathbf{f}$ at the equilibrium point, i.e.,

$$A = \left[\frac{\partial \mathbf{f}}{\partial \mathbf{x}} \right]_{\mathbf{x}=\mathbf{0}}.$$

If the Jacobian matrix is used to define a linear system $\mathbf{A}\mathbf{x}(t)$ then the residual of the linear approximation of $\mathbf{f}(\mathbf{x}(t)$ by this linear function becomes

$$\mathbf{f}_R(\mathbf{x}(t)) = \mathbf{f}(\mathbf{x}(t)) - \mathbf{A}\mathbf{x}(t) \tag{9.148}$$

thus, the nonlinear function can be expressed as a sum of the linear function and the nonlinear residual function as

$$\mathbf{f}(\mathbf{x}(t)) = \mathbf{A}\mathbf{x}(t) + \mathbf{f}_R(\mathbf{x}(t)).$$

From the definition of the Jacobian of f, it follows that

$$\lim_{\|\mathbf{x}(t)\| \to 0} \left[\frac{\|\mathbf{f}_R(\mathbf{x}(t))\|}{\|\mathbf{x}(t)\|} \right] = 0.$$

Therefore, since $\mathbf{f}(0) = 0$, then Equation 9.148 can be regarded as a Taylor expansion of the nonlinear function about the origin and that as $\mathbf{x(t)}$ approaches this origin, the residual function becomes negligible so that the linear function $\mathbf{Ax}(t)$ approximates the nonlinear $\mathbf{f}(\mathbf{x}(t))$, i.e.,

$$\mathbf{f}(\mathbf{x}(t)) = \mathbf{Ax}(t)$$

thus Equation 9.147 can be written as

$$\dot{\mathbf{x}}(t) = \mathbf{Ax}(t),$$

where

$$A = \left[\frac{\partial \mathbf{f}(\mathbf{x}(t))}{\partial \mathbf{x}(t)} \right]_{\mathbf{x}(t)=\mathbf{x}(t_0)=0}.$$

The same approach can be used (with some slight modifications) for non-autonomous systems.

Now the Lyapunov's linearization method defines the conditions under which the linearized model can be used in place of a nonlinear system. It is summarized in the following theorem.

Theorem 9.4 *Consider a nonlinear system $\dot{\mathbf{x}}(t) = \mathbf{f}(\mathbf{x}(t))$ as linearized at the origin by $\dot{\mathbf{x}}(t) = \mathbf{Ax}(t)$ where A is the Jacobian matrix of $\mathbf{f}(\mathbf{x}(t))$ and is bounded. There are three possibilities for the relationship between the nonlinear system and the linear approximation:*

(a) If the linear model is strictly stable, then the equilibrium point of the nonlinear system $f(x(t))$ at $x(t_0) = 0$ is also stable

(b) If the linear model is unstable, then the equilibrium point of the nonlinear system is also unstable.

(c) If the linearized system is marginally stable, then no conclusion can be made on the stability of the equilibrium of the nonlinear system $f(x(t))$.

Based on this theorem, it is evident that one can easily determine the stability of the nonlinear system by studying the linearized model, provided that the linearization is carried out at the origin.

Example 9.1 *Consider the system $\dot{\mathbf{x}}(t) = \mathbf{f}(\mathbf{x}(t))$ where $\mathbf{x} = [x_1\ x_2]^T$ and $\mathbf{f}(\mathbf{x}(t)) = [f_1(\mathbf{x}(t)\ \ f_2(\mathbf{x}(t)]^T$ such that*

$$\dot{x}_1 = f_1(x_1, x_2) = x_1^2 + x_2^2 + x_1 x_2$$

$$\dot{x}_2 = f_2(x_1, x_2) = -x_1 - x_2 - x_1 x_2.$$

It is required to get a linearized model at $\mathbf{x} = [0, 0]^T$ and determine whether the equilibrium $\mathbf{x} = 0$ for the nonlinear system $\mathbf{f}(\mathbf{x}(t))$ is a stable equilibrium. Replace $f_1(x_1, x_2) = -2x_1 + x_2 + x_1 x_2$ and redo the problem.

Solution 9.1 *This is a 2×2 system whose Jacobian matrix is*

$$A(x) = \begin{bmatrix} \dfrac{\partial f_1}{\partial x_1} & \dfrac{\partial f_1}{\partial x_2} \\ \dfrac{\partial f_2}{\partial x_1} & \dfrac{\partial f_2}{\partial x_2} \end{bmatrix}$$

$$= \begin{bmatrix} 2x_1 + x_2 & 2x_2 + x_1 \\ -1 - x_2 & -1 - x_1 \end{bmatrix}$$

so that at $\mathbf{x} = \mathbf{0}$ *it follows that*

$$A = \begin{bmatrix} 0 & 0 \\ -1 & -1 \end{bmatrix}$$

whose eigenvalues are $0, -1$. Clearly, the linearized system will be marginally stable, in which case, one cannot say anything about the stability of the nonlinear system $\mathbf{f}(\mathbf{x}(t))$.

If the same problem is redone using $f_1(x_1, x_2) = -2x_1 + x_2 + x_1 x_2$, then the Jacobian of the nonlinear system at the origin becomes

$$A = \begin{bmatrix} -2 & 1 \\ -1 & -1 \end{bmatrix},$$

whose eigenvalues are $(-\frac{3}{2} + \frac{1}{2}i\sqrt{3}, -\frac{3}{2} - \frac{1}{2}i\sqrt{3})$. Since the real part of the eigenvalues is negative, then this system is asymptotically stable.

9.9.4.3 Lyapunov's Linearization in Feedback Stabilization

In this section the usefulness of Lyapunov's linearization method in the stabilization of nonlinear systems is illustrated. Concentration will be focused on the autonomous system for reasons given earlier. Further details of the material covered in this section are found in later sections on the nonlinear control design. At this time, only an insight into the applicability of the linearization method is given.

Consider an autonomous control system described by

$$\dot{\mathbf{x}}(t) = \mathbf{f}(\mathbf{x}(t), \mathbf{u}(t))$$

where $f : \mathbb{R}^n \times \mathbb{R}^m \to \mathbb{R}^n$. In the control design, the objective is to obtain a feedback control law of the form

$$\mathbf{u}(t) = \mathbf{g}(\mathbf{x}(t))$$

such that the equilibrium of the resulting closed-loop system

$$\dot{\mathbf{x}}(t) = \mathbf{f}(\mathbf{x}(t), \mathbf{g}(\mathbf{x}(t)))$$

becomes asymptotically stable. A simple approach to the solution of this problem is by establishing a linear function of the Jacobians $\mathbf{A}$ and $\mathbf{B}$ with respect to $\mathbf{x}$ and $\mathbf{u}$ respectively at the origin, when no control effort $\mathbf{u}$ is applied, i.e.,

$$\mathbf{A} = \left[\frac{\partial \mathbf{f}}{\partial \mathbf{x}}\right]_{\mathbf{x}=0,\mathbf{u}=0}$$

$$\mathbf{B} = \left[\frac{\partial \mathbf{f}}{\partial \mathbf{u}}\right]_{\mathbf{x}=0,\mathbf{u}=0}$$

It will be seen that, since

$$[d\mathbf{f}] = \left[\frac{\partial \mathbf{f}}{\partial \mathbf{x}}\right] d\mathbf{x} + \left[\frac{\partial \mathbf{f}}{\partial \mathbf{u}}\right] d\mathbf{u}$$

$$= \mathbf{A} d\mathbf{x} + \mathbf{B} d\mathbf{u}$$

(at the origin)

then the matrices $\mathbf{A}$ and $\mathbf{B}$ correspond to the plant and input matrix of a linearized system

$$\dot{\mathbf{x}}(t) = \mathbf{A}\mathbf{x}(t) + \mathbf{B}\mathbf{u}(t)$$

whose control law was studied in Chapter 7, and is

$$\mathbf{u}(t) = -\mathbf{K}\mathbf{x}(t)$$

so that

$$\dot{\mathbf{x}}(t) = (\mathbf{A} - \mathbf{B}\mathbf{K})\mathbf{x}(t).$$

Thus, the only condition required for the stabilization of this system is selection of the gain matrix $\mathbf{K}$ so that the eigenvalues of the closed-loop matrix $(\mathbf{A} - \mathbf{B}\mathbf{K})$ have negative real parts.

9.9.4.4 Lyapunov's Direct Method

The previous section introduced what is known as Lyapunov's indirect method of stability analysis. Basically, as it was shown, this method is based on the power series expansion of the nonlinear function that describes the dynamics of the system in question. Because of the inherent error associated with the truncation of the approximation of nonlinear functions to linear functions, the indirect method does not attract much favor in most applications. This section presents the second Lyapunov method, known as Lyapunov's direct method of nonlinear analysis. The direct method is based on a physical observation that when the total mechanical (or electrical) energy of any physical system is continuously dissipated, the system will eventually settle down to an equilibrium point. Since the energy is a scalar quantity, it turns out that, according to Lyapunov's direct method, the stability of any system can be established by examining the variation of a single scalar function, known as the energy function or the Lyapunov function, which is a generalization of the total mechanical energy of the system. In total, the method is summarized in three basic theorems: the basic stability theorems, asymptotic stability theorems, and instability theorems. Before the theorems are explained in detail, a discussion on the Lyapunov functions is given. As a general rule pertaining to Lyapunov's theory, maturity in mathematics is assumed.

9.9.4.5 Energy Functions

To get a clear meaning of Lyapunov's energy functions and their use in stability analysis, consider a simple second-order nonlinear mechanical system composed of a nonlinear damping element and a nonlinear spring so that its dynamics can be written as

$$\ddot{x} = f(x, \dot{x})$$
$$= -\frac{B}{m}\dot{x}^2 - \frac{k_1}{m}x^2 - \frac{k_0}{m}x \tag{9.149}$$

or simply

$$m\ddot{x} + B\dot{x}^2 + k_1 x^2 + k_0 x = 0. \tag{9.150}$$

The total mechanical energy of such a system, which is the sum of its kinetic energy and potential energy, is given by the equation

$$V(\mathbf{x}) = \frac{1}{2}m\dot{x}^2 + \int_0^x (k_1 y^2 + k_0 y)dy$$
$$= \frac{1}{2}m\dot{x}^2 + \frac{1}{3}k_1 x^3 + \frac{1}{2}k_0 x^2. \tag{9.151}$$

Obviously, by studying this equation, it is possible to establish properties of this system, with knowledge from the previous discussion, that at the equilibrium point ($\dot{x} = 0$, $x = 0$) this energy function will be zero. This observation implies that the asymptotic stability, which is concerned with the convergence of the system when disturbed from the equilibrium point, is related to convergence of this energy function to zero while the exponential stability is related to the rate at which the energy function decreases toward the equilibrium position.

The time derivative of the energy function is given by

$$\dot{V}(\mathbf{x}) = m\dot{x}\ddot{x} + (k_1 x^2 + k_0 x)\dot{x}$$

and by the dynamics equation,

$$m\dot{x}\ddot{x} = -B\dot{x}^3 - k_1 x^2 \dot{x} - k_0 x\dot{x}. \tag{9.152}$$

The rate of energy dissipation becomes

$$\dot{V}(\mathbf{x}) = -B\dot{x}^3 - k_1 x^2 \dot{x} - k_0 x\dot{x} + (k_1 x^2 + k_0 x)\dot{x}$$
$$= -B\dot{x}^3, \tag{9.153}$$

which indicates that, as the system is disturbed from some initial condition, it will dissipate energy through its damping element at a rate $\dot{V}(\mathbf{x}) = -B\dot{x}^3$ until this rate is zero, i.e., $\dot{x} = 0$.

Now, for $\dot{x}$ to be zero, the spring force must be also zero, calling for x to be zero. This simple analysis eventually leads to the conclusion that the system will eventually settle down to an equilibrium point as predicted by asymptotic stability requirements. In short, this has shown that the system is asymptotically stable. Using the same reasoning, the instability of the system can be established if the energy function is such that it is growing up with time.

To some extent, this discussion has not only acted as an eye-opener but also clarified to some extent the meaning and use of energy functions in establishing the stability of the system. Lyapunov's direct method follows an approach that draws conclusions about the behavior of complex systems. Having seen the use of energy functions, the discussion gives in-depth general properties of the energy functions and the selection of such functions in dealing with complex systems.

9.9.4.6 Properties of the Energy Function.

Quick examination of the energy function just developed in the previous section shows two main properties. The first property, which is associated with the system dynamics is that, if the state variables $\dot{x}$, x vary according to the system dynamics equation, then the function will be continuously monotonically decreasing. The second property is concerned with the energy function itself. If the state $\dot{x}$, x is close to zero, then for all $\dot{x}$, x, $V(\mathbf{x})$ will be strictly positive unless $\dot{x}$ and x are zero. These are the two main properties of the energy function that are adopted for use by Lyapunov's direct method. Basing on this eye-opener, the next discussion will concentrate on these two properties.

In simple terms, a positive definite function may be viewed as one that is always positive for any value of its arguments. However, this simplistic definition is not truly enough to explain the definiteness of the function. A more accurate mathematical definition of the positive definiteness of functions classifies them as being either local or global.

Definition 9.1 *A scalar function $V(\mathbf{x})$ where $V : \mathbb{R}^n \to \mathbb{R}$ is said to be* <u>*locally positive definite*</u> *if $V(0) = 0$ and in the neighborhood of $\mathbf{x} = \mathbf{0}$ where $\mathbf{x} \neq \mathbf{0}$ the function is such that $V(\mathbf{x}) > 0$. If, however, when $V(x)$ has such properties as $V(0) = 0$ and that $V(\mathbf{x}) > 0$ for the whole state-space, then $V(\mathbf{x})$ is said to be* <u>*globally positive definite*</u>.

Essentially, this definition means that the function $V(\mathbf{x})$ has a unique minimum at the origin. A geometrical interpretation of this definition can be made by observing the plots of $V(\mathbf{x})$ in the $\mathbf{x}$-space. In a two-dimensional phase plane, the plot of $V(\mathbf{x})$ is presented by level curves that are a set of closed non-intersecting contour curves surrounding the origin, while in a 3-D space, $V(\mathbf{x})$ is presented by an open surface which looks like an upward cup.

Obviously, one can easily treat the contours as the horizontal section of the 3-D cup projected in a 2-D plane. More importantly to note, however, is that this definition of positive definiteness is still not sufficient, as one should indicate the bound of the magnitude of the state vector for which the energy function remains positive definite. This is consolidated into the definition:

Definition 9.2 *If a continuous function $V(\mathbf{x})$ where $V : \mathbb{R}^n \to \mathbb{R}$ is such that $V(\mathbf{0}) = \mathbf{0}$ and $V(\mathbf{x}) > 0, \forall \mathbf{x} \neq \mathbf{0}$, it becomes positive definite function if there exists some constant $r > 0$ such that*

$$\min_{||\mathbf{x}|| > r} V(\mathbf{x}) > 0 \tag{9.154}$$

Together with these concepts of local and global positive definiteness of continuous function are concepts of local and global negative definiteness as well as semi-definiteness. The energy function becomes negative definite if $-V(\mathbf{x})$ is positive definite. Semi-definiteness is a situation in which there is a possibility for this function to be zero.

Example 9.2 *Examine the definiteness of the following function*

$$V(\mathbf{x}) = x_1^2 + x_2^2 + \sin^2 x_3 \tag{9.155}$$

Solution 9.2 *The first condition to be examined is when $\mathbf{x} = 0$ then*

$$V(0) = 0$$

and that for $\mathbf{x} \neq 0$ and $|x_3| < \pi$ it follows that

$$V(\mathbf{x}) > 0.$$

Thus, it can be concluded that $V(\mathbf{x})$ is locally positive definite. However, closer examination shows that when $\mathbf{x} = [0 \quad 0 \quad \pi]^T$ it will also give

$V(0) = 0$ *hence making it not globally positive definite. This concludes that the given function is locally (in the vicinity of the origin) positive definite but not globally positive definite.*

So far, the energy functions that map the state vector into a scalar have been discussed in cases where the system is autonomous. However, the possibility for an energy function $W(t, \mathbf{x})$ that maps both the state vector and time into a scalar for a non-autonomous system, i.e., $W : \mathbb{R}_+ \times \mathbb{R}^n \to \mathbb{R}$. These energy functions can also be defined and treated using similar approaches as for autonomous systems, only with very slight modifications that cater for the presence of the time element in the function.

Definition 9.3 *A continuous function $W(t, \mathbf{x})$ where $W : \mathbb{R}_+ \times \mathbb{R}^n \to \mathbb{R}$ for $x \in \mathbb{R}^n$ is a local positive function if and only if $W(t, 0) = 0$, $\forall t$ and there exists a local positive definite function $V(\mathbf{x})$ where $V : \mathbb{R}^n \to \mathbb{R}$ such that in the neighborhood of $\mathbf{x} = 0$,*

$$W(t, \mathbf{x}) \geq V(\mathbf{x}) \qquad \forall t \geq 0 \tag{9.156}$$

If, however,

$$W(t, \mathbf{x}) \geq V(\mathbf{x}) \qquad \forall t \geq 0 \qquad \forall \mathbf{x} \in \mathbb{R}^n \tag{9.157}$$

the function becomes a globally positive definite function.

This definition shows the close relationship between the energy function for autonomous and non-autonomous systems. In general, energy functions for non-autonomous systems can be regarded as a generalization of an energy function that includes those for autonomous systems

Example 9.3 *Discuss the definiteness of the following function*

$$W(t, \mathbf{x}) = (t^2 + t + 2)(x_1^2 + x_2^2 + x_3^2)$$

Solution 9.3 *It can be seen that if*

$$V(\mathbf{x}) = x_1^2 + x_2^2 + x_3^2$$

then

$$W(t, \mathbf{x}) = (t^2 + t + 2)V(\mathbf{x})$$

so that

$$V(0) = 0$$
$$W(t, 0) = 0.$$

For any $x \in \mathbb{R}^n$ let

$$V(\mathbf{x}) = K$$

where K is some scalar value so that

$$W(t, \mathbf{x}) = (t^2 + t + 2)K.$$

Since

$$(t^2 + t + 2) > 1, \qquad \forall t \geq 0$$

then

$$W(t, \mathbf{x}) > V(\mathbf{x}).$$

Hence, the function is a globally positive definite function.

9.9.4.7 Decreasing Functions and the Derivative of the Energy Function

Since the energy function for a non-autonomous system can equally be treated like an autonomous system, most of the discussion in this section will be based on non-autonomous system energy functions, which have been seen to be the generalization of autonomous energy function. All results obtained under this discussion so far apply well with the autonomous energy functions.

Suppose that the energy function $W(t, \mathbf{x}(t))$ is continuously differentiable with continuous partial derivatives with respect to t, and x_i, $i = 1, 2 \cdots n$. Then the derivative of $W(t, \mathbf{x}(t))$, with respect to t is

$$\frac{dW(t, \mathbf{x}(t))}{dt} = \frac{\partial W(t, \mathbf{x}(t))}{\partial t} + \nabla_x W(t, \mathbf{x}(t)) \dot{\mathbf{x}}(t).$$

Suppose further that $\mathbf{x}(t)$ satisfies the differential equation that describes the system dynamics given as

$$\dot{\mathbf{x}}(t) = f(t, \mathbf{x}(t)) \qquad \forall t \geq 0$$

then the time derivative of $W(t, \mathbf{x}(t))$ with respect to t becomes

$$\frac{dW(t, \mathbf{x}(t))}{dt} = \frac{\partial W(t, \mathbf{x}(t))}{\partial t} + \nabla_x W(t, \mathbf{x}(t)) f(t, \mathbf{x}(t)). \qquad (9.158)$$

This derivative is known as the derivative of the energy function along the system trajectories and is formally contained in the following definition.

Definition 9.4 *If $W : \mathbb{R}_+ \times \mathbb{R}^n \to \mathbb{R}$ is continuously differentiable with respect to all of its arguments, and $\nabla_x W$ is the gradient of W with respect to $\mathbf{x}$, then the function $\dot{W} : \mathbb{R}_+ \times \mathbb{R}^n \to \mathbb{R}$, called the derivative of W along the system trajectories, is defined by*

$$\dot{W}(t, \mathbf{x}(t)) = \frac{\partial W(t, \mathbf{x}(t))}{\partial t} + \nabla_x W(t, \mathbf{x}(t)) f(t, \mathbf{x}(t)).$$

Note that for an autonomous system

$$\frac{\partial W(t, \mathbf{x}(t))}{\partial t} = 0.$$

Hence,

$$\dot{W}(t, \mathbf{x}(t)) = \nabla_x W(t, \mathbf{x}(t)) f(t, \mathbf{x}(t)). \tag{9.159}$$

This derivative reflects the time rate at which the energy function decreases toward the origin. Its use in nonlinear systems analysis will be shown shortly. So far, all energy functions have been treated as Lyapunov functions. In essence, this is an incorrect assumption because not all energy functions are indeed Lyapunov functions. A Lyapunov function is an energy function with decreasing properties, i.e., it should be decreasing with time. As such, an energy function qualifies to be a Lyapunov function only if its time derivative is less than zero, which indicates the decreasing energy. Geometrically, the Lyapunov function can be seen as a family of the energy functions that at some given state will be pointing down toward the inverted cup of the energy surface. The formal definition of the Lyapunov function is given as follows:

Definition 9.5 *If in the vicinity of the equilibrium point, the energy function $W(t, \mathbf{x}(t))$ is positive definite and has continuous partial derivatives, and if its time derivative along any of the state trajectories is negative semi-definite, i.e.,*

$$\dot{W}(t, \mathbf{x}(t)) \leq 0 \tag{9.160}$$

then the energy function $W(t, \mathbf{x}(t)$ is said to be a Lyapunov function for the system.

9.9.4.8   The Concept of Invariant Sets

The derivative of the energy function along the system state trajectory is a very important quantity, as will be shown shortly by the Lyapunov theorems, in determining the asymptotic stability of a system, provided that it is negative definite. However, in some cases, this quantity may become negative semi-definite, in which case it becomes less useful in stability analysis. To be able to get around such situations, the concept of an invariant set is introduced together with the invariant set theorems. In general, the invariant sets are groups of state vectors in the state-space for which the system trajectory is indefinitely confined for some system dynamics.

Definition 9.6 *If a system dynamics is described by a nonlinear differential equation*

$$\dot{\mathbf{x}} = \mathbf{f}(t, \mathbf{x}(t))$$
$$\forall t \geq 0, \quad x \in \mathbb{R}^n$$

with solution $\mathbf{s}(t, \mathbf{x}(t_0))$, *and if there exists a set* $M \subseteq \mathbb{R}^n$ *such that for each* $\mathbf{x}(t_0) \in M$ *there is a* $t_0 \in \mathbb{R}_+$ *for which*

$$\mathbf{s}(t, \mathbf{x}(t_0)) \in M \quad \forall t \geq t_0 \tag{9.161}$$

then the set M *is called an invariant set of the differential equation that describes the system dynamics.*

In simple words, this definition implies that if M is an invariant set, then every system trajectory that starts from some point in M remains there for all future time. Invariant sets represent a collection of state vectors that are solutions (or approximate solutions) of the nonlinear differential equation of the system. With this view, every equilibrium point or limit cycle, and even the whole state-space can be viewed as invariant sets, however, for analytical purposes, the whole state-space does not lead to useful conclusions, hence it is known as a trivial invariant set.

As mentioned earlier, invariant sets are very powerful in nonlinear system analysis when the energy function fails. The application of invariant sets in such problems is contained in the popular invariant set theorems, which can either be local or global. The local invariant set theorem will be discussed in more detail before extending the concepts therein to the global invariant set theorem.

Local Invariant Set Theorem.

- *For a continuous nonlinear system*

$$\dot{\mathbf{x}} = \mathbf{f}(t, \mathbf{x}(t))$$
$$\forall t \geq 0, \quad x \in \mathbb{R}^n,$$

  let $W(t, x(t))$ *be associated scalar energy function with continuous partial derivatives. If for some* $r > 0$ *the region* B_r *defined by*

$$W(t, \mathbf{x}(t)) < r$$
$$\forall t \geq 0$$

is bounded and

$$\dot{W}(t, \mathbf{x}(t)) \leq 0$$
$$\forall \mathbf{x} \in B_r$$

and $R \subseteq B_r$ is a set of all points within B_r where $\dot{W}(t, x(t)) = 0$ and if M is the largest invariant set in R (i.e., the union of all invariant sets in R) then every solution $x(t)$ origination in B_r tends to M as $t \to \infty$

This local invariant set theorem presents an alternative description of the system asymptotic stability if the set M consists only of the equilibrium point, as will be shown shortly. One of the major concerns that could attract some attention in the application of invariant sets in Lyapunov's theorems is the method by which to establish invariant sets for a given system. The following discussion explains this problem by first defining the limit sets.

Definition 9.7 *Suppose that a nonlinear system*

$$\dot{\mathbf{x}} = \mathbf{f}(t, \mathbf{x}(t))$$
$$\forall t \geq 0, \quad x \in \mathbb{R}^n$$

has a solution trajectory $\mathbf{s}(t, \mathbf{x}(t_0))$ where $x(t_0) \in \mathbb{R}^n$, $t_0 \in \mathbb{R}_+$. If there exists a sequence $\{t_i\}_{i=1}^{\infty} \in [t_0, \infty] \ni t_i \to \infty$ and for some point $\mathbf{p} \in \mathbb{R}^n$

$$\lim_{t_i \to \infty} \|\mathbf{p} - \mathbf{s}(t_i, \mathbf{x}(t_0))\| = 0 \tag{9.162}$$

then $\mathbf{p}$ is called the limit point of the trajectory. A set of such limit points $\Omega(\mathbf{x}(t_0))$ is called the limit set of the trajectory.

This definition simply states that, for some solution trajectories, there exists some points $\mathbf{p}$ called the limit points such that for some number $\varepsilon > 0$, there is a $t \in [0, \infty]$ such that

$$\|\mathbf{p} - \mathbf{s}(t, \mathbf{x}(t_0))\| < \varepsilon,$$

which means that the distance of $\mathbf{p}$ from the trajectory is bounded. The limit set for a given system can be either bounded or unbounded, depending on the trajectory itself. If the trajectory is bounded, then the limit set will also be bounded, otherwise it becomes unbounded. If the limit set is bounded and the distance between a point $\mathbf{x}$ and the limit set Ω is defined, such that

$$d(\mathbf{x}, \Omega) = \min_{\mathbf{y} \in \Omega} \|\mathbf{x} - \mathbf{y}\|$$

then the trajectory $\mathbf{s}(.,\mathbf{x}(t_o))$ is bounded for $\mathbf{x}(t_0) \in \mathbb{R}^n$, $\quad t_0 \in \mathbb{R}_+$ then

$$\lim_{t \to \infty} d[\mathbf{s}(t,\mathbf{x}(t_0)), \Omega(\mathbf{x}(t_0))] = 0, \qquad (9.163)$$

which simply means that the distance between the trajectory and the limit set converges to zero, or the trajectory falls in the limit set.

Having clarified the relationships between the system trajectories, it is possible to establish the invariant sets for a given system. Generally, if the nonlinear system

$$\dot{\mathbf{x}} = \mathbf{f}(t,\mathbf{x}(t))$$
$$\forall t \geq 0, \quad x \in \mathbb{R}^n$$

is periodic (i.e., has limit cycles) and the trajectory $\mathbf{s}(.,\mathbf{x}(t_o))$ is bounded, then the limit set $\Omega(\mathbf{x}(t_0))$ is an invariant set of this system. For autonomous systems of the form

$$\dot{\mathbf{x}} = \mathbf{f}(\mathbf{x}(t)) \qquad (9.164)$$

the invariant sets enclose the system domains of attraction, i.e., the area in the state-space in which the system trajectories converge, if the system has any.

Definition 9.8 *If 0 is an attractive equilibrium of the autonomous system, the domain of attraction $D(0)$ is defined as*

$$D(\mathbf{0}) = \{\mathbf{x}(0) \in \mathbb{R}^n : \lim_{t \to \infty} \mathbf{s}(t,\mathbf{x}(0)) \to 0\} \qquad (9.165)$$

It will be recalled that domains of attraction are the areas that contain the stable equilibria for the system trajectory.

9.10 Design of Nonlinear Control Systems

9.10.1 Feedback Linearization Methods

These methods are an extension of the linear feedback methods. In this section the basic principles of system linearization are discussed. The emphasis will be on the methods for system linearization about equilibrium points so that the linear feedback control can be applied. The approach begins with the more familiar linearization techniques.

9.10.2   Introduction to Adaptive Control

In some control systems, certain parameters are either not constant or vary in an unknown manner. One way of minimizing the effects of such contingencies is by designing for minimum sensitivity. If, however, parameter variations are large or very rapid, it may be desirable to design for the capability of continuously measuring them and changing the compensation so that the system performance criteria are always satisfied. This is called adaptive control design.

Some nonlinear systems are difficult to model for several technical reasons. Modern approaches to such systems and even to those that can be modeled with uncertainty is to use adaptive control strategies. Under such strategies the controller is made to tune itself with the nonlinear and even time-varying plant dynamics producing the required control action. The most popular methods under this scheme are the Self Tuning Regulator (STR) and the Model Reference Adaptive Controller (MRAC). This section discusses the basics of these adaptive control strategies. The Model Reference Adaptive Controller requires the presence of the model that represents the required dynamics. The controller works by adapting the parameters in accordance with adaptation of the difference between the actual system dynamics and the model reference.

9.10.3   Introduction to Neural Control

The increasing demand for high precision control over a wide range of operating regions and the emergence of new computational methods using neural networks that abstract parallel information-handling features of the human brain with massively interconnected processing elements, neural networks are emerging as modern adaptive controllers. The main attractive features of artificial neural networks include the self-learning and distributed memory capabilities. The self learning features of neural networks are applied in learning the system dynamics, thereby tuning the controller parameters accordingly. Though most of the theory of neural controllers is still in its infancy, this section will provide an overview and the basic ideas behind neurocontrol.

9.10.4   Introduction to Robust Control

A control system designed using the methods and concepts of the preceding chapters assumes knowledge of the model of the plant and controller and constant parameters. The plant model will always be an inaccurate representation of the actual physical system because of the following issues:

- Parameter changes

- Unmodeled dynamics

- Unmodeled time delays

- Changes in the equilibrium point (operating point).

- Sensor noise

- Unpredicted disturbance inputs

Robust control is an approach to handling model uncertainties associated with time varying systems, both linear and nonlinear. This strategy happens to be a special case of adaptive control, the robust. The goal of robust systems design is to retain assurance of system performance in spite of model inaccuracies and changes. A system is robust when it has acceptable changes in performance due to model changes or inaccuracies. A robust control system exhibits the desired performance despite the presence of significant plant (process) uncertainty.

The system structure that incorporates potential system uncertainties is shown in Figure 9.30 This model includes the sensor noise $N(s)$, the unpredicted disturbance input $D(s)$, and a plant $G(s)$ with potentially unmodeled dynamics or parameter changes. The unmodeled dynamics and parameter changes may be significant or very large, and for these systems, the challenge is to create a design that retains that desired performance.

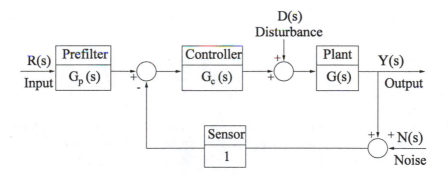

FIGURE 9.30
Robust Control System Design

Designing highly accurate systems in the presence of significant plant uncertainty is a classical feedback design problem. The theoretical bases for the solution of this problem dates back to the work of H.S. Black and H.W. Bode in the early 1930s, when this problem was referred to as the sensitivities design problem. A significant amount of literature has been published since then regarding the design of systems subject to large plant uncertainty. The designer seeks to obtain a system that performs adequately over a large range of uncertain parameters. A system is said to be robust when it is durable, hardy, and resilient.

A control system is robust when (i) it has low sensitivities, (ii) it is stable over the range of parameter variations. and (iii) the performance continues to meet the specifications in the presence of a set of changes in the system parameters. Robustness is the sensitivity to effects that are not considered in the analysis and design phase-for example, disturbances, measurement noise, and unmodeled dynamics. The system should be able to withstand these neglected effects when performing the tasks for which it was designed.

For small parameter perturbation the differential sensitivities and root sensitivity are used as a measure of robustness. System sensitivity is defined as

$$S_\alpha^T = \frac{\partial T/T}{\partial \alpha/\alpha}$$

where α is the parameter, and T the transfer function of the system. Root sensitivity is defined as

$$S_\alpha^{r_i} = \frac{\partial r_i}{\partial \alpha/\alpha}$$

When the zeros of $T(s)$ are independent of the parameter α, it can be shown that

$$S_\alpha^T = -\sum_{i=1}^{n} S_\alpha^{r_i} \frac{1}{(s+r_i)},$$

for an nth-order system. For example, if there is a closed-loop system, as shown in Figure 9.31, where the variable parameter is α, then $T(s) = 1/[s + (\alpha + 1)]$, and

$$S_\alpha^T = \frac{-\alpha}{s + \alpha + 1}.$$

Furthermore, the root is $r_1 = +(\alpha + 1)$, and

$$-S_\alpha^{r_i} = -\alpha$$

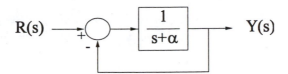

FIGURE 9.31
Closed-Loop System (Variable Parameter)

9.11   Problems

Problem 9.1 *A nonlinear system is described by the following equations,*

$$\dot{x}_1 = 2x_1 + 3x_1^2 - 2x_2 + u$$

$$\dot{x}_2 = x_2 - 0.25(\sin x_1 + \cos x_2) + u.$$

(a) With $u = 0$ plot the state-space trajectories in the neighborhood of the origin defined by $|x_1| \leq 1$, $|x_2| \leq 2$.

(b) Using Lyapunov's first method, linearize the above system about $x_1 = x_2 = u = 0$, and assess the stability of this linearized system.

Problem 9.2 *Investigate whether the following functions $V(x)$ are positive definite.*

(a) $V(x) = 2x_1^2 + 2x_1x_2 + 4x_2^2 - 4x_1x_3 + 5x_3^2$

(b) $V(x) = x_1^2 + 4x_1x_2 + 5x_2^2 - 6x_1x_3 - 8x_2x_3 + 8x_3^2$

Problem 9.3 *Construct a Lyapunov function for the system*

$$\dot{x}_1 = x_1^2 + x_2^2 - x_1$$

$$\dot{x}_2 = x_1^2 - x_2^2 - x_2$$

and use it to investigate the stability at the origin. State the domain of attraction if the system is asymptotically stable.

Problem 9.4 *A nonlinear system is described by*

$$\dot{x}_1 = x_1 + x_2 - (x_1^3 + x_1x_2^2)$$

$$\dot{x}_2 = x_2 - x_1 - (x_2^3 + x_1x_2^2)$$

if $r^2 = x_1^2 + x_2^2$, show that the system has a limit cycle at $r = 1$.

Problem 9.5 *For the following system, determine the equilibrium points and discuss the stability of each equilibrium using Lyapunov's (indirect) linearization method.*

$$\dot{x}_1 = x_1 - x_1^2 - \frac{1}{2}x_1 x_2$$

$$\dot{x}_2 = x_2 - x_2^2 + \frac{1}{2}x_1 x_2.$$

Sketch the flow of trajectories in the (x_1, x_2) phase plane and classify each equilibrium point (node, saddle, etc.)

Problem 9.6 *Sketch the isoclines for the equations:*

$$(a) \qquad\qquad\qquad \frac{dy}{dx} = xy(y - 2)$$

$$(b) \qquad\qquad \ddot{x} + \omega_n^2 x + \mu x^3 = 0$$

$$(c) \qquad \ddot{x} - \dot{x}(1 - x^2) + x = 0.$$

Problem 9.7 *The equations describing the motion of an earth satellite in the orbit plane are*

$$r\frac{d^2\theta}{dt^2} + 2\frac{dr}{dt}\frac{d\theta}{dt} = 0$$

$$\frac{d^2r}{dt^2} - r\left(\frac{d\theta}{dt}\right)^2 = -\frac{k^2}{pr^2}.$$

A satellite is nearly in a circular orbit determined by r and $\dfrac{d\theta}{dt} \equiv \omega$. An exactly circular orbit is defined by

$$r = r_o = constant, \quad \omega = \omega_o = constant.$$

Since $\dfrac{dr_o}{dt} = 0$ and $\dfrac{d\omega_o}{dt} = 0$, the first differential equation is eliminated for a circular orbit. The second equation reduces to $r_o^2 \omega_o^2 = \dfrac{k^2}{pr_o^2}$. Find a set of linear equations that approximately describes the differences.

Problem 9.8 *Show that the describing function for the saturation element in the previous problem is given by*

$$\frac{B_1}{A}e^{j\phi_1} = \frac{2}{\pi}\left[\sin^{-1}\left(\frac{1}{A}\right) + \frac{1}{A}\cos\left(\sin^{-1}\left(\frac{1}{A}\right)\right)\right].$$

Problem 9.9 *Show that equation* $\dfrac{d^2x}{dt^2} = f\left(x, \dfrac{dx}{dt}\right)$ *can be equivalently described by a pair of first-order differential equations.*

Problem 9.10 *Show that the phase plane trajectory of the solution of the differential equation*

$$\frac{d^2x}{dt^2} + x = 0,$$

with initial conditions $x(0) = 0$ *and* $\dot{x}(0) = 1$, *is a circle of unit radius centered at the origin.*

Problem 9.11 *Determine the equation of the phase plane trajectory of the differential equation*

$$\frac{d^2x}{dt^2} + \frac{dx}{dt} = 0,$$

with the initial conditions $x(0) = 0$ *and* $\dot{x}(0) = 1$.

Appendix A

Laplace and z-Transforms

A.1 Properties of Laplace Transforms

Number	Laplace Transform $F(s)$	Time Function $f(t)$	Comment Transform pair
1	$\alpha F_1(s) + \beta F_2(s)$	$\alpha f_1(t) + \beta f_2(t)$	Superposition
2	$F(s)e^{-s\lambda}$	$f(t - \lambda)$	Time delay
3	$\dfrac{1}{|a|}F\left(\dfrac{s}{a}\right)$	$f(at)$	Time scaling
4	$F(s + a)$	$e^{-at}f(t)$	Frequency shift
5	$s^m F(s) - s^{m-1}f(0)$ $-s^{m-2}\dot{f}(0) - ... - f^{(m-1)}(0)$	$f^{(m)}(t)$	Differentiation
6	$\dfrac{1}{s}F(s)$	$\int f(\xi)d\xi$	Integration
7	$F_1(s)F_2(s)$	$f_1(t) * f_2(t)$	Convolution
8	$\lim\limits_{s\to\infty} sF(s)$	$f(0^+)$	IVT
9	$\lim\limits_{s\to 0} sF(s)$	$\lim\limits_{t\to\infty} f(t)$	FVT
10	$\dfrac{1}{2\pi j}\int_{c-j\infty}^{c+j\infty} F_1(\zeta)F_2(s - \zeta)d\zeta$	$f_1(t)f_2(t)$	Time product
11	$-\dfrac{d}{ds}F(s)$	$tf(t)$	Multiply by time

A.2   Table of Laplace Transforms

Number	$F(s)$	$f(t), t \geq 0$
1	1	$\delta(t)$
2	$1/s$	$1(t)$
3	$1/s^2$	t
4	$2!/s^3$	t^2
5	$3!/s^4$	t^3
6	$m!/s^{m+1}$	t^m
7	$\dfrac{1}{s+a}$	e^{-at}
8	$\dfrac{1}{(s+a)^2}$	te^{-at}
9	$\dfrac{1}{(s+a)^3}$	$\dfrac{1}{2!}t^2 e^{-at}$
10	$\dfrac{1}{(s+a)^m}$	$\dfrac{1}{(m-1)!}t^{m-1}e^{-at}$
11	$\dfrac{a}{s(s+a)}$	$1 - e^{-at}$
12	$\dfrac{a}{s^2(s+a)}$	$\dfrac{1}{a}\left(at - 1 + e^{-at}\right)$
13	$\dfrac{b-a}{(s+a)(s+b)}$	$e^{-at} - e^{-bt}$
14	$\dfrac{s}{(s+a)^2}$	$(1 - at)e^{-at}$
15	$\dfrac{a^2}{s(s+a)^2}$	$1 - e^{-at}(1 + at)$
16	$\dfrac{(b-a)s}{(s+a)(s+b)}$	$be^{-bt} - ae^{-at}$
17	$\dfrac{a}{s^2 + a^2}$	$\sin at$
18	$\dfrac{s}{s^2 + a^2}$	$\cos at$
19	$\dfrac{s+a}{(s+a)^2 + b^2}$	$e^{-at}\cos bt$
20	$\dfrac{b}{(s+a)^2 + b^2}$	$e^{-at}\sin bt$
21	$\dfrac{a^2 + b^2}{s\left[(s+a)^2 + b^2\right]}$	$1 - e^{-at}\left(\cos bt + \dfrac{a}{b}\sin bt\right)$

A.3 Properties of $\mathcal{Z}$-Transforms

1. Definition.
$$F(z) = \Sigma_{k=0}^{\infty} f(kh) z^{-k}$$

2. Inversion.
$$f(kh) = \frac{1}{2\pi i} \oint F(z) z^{k-1} dz$$

3. Linearity
$$\mathcal{Z}\{\alpha f + \beta g\} = \alpha \mathcal{Z} f + \beta \mathcal{Z} g$$

4. Time shift
$$\mathcal{Z}\{q^{-n} f\} = z^{-n} F$$
$$\mathcal{Z}\{q^{n} f\} = z^{n}(F - F_1) \text{ where } F_1(z) = \Sigma_{j=0}^{n-1} f(jh) z^{-j}$$

5. Initial-value theorem
$$f(0) = \lim_{z \to \infty} F(z)$$

6. Final-value theorem
If $(1 - z^{-1})F(z)$ does not have any poles on or outside the unit circle, then
$$\lim_{k \to \infty} f(kh) = \lim_{z \to 1}(1 - z^{-1})F(z)$$

7. Convolution
$$\mathcal{Z}\{f \times g\} = \mathcal{Z}\left\{\Sigma_{n=0}^{k} f(n)g(k-n)\right\} = (\mathcal{Z}f)(\mathcal{Z}g)$$

A.4   Table of $\mathcal{Z}$-Transforms

$F(s) \equiv$ Laplace transform of $f(t)$, $F(z) \equiv \mathcal{Z}$-transform of $f(kT)$ and $f(t) = 0$ for $t = 0$

Number	$F(s)$	$f(kT)$	$F(z)$
1		$1,\, k = 0;\, k \neq 0$	1
2		$1,\, k = k_o;\, k \neq k_o$	z^{-k_o}
3	$\dfrac{1}{s}$	$1(kT)$	$\dfrac{z}{z-1}$
4	$\dfrac{1}{s^2}$	kT	$\dfrac{Tz}{(z-1)^2}$
5	$\dfrac{1}{s^3}$	$\dfrac{1}{2!}(kT)^2$	$\dfrac{T^2}{2}\left[\dfrac{z(z+1)}{(z-1)^3}\right]$
6	$\dfrac{1}{s^4}$	$\dfrac{1}{3!}(kT)^3$	$\dfrac{T^3}{6}\left[\dfrac{z(z^2+4z+1)}{(z-1)^4}\right]$
7	$\dfrac{1}{s^m}$	$\displaystyle\lim_{a\to 0}\dfrac{(-1)^{m-1}}{(m-1)!}\times\left(\dfrac{\partial^{m-1}}{\partial a^{m-1}}e^{-akT}\right)$	$\displaystyle\lim_{a\to 0}\dfrac{(-1)^{m-1}}{(m-1)!}\times\left(\dfrac{\partial^{m-1}}{\partial a^{m-1}}\dfrac{z}{z-e^{-aT}}\right)$
8	$\dfrac{1}{s+a}$	e^{-akT}	$\dfrac{z}{z-e^{-aT}}$
9	$\dfrac{1}{(s+a)^2}$	kTe^{-akT}	$\dfrac{Tze^{-aT}}{(z-e^{-aT})^2}$

A.5 Table of $\mathcal{Z}$-Transforms (contd.)

No.	$F(s)$	$f(kT)$	$F(z)$
10	$\dfrac{1}{(s+a)^3}$	$\dfrac{1}{2}(kT)^2 e^{-akT}$	$\dfrac{T^2}{2}e^{-aT}z\dfrac{(z+e^{-aT})}{(z-e^{-aT})^3}$
11	$\dfrac{1}{(s+a)^m}$	$\dfrac{(-1)^{m-1}}{(m-1)!}\times$ $\left(\dfrac{\partial^{m-1}}{\partial a^{m-1}}e^{-akT}\right)$	$\dfrac{(-1)^{m-1}}{(m-1)!}\left(\dfrac{\partial^{m-1}}{\partial a^{m-1}}\dfrac{z}{z-e^{-aT}}\right)$
12	$\dfrac{a}{s(s+a)}$	$1-e^{-akT}$	$\dfrac{z(1-e^{-aT})}{(z-1)(z-e^{-aT})}$
13	$\dfrac{a}{s^2(s+a)}$	$\dfrac{1}{a}(akT-1+e^{-akT})$	$\dfrac{z\,[Az+B]}{a(z-1)^2(z-e^{-aT})}$ $A=(aT-1+e^{-aT})$ $B=(1-e^{-aT}-aTe^{-aT})$
14	$\dfrac{b-a}{(s+a)(s+b)}$	$e^{-akT}-e^{-bkT}$	$\dfrac{(e^{-aT}-e^{-bT})z}{(z-e^{-aT})(z-e^{-bT})}$
15	$\dfrac{s}{(s+a)^2}$	$(1-akT)e^{-akT}$	$\dfrac{z\left[z-e^{-aT}(1+aT)\right]}{(z-e^{-aT})^2}$
16	$\dfrac{a^2}{s(s+a)^2}$	$1-e^{-akT}(1+akT)$	$\dfrac{z\,[Az+B]}{(z-1)(z-e^{-aT})^2}$ $A=(1-e^{-aT}-aTe^{-aT})$ $B=e^{-2aT}-e^{-aT}+aTe^{-aT}$
17	$\dfrac{(b-a)\,s}{(s+a)(s+b)}$	$be^{-bkT}-ae^{-akT}$	$\dfrac{z\left[z(b-a)-(be^{-aT}-ae^{-bT})\right]}{(z-e^{-aT})(z-e^{-bT})}$
18	$\dfrac{a}{s^2+a^2}$	$\sin akT$	$\dfrac{z\sin aT}{z^2-(2\cos aT)z+1}$
19	$\dfrac{s}{s^2+a^2}$	$\cos akT$	$\dfrac{z(z-\cos aT)}{z^2-(2\cos aT)z+1}$
20	$\dfrac{s+a}{(s+a)^2+b^2}$	$e^{-akT}\cos bkT$	$\dfrac{z(z-e^{-aT}\cos bT)}{z^2-2e^{-aT}(\cos bT)z+e^{-2aT}}$
21	$\dfrac{b}{(s+a)^2+b^2}$	$e^{-akT}\sin bkT$	$\dfrac{ze^{-aT}\sin bT}{z^2-2e^{-aT}(\cos bT)z+e^{-2aT}}$
22	$\dfrac{a^2+b^2}{s\left[(s+a)^2+b^2\right]}$	$1-e^{-akT}\times(\cos bkT$ $+\dfrac{a}{b}\sin bkT)$	$\dfrac{z(Az+B)}{(z-1)\left[z^2-Cz+e^{-2aT}\right]}$ $A=1-e^{-aT}\cos bT$ $-\dfrac{a}{b}e^{-aT}\sin bT$ $B=e^{-2aT}+\dfrac{a}{b}e^{-aT}\sin bT$ $-e^{-aT}\cos bT$ $C=2e^{-aT}(\cos bT)$

Appendix B

MATLAB: Basics and Exercises

MATLAB is an interactive high-level programming language for numerical computation and data visualization. It is used extensively for the design and analysis of control systems. There are many different toolboxes available that extend the basic functions of MATLAB into different application areas. In this Appendix the basic commands will be introduced and the reader will be familiarized with MATLAB. In addition, a number of exercises involving dynamic systems are provided. MATLAB is supported on Unix, Macintosh, and Windows environments. There are number of MATLAB websites that can be used to supplement this appendix. University of Michigan and Carnegie Mellon University support a very effective MATLAB tutorial through the following websites:

http://www.engin.umich.edu/group/ctm/index.html

http://hpme12.me.cmu.edu/matlab/html

B.1 Getting Started

MATLAB is invoked by entering the command *"matlab"* at the computer system prompt or by clicking on the MATLAB *icon*, depending on the type of machine being used. Once started, MATLAB will clear the screen, provide some introductory remarks, and produce the MATLAB *command prompt* >>. For the most part, MATLAB commands are independent of the type of machine and operating system being used. However, the way that MATLAB interfaces with the computing environment varies dramatically from machine to machine. As a result, use of features such as printing and command line editing are machine dependent. In order to exit from MATLAB type *"quit"* or *"exit "* at the MATLAB prompt, followed by the return or enter key.

B.2 Creating MATLAB Files

It is much more convenient to use MATLAB *script files* than to enter commands line by line at the MATLAB command prompt. A script file is an ASCII file (regular text file) that contains a series of commands written just as they would be entered at the MATLAB command prompt. Statements beginning with % are considered to be comments and are hence, ignored during execution. Each script file should have a name that ends with a ".*m*" extension The script file (MATLAB program file) is then executed by typing the name of the script file without the ".*m*" extension at the MATLAB command prompt. For example, if the script file *lab*1.*m* contains the following commands used to plot a sine curve:

%   *MATLAB* (*Exercise*1)
%   *Plotting a simple sine curve*
$t \ = \ 0 : 0.1 : 10;$
$y \ = \ sin(2 * t);$
$plot(t, y)$

Typing *lab*1 at the MATLAB command prompt will plot the sine curve. The file should be in the same directory as the MATLAB prompt. To verify, type *ls* at the MATLAB prompt to see that the file exists in the same directory. If not, then type *cd directory_name* to go the right directory. Once the MATLAB prompt is set, open another window to edit the text (script) file, i.e., enter the commands of the program that needs to be executed. The text file should be saved before running it at the MATLAB prompt.

B.3   Commands

MATLAB has many commands and a few are listed below along with their syntax. They are supplemented with examples and illustrations.

B.3.1   Vectors

For a row vector enter each element of the vector (separated by a space) between brackets, and set it equal to a variable. For example, to create a row vector **a**, enter the following in a MATLAB command window:

$>> \mathbf{a} = \begin{bmatrix} 1 \ 2 \ 3 \ 4 \ 5 \ 6 \ 7 \end{bmatrix}$
$\mathbf{a} =$
        1 2 3 4 5 6 7

For a proper vector **d,** enter each element of the vector (separated by a semicolon)

$>>$ d $= \begin{bmatrix} 1; \, 2; \, 3; \end{bmatrix}$

d $=$

    1
    2
    3

Alternatively the following transpose command can be used to obtain a proper vector.

$>>$ d $= \begin{bmatrix} 1 \ 2 \ 3 \end{bmatrix}'$

To create a row vector with elements between 0 and 20 evenly spaced in increments of 2 the following command can be used (this method is frequently used to create a time vector):

$>>$ **t** $=$ **0:2:20**

t $=$

    0 2 4 6 8 10 12 14 16 18 20

Manipulating vectors is as easy as creating them. To add 2 to each element of vector **a,** the command takes the form:

$>>$ **b** $=$ **a+2**

b $=$

    3 4 5 6 7 8 9

Now suppose, the sum of two vectors is required. The two vectors have to be of the same length, and the corresponding elements are simply added as shown below:

$>>$ **c** $=$ **a+b**

c $=$

    4 6 8 10 12 14 16

Subtraction of vectors of the same length works in exactly the same way.

B.3.2   Functions

MATLAB includes many standard functions. Each function is a block of code that accomplishes a specific task. Such MATLAB functions include the following: *sin, cos, log, exp,* and *sqrt.* Commonly used constants such as *pi,* and *i* or *j* for the square root of *-1,* are also incorporated.

$>>$ **sin(pi/4)**

ans $=$

    0.7071

To determine the usage of any function, type *help function_name* at the MATLAB command window. The *function* command facilitates the creation of new functions by the user.

B.3.3   Plotting

It is easy to create plots in MATLAB. Suppose the task involves plotting a sine curve as a function of time. The task involves making a time vector and then computing the value of the *sin* function at each time value. It is important to note that a *semicolon* (;) after a command instructs the MATLAB engine not to display the results of that particular command. The sine curve is plotted as follows:

>> **t = 0:0.25:7;**
>> **y = sin(t);**
>> **plot(t,y)**

MATLAB will return the following plot.

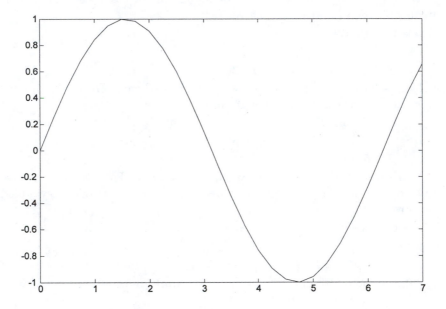

The plot command has extensive add-on capabilities. The title of the plot, labels of its axes, text and grids can be included to make referencing easy. The following commands can be used inside a MATLAB script file *(file_name.m),* and when the file is run, the curve with its new features are produced:

   plot(t,y)
   title('Plot name, e.g., System Response')
   xlabel('Time [sec]')
   ylabel('y [m]')
   grid
   gtext('Put text on graph, e.g., sin(t)')

MATLAB will return the following plot.

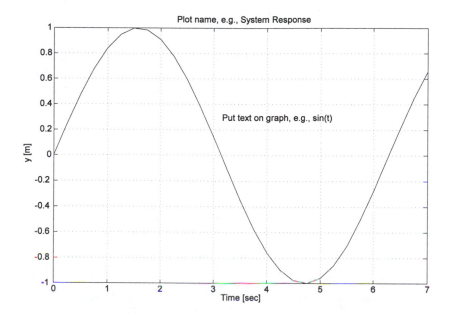

The *grid* command produces grid lines on the graph, and these can be used to compare plots. The command *gtext* allows text to be put on the graph. MATLAB provides crosswires in the graphics window that can be moved by the computer mouse to the desired location for the text label. It is important to note that the *opening apostrophe* in MATLAB commands must be like a *9* and not a *6* as in normal sentence construction. This is part of MATLAB syntax.

B.3.4   Polynomials

In MATLAB, a polynomial is represented by a row vector. To create a polynomial in MATLAB, each coefficient of the polynomial is entered into the vector in descending order. For instance, consider the following polynomial,

$$s^4 + 3s^3 - 15s^2 - 2s + 9.$$

The polynomial is entered in the following manner:
```
>> x = [1 3 −15 −2 9]
x =
    1 3 −15 −2 9
```
MATLAB interpret a vector of length $(n+1)$ as an nth-order polynomial. Thus, if the polynomial is missing any coefficients, zeros must be entered at the appropriate places in the vector. For example,

$$s^4 + 1,$$

would be represented as:

  >> y = $\begin{bmatrix} 1 & 0 & 0 & 0 & 1 \end{bmatrix}$

The value of a polynomial can be found by using the *polyval* function. For example, to find the value of the above polynomial at $s = 2$, the following command is used:

  >> z = **polyval([1 0 0 0 1], 2)**

  z =

      17

The roots of a polynomial can also be extracted. This is particularly useful for high-order polynomials such as

$$s^4 + 3s^3 - 15s^2 - 2s + 9.$$

The roots are obtained by entering the command:

  >> **roots([1 3 -15 -2 9])**

  ans =

      -5.5745
      2.5836
      0.7860
      -0.7951

The product of two polynomials is found by taking the convolution of their coefficients. The function *conv* is engaged as follows:

  >> x = $\begin{bmatrix} 1 & 2 \end{bmatrix}$;
  >> y = $\begin{bmatrix} 1 & 4 & 8 \end{bmatrix}$;
  >> z = **conv(x,y)**

  z =

      1 6 16 16

Dividing two polynomials is achieved by using the function *deconv*. It will return the result including the remainder. For example dividing z by y from above and leads to:

  >> [**xx, R**] = **deconv(z,y)**

  xx =

      1 2

  R =

      0 0 0 0

As expected, **xx** = **x**, and the remainder is **0**. Addition of two polynomials, that have the same order, is achieved by adding their corresponding row vectors. For example, if two polynomials **x** and **y** have the same order, then their sum can be obtained as follows

  >> z = **x** + **y**

B.3.5   Matrices

Entering matrices into MATLAB is similar to entering vectors except that each row of elements is separated by a semicolon (;) or a return key.

```
>> A = [1 2; 3 4]
```
yields
```
A =
    1      2
    3      4
>> A = [1,2
   3,4]
```
produces the same result.

Matrices in MATLAB can be manipulated in many ways. The transpose of a matrix is obtained using the apostrophe key.
```
>> C = A'
C =
    1      3
    2      4
```
It should be noted that if **A** had been complex, the apostrophe would have produced the complex conjugate transpose. The transpose can also be obtained by just typing **A'** at the MATLAB prompt.

Two matrices **B** and **C** can be multiplied together if the number of columns in **B** is equal to the number of rows in **C**. It is also essential to note that the order of multiplication for matrices matters.
```
>> B = [1 0; 0 1];
>> A*B
ans =
    1      2
    3      4
```
A square matrix can be multiplied by itself many times by raising it to the necessary power. For example,
```
>> A = [1 2; 3 4];
>> A^3
ans =
    37      54
    81     118
```
The inverse of a matrix is obtained as follows:
```
>> inv(A)
ans =
   -2.0000     1.0000
    1.5000    -0.5000
```

B.4   Printing

Printing in MATLAB is machine dependent.

Macintosh

To print a plot or a *m-file* from a Macintosh, just click on the plot or *m-file*, select *Print* under the File menu, and hit return.

Windows

To print a plot or a *m-file* from a computer running Windows, just select *Print* from the File menu in the window of the plot or *m-file*, and hit return.

Unix

To print a file on a Unix workstation the following command can be entered:

> **lp -P<printer_name> file_name**

For example, if the name of the printer is *lexlab2* and the file is *test.m,* then,

> **lp -Plexlab2 test.m**

Plots

Plots can be printed by going to the file menu of the plot and clicking the print option. Another window will pop up, and the print option can be selected. Thus, the file will send to be printed by the default printer. If the objective is to save the plot and print it later, either of the following two commands can be used in the MATLAB command window soon after producing the plot.

>> **print plot_name.ps**

>> **print -deps plot_name.eps**

Once saved the plot can be printed later or included as part of a text document. For example, in Unix the plot can be printed as follows:

> **lp -P<printer_name> plot_name.ps**

B.5   Using M-files in MATLAB

In handling MATLAB files *(m-files)*, there are slight differences between the machine platforms.

Macintosh

There is a built-in editor for m-files; The *"New M-file"* option can be chosen from the *File menu.* Other editors can be used but its important to save the files in text format and load them when MATLAB is invoked.

PC Windows

Running MATLAB from Windows is very similar to running it on a Macintosh. However, its important to note that the *m-file* will be saved in the clipboard. Therefore, it must be saved as *file_name.m*

Unix

Two Unix windows must be opened: the MATLAB command window, and the file editing window. Both windows must be operating from the same directory. All the MATLAB commands are written into the script file *"file_name.m"* opened from the editing window. Unix editors include such programs as *emacs* and *pico*. For example a file can be opened and edited by first typing the following command in the editing Unix window:

> **emacs file_name.m**

When all the MATLAB commands have been written into *file_name.m*, the file can then be executed by typing the following command from the MATLAB command window:

>> **file_name**

B.6 Saving Workspace

As has been discussed already, MATLAB commands can be typed directly in a MATLAB command window, or they can be edited into an *m-file* and then the file executed afterwards. When typing commands directly in the MATLAB window, the work being carried out can be saved into a file *file_name.m* by using the *diary* command as follows:

>> **diary file_name.m**
>> **A = [1 2; 3 4]**
A =
    1   2
    3   4
>> **B = [1 0; 0 1]**

B =
    1   0

    0   1
>> **diary**

The command *diary file_name.m* causes all subsequent prompt inputs and their outputs to be written into *file_name.m*. The command *diary off* suspends that facility.

B.7   Getting Help in MATLAB

MATLAB has fairly good on-line help that can be accessed as follows:
>> **help command_name**

It also important to notice that the value of a particular variable can be obtained at any time by typing its name. For example,
>> **A**
A =
    1   2
    3   4

Also more than one command statement can be placed on a single line, by separating the statements by a semicolon or comma. If a variable is not assigned to a specific operation or result, MATLAB will store it in a temporary variable called *"ans."*

B.8   Control Functions

MATLAB has built-in functions to help in the design and analysis control systems. Given below are some of the functions

(a) *step*

This function produces the step response of a system, i.e., the system output due to a step input. It takes as arguments the state-variable matrices (**A,B,C,D**) from the linear system representation,

$$\dot{\mathbf{x}} = \mathbf{A}\mathbf{x} + \mathbf{B}\mathbf{u}$$
$$\mathbf{y} = \mathbf{C}\mathbf{x} + \mathbf{D}\mathbf{u}.$$

The syntax takes the form
>> **step(A, u*B, C, u*D, iu, t);**

The term, **u,** is the size of the step input (a constant), and **iu** indicates the number of inputs. The quantity, **t,** is a user-supplied time vector that specify the time interval over which the step response should be calculated. If it is not specified a default time interval is used. Given below is the MATLAB program that can be used to produce the step response for a car's cruise-control system. The input is a step force, $u = 500N$.

A = [0 1; 0 -0.05];
B = [0; 0.001];
C = [0 1];
D = 0;
step(A, 500*B, C, D, 1)
title('Cruise Control Step Response')
This returns the plot shown below.

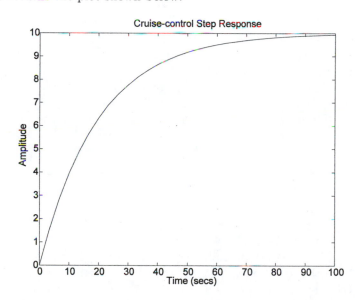

(b) *impulse*

This function produces the step response of a system, i.e., the system output due to a step input. It takes as arguments the state-variable matrices (**A,B,C,D**) from the linear system representation,

$$\dot{\mathbf{x}} = \mathbf{Ax} + \mathbf{Bu}$$
$$\mathbf{y} = \mathbf{Cx} + \mathbf{Du}.$$

The syntax takes the form
    >> **impulse(A, B, C, D, iu, t);**

where the input is an impulse and **iu** indicates the number of inputs. The vector **t** is a user-supplied time vector that specify the time interval over which the step response should be calculated. Given below is example of MATLAB program that produces an impulse response:

A = [-3 -2 ; 1 0];
B = [1;0];
C = [2 1];
D = 0;
impulse(A,B,C,D,1)
title('Impulse Response')
The following plot is obtained.

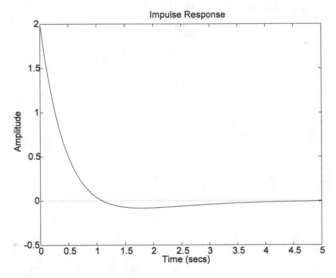

As shown in preceding examples, the state-variable matrices must be entered before the *step* and *impulse* functions are used. Alternatively, the system transfer functions can be used to obtain system responses. MATLAB also has built-in programs to convert between different models of dynamic systems, such as the *state-variable matrix* form and the *transfer function* form. For example,

>> **[num,den] = ss2tf(A,B,C,D)**

converts the system from the state-variable matrix form to the transfer function form. The impulse and step responses are then obtained as follows:
>> **step(num,den)**
>> **impulse(num,den)**

Other commands that convert between models include the following

>> **[A,B,C,D] = tf2ss(num,den)**

>> [z,p,k] = ss2zp(A,B,C,D)

>> [A,B,C,D] = zp2ss(z,p,k)

>> [A1,B1,C1,D1] = ss2ss(A,B,C,D)

B.9 More Commands

1. *inv*($\mathbf{A}$) produces the inverse of a square matrix $\mathbf{A}$. A warning message is printed if $\mathbf{A}$ singular or badly scaled

2. *loops* repeat statements a specific number of times.

   The general form of a *for* statement is:

   for variable = expression, statement, ..., statement *end*

   The columns of the expression are stored one at a time in the variable and then the following statements, up to the *end*, are executed. The expression is often of the form $X{:}Y$, in which case its columns are simply scalars. For example:

   N =10

   for I = 1:N;

       for J = 1:N;

       A(I,J) = 1/(I+J-1);

       end

   end

3. *eye* (identity matrices)

   eye(N) produces an N-by-N identity matrix.

   eye(M, N) or **eye([M, N])** produces an M-by-N matrix with 1's on the diagonal and zeros elsewhere.

   eye(size(A)) produces an identity matrix with the same size as A.

4. *rank* (linear independence)

   If X is a matrix then

   k = rank(X)

   gives the number of linearly independent rows or columns of the matrix.

B.10   LABWORK I

1. Consider $A = \begin{bmatrix} -5 & 1 & 0 \\ 0 & -2 & 1 \\ 0 & 0 & 1 \end{bmatrix}$, $b = \begin{bmatrix} 0 \\ 0 \\ 1 \end{bmatrix}$, $c = \begin{bmatrix} -1 & 1 & 0 \end{bmatrix}$

   (a) Suppose $Ax = b$, find x.

   (b) Suppose $yA = c$, find y.

   (c) $G(s) = c(sI - A)^{-1}b$, find $G(0)$ and $G(1)$.

   (d) Define $C_M = \begin{bmatrix} b & Ab & A^2b \end{bmatrix}$. Find rank of C_M

   (e) Now consider an arbitrary $n \times n$ matrix A and $n \times 1$ vector b.

   Let $C_M = \begin{bmatrix} b & Ab & A^2b & \dots & A^{n-1}b \end{bmatrix}$. Write the script file that computes the rank of C_M.

2. Consider the function

$$H(s) = \frac{\eta(s)}{d(s)}$$

   where

$$\eta(s) = s^3 + 6.4s^2 + 11.29s + 6.76$$
$$d(s) = s^4 + 14s^3 + 46s^2 + 64s + 40$$

   (a) Find $\eta(-12), \eta(-10), \eta(-8)$.

   (b) Find $d(-12), d(-10), d(-8)$.

   (c) Find $H(-12), H(-10), H(-8)$.

3. Let $A1$ be any $n \times m$ matrix and $A2$ be any $p \times q$ matrix. Create a function **block(A1,A2)** that generates the $(n+p) \times (m+q)$ block diagonal matrix

$$\begin{bmatrix} A1 & 0 \\ 0 & A2 \end{bmatrix},$$

   where the of diagonal blocks have all elements equal to zero. Use the **zeros** command.

4. For designing an automobile suspension, a two-mass system can be used for modeling as shown in the following diagram.

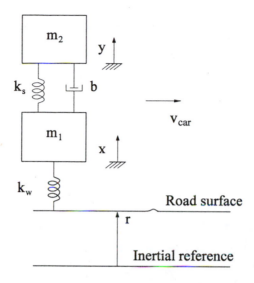

The Quarter-Car Model

This is called a quarter-car model because it comprises one of the four wheel suspensions. The car and wheel positions are denoted by $y(t)$ and $x(t)$ respectively. These displacements are from static equilibrium which corresponds to no inputs except gravity.

(a) Draw the free-body diagram of this system, assuming one-dimensional vertical motion of the mass above wheel.

(b) Write down the equations of motion for the automobile.

(c) Express these equations in a state-variable matrix form $(\mathbf{A},\mathbf{B},\mathbf{C},\mathbf{D})$ using the following state-variable vector,

$$x(t) = [x \ \dot{x} \ y \ \dot{y}]^T,$$

and justify this choice of state variables. Note that the car and wheel positions, $y(t)$ and $x(t)$, are the two outputs of the car system while the input is the unit step bump $r(t)$.

(d) Plot the position of the car and the wheel after the car hits a "unit bump" (i.e. $r(t)$ is a unit step) using MATLAB. Assume $m_1 = 10kg$, $m_2 = 250kg$, $k_w = 500,000N/m$, $k_s = 10,000N/m$. Find the value of b that you would prefer if you were a passenger in the car.

B.11 LABWORK II

1. Given the matrices

$$A = \begin{bmatrix} 1\,0\,2\,1 \\ 0\,5\,3\,2 \\ 2\,1\,1\,4 \\ 4\,2\,2\,3 \end{bmatrix} ; \quad B = \begin{bmatrix} 2\,5 \\ 3\,3 \\ 6\,2 \\ 0\,1 \end{bmatrix} ; \quad C = \begin{bmatrix} 1\,0 \\ 1\,2 \\ 3\,3 \\ 1\,3 \end{bmatrix}$$

$$x = \begin{bmatrix} 1\,3\,5\,1 \end{bmatrix} ; \quad y = \begin{bmatrix} 2\,1\,4\,3 \end{bmatrix} ; \quad z = \begin{bmatrix} 0\,7\,8\,9 \end{bmatrix} ;$$

evaluate

   (a) $AB + AC$

   (b) $A(B + C)$ and verify that $A(B + C) = AB + AC$

   (c) $x + y$

   (d) $3x + 4y + z$

2. Plot the following curves for the specified ranges:

   (a) $y = \sin 3t$ for $t = 0 : 10$

   (b) $y = \cos 2t$ for $t = 0 : 10$

   (c) $y = 3\sin t + 4\cos t$ for $t = -5 : 15$

   (d) $y = e^{-3t}$ for $t = -1 : 5$

B.12 LABWORK III

1. Consider the mechanical system shown below.

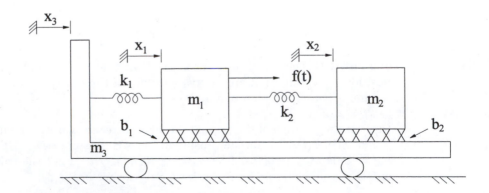

The forces exerted by the springs are zero when $x_1 = x_2 = x_3 = 0$. The input force is $f(t)$ and the absolute displacements of m_1, m_2 and m_3 are x_1, x_2 and x_3, respectively. The output is the spring force in k_2. The differential equations which represent the system can be expressed in the state-variable form (A, B, C, D). This can be done by choosing the vector of state variables as

$$\mathbf{x} = \begin{bmatrix} x_1 & v_1 & x_2 & v_2 & x_3 & v_3 \end{bmatrix}^T.$$

The state-variable matrix form is then given by

$$\begin{bmatrix} \dot{x}_1 \\ \dot{v}_1 \\ \dot{x}_2 \\ \dot{v}_2 \\ \dot{x}_3 \\ \dot{v}_3 \end{bmatrix} = \begin{bmatrix} 0 & 1 & 0 & 0 & 0 & 0 \\ -\dfrac{(k_1+k_2)}{m_1} & -\dfrac{b_1}{m_1} & \dfrac{k_2}{m_1} & 0 & \dfrac{k_1}{m_1} & \dfrac{b_1}{m_1} \\ 0 & 0 & 0 & 1 & 0 & 0 \\ \dfrac{k_2}{m_2} & 0 & -\dfrac{k_2}{m_2} & -\dfrac{b_2}{m_2} & 0 & \dfrac{b_2}{m_2} \\ 0 & 0 & 0 & 0 & 0 & 1 \\ \dfrac{k_1}{m_3} & \dfrac{b_1}{m_3} & 0 & \dfrac{b_2}{m_3} & -\dfrac{k_1}{m_3} & -\dfrac{(b_1+b_2)}{m_3} \end{bmatrix} \begin{bmatrix} x_1 \\ v_1 \\ x_2 \\ v_2 \\ x_3 \\ v_3 \end{bmatrix} + \begin{bmatrix} 0 \\ \dfrac{1}{m_1} \\ 0 \\ 0 \\ 0 \\ 0 \end{bmatrix} f(t)$$

$$y = \begin{bmatrix} k_2 & 0 & -k_2 & 0 & 0 & 0 \end{bmatrix} \begin{bmatrix} x_1 \\ v_1 \\ x_2 \\ v_2 \\ x_3 \\ v_3 \end{bmatrix} + [0]\, f(t)$$

For the following data

$$m_1 = 0.5 kg$$
$$m_2 = 0.75 kg$$
$$m_3 = 2.0 kg$$
$$b_1 = 1000 Ns/m$$
$$b_2 = 5 Ns/m$$
$$k_1 = 50 N/m$$
$$k_2 = 100 N/m$$

(a) Find the output response $y(t)$ to an input force of $50N$.

(b) Find the response to an impulse force input.

(c) Choose values of b_1 and b_2 that will reduce oscillations, give a practical settling time, and produce smooth responses.

(d) Explain why the steady state values of the MATLAB plots are in agreement with what is expected from analyzing the mechanical system.

(e) Is the matrix A invertible, and why is this the case?

(suggestion: For the responses, specify the time for which the plot is to be graphed, e.g., from 0sec to 3sec)

2. If the output was the platform displacement is x_3 then the matrices **C** and **D** are obtained as follows

$$y = \begin{bmatrix} 0\ 0\ 0\ 0\ 1\ 0 \end{bmatrix} \begin{bmatrix} x_1 \\ v_1 \\ x_2 \\ v_2 \\ x_3 \\ v_3 \end{bmatrix} + [0]\, f(t).$$

(a) Find the output response $y(t)$ to an input force of $50N$

(b) Find the response to an impulse force input.

(c) Explain why these MATLAB plots are in agreement with what is expected from analyzing the mechanical system.

3. For the translational mechanical system under consideration, the vector of independent variables (minimum number of state variables) can be chosen as

$$\mathbf{x}(t) = \begin{bmatrix} x_{R_1} & v_1 & x_{R_2} & v_2 & v_3 \end{bmatrix}^T,$$

where

$$x_{R_1} = (x_1 - x_3)$$
$$x_{R_2} = (x_1 - x_2)$$

The state-variable matrix form is then given by,

$$
\begin{bmatrix} \dot{x}_{R_1} \\ \dot{v}_1 \\ \dot{x}_{R_2} \\ \dot{v}_2 \\ \dot{v}_3 \end{bmatrix} = \begin{bmatrix} 0 & 1 & 0 & 0 & -1 \\ -\dfrac{k_1}{m_1} & -\dfrac{b_1}{m_1} & -\dfrac{k_2}{m_1} & 0 & \dfrac{b_1}{m_1} \\ 0 & 1 & 0 & -1 & 0 \\ 0 & 0 & \dfrac{k_2}{m_2} & -\dfrac{b_2}{m_2} & \dfrac{b_2}{m_2} \\ \dfrac{k_1}{m_3} & \dfrac{b_1}{m_3} & 0 & \dfrac{b_2}{m_3} & -\dfrac{(b_1+b_2)}{m_3} \end{bmatrix} \begin{bmatrix} x_{R_1} \\ v_1 \\ x_{R_2} \\ v_2 \\ v_3 \end{bmatrix} +
$$

$$
\begin{bmatrix} 0 \\ 1 \\ \dfrac{m_1}{0} \\ 0 \\ 0 \end{bmatrix} f(t)
$$

$$
y = \begin{bmatrix} 0 & 0 & k_2 & 0 & 0 \end{bmatrix} \begin{bmatrix} x_{R_1} \\ v_1 \\ x_{R_2} \\ v_2 \\ v_3 \end{bmatrix} + [0]\, f(t).
$$

(a) Show that this representation produce the same results as the six state-variable matrix system by repeating 1(a), (b) and (c) using the five state-variable matrix system.

(b) Is the matrix A invertible, and why is this the case ?

B.13 LABWORK IV

1. Consider the 6×6 matrix state-space system described in LABWORK III.

   The same system can be represented in transfer function form. In MATLAB one can convert from state-space to transfer function form by using the following command

   $$[\mathbf{num}, \mathbf{den}] = \mathbf{ss2tf(A, B, C, D)}.$$

   (a) Use the above command to get the pair (num, den) for the above system. Plot the step and impulse responses using num, den.

The syntax for plotting step and impulse function using transfer function is given as

$$\text{step}(\mathbf{num}, \mathbf{den}, \mathbf{t})$$
$$\text{impulse}(\mathbf{num}, \mathbf{den}, \mathbf{t}).$$

Compare these plots with those obtained earlier using the state-space matrices $(\mathbf{A}, \mathbf{B}, \mathbf{C}, \mathbf{D})$ and explain why the responses are in agreement with what is expected.

(b) Similar to the *ss2tf* function MATLAB also has *tf2ss* function which converts the transfer function to the state-space form. The syntax for this function is given by

$$[\mathbf{A1}, \mathbf{B1}, \mathbf{C1}, \mathbf{D1}] \;=\; \text{tf2ss}(\mathbf{num}, \mathbf{den}).$$

Using this function to transform the *num, den* obtained in part (a), to state-space matrices $(\mathbf{A1}, \mathbf{B1}, \mathbf{C1}, \mathbf{D1})$. Compare these matrices $(\mathbf{A1}, \mathbf{B1}, \mathbf{C1}, \mathbf{D1})$ with the state-space matrices above $(\mathbf{A}, \mathbf{B}, \mathbf{C}, \mathbf{D})$ and comment on the results so obtained. Is the matrix $\mathbf{A1}$ invertible?

2. Repeat the same problem 1 for the following 5×5 matrix state-space system described in LAB III

Compare the denominator of the transfer function obtained for matrices above with the denominator obtained for the above problem (problem 1).

3. Apart from the step and impulse inputs we can have may other kinds of inputs e.g., sinusoidal, triangular wave etc. To obtain the response for such inputs MATLAB has a built in function *lsim*. The usage for *lsim* is shown below

$$\text{lsim}(\mathbf{num}, \mathbf{den}, \mathbf{u}, \mathbf{t})$$
$$\text{lsim}(\mathbf{A}, \mathbf{B}, \mathbf{C}, \mathbf{D}, \mathbf{u}, \mathbf{t})$$

where u is the input.

Find the response of the system described in problem 1 (using both $\mathbf{A}, \mathbf{B}, \mathbf{C}, \mathbf{D}$ and $\mathbf{num}, \mathbf{den}$) to the following inputs:

(a) $u = 2t$

(b) $u = \sin 5t$

What would happen to the response if the frequency in part (b) is increased from 5 to 10?

Use *subplot* to plot the four curves on one sheet and compare them.

4. Repeat the above problem 3 for the system described in problem 2 (using both $\mathbf{A}, \mathbf{B}, \mathbf{C}, \mathbf{D}$ and $\mathbf{num}, \mathbf{den}$).

Data:

$$m_1 = 0.5kg$$
$$m_2 = 0.75kg$$
$$m_3 = 2kg$$
$$b_1 = 1000Ns/m$$
$$b_2 = 5Ns/m$$
$$k_1 = 50N/m$$
$$k_2 = 100N/m$$

B.14 LABWORK V

1. Consider the system described by the following state-space matrices:

$$\mathbf{A} = \begin{bmatrix} 0 & 1 \\ 0 & -\dfrac{1}{20} \end{bmatrix}, \quad \mathbf{B} = \begin{bmatrix} 0 \\ \dfrac{1}{20} \end{bmatrix}$$
$$\mathbf{C} = \begin{bmatrix} 1 & 0 \end{bmatrix}, \quad \mathbf{D} = [0]$$

Find the transfer function of the system using the MATLAB function **ss2tf**.

2. The input-output differential equation of a system with zero initial conditions is given below:

$$\ddot{y} + 6\dot{y} + 25y = 9u + 3\dot{u}$$

(a) Find the transfer function of the system by using Laplace transforms.

(b) Using the transfer function (**num, den**) obtained in part (a) find the state-variable matrices (**A, B, C** and **D**). (Use MATLAB function **tf2ss**.)

3. The MATLAB function

$$[z, p, k] = ss2zp(\mathbf{A}, \mathbf{B}, \mathbf{C}, \mathbf{D})$$

finds the zeros and poles of the system described by (**A, B, C** and **D**). Find the zeros and poles of the system described in problem 2 using the state-space description $(\mathbf{A}, \mathbf{B}, \mathbf{C}$ and $\mathbf{D})$. Also construct the transfer function using the zeros and poles, and compare with the transfer function obtained in problem 2 (a).

B.15   LABWORK VI

1. Translational, rotational and electrical systems can be shown to manifest the same dynamic behavior and hence their models can be used interchangeably. Consider the four second order systems shown in following diagrams:

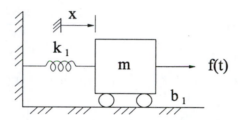

FIGURE B.1
A Translational Mechanical System

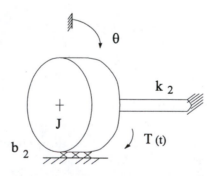

FIGURE B.2
A Rotational Mechanical System

Using free-body diagrams and circuit laws (current and voltage) it can be shown that the input-output differential equations for the four

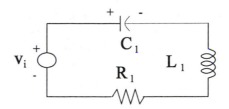

FIGURE B.3
An RLC Series Circuit

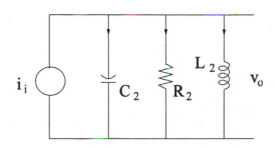

FIGURE B.4
An RLC Parallel Circuit

systems are given by:

$$\ddot{x} + \frac{b_1}{m}\dot{x} + \frac{k_1}{m}x = \frac{1}{m}f(t)$$

$$\ddot{\theta} + \frac{b_2}{J}\dot{\theta} + \frac{k_2}{J}\theta = \frac{1}{J}T(t)$$

$$\ddot{i} + \frac{R_1}{L_1}\dot{i} + \frac{1}{L_1C_1}i = \frac{1}{L_1}\dot{v}_i(t)$$

$$\ddot{v}_o + \frac{1}{R_2C_2}\dot{v}_o + \frac{1}{L_2C_2}v_o = \frac{1}{C_2}\dot{i}(t).$$

2. By comparing these models with the standard second order model

$$\ddot{y} + 2\xi\omega_n\dot{y} + \omega_n^2 y = f(t),$$

find the expressions for the respective natural frequency (ω_n) and damping ratio (ξ) for each of the four systems.

Use the following data for the four dynamic systems for the rest of the problems:

$$b_1 = 15Ns/m$$
$$b_2 = 15Nms$$
$$k_1 = 100N/m$$
$$k_2 = 100Nm$$
$$m = 1kg$$
$$J = 1kgm^2$$
$$R_1 = 1.5\Omega$$
$$R_2 = \frac{2}{3}\Omega$$
$$C_1 = C_2 = 0.1F$$
$$L_1 = L_2 = 0.1H$$

3. (a) Find the differential equation for each of the four systems (hand calculations).

   (b) Find the transfer function for each of the four systems (hand calculations).

   (c) Compare the four characteristic equations.

   (d) Compare the four transfer functions.

4. Use MATLAB to obtain the following:

   (a) the state-variable matrix models (**A, B, C and D**) for each of the four systems.

   (b) the unit step response for each of the four systems.

   (c) the impulse response for each of the four systems.

5. (a) Compare the system responses (behavior) of the two mechanical systems.

   (b) Compare the system responses (behavior) of the two electrical systems.

6. Use the *FVT* and *IVT* to confirm the eight MATLAB plots.

7. Find the DC gain for each of the four systems.

Bibliography

[1] Astrom, K.J. and B. Wittenmark. *Computer Control Systems: Theory and Design.* Prentice-Hall, 1994.

[2] Astrom, K.J. and B. Wittenmark. *Adaptive Control Systems: Theory and Design.* Addison-Wesley, 1995.

[3] Bar-Shalom, Y. and T.E. Fortmann. *Tracking and Data Association.* Academic Press, 1988.

[4] Bar-Shalom, Y. and X. Li. *Estimation and Tracking.* Artech House, 1993.

[5] Close, C.M. and D.K. Frederick. *Modeling and Analysis of Dynamic Systems.* Addison-Wesley, 1995.

[6] Cochin, I. and W. Cadwallender. *Analysis and Design of Dynamic Systems.* Addison-Wesley, 1997.

[7] Dorf, R.C. and R.H. Bishop. *Modern Control Systems.* Addison-Wesley, 1998.

[8] Franklin, G.F., J.D. Powell, and A. Emami-Naeini. *Feedback Control of Dynamic Systems.* Addison-Wesley, 1994.

[9] Franklin, G.F., J.D. Powell, and M.L. Workman. *Digital Control of Dynamic Systems.* Addison-Wesley, 1998.

[10] Friedland, F. *Control System Design: Introduction to State-Space Methods.* McGraw-Hill, 1987.

[11] Jacobs, O.L.R. *Introduction to Control Theory.* Oxford University Press, 1993.

[12] Kuo, B.C. *Automatic Control Systems.* Addison-Wesley, 1995.

[13] Maybeck, P.S. *Stochastic Models, Estimation and Control, Vol. I.* Academic Press, 1979.

[14] Maybeck, P.S. *Stochastic Models, Estimation and Control, Vol. 3.* Academic Press, 1982.

[15] Mutambara, A.G.O. *Decentralized Estimation and Control for Multi-sensor Systems.* CRC Press, 1998.

[16] Nise, N.S. *Control Systems Engineering.* Benjamin-Cummings, 1995.

[17] Ogata, K. *Modern Control Engineering.* Prentice-Hall, 1997.

[18] Ogata, K. *System Dynamics.* Prentice Hall, 1998.

[19] Skogestad, S. and I. Postlethwaite. *Multivariable Feedback Control.* John Wiley and Sons, 1996.

[20] Vu, H.V. and R.S. Esfandiari. *Dynamic Systems: Modeling and Analysis.* McGraw-Hill, 1997.

Index